# Intermittency and Self-Organisation in Turbulence and Statistical Mechanics

# Intermittency and Self-Organisation in Turbulence and Statistical Mechanics

Special Issue Editor

**Eun-jin Kim**

MDPI • Basel • Beijing • Wuhan • Barcelona • Belgrade

*Special Issue Editor*
Eun-jin Kim
University of Sheffield
UK

*Editorial Office*
MDPI
St. Alban-Anlage 66
4052 Basel, Switzerland

This is a reprint of articles from the Special Issue published online in the open access journal *Entropy* (ISSN 1099-4300) from 2018 to 2019 (available at: https://www.mdpi.com/journal/entropy/special_issues/Turbulence_Mechanics)

For citation purposes, cite each article independently as indicated on the article page online and as indicated below:

| |
|---|
| LastName, A.A.; LastName, B.B.; LastName, C.C. Article Title. *Journal Name* **Year**, *Article Number*, Page Range. |

**ISBN 978-3-03921-108-1 (Pbk)**
**ISBN 978-3-03921-109-8 (PDF)**

# Contents

# About the Special Issue Editor

**Eun-jin Kim** (Dr., Prof.) Dr. Kim obtained her BSc in Physics from Yonsei University in Seoul, Korea, and PhD in Physics from the University of Chicago, USA. She held postdoctoral positions at the Universities of Leeds and Exeter in UK, High-Altitude Observatory in Boulder, USA and University of California, San Diego, USA. She is currently an Associate Professor at the University of Sheffield, UK. Dr Kim is interested in complexity, self-organisation and non-equilibrium processes, and has a unique track record in multidisciplinary research, with applications to astrophysical and laboratory fluids/plasmas and biosystems. In particular, Dr. Kim is keen on the information theory (information length) to model complexity and self-organisation in nonlinear dynamical systems, fluid/plasma turbulence, and biosystems. She is a holder of a Leverhulme Trust Research Fellowship. She published over 110 refereed journal papers (51 as first author).

*entropy*

MDPI

*Editorial*

# Intermittency and Self-Organisation in Turbulence and Statistical Mechanics

**Eun-jin Kim**

School of Mathematics and Statistics, University of Sheffield, Sheffield S3 7RH, UK; e.kim@sheffield.ac.uk

Received: 4 June 2019; Accepted: 6 June 2019; Published: 6 June 2019

**Keywords:** turbulence; statistical mechanics; intermittency; coherent structure; multi-scale problem; self-organisation; bifurcation; non-locality; scaling; multifractal

---

There is overwhelming evidence, from laboratory experiments, observations, and computational studies, that coherent structures can cause intermittent transport, dramatically enhancing transport. A proper description of this intermittent phenomenon, however, is extremely difficult, requiring a new non-perturbative theory, such as statistical description. Furthermore, multi-scale interactions are responsible for inevitably complex dynamics in strongly non-equilibrium systems, a proper understanding of which remains one of the main challenges in classical physics. However, as a remarkable consequence of multi-scale interaction, a quasi-equilibrium state (so-called self-organisation) can be maintained.

This Special Issue presents different theories of statistical mechanics to understand this challenging multiscale problem in turbulence. The 14 contributions to this Special Issue focus on the various aspects of intermittency, coherent structures, self-organisation, bifurcation and nonlocality. Given the ubiquity of turbulence, the contributions cover a broad range of systems covering laboratory fluids (channel flow, the Von Kármán flow), plasmas (magnetic fusion), laser cavity, wind turbine, air flow around a high-speed train, solar wind and industrial application. The following is a short summary of each contribution.

Mathur et al. [1] address the importance of structures in the transient behaviour of a channel flow at high Reynolds number Re. Large-eddy simulations of turbulent channel flow subjected to a step-like acceleration reveal the transition of transient channel flow comprised of a three-stage response similar to that of the bypass transition of boundary layer flows; the effect of the structures (the elongated streaks) becomes more important in the transition for large Re. Their analysis employing conditionally-averaged turbulent statistics elucidates the interplay between structures and active/inactive regions of turbulence depending on Re.

Chliamovitch and Thorimbert [2] present a new method of dealing with non-locality of turbulence flows through the formulation of the bilocal kinetic equation for pairs of particles. Based on a maximum-entropy-based generalisation of Boltzmann's assumption of molecular chaos, they utilise the two-particle kinetic equations and derive the balance equations from the bilocal invariants to close their kinetic equations. The end product of their calculation is non-viscous hydrodynamics, providing a new dynamical equation for the product of fluid velocities at different points in space.

Jacquet et al. [3] address the formation of coherent structures and their self-organisation in a reduced model of turbulence. They present the transient behaviour of self-organised shear flows by solving the Fokker–Planck equation for time-dependent Probability Density Functions (PDFs) and model the formation of self-organisation shear flows by the emergence of a bimodal PDF with the two peaks for non-zero mean values of a shear flow. They show that the information length—The total number of statistically different states that a system passes through in time—is a useful statistical measure in understanding attractor structures and the time-evolution out of equilibrium.

Xu et al. [4] deal with an unsteady flow in wind turbines and show the importance of structures (turbulent winds/wind shears) on the stability of the floating wind. Based on the vortex theory for the wake flow field of the wind turbine, they invoke the Free Vortex Wave method to calculate the rotor power of the wind turbine. Depending on the turbulent wind, wind shear, and the motions of the floating platform, they put forward a trailing-edge flap control strategy to reduce rotor power fluctuations of a large-scale offshore floating wind turbine. Their proposed strategy is shown to improve the stability of the output rotor power of the floating wind turbine under the turbulent wind condition.

Anderson et al. [5] model anomalous diffusion and non-local transport in magnetically confined plasmas by using a non-linear Fractional Fokker–Planck (FFP) equation with a fractional velocity derivative. Their model is based on the Langevin equation with a nonlinear cubic damping and an external additive forcing given by a Lévy-stable distribution with the fractality index $\alpha$ ($0 < \alpha < 2$). By varying $\alpha$, they numerically solve the stationary FFP equation and analyse the statistical properties of stationary distributions by using the Boltzmann–Gibbs entropy, Tsallis' q-entropy, q-energies, and generalised diffusion coefficient, and show the significant increase in transport for smaller $\alpha$.

Saini et al. [6] highlight key challenges in modelling high Reynolds number unsteady turbulent flows due to complex multi-scale interactions and structures (e.g., near wall) and discuss different advanced modelling techniques. Given the limitation of the traditional Reynolds-Averaged Navier–Stokes (RANS) based on stationary turbulent flows, they access the validity of the Improved Delayed Detached Eddy Simulation (IDDES) methodology using two different unsteady RANS models. By investigating different types of flows including channel (fully attached) flow and periodic hill (separated) flow at different Reynolds numbers, they point out the shortcomings of the IDDES methodology and call for future work.

Barbay et al. [7] address the formation of oscillatory patterns (structures), bifurcations and extreme events in an extended semiconductor microcavity laser. Experimentally, as an example of self-pulsing spatially extended systems, they consider vertical-cavity surface emitting lasers with an integrated saturable absorber and study the complex dynamics and extreme events accompanied by spatiotemporal chaos. Theoretically, by employing the Ginzburg–Landau model, they characterize intermittency by the Lyapunov spectrum and Kaplan–Yorke dimension and show the chaotic alternation of phase and amplitude turbulence, extreme events induced by the alternation of defects and phase turbulence.

Wang et al. [8] investigate the effect of streaks (structures) on wall-bounded turbulence at low-to-moderate Reynolds number by using 2D Particle Image Velocimetry measurement and direct numerical simulations. To understand the spanwise spacing of neighbouring streaks, they present a morphological streak identification analysis and discuss wall-normal variation of the streak spacing distributions, fitting by log-normal distributions, and Re-(in)dependence. They then reproduce part of the spanwise spectra by a synthetic simulation by focusing on the Re-independent spanwise distribution of streaks. Their results show the important role of streaks (structures) in determining small-scale velocity spectra beyond the buffer layer.

Van Milligen et al. [9] address the importance of self-organisation and structures in transport in magnetically confined fusion plasmas far from equilibrium by studying the radial heat transport in strongly heated plasmas. By using the transfer entropy, they identify the formation of weak transport barriers near rational magnetic surfaces most likely due to zonal flows (structures) and show that jumping over transport barriers is facilitated with the increasing heating power. The behaviour of three different magnetic confinement devices is shown to be similar. They invoked a resistive magneto-hydrodynamic (fluid) model and continuous-time random walk to understand the experiment results.

He et al. [10] address turbulence in the air over a high-speed train and the formation of a coherent structure near the vent of a train, which plays an important role in the dissipated energy through the skin friction. By modelling the ventilation system of a high-speed train by a T-junction duct with vertical blades, they calculate the velocity signal of the cross-duct in three different sections (upstream, mid-center and downstream), and analyse the coherent structure of the denoised signals by using the continuous wavelet transform. Results show that the skin friction of the train decreases with the increasing ratio of the suction velocity of ventilation to the velocity of the train.

Alberti et al. [11] discuss turbulence, intermittency and structure in the solar wind by using fluid (magnetohydrodynamic) and kinetic approaches. By analysing solar wind magnetic field measurements from the ESA Cluster mission and by using the empirical mode decomposition based multi-fractal analysis and a chaotic approach, they investigate self-similarity properties of solar wind magnetic field fluctuations at different timescales and the scaling relation of structure functions at different orders. The main results include multi-fractal and mono-fractal scalings in the inertial range and the kinetic/dissipative range, respectively.

Geneste et al. [12] address intermittency in high Reynolds number turbulence by studying the universality of the multi-fractal scaling of structure function of the Eulerian velocity. Experimentally, they measure the radial, axial and azimuthal velocity in a Von Kármán flow, using the Stereoscopic Particle Image Velocimetry technique at different resolutions while performing direct numerical simulations of the Navier-Stokes equations. They demonstrate a beautiful log-universality in structure functions, link it to multi-fractal free energy based on the analogy between multi-fractal and classical thermodynamics and invoke a new idea of a phase transition related to fluctuating dissipative time scale.

Podgórska [13] discuss the effect of internal (fine-scale) intermittency due to vortex stretching on liquid–liquid dispersions in a turbulent flow with applications to industry. The internal intermittency is related to a strong local and instantaneous variability of the energy dissipation rate, and the k-$\varepsilon$ model and multifractal formalism are used to understand turbulence properties and internal intermittency in droplet breakage and coalescence. By solving the population balance equation and CFD simulations, they elucidate the effects of the impeller type—six-blade Rushton turbine and three-blade high-efficiency impeller—and droplet breakage coalescence (dispersion) on drop size distribution.

De Divitiis [14] review their previous works on homogenous isotropic turbulence for incompressible fluids and a specific (non-diffusive) Lyapunov theory for closing the von Kármán–Howarth and Corrsin equations without invoking the eddy-viscosity concepts. In particular, they show that the bifurcation rate of the velocity gradient along fluid particle trajectories exceeds the largest Lyapunov exponent and that the statistics of finite-time Lyapunov exponent of the velocity gradient follows normal distributions. They also discuss the statistics of velocity and temperature difference by utilising a statistical decomposition based on extended distribution functions and the Navier–Stokes equations.

**Acknowledgments:** We express our thanks to the authors of the above contributions, and to the journal Entropy and MDPI for their support during this work.

**Conflicts of Interest:** The author declares no conflict of interest.

## References

1. Mathur, A.; Seddighi, M.; He, S. Transition of Transient Channel Flow with High Reynolds Number Ratios. *Entropy* **2018**, *20*, 375. [CrossRef]
2. Chliamovitch, G.; Thorimbert, Y. Turbulence through the Spyglass of Bilocal Kinetics. *Entropy* **2018**, *20*, 539. [CrossRef]
3. Jacquet, Q.; Kim, E.; Hollerbach, R. Time-Dependent Probability Density Functions and Attractor Structure in Self-Organised Shear Flows. *Entropy* **2018**, *20*, 613. [CrossRef]
4. Xu, B.; Feng, J.; Wang, T.; Yuan, Y.; Zhao, Z.; Zhong, W. Trailing-Edge Flap Control for Mitigating Rotor Power Fluctuations of a Large-Scale Offshore Floating Wind Turbine under the Turbulent Wind Condition. *Entropy* **2018**, *20*, 676. [CrossRef]
5. Anderson, J.; Moradi, S.; Rafiq, T. Non-Linear Langevin and Fractional Fokker–Planck Equations for Anomalous Diffusion by Lévy Stable Processes. *Entropy* **2018**, *20*, 760. [CrossRef]
6. Saini, R.; Karimi, N.; Duan, L.; Sadiki, A.; Mehdizadeh, A. Effects of Near Wall Modeling in the Improved-Delayed-Detached-Eddy-Simulation (IDDES) Methodology. *Entropy* **2018**, *20*, 771. [CrossRef]

7. Barbay, S.; Coulibaly, S.; Clerc, M. Alternation of Defects and Phase Turbulence Induces Extreme Events in an Extended Microcavity Laser. *Entropy* **2018**, *20*, 789. [CrossRef]
8. Wang, W.; Pan, C.; Wang, J. Wall-Normal Variation of Spanwise Streak Spacing in Turbulent Boundary Layer with Low-to-Moderate Reynolds Number. *Entropy* **2019**, *21*, 24. [CrossRef]
9. Van Milligen, B.; Carreras, B.; García, L.; Nicolau, J. The Radial Propagation of Heat in Strongly Driven Non-Equilibrium Fusion Plasmas. *Entropy* **2019**, *21*, 148. [CrossRef]
10. He, J.; Wang, X.; Lin, M. Coherent Structure of Flow Based on Denoised Signals in T-junction Ducts with Vertical Blades. *Entropy* **2019**, *21*, 206. [CrossRef]
11. Alberti, T.; Consolini, G.; Carbone, V.; Yordanova, E.; Marcucci, M.; De Michelis, P. Multifractal and Chaotic Properties of Solar Wind at MHD and Kinetic Domains: An Empirical Mode Decomposition Approach. *Entropy* **2019**, *21*, 320. [CrossRef]
12. Geneste, D.; Faller, H.; Nguyen, F.; Shukla, V.; Laval, J.; Daviaud, F.; Saw, E.; Dubrulle, B. About Universality and Thermodynamics of Turbulence. *Entropy* **2019**, *21*, 326. [CrossRef]
13. Podgórska, W. The Influence of Internal Intermittency, Large Scale Inhomogeneity, and Impeller Type on Drop Size Distribution in Turbulent Liquid-Liquid Dispersions. *Entropy* **2019**, *21*, 340. [CrossRef]
14. De Divitiis, N. Statistical Lyapunov Theory Based on Bifurcation Analysis of Energy Cascade in Isotropic Homogeneous Turbulence: A Physical–Mathematical Review. *Entropy* **2019**, *21*, 520. [CrossRef]

*entropy*

MDPI

*Article*

# Transition of Transient Channel Flow with High Reynolds Number Ratios

**Akshat Mathur [1], Mehdi Seddighi [1,2] and Shuisheng He [1,*]**

1 Department of Mechanical Engineering, University of Sheffield, Sheffield S1 3JD, UK; akshatm@gmail.com (A.M.); M.Seddighi@ljmu.ac.uk (M.S.)

2 Department of Maritime and Mechanical Engineering, Liverpool John Moores University, Liverpool L3 3AF, UK

* Correspondence: s.he@sheffield.ac.uk; Tel.: +44-114-222-7756; Fax: +44-114-222-7890

Received: 24 March 2018; Accepted: 15 May 2018; Published: 17 May 2018

**Abstract:** Large-eddy simulations of turbulent channel flow subjected to a step-like acceleration have been performed to investigate the effect of high Reynolds number ratios on the transient behaviour of turbulence. It is shown that the response of the flow exhibits the same fundamental characteristics described in He & Seddighi (J. Fluid Mech., vol. 715, 2013, pp. 60–102 and vol. 764, 2015, pp. 395–427)—a three-stage response resembling that of the bypass transition of boundary layer flows. The features of transition are seen to become more striking as the Re-ratio increases—the elongated streaks become stronger and longer, and the initial turbulent spot sites at the onset of transition become increasingly sparse. The critical Reynolds number of transition and the transition period Reynolds number for those cases are shown to deviate from the trends of He & Seddighi (2015). The high Re-ratio cases show double peaks in the transient response of streamwise fluctuation profiles shortly after the onset of transition. Conditionally-averaged turbulent statistics based on a λ_2-criterion are used to show that the two peaks in the fluctuation profiles are due to separate contributions of the active and inactive regions of turbulence generation. The peak closer to the wall is attributed to the generation of "new" turbulence in the active region, whereas the peak farther away from the wall is attributed to the elongated streaks in the inactive region. In the low Re-ratio cases, the peaks of these two regions are close to each other during the entire transient, resulting in a single peak in the domain-averaged profile.

**Keywords:** pipe flow boundary layer; turbulent transition; large eddy simulation; channel flow

---

## 1. Introduction

Unsteady turbulent flow remains a topic of interest to researchers for many years. The transient response of turbulence to unsteady flow conditions exhibits interesting underlying physics that are not generally observed in steady turbulent flows. It has the potential to give insight into the fundamental physics of turbulence, as well as holds practical importance in engineering applications and turbulence modelling. Unsteady flows are generally classified as periodic and non-periodic flows. Turbulent periodic flows have been investigated extensively over the years, both experimentally and computationally. Examples of such studies include Tu and Ramaprian [1], Shemer et al. [2], Brereton et al. [3], Tardu et al. [4], Scotti and Piomelli [5] and He and Jackson [6]. The focus of the present paper is non-periodic turbulent flows, especially concerning accelerating (or ramp-up) flows, the work of which is reviewed below.

Maruyama et al. [7] presented one of the earliest experimental investigations on the transient response of turbulence following a step change in flow. It was reported that the generation and propagation of "new" turbulence are the dominant processes in the step-increase flow cases, whereas, the decay of "old" turbulence is the dominant process in step-decrease case. He and Jackson [8] presented a comprehensive experimental investigation of linearly accelerating and decelerating pipe flows, with initial and final Reynolds numbers ranging from 7000 to 45,200 (based on bulk velocity and pipe diameter). Consistent with the earlier studies, the authors concluded that turbulence responds first in the near-wall region and then propagates to the core of the flow. It was further reported that the streamwise velocity is the first to respond in the wall region followed by the transverse components, while all components responded approximately at the same time in the core region. Overall, turbulence was shown to produce a two-stage response—an initial slow response followed by a rapid one. The behaviour of turbulence was explained by the delays associated with turbulence production, energy redistribution and propagation processes. Experimental investigation with much higher initial and final Reynolds numbers (i.e., 31,000 and 82,000, respectively, based on bulk velocity and pipe diameter) and higher acceleration rates was presented by Greenblatt and Moss [9]. It was reported that the results were in agreement with the earlier studies. In addition, the authors reported a second peak of turbulence response in a region away from the wall (at $y^+ \sim 300$). Other notable reports on the transient response of turbulence include the experimental study of He et al. [10], and the computational investigations of Chung [11], Ariyaratne et al. [12], Seddighi et al. [13] and Jung and Chung [14].

Recent numerical studies of He and Seddighi [15,16] and Seddighi et al. [17] have proposed a new interpretation of the behaviour of transient turbulent flow. It was reported that the transient flow following a rapid increase in flow rate of turbulent flow is effectively a laminar-turbulent transition similar to bypass transition in a boundary layer. With an increase in flow rate, the flow does not progressively evolve from the initial turbulent flow to a new one, but undergoes a process with three distinct phases of pre-transition (laminar in nature), transition and fully-turbulent. These resemble the three regions of boundary layer bypass-transition, namely, the buffeted laminar flow, the intermittent flow and fully developed regions, respectively. The turbulent structures present at the start of the transient, like the "free-stream turbulence" in boundary layer flows, act as a perturbation to a time-developing laminar boundary layer. Elongated streaks of high and low streamwise velocities are formed, which remain stable in the pre-transition period. In the transition period, isolated turbulent spots are generated which eventually grow in both streamwise and spanwise directions and merge with one another occupying the entire wall surface. Seddighi et al. [17] further reported that a slow ramp-type accelerating flow also shows a transitional response despite having quantitative differences in its mean and instantaneous flow. Jung and Kim [18] conducted a more comprehensive study on the effects of changing the acceleration rate and the final/initial Reynolds number ratio by systematically varying these parameters in a direct numerical simulation (DNS) study. They noted that when the increase of the Reynolds number is small or when the acceleration is mild, transition could not be clearly identified through visualisation, which was consistent with the observation by He and Seddighi [16]. The authors went further and attempted to develop a criterion for when transition could be clearly observed.

More recently, the transition nature of a transient turbulent flow starting from a turbulent flow has been demonstrated experimentally by Mathur et al. [19] in a channel, and Sundstrom and Cervantes [20,21] in a circular pipe. The former focused on the transition physics, especially the abrupt changes in the length and time scales of turbulence as the transition occurs. Their experiments were accompanied by large eddy simulations (LES) of the experiments and an analytical solution based on the extended Stokes first problem solutions for the early stages of the flow. Sundstrom and Cervantes [20] obtained an analytical solution for the pre-transition phase of an accelerating flow and demonstrated that the velocity profile possess a self-similarity during the early stages. Sundstrom and Cervantes [21] on the other hand compared experimental results of accelerating and pulsating flows. They have found that, like accelerating flows, the accelerating phase of the pulsating flow also demonstrated distinct staged development, namely, a laminar-like development followed by rapid generation of turbulence.

The DNS study presented by He and Seddighi [16] (HS15, hereafter) covered a Reynolds number range from 2800 to 12,600 (i.e., a maximum Reynolds number ratio of 4.5). The initial turbulence intensity, $Tu_0$, equivalent to 'free-stream turbulence' of boundary layer flows was thus defined by HS15, by using peak turbulence following the commencement of the transient:

$$Tu_0 = \frac{\left(u'_{rms,0}\right)_{max}}{U_{b1}} \approx 0.375 \frac{U_{b0}}{U_{b1}} (Re_0)^{-0.1} \quad (1)$$

where $\left(u'_{rms,0}\right)_{max}$ is the peak r.m.s. streamwise fluctuating velocity of the initial flow; $U_{b0}$ and $U_{b1}$ are the initial and final bulk velocities, respectively; and $Re_0$ is the initial Reynolds number ($Re_0 = U_{b0}\delta/\nu$, where $\delta$ is the channel half-height and $\nu$ denotes the fluid kinematic viscosity). The "turbulence intensity" range covered by HS15 was 15.4% down to 3.8%. The purpose of the present study is to extend the range of turbulence intensity or Reynolds number ratio using large eddy simulations. The present paper increases the final flow to a Reynolds number of 45000; thereby increasing the Reynolds number ratio to ~19 and decreasing the turbulence intensity to 0.9%. The effect of high *Re*-ratio on the overall transition process, the transitional Reynolds number and the turbulent fluctuations is presented here. The simulations are also performed on different domain sizes to investigate the effect of domain length.

## 2. Methodology

Large-eddy simulations of unsteady turbulent channel flow are performed using an *in-house* code, developed by implementing subgrid calculations on the base DNS code, *CHAPSim* [15,22]. The resulting filtered governing equations in dimensionless form read:

$$\frac{\partial \overline{u}_i}{\partial t} + \frac{\partial}{\partial x_j}\left(\overline{u}_i \overline{u}_j\right) = -\frac{\partial \overline{P}}{\partial x_i} + \frac{1}{Re_c}\frac{\partial^2 \overline{u}_i}{\partial x_j \partial x_j} - \frac{\partial \tau_{ij}}{\partial x_j} \quad (2)$$

$$\frac{\partial \overline{u}_i}{\partial x_i} = 0 \quad (3)$$

where the overbar (-) denotes a spatially-filtered variable, $Re_c$ is Reynolds number based on characteristic velocity ($Re_c = U_c\delta/\nu$) and $\tau_{ij}$ represents the residual (or subgrid-scale) stress:

$$\tau_{ij} = \overline{u_i u_j} - \overline{u}_i \overline{u}_j \quad (4)$$

Here, the governing equations are non-dimensionalised using the channel half-height ($\delta$), characteristic velocity ($U_c$), time scale ($\delta/U_c$) and pressure-scale ($\rho U_c^2$). $x_1$, $x_2$, $x_3$ and $u_1$, $u_2$, $u_3$ stand for streamwise, wall-normal and spanwise coordinates and velocities, respectively. Although the characteristic velocity ($U_c$) used in the simulations was the centreline velocity of the laminar Poiseuille flow at the initial flow rate, the results presented here are re-scaled using the initial bulk velocity ($U_{b0}$) as the characteristic velocity. The governing Equations (2) and (3) are spatially discretized using second-order central finite-difference scheme. An explicit third-order Runge-Kutta scheme is used for temporal discretization of the non-linear terms, and an implicit second-order Crank-Nicholson scheme for the viscous terms. In addition, the continuity equation is enforced using the fractional-step method (Kim and Moin [23]; Orlandi [24]). The Poisson equation for the pressure is solved by an efficient 2-D fast Fourier transform (FFT, Orlandi [24]). Periodic boundary conditions are applied in the streamwise and spanwise directions and a no-slip boundary condition on the top and bottom walls. The code is parallelized using the message-passing interface (MPI) for use on a distributed-memory computer cluster. Detailed information on the numerical methods and discretization schemes used in the code, and its validation can be found in Seddighi [22] and He and Seddighi [15]. The subgrid-scale stress is modelled using the Boussinesq eddy viscosity assumption:

$$\tau_{ij} - \frac{1}{3}\tau_{kk}\delta_{ij} = 2\nu_{sgs}\overline{S}_{ij} \quad (5)$$

where $\delta_{ij}$ is Kronecker delta, $\nu_{sgs}$ is the *subgrid-scale viscosity* and $\overline{S}_{ij}$ is the resolved strain rate. The subgrid-scale viscosity is modelled using the WALE model of Nicoud and Ducros [25]:

$$\nu_{sgs} = (C_w\Delta)^2 \frac{\left(\overline{S}_{ij}^d \overline{S}_{ij}^d\right)^{3/2}}{\left(\overline{S}_{ij}\overline{S}_{ij}\right)^{5/2} + \left(\overline{S}_{ij}^d \overline{S}_{ij}^d\right)^{5/4}} \tag{6}$$

where $\overline{S}_{ij}^d$ is the traceless symmetric part of the square of the filtered velocity gradient tensor, $\overline{S}_{ij}$ is the filtered strain rate tensor, $C_w$ is the model constant and $\Delta$ is the filter width which is defined as $(\Delta x_1.\Delta x_2.\Delta x_3)^{1/3}$. As the above model invariant is based on both local strain rate and rotational rate of the flow, the model is said to account for all turbulent regions and is shown to even reproduce transitional flows [25].

For validation purpose, the results of the present code have been compared with DNS results. In Figure 1, steady turbulent channel flow statistics for the present code at $Re_\tau \sim 950$ have been compared with those of Lee and Moser [26] at $Re_\tau \sim 1000$ ($Re_\tau = u_\tau\delta/\nu$, is the frictional Reynolds number defined using the friction velocity, $u_\tau$, and channel half-height). It can be seen that the LES profiles are in agreement with those of DNS. It should be noted that the peak streamwise turbulent fluctuation is predicted fairly accurately by the LES, even though the predictions are less accurate away from the wall-region. A further validation of the present LES code for unsteady flow is presented in Figure 2, where two DNS accelerating flow cases of He and Seddighi [15,16] are reproduced. It is clear from the figure that the transient response of friction factor predicted by LES follows very closely that of DNS. Although the final steady value of LES is slightly higher than that of DNS (i.e., turbulence shear is slightly over-predicted), the timing of the minimum friction factor and the recovery periods are accurately predicted by the LES.

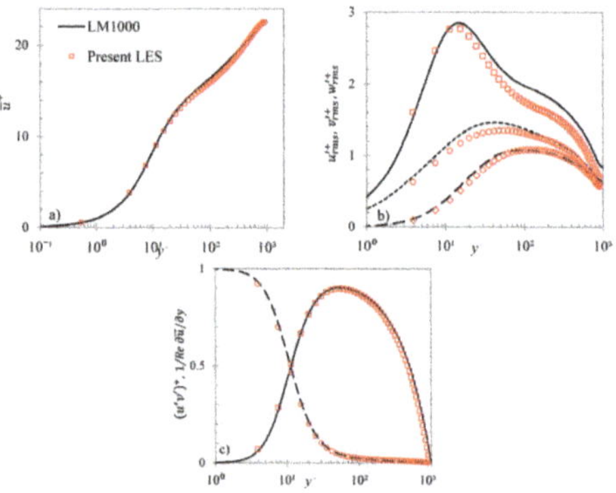


**Figure 1.** Comparison of present LES of steady channel flow at $Re_\tau \sim 950$ with DNS of Lee & Moser (2015) $Re_\tau \sim 1000$. (**a**) mean velocity in wall coordinates; (**b**) r.m.s. velocity fluctuations in wall coordinates (DNS: — $u'^+_{rms}$, - - $v'^+_{rms}$, - $w'^+_{rms}$; LES: □ $u'^+_{rms}$, ◇ $v'^+_{rms}$, ○ $w'^+_{rms}$); and (**c**) Reynolds and viscous stresses in wall coordinates (DNS: — $(u'v')^+$, - - $1/Re\ \partial\overline{u}/\partial y$; LES: □ $(u'v')^+$, ○ $1/Re\ \partial\overline{u}/\partial y$ ).

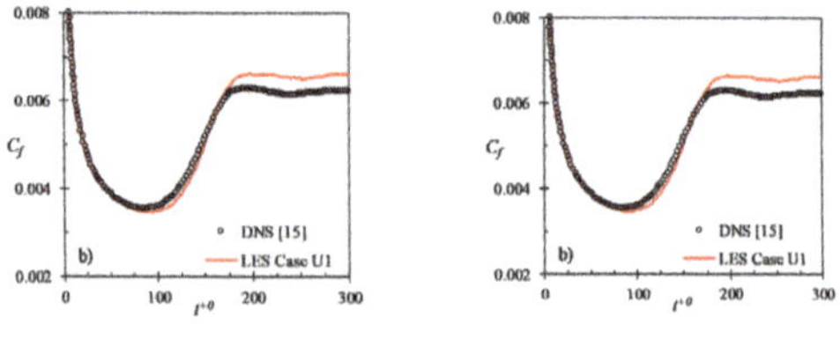


**Figure 2.** Present LES validation cases, U1 and U2, compared with the DNS cases of He & Seddighi (2013) [15].

## 3. Results and Discussion

Simulations are performed for a spatially fully developed turbulent channel flow subjected to a step-like linear acceleration using large eddy simulations. Two cases (U1 and U2), as described above, have been used to validate the LES spatial resolution with that of the DNS results of He and Seddighi [15,16]. Further four cases have been designed with Reynolds number ratios up to 19. The present cases have been described in Table 1. The spatial resolution provided in the table is in wall units of the final flow. Multiple realizations have been performed for each case, each starting from a different initial flow field. The spatial resolution of the cases U3–U5 resembles that of the LES validation cases, U1 and U2. However, due to limited computational resources, the resolution of the case U6 has been restricted to lower values. It is expected that the basic physical phenomena and trend of 'transition' has been captured despite the lower spatial resolution. Cases U3–U6 have also been repeated with different domain lengths to ensure that there is a minimal effect of the domain length on the physical process.

**Table 1.** Present accelerating flow cases with the DNS cases of He & Seddighi (2013, 2015) for comparison.

| Case | $Re_0$ | $Re_1$ | $\frac{Re_1}{Re_0}$ | $Tu_0$ | Grid | $L_x$/ffi | $L_z$/ffi | $\Delta x^{+1}$ | $\Delta z^{+1}$ | $\Delta y_c^{+1}$ |
|---|---|---|---|---|---|---|---|---|---|---|
| HS13 [15] | 2825 | 7404 | 2.6 | 0.065 | 512 × 200 × 200 | 12.8 | 3.5 | 11 | 7 | 7 |
| HS15 [16] | 2800 | 12,600 | 4.5 | 0.038 | 1024 × 240 × 480 | 18 | 5 | 12 | 7 | 10 |
| U1 | 2825 | 7400 | 2.6 | 0.065 | 192 × 128 × 160 | 12.8 | 3.5 | 28 | 9 | 13 |
| U2 | 2825 | 12,600 | 4.5 | 0.038 | 450 × 200 × 300 | 18 | 5 | 26 | 11 | 13 |
| U3 | 2825 | 18,500 | 6.5 | 0.026 | 1200 × 360 × 540 | 24 | 5 | 19 | 9 | 10 |
| U4 | 2825 | 25,000 | 8.8 | 0.019 | 2400 × 360 × 360 | 48 | 3 | 24 | 10 | 13 |
| U5 | 2825 | 35,000 | 12.4 | 0.014 | 2400 × 360 × 360 | 48 | 3 | 32 | 13 | 18 |
| U6 | 2333 | 45,000 | 19.3 | 0.009 | 2400 × 360 × 360 | 72 | 3 | 60 | 17 | 22 |

### 3.1. Instantaneous Flow Features

The flow structures at several time instants during the transient period for cases U3 and U6 are presented in Figure 3, using the isosurface plots of $u'/U_{b0}$ and $\lambda_2/(U_{b0}/\delta)^2$. Here, the blue and green isosurfaces are the positive and negative streamwise velocity fluctuations, $u'(= u - \overline{u})$; and red iso-surfaces are vortical structures represented by $\lambda_2$, where $\lambda_2$ is the second largest eigenvalue of the symmetric tensor $S^2 + \Omega^2$, $S$ and $\Omega$ are the symmetric and anti-symmetric velocity gradient tensor $\nabla u$. Figure 3a shows instantaneous plots in the entire domain size ($24\delta \times 5\delta$ in X–Z direction) for case U3. However, due to space constraints, only one-third of the domain length ($24\delta \times 3\delta$ in X–Z directions) is presented for case U6 in Figure 3b. Also presented in the inset is the development of the friction coefficient for the corresponding wall for a single realization. The symbols indicate the time instants for which the instantaneous plots are shown. The critical times of onset and completion of transition are clearly identifiable from the development of the friction coefficient (He and Seddighi [15]). The time of minimum friction coefficient approximately corresponds to the appearance of first turbulent spots and, hence, the onset of transition; while the time of first peak corresponds to a complete coverage of wall with newly generated turbulence and, hence, the completion time.

It is seen that the response of the transient flow is essentially the same as that described in He and Seddighi [15,16]—a three stage response resembling the bypass transition of boundary layer flows. In the initial flow (at $t^{+0} = 0$), patches of high- and low-speed fluctuating velocities and vortical structures are seen, representative of a typical turbulent flow. In the early period of the transient (at $t^{+0} = 20$), elongated streaks are formed, represented by alternating tubular structures of isosurfaces of positive and negative $u'/U_{b0}$. These structures are similar to those found in the pre-transition regions of the boundary layer flow (Jacobs and Durbin [27]; Matsubara and Alfredsson [28]). The number of vortical structures is also seen to reduce during this stage. Further at $t^{+0} = 40$, it seen that the streak structures are further stretched and become stronger. It is noted that in the higher Reynolds number-ratio case, the streaks appear stronger and longer; and the vortical structures appear to reduce by a greater extent—a trend also reported in HS15. New vortical structures start to appear at $t^{+0} = 65$, representing burst of turbulent spots which trigger the onset of transition. Afterwards, these turbulent spots grow with time to occupy more wall surface and eventually cover the entire domain signifying the completion of transition. It is again observed that the number of the initial turbulent spots seem to be more scarce for case U6 and some of the streaks extend nearly the entire domain length. Thus, the present domain lengths are sufficiently increased to reduce any effect of the domain size in the higher Reynolds-number ratio cases. This is further demonstrated later in the next section.

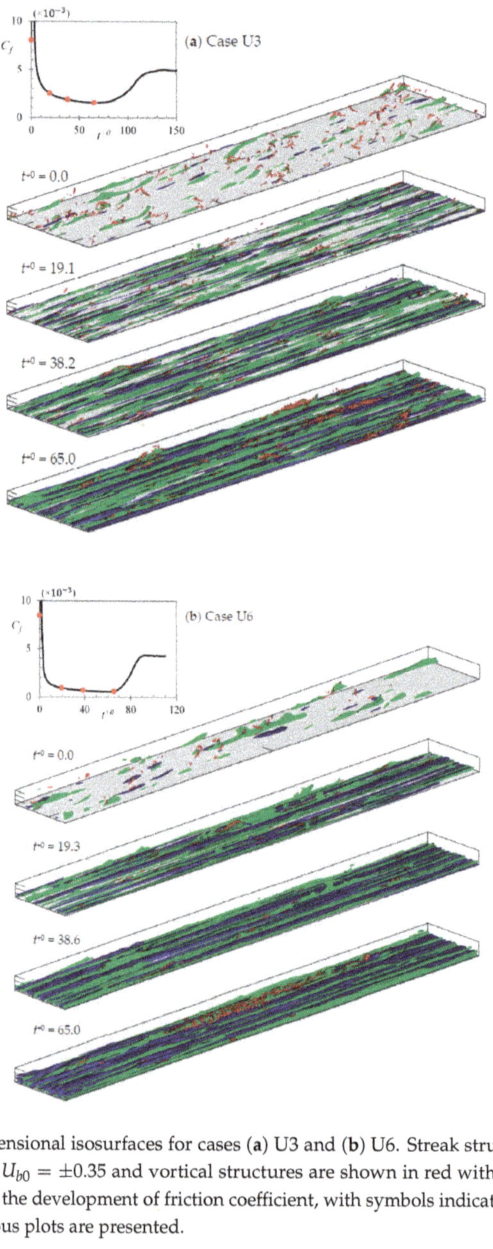


**Figure 3.** Three dimensional isosurfaces for cases (**a**) U3 and (**b**) U6. Streak structures are shown in blue/green with $u'/U_{b0} = \pm 0.35$ and vortical structures are shown in red with $\lambda_2/(U_{b0}/\delta)^2 = -5$. The inset plot shows the development of friction coefficient, with symbols indicating the time instants at which instantaneous plots are presented.

In order to visualise the instability and breakdown occurring in the low-speed streak, the site of the initial turbulent spot for case U3 is traced back in time; and a *sliding window* (of size $3\delta \times 1\delta$ in the X-Z direction) is used to follow the event in the domain during the late pre-transition and early transitional period, moving roughly a distance of 1δ downstream per two initial wall-units of time ($\Delta L_x / \Delta t^{+0} \sim 0.5\delta$). Visualisations of 3D isosurface structures inside this window are presented in Figure 4 at several time instants during this period. It is seen that for the most part of the pre-transition period (up to $t^{+0} = 49.7$) the streaks undergo elongation and enhancement. At about halfway during pre-transition period, the low-speed streak begins to develop an instability, similar to the sinuous instability of boundary-layer transitional flows (Brandt et al. [29–31]; Schlatter et al. [32]). This type of instability is reported to be driven by the spanwise inflections of the streamwise velocity and is characterised by antisymmetric spanwise oscillations of the low-speed streak (Swearingen and Blackwelder [33]). In the late pre-transitional period (about $t^{+0} = 57.3$), the streak appears to break down accompanying the generation of some vortical structures. Afterwards, bursts of turbulent structures appear surrounding the low-speed streak site, which continue to grow in size and soon outgrow the size of the window.

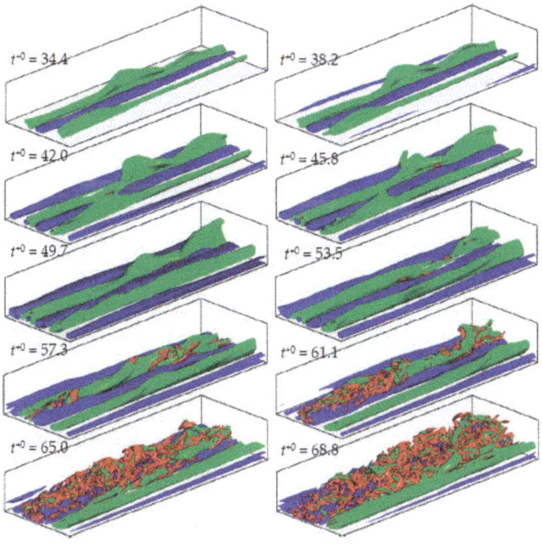


**Figure 4.** Visualization of streak instability and breakdown in case U3 using a sliding window. 3D iso-surface streak structures are shown in blue/green with $u'/U_{b0} = \pm 0.65$, and vortical structures are shown in red with $\lambda_2/(U_{b0}/\delta)^2 = -80$.

Overall, it is seen that the features of the transition process become more striking in case U6 than that in U3. The quantitative information about streaks can be obtained by the correlations of the streamwise velocity ($R_{11}$). Correlations in the streamwise direction provide a measure of the length of the streaks, whereas those in the spanwise direction measure the strength and the spacing between streaks. Figure 5 presents these correlations for case U3 (a,b) and U6 (c,d) in the streamwise (a,c) and spanwise directions (b,d). It can be seen from the initial flows (at $t^{+0} = 0$) of both cases that the length of the streaks (given by the streamwise correlations) is about 800 wall units (based on the initial flow) and the location of minimum spanwise correlations is about 50 wall units, implying that the spacing of streaks is about 100 wall units. This is representative of a typical turbulent flow. After the start of the transient, these streaks are stretched in the streamwise direction. It is seen that until the end of the pre-transitional period (at $t^{+0} = 70 - 80$), the streaks are stretched to a maximum of 1200 wall units in case U3, whereas to 3000 wall units in case U6. During this time, the spacing between the streaks is reduced to about 75 wall units in case U3, and to 56 wall units in case U6. The minimum value of the spanwise correlations provides a measure of strength of the streaks. It is clearly seen that this value is lower for case U6 in comparison to that in U3. Thus, the streaks in the pre-transitional stage of case U6 are much longer, stronger and more densely packed than those in case U3.

To further illustrate the development of the flow structures during pre-transition period, the variations of the integral length scales ($L = \int_0^{x_0} R_{11} dX$, where $x_0$ is the location when $R_{11}$ first reaches zero) in U3 and U6 are shown in Figure 6. It can be seen that the integral length scale increases significantly during the pre-transition period, reaching a peak at the time around the onset of transition. The peak value is over doubled that of its initial value in U3 but around 8 times in U6. This trend is clearly consistent with the streaks observed in Figure 3 and the correlations shown in Figure 5.

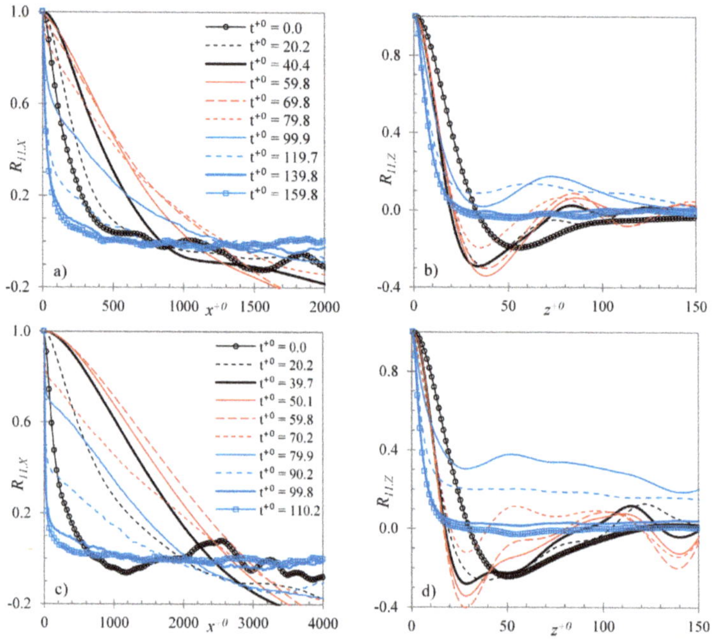


**Figure 5.** Streamwise velocity autocorrelations at several time instants during the transient for case U3 (**a**,**b**) and U6 (**c**,**d**) in the streamwise (**a**,**c**) and spanwise directions (**b**,**d**) at $y^{+0} = 10$.

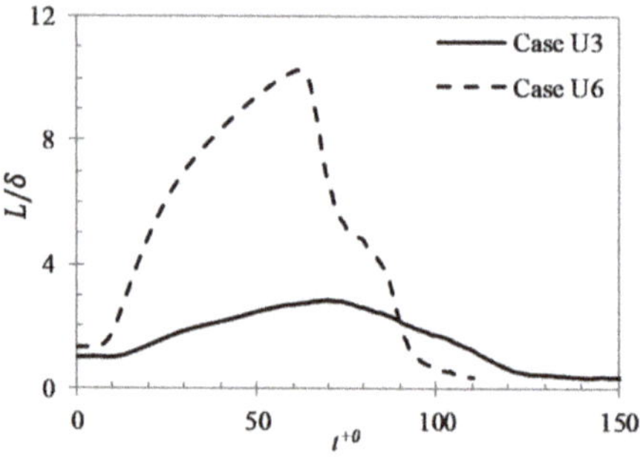


**Figure 6.** Development of the integral length scale of the flow in U3 and U6.

The near wall vortical structures were visualised by the $\lambda_2$-criterion in Figures 3 and 4 earlier. The same criterion can also be used to get some quantitative information about these structures. Jeong and Hussain [34] noted that $\lambda_2$ is positive everywhere outside a vortex core and can assume values comparable to the magnitudes of the negative $\lambda_2$ values inside the vortices. Jeong et al. [35] showed that due to significant cancellation of negative and positive regions of $\lambda_2$ in the buffer region, a spatial mean $\langle\lambda_2\rangle$ was an ineffective indicator of the vortical events. It was reported that the r.m.s. fluctuation of $\lambda_2$, $\lambda'_{2,rms}$, shows a peak value at $y^+ \sim 20$, indicating prominence of vortical structures in the buffer region. Hence, the maximum value of $\lambda'_{2,rms}$ can be used to compare the relative strength of these structures in the flow. Figure 7 shows the variation of $(\lambda'_{2,rms})_{max}$ during the transient for the cases U3 and U6. Here, $(\lambda'_{2,rms})_{max}$ is normalised by $U_{b0}/\delta$. It can be seen that in the early period of the transient, the value of $(\lambda'_{2,rms})_{max}$ increases abruptly during the excursion of the flow acceleration (till $t^{+0} \sim 3$). This is attributed to the straining of near-wall velocity due to the imposed flow acceleration, resulting in distortion of the pre-existing vortical structures and, hence, high fluctuations of $\lambda_2$. After the end of the acceleration, the values are seen to gradually reduce, which signify a breakdown of the equilibrium between the near-wall turbulent structures and the mean flow. The formation of high shear boundary layer due to the imposed acceleration causes the high-frequency disturbances to damp and shelters the small structures from the free-stream turbulence. This phenomenon of disruption of the near-wall turbulence is referred to as *shear sheltering* [36]. Later in the late pre-transition stage, $(\lambda'_{2,rms})_{max}$ begins to increase gradually as the new structures begin to form. At the onset of transition, this value increases rapidly due to burst of turbulent spots and generation of new turbulent structures in the flow. The rate of increase of $(\lambda'_{2,rms})_{max}$ can be used to indicate the strength of turbulence generation. It is clearly seen that the rate is higher for case U6, implying a stronger rate of turbulence generation in comparison to case U3.

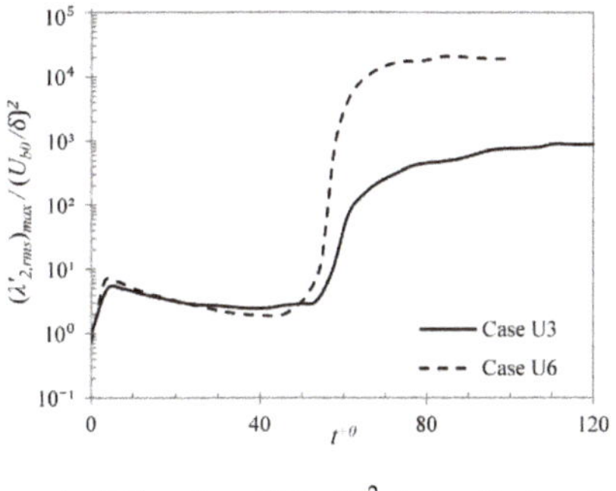


**Figure 7.** Time development of $(\lambda'_{2,rms})_{max}/(U_{b0}/\delta)^2$ during the transient for cases U3 and U6.

This trend is similar to that observed in HS15. Therein, the highest Reynolds number ratio case showed a distinct and clear transition process, but the transition of in the lowest ratio case was indiscernible from the instantaneous visualisations. Here, it is seen that as the Reynolds number ratio is increased further (larger than those in HS15), the features of the transition appear to be more striking and prominent. The streaks in the pre-transitional stage are longer and stronger, and are more densely packed, and after the onset of transition the generation of turbulence is stronger.

### 3.2. Correlations of Transition

The onset of transition can be clearly identified using the minimum friction factor during the transient [15]. Thus, a critical time of onset of transition ($t_{cr}$) can be obtained and used to calculate an equivalent critical Reynolds number, $Re_{t,cr} = t_{cr}U_{b1}^2/\nu$, where $U_{b1}$ is the bulk velocity of the final flow. Here, the equivalent Reynolds number ($Re_t$) can be considered analogous to the Reynolds number ($Re_x = xU_\infty/\nu$, where is $x$ the distance from the leading edge and $U_\infty$ is the free stream velocity) used in the boundary layer flows. It was demonstrated by HS15 that although these two

Reynolds numbers cannot be quantitatively compared, $Re_t$ has the same significance in the channel flow transition as $Re_x$ has in boundary layer transition.

Similar to that in boundary layer transition, the critical Reynolds number here is closely dependent on the initial 'free-stream turbulence' and can be represented by:

$$Re_{t,cr} = 1340\ Tu_0^{-1.71} \tag{7}$$

Figure 8 shows the relation between the equivalent critical Reynolds number and the initial turbulence intensity for the present LES cases and the DNS cases of HS15 for comparison. The present data follows the Equation (7) established from the higher turbulence intensity cases (U1–U4). However, the lower turbulent intensity cases, namely cases U5 and U6, are seen to diverge from this relation, with transition occurring at higher $Re_t$ values.

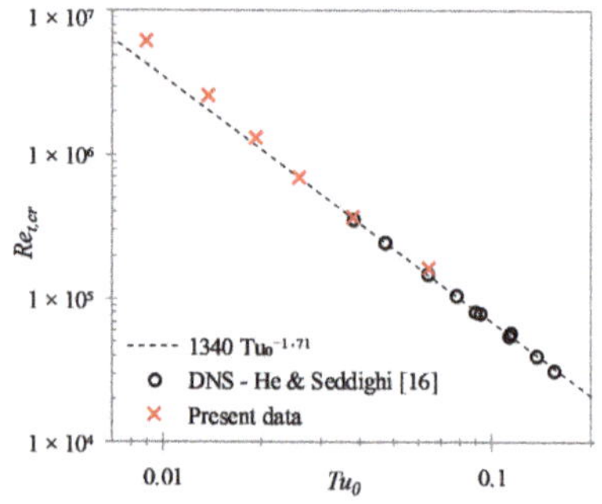


**Figure 8.** Dependence of equivalent critical Reynolds number on initial turbulence intensity.

Similar to onset of transition, friction factor can also be used to determine the time of completion of the transition process ($t_{turb}$). By assuming that the transition is complete when the friction factor reaches its first peak, a transition period can thus be obtained ($\Delta t_{cr} = t_{turb} - t_{cr}$). The relation between the equivalent transition period Reynolds number ($\Delta Re_{t,cr} = \Delta t_{cr} U_{b1}^2 / \nu$) and the critical Reynolds number is presented in Figure 9. Also shown in the figure is the power-relation for transition length of boundary layer flows by Narasimha et al. [37], and the linear-relation between the same by Fransson et al. [38]. It should be noted that $Re_{cr}$ in the figure denotes $Re_{t,cr}$ and $Re_{x,cr}$ for the boundary layer flow and the transient channel flow, respectively. It is seen that, similar to the findings of HS15, the presented data is reasonably well predicted by the boundary layer correlations if a factor of 0.5 is applied to the present $\Delta Re_{t,cr}$. However, the present data seem to suggest a power-relation between $\Delta Re_{t,cr}$ and $Re_{t,cr}$, similar to that of Narasimha et al. [37].

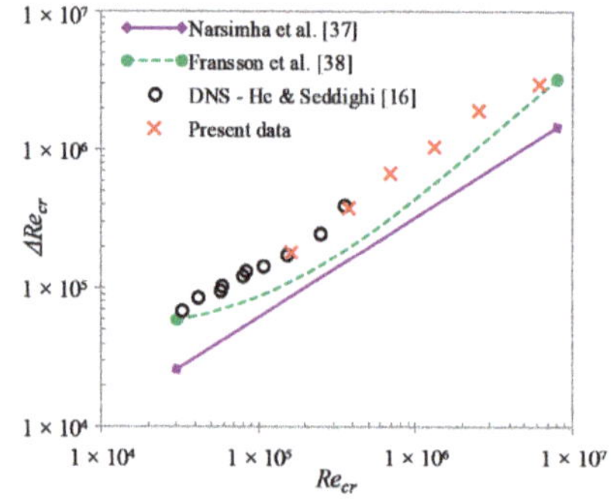


**Figure 9.** Relationship between transition period Reynolds number and critical Reynolds number.

The critical Reynolds number discussed above is naturally a statistical concept. In each flow realisation, the generation of turbulence spots and transition to turbulence may vary significantly around the "mean" $Re_{t,cr}$. The generation of turbulent spots is to some extent dependent on the initial flow structures. Due to this, the time and spatial position at which the generation of turbulent spot occurs can vary with different initial flow fields. Thus, several simulations have been run for each case, each starting from a different initial flow field to arrive at an average critical and transition period Reynolds numbers. It is observed that there are large deviations in the critical Reynolds number for different realizations, and for the top and bottom walls of a single realization for the present cases. Friction factor histories for both walls of different realizations for cases U3 and U6 are presented in Figure 10. It is seen that the deviations in the critical time are larger in case U6 than those in case U3. The degree of the scatters of the critical Reynolds number for the present cases is found to be linearly proportional to the average value. As shown in Figure 11, the r.m.s. of fluctuation of the critical Reynolds numbers are roughly 10% of the average value.

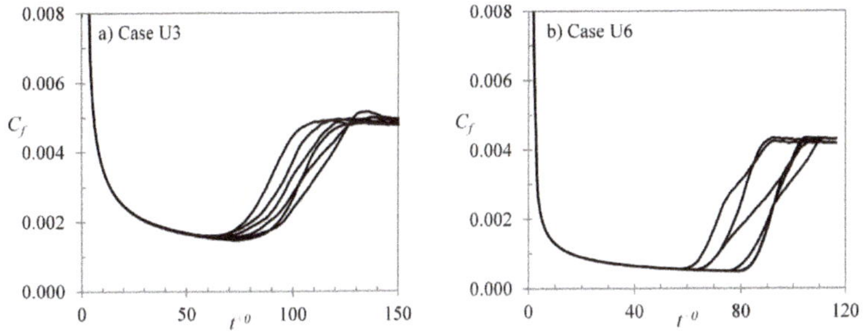


**Figure 10.** Deviations in different realizations for cases (**a**) U3; and (**b**) U6.

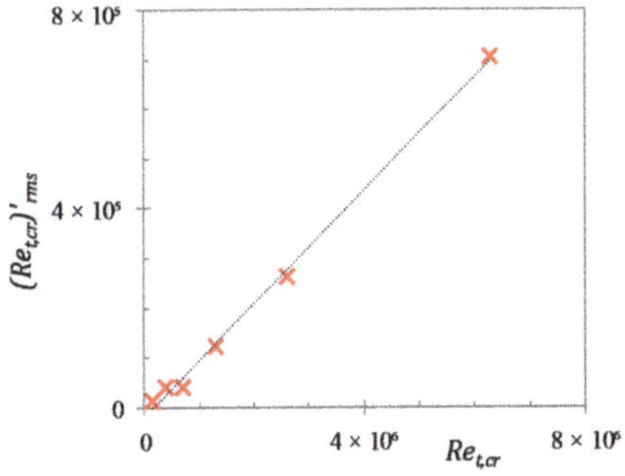


**Figure 11.** Deviations observed in the equivalent critical Reynolds number for the present cases.

The present higher Reynolds number ratio cases (namely, case U3–U6) were also simulated with different domain lengths to see its effect on the onset of transition and the deviations observed in its predicted critical time. Case U3 was performed with two different domain lengths—18δ and 24δ; cases U4 and U5 each with three lengths—18δ, 24δ and 48δ; whereas, case U6 with four different lengths—18δ, 24δ, 48δ and 72δ. It should be noted that the spatial resolution for different domain lengths of each case was kept roughly the same so that an appropriate comparison can be made. Figure 12 presents the friction factor histories for both walls of every realization for cases U3 and U6. It is observed that as the domain length is increased, the spread of deviations of $Re_{t,cr}$ for multiple realizations is slightly decreased. For case U6, the spread of deviations for the two larger domain lengths is almost identical. Hence, it can be deduced that the effect of domain lengths is very small for the two larger domains. The average critical Reynolds numbers and their r.m.s. deviations, for different domain lengths of cases U3–U6 are presented in Figure 13a,b, respectively. It is clearly seen that the critical Reynolds numbers obtained using different domain lengths for U3 to U5 are largely the same in each case, hence demonstrating the smallest domain size is adequate in capturing the transition time. It is also seen that the larger the domain or the smaller the Reynolds number ratio, the smaller the r.m.s. of $Re_{t,cr}$ suggesting less realisations are needed for such cases to obtained a reliable $Re_{t,cr}$. For case U6, the critical Reynolds number observed decreases slightly as the domain length is increased even for the largest domain sizes (Figure 13a). The streaks are very long and the initial turbulence spots generated are spares in a high *Re*-ratio flow, and hence a larger domain is required.

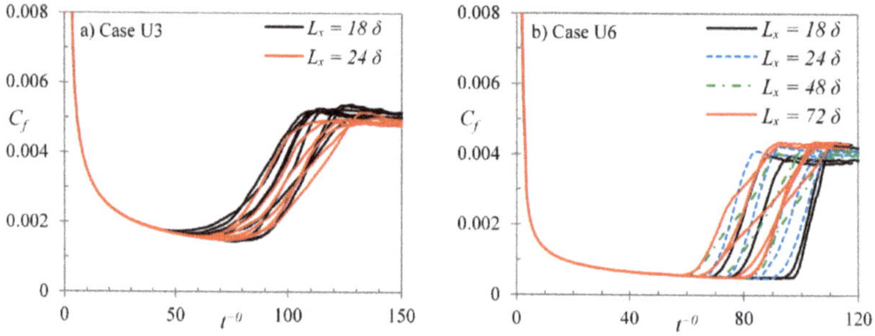


**Figure 12.** Friction factor developments using different domain lengths for cases (**a**) U3; and (**b**) U6.

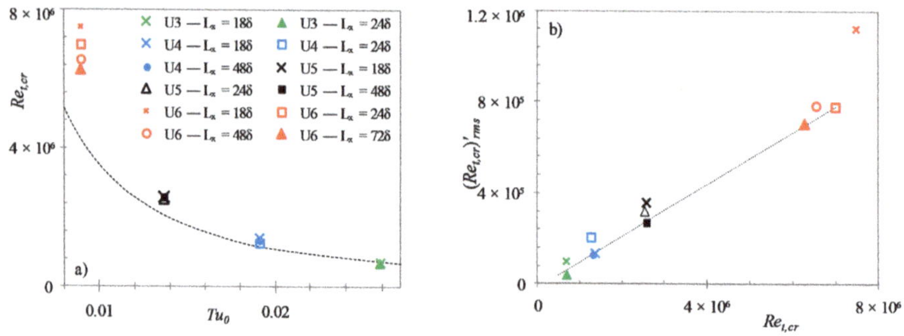


**Figure 13.** Effect of domain length on (**a**) the critical Reynolds number; and (**b**) r.m.s. fluctuation of critical Reynolds number. Here, the largest domain length in each case is marked with a solid/filled symbol.

### 3.3. Turbulent Fluctuations

Figure 14 presents the development of r.m.s. fluctuating velocity profiles for cases U3 and U6. As shown earlier in Figure 3, the critical time for both cases is approximately $t^{+0} = 65$, while the completion time for U3 and U6 are roughly $t^{+0} = 120$ and 85, respectively. It can be seen that following the start of the transient, $u'_{rms}$ progressively increases in the wall region and maintains this trend until the onset of transition. On the other hand, the transverse components ($v'_{rms}$ and $w'_{rms}$) reduce slightly from the initial values and remain largely unchanged until the onset of transition. The Reynolds stress increases very slightly during this period, exhibiting a behaviour that is closer to that of the transverse components than to that of the normal component. During the transition period, $u'_{rms}$ further increases rapidly in the near wall region. It is interesting to note that case U6 clearly shows formation of two peaks of $u'_{rms}$ during this period ($t^{+0} = 67 - 85$), however, case U3 shows a single peak. Similar double-peaks are also observed in cases U4 and U5 (not shown). The first peak, very close to the wall, is formed rapidly during the transitional period, increasing from very low initial values; whereas, the second peak, farther from the wall, is only slightly higher than that at the point of onset of transition. At the end of the transitional period, $u'_{rms}$ reduces and approaches its final steady value. During the transition period the transverse components increase rapidly and monotonically to peak values, showing a slight overshoot towards the end of the transient. The feature of two peaks is not shown by these components.

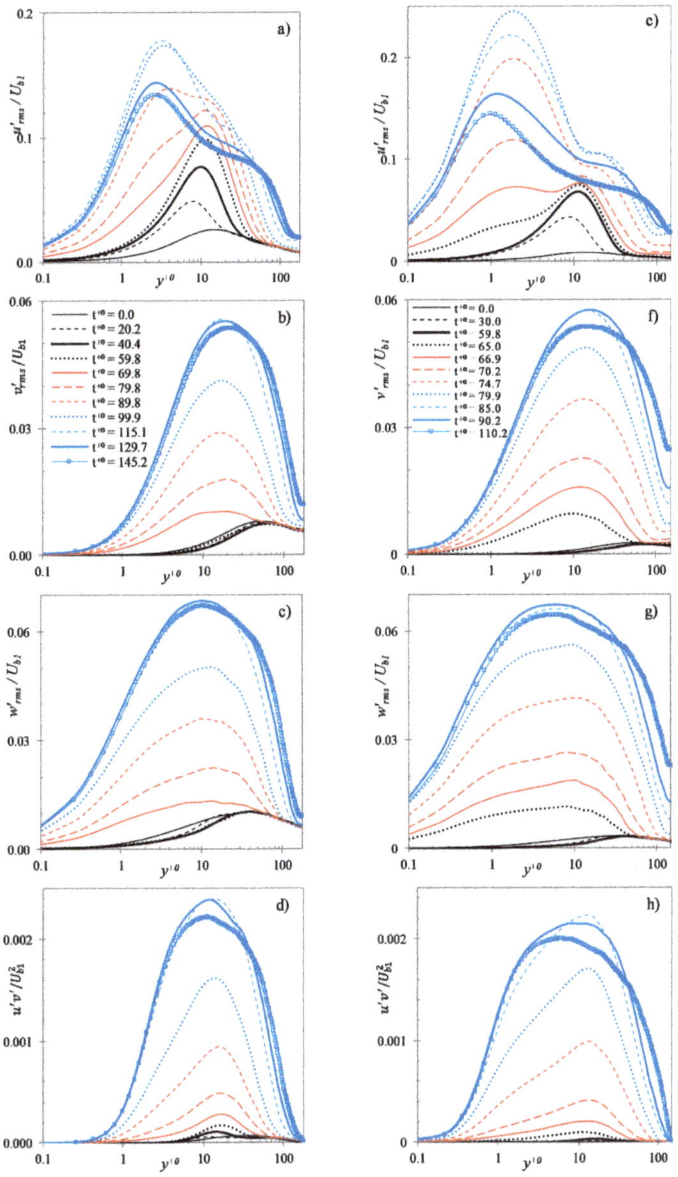

**Figure 14.** R.M.S. fluctuating velocities and Reynolds stress at several time instants during the transient in cases U3 (**a–d**) and U6 (**e–h**).

To further analyse the origin and location of the two peaks in the present cases, the *conditional sampling* technique of Jeong et al. [35] and Talha [39] is used. Here, the r.m.s. fluctuation of $\lambda_2$, $\lambda'_{2,rms}$, is used to distinguish the 'active areas' of turbulent generation from the 'inactive areas'. It should be noted that this technique is performed to separate the active areas of turbulence generation in the *x-z* domain, rather than in the wall-normal direction. The criterion is based on the comparison of a local r.m.s. fluctuation of $\lambda_2$ with a *base* value. The base value chosen here is the $\lambda'_{2,rms}$ of the entire *x*–*z* plane at the critical time of onset of transition. Similar to that used by Jeong et al. [35], a window of size ($\Delta x^+$, $\Delta z^+$) = (120, 50) is used to determine the local r.m.s. fluctuation. The r.m.s. fluctuation is computed in the *x-z* direction and, thus, is a function of *y*. The values are then summed in the wall-normal direction for 50 wall units and compared with each other. The criterion for determining active area reads:

$$\sum_{j=1}^{N_y} \widetilde{\lambda'_{2,rms}} \geq 0.1 \sum_{j=1}^{N_y} \lambda'_{2,rms,cr} \tag{8}$$

where $\widetilde{\lambda'_{2,rms}}$ is the local r.m.s. fluctuation value within the window, $\lambda'_{2,rms,cr}$ is the r.m.s. fluctuation value of the entire *x*–*z* plane at the onset of transition, and $N_y$ is the number of control volumes in the wall region of $y^+ < 50$. It should be noted that the wall units are based on the average friction velocity of all active areas in the domain. Hence, the determination of the window size is an iterative process. Number of iterations was kept such that the change in active area determination for successive iterations was less than 0.1%. It is seen in Figure 7 that the value of $(\lambda'_{2,rms})_{max}$ at the onset of transition ($t^{+0} = 65$) reaches close to the fully turbulent value. Thus, the criterion (Equation (8)) distinguishes the areas of *newly* generated turbulence in the transitional period. For any time before the onset of transition or after the completion of transition, the criterion gives 0% or 100% (of *x*–*z* domain), respectively, as active areas of turbulence generation.

The above scheme is used to distinguish the active areas of turbulent generation for all the present cases. At the beginning of the transient, the entire wall surface is classified as inactive region. At the onset of transition, the active region emerges at the location of the turbulent spot burst. During the transitional period, the active area grows in size and eventually covers the entire wall surface at the end of transitional period. To validate the above criterion, the instantaneous flow for case U3 during transitional period (at $t^{+0} = 89.8$) is presented in Figure 15. The instantaneous 3D iso-structures of $u'$ and $\lambda_2$ are presented in Figure 15a,b, respectively. Figure 15c shows the instantaneous contours of $u'$ at $y^{+0} = 5$, and Figure 15d shows the approximation of the active wall surface determined using Equation (7). It is clearly seen that the present scheme is suitable to capture the active areas of turbulent production during the transition. Although the edges of active regions may be smeared somewhat, any uncertainties caused to the active/inactive areas are negligible.

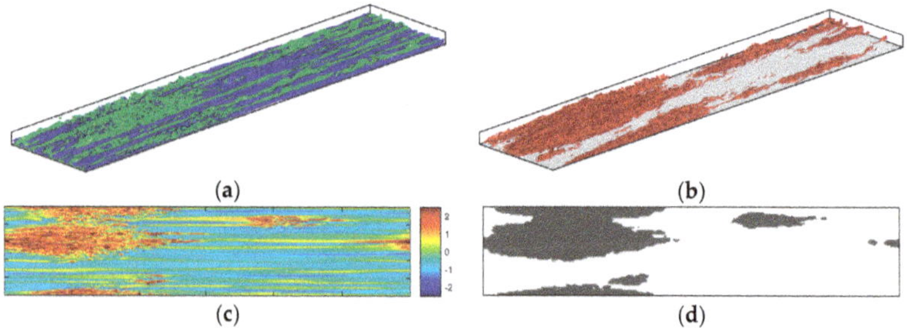


**Figure 15.** Instantaneous flow for case U3 at $t^{+0} = 89.8$ (**a**) isosurface structures of $u'/U_{b0} = \pm 0.35$; (**b**) isosurface structures of $\lambda_2/(U_{b0}/\delta)^2 = -5$; (**c**) contours of streamwise fluctuating velocity $u'/U_{b0}$ at $y^{+0} = 5$; (**d**) active region of turbulence production (shown in gray) determined using Equation (7).

Conditionally-averaged turbulent statistics for the active and inactive areas thus obtained are used to investigate the turbulent intensity contributions from each region. First, the statistics for case U6 at $t^{+0} = 67.5$ are presented where the double peak first seems to emerge. At this instant, active region constitutes only 5% of the wall surface. Figure 16 presents the conditionally-averaged velocity profiles, $\overline{u}_a$ and $\overline{u}_i$ for the active and inactive regions, respectively, along with the domain-averaged velocity profile, $\overline{u}_d$. It can be seen that the profiles of the two regions are very different. The inactive region profile resembles that of the pre-transition period, exhibiting a plug-like response to the acceleration, with profile flat in the core. The active region profile, however, has developed farther away from the wall and the near-wall shear resembles that of the final steady flow. The conditionally-averaged streamwise velocity fluctuation profiles at this time are presented in Figure 17. The contributions of fluctuation energy ($\overline{u'^2}$) from active/inactive regions to the domain-averaged profile are shown in Figure 17a, whereas, the conditionally-averaged r.m.s. fluctuation profiles ($u'_{rms}$) within these regions are shown in Figure 17b. It is clear from Figure 17a that the double peaks in the streamwise fluctuations is the net effect of two separate peaks from two separate regions of the flow, i.e., the active and inactive regions. The near-wall peak originates from the active region whereas that the peak further away from the wall originates from the inactive region. The former (located at $y^{+0} \sim 1.2$ or $y^{+1} \sim 15$) is attributed to the burst of new turbulent structures in the active region with its $y$-location consistent with that of the final steady flow, whereas, the latter (located at $y^{+0} \sim 12$) is the contribution of the elongated streaks in the inactive region. It should be noted that active area profile, $u_a'^2$, in Figure 17a too has a local second peak further away from the wall (around $y^{+0} \sim 20$). This is merely a numerical feature due to the method employed in the calculation, where the fluctuation is calculated with respect to the domain-averaged mean profile i.e., $u_a'^2 = \langle (u_a - \overline{u}_d)^2 \rangle$ and $u_i'^2 = \langle (u_i - \overline{u}_d)^2 \rangle$, where $\langle \, \rangle$ denotes a spatial average in the homogeneous ($x$–$z$) plane. This, however, is not an appropriate representation of the conditionally-averaged fluctuation energy because the domain-averaged profile varies from the conditionally-averaged profiles of the active and inactive regions (as seen in Figure 16). To further support this statement, conditionally-averaged r.m.s. fluctuation profiles within these two regions are presented separately in Figure 17b. Here, the velocity fluctuation is calculated with respect to the conditionally-averaged mean flow, i.e., $u'_{a,rms} = (u_a - \overline{u}_a)_{rms}$ and $u'_{i,rms} = (u_i - \overline{u}_i)_{rms}$. It is clear that the active region profile, here, shows a single peak consistent with the final steady profile.

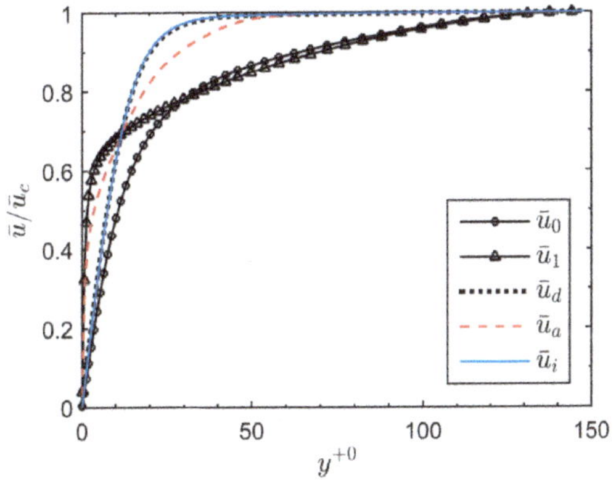


**Figure 16.** Conditionally-averaged velocity profiles of the active ($\overline{u}_a$) and inactive regions ($\overline{u}_i$), along with the domain-averaged ($\overline{u}_d$) for case U6 at $t^{+0} = 67.5$. Also shown are the initial ($\overline{u}_0$) and final ($\overline{u}_1$) steady flow profiles, for comparison.

Now, the development of these conditionally-averaged r.m.s. fluctuation profiles during the transient is presented in Figure 18. As shown earlier in Figure 3, the critical times of onset and completion of transition for case U6 are roughly $t^{+0} = 65$ and 85, respectively. It is seen that the inactive region profiles increase monotonously from the beginning of the transient until the end of the transitional period. The peak of the profile originates at $y^{+0} \sim 5$ and moves further away from the wall during the transient, reaching $y^{+0} \sim 12$ until the end of the transitional period. On the other hand, the active region profile is generated at the point of onset of transition which thereafter reduced gradually during the transitional period. The peak of this profile originates at $y^{+0} \sim 1.3$ ($y^{+1} \sim 20$) at the onset of transition and only moves slightly towards the wall during the transitional period and the post-transition period until it settles to the final steady value at $y^{+0} \sim 1$ ($y^{+1} \sim 14$).

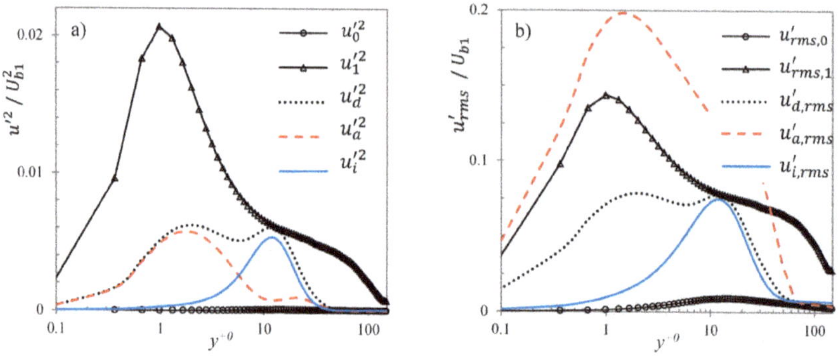


**Figure 17.** (**a**) Domain-averaged velocity fluctuation energy ($u_d'^2$), with contributions from the active ($u_a'^2$) and inactive ($u_i'^2$) regions for case U6 at $t^{+0} = 67.5$, and (**b**) conditionally-averaged velocity fluctuations of the active ($u'_{a,rms}$) and inactive regions ($u'_{i,rms}$), along with the domain average ($u'_{d,rms}$). Also shown in each plot are the domain-averaged initial (subscript 0) and final (subscript 1) steady profiles.

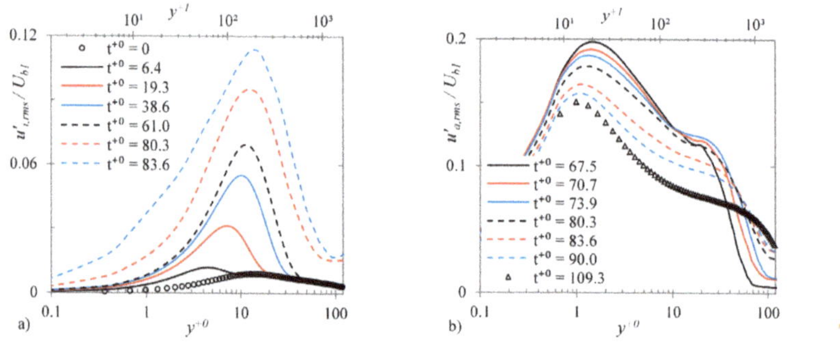


**Figure 18.** R.M.S. streamwise fluctuating velocity profiles at several time instants during the transient for (**a**) inactive and (**b**) active regions for case U6.

The maximum streamwise energy growth, $u_{rms,max}'^2 (= max_y\{u'_{rms}\}^2)$, and the $y$-location of its peak for the two different regions of case U6 is presented in Figure 19a,b, respectively. The domain-averaged energy, $(u'_{d,rms})^2$, similar to that in DNS cases of HS15, exhibits an initial delay following the start of the transient which is attributed to an early receptivity stage [38]. During the pre-transitional period, the energy increases linearly with time until the onset of transition. At this point, the energy increases rapidly owing to the burst of 'new' turbulence, overshooting the final steady value and reaching a peak around the end of the transitional period and thereafter reducing to reach the final steady value. It is seen that the energy growth in the inactive region, $(u'_{i,rms})^2$, grows linearly even after the onset of transition and continues to do so until the end of the transitional period. This is expected as the burst of turbulence generation occurs only in the active region, while the inactive region is dominated by the stable streaky structures which continue to develop further. Energy in the active region $(u'_{a,rms})^2$, on the other hand, is generated at the onset of transition at a value much higher than the final steady value which gradually reduces until the end of the transitional period and reaches the final steady value. It is worth noting that the sharp increase and the high peak observed in the maximum domain-averaged energy during the transitional period is only a numerical feature arising due to the method of statistical calculation. The domain-averaged energy comprises of the turbulent fluctuations from both the active and inactive regions calculated with respect to the domain-averaged mean velocity, resulting in high values of fluctuations. A more suitable representation during the transitional period is a weighted-average of the fluctuation energy, $(u'_{rms})^2_w = \alpha \cdot (u'_{a,rms})^2 + (1-\alpha) \cdot (u'_{i,rms})^2$, where subscript '$w$' denotes the *weighted-average*, and $\alpha$ is the *active* fraction of wall surface (plotted in Figure 19a). It is clear that the average energy of the streamwise fluctuations show only a slight overshoot during the transitional period. The overshoot is attributed to the increasingly dominant effect of the active region during this period, while the slight decrease towards the end of the transitional period is attributed to the redistribution of streamwise energy to transverse components.

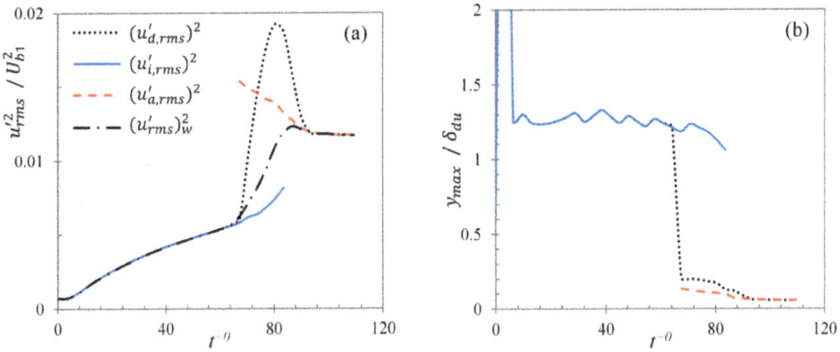


**Figure 19.** Conditionally-averaged (**a**) maximum energy growth and (**b**) the $y$-location of its peak, for case U6.

The $y$-location of the peak of streamwise energy, normalised by the displacement thickness of the velocity field ($\delta_u$), are shown in Figure 19b. It should be noted that conditionally-averaged peak energy location is normalised by $\delta_u$ of respective conditionally-averaged profile. Immediately after the commencement of the transient, a sharp increase is seen in $y/\delta_u$ value of the peak location in the inactive region. This is attributed to the formation of a new thin boundary layer of high shear due to the imposed acceleration, and hence a smaller boundary layer thickness. Further in the pre-transition period the peak of the energy profile is seen to scale with the displacement thickness, rather than the inner scaling, which is atypical of turbulent flows. The location of the peak maintains at ~$1.25\delta_u$ up until the onset of transition, implying that the streamwise energy grows with the growth of the time-developing boundary layer—a feature observed in bypass transitional flow. The peak in the inactive region is seen to largely maintain its location after the onset of transition showing only a slight decrease towards the end of the transitional period. The peak in the active region appears very close to the wall, typical of high Reynolds number turbulent flows. The displacement thickness of turbulent boundary layer in the active region increases with time as it becomes fully developed. Thus, the peak of the streamwise energy appears to move from ~$0.12\delta_u$ at the point of onset of transition to ~$0.06\delta_u$ at the end of the transient. During the pre-transitional period, the entire wall surface is inactive region, thus the domain-averaged peak follows the same trend as that in the inactive region. At the onset of transition, the active region peak, which appears much closer to the wall, has a much higher value than that in inactive region. At this point, the domain-averaged peak is dominated by the active region energy, and seems to follow the location of the active region peak. From the point of onset of transition until the end of transitional region, both active and inactive regions co-exist and exhibit separate developments of their respective streamwise energies. At the onset of transition, there is a large difference between the peak energy of the active region and that in the inactive region. Thus, even though the active region covers only a small fraction of the wall surface, the domain-averaged energy shows a dominant contribution from active region in the near-wall region. The difference between wall normal locations of the peak energies for the two regions also plays a role in enhancing the difference between two separate contributions. The domain-averaged profile, thus, shows the net effect of two peaks. The peak closer to the wall is attributed to the turbulent spots generated at the onset of transition, whereas, the one further away from the wall is attributed to the elongated streaks. In the late transitional period, most of the wall surface is covered with the new turbulence, thus reducing the area of the inactive region. This results in a decreasing contribution of the inactive region, until the inactive region energy is completely masked by the active region energy. At the end of the transitional period, the entire wall becomes the active region with only a single peak in the entire domain. Thus, from the late-transitional period until the end of the transient, the domain-averaged profile shows only a single peak (i.e., peak associated with the generation of 'new' turbulence in the active region). Separate developments of active and inactive regions exist in all the present cases (U1–U6). However, the feature of double-peaks is clearly visible only in cases U4–U6.

Figure 20a,b show the maximum streamwise fluctuations and the $y$-location of the peaks for the cases U1–U5, respectively. Here, the dotted lines represent the domain-averaged values, and the solid and dashed lines represent the conditionally-averaged inactive and active region values, respectively. It can be seen that at the onset of transition (time at which active region value appears), the difference between the maximum fluctuations of the active and inactive regions is very small for cases U1–U3. The resulting active region contribution to the domain-averaged value in the near-wall region is also less than that of the inactive region. Thus, the net effect in the domain-averaged value for these cases shows only a single peak during the transitional period—the peak corresponding to the inactive region; while the active region peak is masked by the inactive region fluctuations. Later in the transitional period, when the active region grows in size, its contribution becomes comparable to that of the inactive region. However, due to close proximity of the two peaks, the domain-averaged profile appears as a single peak. Again, in the late transitional period, the area occupied by the inactive region becomes increasingly small and its contribution to the calculation of turbulent quantities diminishes. The area is then dominated by 'new' turbulence in the active region. Thus, these cases show a single peak in the streamwise fluctuation during the entire transient period.

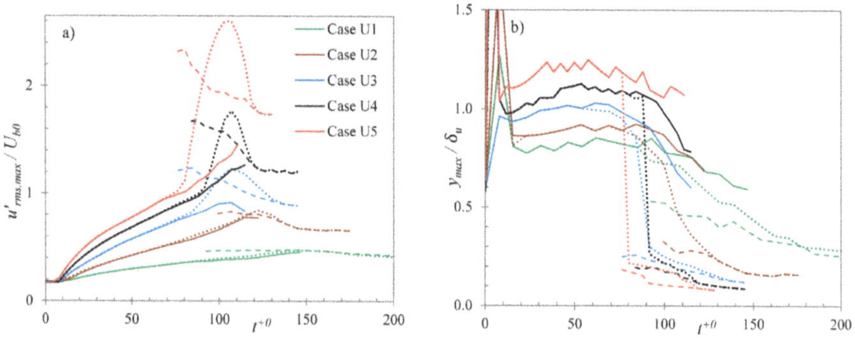


**Figure 20.** Domain- and conditionally-averaged (**a**) maximum streamwise fluctuations; and (**b**) the $y$-locations of their peaks, for cases U1–U5 (Dotted: domain-averaged; solid: inactive region; dashed: active region).

The two peaks shown by the streamwise component during the transient of high Re-ratio cases are very similar to the experimental results of Greenblatt and Moss [9]. However, in their case the peaks farther from the wall were formed at $y^{+0} = 300$, which persisted until the end of the unsteady flow period. Due to limitations in their near-wall velocity data, the full magnitude and location of the near-wall peak was not captured. Although the present results do show two peaks, a direct comparison of these with the two peaks of Greenblatt and Moss [9] might not be appropriate due to the large differences in the initial and final Reynolds numbers. It is possible that their peak farther from the wall (at $y^{+0} = 300$) is a high Reynolds number effect.

## 4. Conclusions

LES has been performed for step-like accelerating channel flow with a Reynolds number ratio up to ~19 (or $Tu_0$ of 0.9%). Similar to the findings of HS15, the present cases with higher Reynolds number ratio also show a three-stage response resembling that of the bypass transition in boundary layer flows. However, the features of transition become more striking when the Reynolds number ratio increases—the elongated streaks in the pre-transitional period become increasingly longer and stronger, and the turbulent spots generated at the initial stage at the onset of transition become increasingly sparse. For the lower turbulence intensity cases, the critical Reynolds number of transition is seen to diverge from the DNS trend of HS15. It was observed that there are large deviations of the critical Reynolds number for different realizations of each case. For the present cases, these deviations increase linearly with the mean value. It is noted that the length of the domain needs to be sufficiently large to accurately capture the transition time when the Reynolds number ratio is high. The present cases are performed using different domain lengths to verify the adequacy of the domain lengths.

The higher Reynolds number ratio cases are found to show double peaks in the transient response of streamwise fluctuations profiles shortly after the onset of transition. A conditional sampling technique is used to further investigate the streamwise fluctuations in all the cases. The wall surface is classified into active and inactive regions of turbulence generation based on a $\lambda_2$-criterion. Conditionally-averaged turbulent statistics, thus obtained, are used to show that the fluctuation energies in the two regions undergo separate developments during the transitional period. For the high-Reynolds number ratio cases, the two peaks in the domain-averaged fluctuation profiles originate from the separate contributions of the active and inactive regions. The peak close to the wall is attributed to the generation of 'new' turbulence in the active region; whereas the peak further away from the wall is attributed to the elongated streaks in the inactive region. In the low-Reynolds number ratio cases, the peaks of the two regions are masked by each other during the entire transient, resulting in a single peak in the domain-averaged profile.

**Author Contributions:** S.H initiated the research. M.S. wrote the DNS code. A.M. together with M.S. implemented LES in the code. A.M. conducted the LES simulations. All authors analysed the results. A.M. led the writing of the manuscript, with contributions from M.S. and S.H.

**Acknowledgments:** We gratefully acknowledge that the work reported herein was partially funded by UK Engineering and Physical Science Research Council (grant no. EP/G068925/1). Some earlier work was carried out making use of the UK national supercomputer ARCHER, access to which was provided by UK Turbulence Consortium funded by the Research Council (grant no. EP/L000261/1). We also acknowledge that some of the data were presented at Turbulence, Heat and Mass Transfer 8, 15–18 September 2015, Sarajevo, Bosnia and Herzegovina and a short paper was included in the conference proceedings.

**Conflicts of Interest:** The authors declare no conflicts of interest.

## References

1. Tu, S.W.; Ramaprian, B.R. Fully developed periodic turbulent pipe flow. Part 1—Main experimental results and comparison with predictions. *J. Fluid Mech.* **1983**, *137*, 31–58. [CrossRef]
2. Shemer, L.; Wygnanski, I.; Kit, E. Pulsating flow in a pipe. *J. Fluid Mech.* **1985**, *153*, 313–337. [CrossRef]
3. Brereton, G.J.; Reynolds, W.C.; Jayaraman, R. Response of a turbulent boundary layer to sinusoidal free-stream unsteadiness. *J. Fluid Mech.* **1990**, *221*, 131–159. [CrossRef]

4. Tardu, S.F.; Binder, G.; Blackwelder, R.F. Turbulent channel flow with large-amplitude velocity oscillations. *J. Fluid Mech.* **1994**, *267*, 109–151. [CrossRef]
5. Scotti, A.; Piomelli, U. Numerical simulation of pulsating turbulent channel flow. *Phys. Fluids* **2001**, *13*, 1367–1384. [CrossRef]
6. He, S.; Jackson, J.D. An experimental study of pulsating turbulent flow in a pipe. *Eur. J. Mech. B Fluids* **2009**, *28*, 309–320. [CrossRef]
7. Maruyama, T.; Kuribayashi, T.; Mizushina, T. The structure of the turbulence in transient pipe flows. *J. Chem. Eng. Jpn.* **1976**, *9*, 431–439. [CrossRef]
8. He, S.; Jackson, J.D. A study of turbulence under conditions of transient flow in a pipe. *J. Fluid Mech.* **2000**, *408*, 1–38. [CrossRef]
9. Greenblatt, D.; Moss, E.A. Rapid temporal acceleration of a turbulent pipe flow. *J. Fluid Mech.* **2004**, *514*, 65–75. [CrossRef]
10. He, S.; Ariyaratne, C.; Vardy, A.E. Wall shear stress in accelerating turbulent pipe flow. *J. Fluid Mech.* **2011**, *685*, 440–460. [CrossRef]
11. Chung, Y.M. Unsteady turbulent flow with sudden pressure gradient changes. *Int. J. Numer. Methods Fluids* **2005**, *47*, 925–930. [CrossRef]
12. Ariyaratne, C.; He, S.; Vardy, A.E. Wall friction and turbulence dynamics in decelerating pipe flows. *J. Hydraul. Res.* **2010**, *48*, 810–821. [CrossRef]
13. Seddighi, M.; He, S.; Orlandi, P.; Vardy, A.E. A comparative study of turbulence in ramp-up and ramp-down unsteady flows. *Flow Turbul. Combust.* **2011**, *86*, 439–454. [CrossRef]
14. Jung, S.Y.; Chung, Y.M. Large-eddy simulation of accelerated turbulent flow in a circular pipe. *Int. J. Heat Fluid Flow* **2012**, *33*, 1–8. [CrossRef]
15. He, S.; Seddighi, M. Turbulence in transient channel flow. *J. Fluid Mech.* **2013**, *715*, 60–102. [CrossRef]
16. He, S.; Seddighi, M. Transition of transient channel flow after a change in Reynolds number. *J. Fluid Mech.* **2015**, *764*, 395–427. [CrossRef]
17. Seddighi, M.; He, S.; Vardy, A.E.; Orlandi, P. Direct Numerical Simulation of an accelerating channel flow. *Flow Turbul. Combust.* **2014**, *92*, 473–502. [CrossRef]
18. Jung, S.Y.; Kim, K. Transient behaviors of wall turbulence in temporally accelerating channel flows. *Int. J. Heat Fluid Flow* **2017**, *67*, 13–26. [CrossRef]
19. Mathur, A.; Gorji, S.; He, S.; Seddighi, M.; Vardy, A.E.; O'Donoghue, T.; Pokrajac, D. Temporal acceleration of a turbulent channel flow. *J. Fluid Mech.* **2018**, *835*, 471–490. [CrossRef]
20. Sundstrom, L.R.J.; Cervantes, M.J. The self-similarity of wall-bounded temporally accelerating turbulent flows. *J. Turbul.* **2018**, *1*, 49–60. [CrossRef]
21. Sundstrom, L.R.J.; Cervantes, M.J. On the similarity of pulsating and accelerating turbulent pipe flows. *Flow Turbul. Combust.* **2018**, *100*, 417–436. [CrossRef]
22. Seddighi, M. Study of Turbulence and Wall Shear Stress in Unsteady Flow over Smooth and Rough Wall Surfaces. Ph.D. Thesis, University of Aberdeen, Aberdeen, UK, 2011.
23. Kim, J.; Moin, P. Application of a fractional-step method to incompressible Navier-Stokes equations. *J. Comput. Phys.* **1985**, *59*, 308–323. [CrossRef]
24. Orlandi, P. *Fluid Flow Phenomena: A Numerical Toolkit*; Kluwer Academic Publishers: Dordrecht, The Netherlands, 2000.
25. Nicoud, F.; Ducros, F. Subgrid-scale stress modelling based on the square of the velocity gradient tensor. *Flow Turbul. Combust.* **1999**, *62*, 183–200. [CrossRef]
26. Lee, M.; Moser, R.D. Direct numerical simulation of turbulent channel flow up to $Re\tau \approx 5200$. *J. Fluid Mech.* **2015**, *774*, 395–415. [CrossRef]
27. Jacobs, R.G.; Durbin, P.A. Simulations of bypass transition. *J. Fluid Mech.* **2001**, *428*, 185–212. [CrossRef]
28. Matsubara, M.; Alfredsson, P.H. Disturbance growth in boundary layers subjected to free-stream turbulence. *J. Fluid Mech.* **2001**, *430*, 149–168. [CrossRef]
29. Brandt, L. Numerical studies of the instability and breakdown of a boundary-layer low-speed streak. *Eur. J. Mech. B Fluids* **2007**, *26*, 64–82. [CrossRef]
30. Brandt, L.; Henningson, D.S. Transition of streamwise streaks in zero-pressure-gradient boundary layers. *J. Fluid Mech.* **2002**, *472*, 229–261. [CrossRef]

31. Brandt, L.; Schlatter, P.; Henningson, D.S. Transition in boundary layers subject to free-stream turbulence. *J. Fluid Mech.* **2004**, *517*, 167–198. [CrossRef]
32. Schlatter, P.; Brandt, L.; Lange, H.C.D.; Henningson, D.S. On streak breakdown in bypass transition. *Phys. Fluids* **2008**, *20*, 101505. [CrossRef]
33. Swearingen, J.D.; Blackwelder, R.F. The growth and breakdown of streamwise vortices in the presence of a wall. *J. Fluid Mech.* **1987**, *182*, 255–290. [CrossRef]
34. Jeong, J.; Hussain, F. On the identification of a vortex. *J. Fluid Mech.* **1995**, *285*, 6–94. [CrossRef]
35. Jeong, J.; Hussain, F.; Schoppa, W.; Kim, J. Coherent structures near the wall in a turbulent channel flow. *J. Fluid Mech.* **1997**, *332*, 185–214. [CrossRef]
36. Zaki, T.A.; Saha, S. On shear sheltering and the structure of vortical modes in single- and two-fluid boundary layers. *J. Fluid Mech.* **2009**, *626*, 111–147. [CrossRef]
37. Narasimha, R.; Narayanan, M.A.B.S.C. Turbulent spot growth in favourable pressure gradients. *AIAA J.* **1984**, *22*, 837–839. [CrossRef]
38. Fransson, J.H.M.; Matsubara, M.; Alfredsson, P.H. Transition induced by free-stream turbulence. *J. Fluid Mech.* **2005**, *527*, 1–25. [CrossRef]
39. Talha, T. A Numerical Investigation of Three-Dimensional Unsteady Turbulent Channel Flow Subjected to Temporal Acceleration. Ph.D. Thesis, University of Warwick, Coventry, UK, 2012.

*entropy* MDPI

*Article*

# Turbulence through the Spyglass of Bilocal Kinetics

**Gregor Chliamovitch * and Yann Thorimbert**

Department of Computer Science, University of Geneva, Route de Drize 7, 1227 Geneva, Switzerland; Yann.Thorimbert@unige.ch
* Correspondence: Gregor.Chliamovitch@unige.ch

Received: 13 June 2018; Accepted: 16 July 2018; Published: 20 July 2018

**Abstract:** In two recent papers we introduced a generalization of Boltzmann's assumption of molecular chaos based on a criterion of maximum entropy, which allowed setting up a bilocal version of Boltzmann's kinetic equation. The present paper aims to investigate how the essentially non-local character of turbulent flows can be addressed through this bilocal kinetic description, instead of the more standard approach through the local Euler/Navier–Stokes equation. Balance equations appropriate to this kinetic scheme are derived and closed so as to provide bilocal hydrodynamical equations at the non-viscous order. These equations essentially consist of two copies of the usual local equations, but coupled through a bilocal pressure tensor. Interestingly, our formalism automatically produces a closed transport equation for this coupling term.

**Keywords:** kinetic theory; fluid dynamics; turbulence

---

## 1. Introduction

The study of turbulent flows has to face two main difficulties, namely non-linearity, which arises from the advective term in the Euler/Navier–Stokes transport equation; and non-locality, which stems from the fact that the theory of complex flows relies to a large extent [1,2] on the correlation function $Q_{ij} = \langle u'_i(\mathbf{x})u'_j(\mathbf{y})\rangle$—that is the average product of the fluctuating component of the velocities of fluid elements at two distant points in space. As such, $Q_{ij}$ is a fundamentally bilocal object.

These two issues are logically disjoint, and the present paper does not bring any new insight regarding the former, focusing instead exclusively on non-locality. The problem raised by bilocality is that turbulence is usually considered from the standpoint of the Navier–Stokes equation (or Euler equation in the non-viscous case), which in turn is derived from the local considerations of kinetic theory (see for instance [3–6] for a few milestones in this direction). Thus, it appears somewhat paradoxical to expect strictly local considerations to lead to a complete picture of a fundamentally bilocal phenomenon.

A different approach would be to start from kinetic theory considered from a bilocal standpoint and then on top of that build a hydrodynamics model that incorporates bilocal features from scratch. The viability of this more sensible approach crucially depends on the possibility of deriving a coherent bilocal kinetic theory of gases, which, technically speaking, amounts to obtaining a closed kinetic equation for the distribution function $f_2$ that describes the distribution of pairs of particles [7,8].

## 2. Two-Particle Kinetics

### 2.1. Generalized Molecular Chaos

Among the existing schemes for setting up a coherent equation for $f_2$, the authors and co-workers recently proposed an approach that relies on a maximum-entropy-based generalization of Boltzmann's assumption of molecular chaos [9,10]. The key observation is that the *Stosszahlansatz*, namely the substitution $f_2(\xi_1, \xi_2) \to f_1(\xi_1)f_1(\xi_2)$ (introducing for convenience the aggregated variable

$\xi_i = (\mathbf{q}_i, \mathbf{p}_i)$) before a collision, can be interpreted either as an assertion regarding the physical state of pre-colliding particles (regarding the range of validity of the *Stosszahlansatz*, see for instance [11,12]), or as a heuristic assumption which substitutes the unknown pre-collisional distribution $f_2$ for its least biased approximation, since the factorized distribution is precisely the distribution that maximizes entropy while being consistent with imposed marginal distributions [13] (the fact that maximum entropy distributions do not require a subjective interpretation and can be assigned an objective meaning is discussed at length in [14]).

The added value of this re-interpretation of molecular chaos is that it lends itself nicely to generalization, and in [9] it was shown how to derive a kinetic equation for the two-particle distribution. This makes it necessary to close the second-order BBGKY equation, whose collision term involves the three-particle distribution $f_3$. The procedure thus requires the substitution of the pre-collisional three-particle distribution with its maximum entropy approximation which is compatible with the $f_2$ appearing in the streaming term. The general result to keep in mind here [13] is that the maximum entropy approximation we can make on the three-particle repartition function under constraints on the bivariate marginals can be expressed as a product of bivariate functions, so that we should make

$$f_3(\xi_1, \xi_2, \xi_3) \to G_1(\xi_1, \xi_2) G_2(\xi_1, \xi_3) G_3(\xi_2, \xi_3). \tag{1}$$

Though elegant, this result is of limited practical scope unless one can obtain extra knowledge about the functions $G_{1,2,3}$. Fortunately, classical particle repartition functions have the peculiarity of being symmetric under exchange of particles, which implies that $G_1 = G_2 = G_3$. Hence, before collision, we are led to the ansatz

$$f_3(\xi_1, \xi_2, \xi_3) \to G(\xi_1, \xi_2) G(\xi_1, \xi_3) G(\xi_2, \xi_3) \tag{2}$$

for some function $G$ which is implicitly related to $f_2$ through

$$f_2(\xi_1, \xi_2) = \int d\xi_3 G(\xi_1, \xi_2) G(\xi_1, \xi_3) G(\xi_2, \xi_3). \tag{3}$$

Note that compared to other closure schemes to be found in the literature, this scheme has the two-fold advantage of being constructive, and of yielding a standalone kinetic equation for $f_2$ and *not* a coupled system of equations for $f_1$ and $f_2$ (or possibly another function encapsulating the dependence between particles, cf. [15]).

### 2.2. Two-Particle Kinetic Equation

Once we have this ansatz at hand, the steps that usually lead to the one-particle Boltzmann equation can be replicated almost exactly in the case of the two-particle distribution. Throughout this work, we shall retain the usual assumptions of kinetic theory [7,8,16], leading us to neglect triple collisions. The streaming term for the two-particle distribution characterizing particles '1' and '2' will thus be altered by (1) binary collisions between '1' and another particle with '2' being a spectator, and (2) binary collisions between '2' and another particle with '1' being a spectator. Particles interact through either a hard-sphere contact interaction or a short-range, repulsive central force field [17,18].

A binary interaction is defined as occurring when two particles meet in a ball $B$ of radius $R$. Defining ternary interactions is more subtle, since inasmuch as the interaction potential is the same regardless of the order of the interaction, it seems artificial to introduce a specific cutoff. We shall therefore define the range of triple collisions as the lenticular overlap of balls $B_R^{(1)}$ and $B_R^{(2)}$ characterizing the domain of interaction with '1' and '2', respectively. Neglecting triple collisions thus amounts to assuming that $|\mathbf{q}_1 - \mathbf{q}_2| > 2R$. Note that it is particularly important to stick tightly to the assumptions made in one-particle theory in order to guarantee that any new prediction arising in the present bilocal description can be ascribed to the statistical description considered, and not to the introduction of new physical assumptions (even though the framework presented here might eventually find its greatest relevance in systems where correlation is known to be important (e.g., granular gases [19]), in which case the assumptions made here should be relaxed and generalized).

This line of reasoning allows us to write a self-standing equation for the function $f_2$ describing the joint distribution of particles '1' and '2', which was found to be [9]

$$\begin{aligned}
&\left(\frac{\partial}{\partial t} + \frac{\mathbf{p}_1}{m}\cdot\nabla_{\mathbf{x}} + \frac{\mathbf{p}_2}{m}\cdot\nabla_{\mathbf{y}}\right) f_2(\mathbf{x},\mathbf{p}_1;\mathbf{y},\mathbf{p}_2;t) \\
&= \int d\mathbf{p}_3 d\omega \frac{|\mathbf{p}_3-\mathbf{p}_1|}{m}\left(G^{\mathbf{x},\mathbf{y}}_{\mathbf{p}_1',\mathbf{p}_2'} G^{\mathbf{x},\mathbf{x}}_{\mathbf{p}_1',\mathbf{p}_3'} G^{\mathbf{y},\mathbf{x}}_{\mathbf{p}_2',\mathbf{p}_3'} - G^{\mathbf{x},\mathbf{y}}_{\mathbf{p}_1,\mathbf{p}_2} G^{\mathbf{x},\mathbf{x}}_{\mathbf{p}_1,\mathbf{p}_3} G^{\mathbf{y},\mathbf{x}}_{\mathbf{p}_2,\mathbf{p}_3}\right) \\
&+ \int d\mathbf{p}_4 d\omega \frac{|\mathbf{p}_4-\mathbf{p}_2|}{m}\left(G^{\mathbf{x},\mathbf{y}}_{\mathbf{p}_1',\mathbf{p}_2'} G^{\mathbf{x},\mathbf{y}}_{\mathbf{p}_1',\mathbf{p}_4'} G^{\mathbf{y},\mathbf{y}}_{\mathbf{p}_2',\mathbf{p}_4'} - G^{\mathbf{x},\mathbf{y}}_{\mathbf{p}_1,\mathbf{p}_2} G^{\mathbf{x},\mathbf{y}}_{\mathbf{p}_1,\mathbf{p}_4} G^{\mathbf{y},\mathbf{y}}_{\mathbf{p}_2,\mathbf{p}_4}\right),
\end{aligned} \tag{4}$$

with $\mathbf{p}_{1,2,3,4}$ and $\mathbf{p}'_{1,2,3,4}$ denoting the momenta before and after the collision, respectively. For notational convenience, we have put $\mathbf{q}_1 = \mathbf{q}_3 = \mathbf{x}$ and $\mathbf{q}_2 = \mathbf{q}_4 = \mathbf{y}$, as well as the shortcut $G^{\mathbf{x},\mathbf{y}}_{\mathbf{p}_1,\mathbf{p}_2} = G(\mathbf{x},\mathbf{p}_1;\mathbf{y},\mathbf{p}_2;t)$.

The first term on the r.h.s. corresponds to the contribution of the collisions possibly undergone at position $\mathbf{x}$ by particle '1' with some particle '3', while the second term accounts for the contribution of the collisions possibly undergone at position $\mathbf{y}$ by particle '2' with some particle '4'. It must be emphasized that the same usual assumptions on density that allow neglecting triple collisions also imply that a binary collision occurs *either* at $\mathbf{x}$ *or* $\mathbf{y}$, but not simultaneously at both places—this will turn out to be important when discussing the appropriate collisional invariants.

### 2.3. Collisional Invariants

Despite its un-glamorous aspect, the structure of Equation (4) is similar to the structure of the one-particle Boltzmann equation, except that the function $G$ appearing in the collision integral, which comes directly from the maximum entropy formulation of the generalized *Stosszahlansatz*, is not $f_2$ itself but an implicit function of $f_2$. Our point in [10] was that although $f_2$ does not appear explicitly in the collision integral, this does not preclude the kind of manipulations usually performed on the Boltzmann equation, and we managed to derive appropriate collisional invariants and the bilocal equilibrium they give rise to. (Nevertheless, it seems that the standard derivation of the $H$-theorem for $f_1$ cannot be generalized in a straightforward way to $f_2$ in our formalism, even though there is no reason to believe that the two-particle entropy $H_2 = -\int f_2 \ln f_2$ does not increase over time.) The salient point in our analysis was that the formulation of local collisions in bilocal terms makes it necessary to consider a collisional invariant other than mass, momentum and kinetic energy; in particular, it happened that defining a bilocal invariant $\chi$ through the relation

$$\chi(\mathbf{p}_1',\mathbf{p}_2') + \chi(\mathbf{p}_3',\mathbf{p}_4') = \chi(\mathbf{p}_1,\mathbf{p}_2) + \chi(\mathbf{p}_3,\mathbf{p}_4) \tag{5}$$

makes it necessary to retain $\chi_1 = 1$, $\chi_2 = (p_1^i + p_2^i)$, $\chi_3 = (\mathbf{p}_1^2 + \mathbf{p}_2^2)$, but also, more interestingly,

$$\chi_4 = p_1^i p_2^j. \tag{6}$$

in [10] we considered only the invariant $\chi_4 = \mathbf{p}_1 \cdot \mathbf{p}_2$, but (6) is more general. This is due to the fact that, as mentioned above, the collision occurs at either $\mathbf{x}$ or $\mathbf{y}$. In the former case, definition (5) with Equation (6) becomes

$$(p'^i_1 + p'^i_3)p_2^j = (p_1^i + p_3^i)p_2^j \tag{7}$$

while in the latter it becomes

$$(p'^j_2 + p'^j_4)p_1^i = (p_2^j + p_4^j)p_1^i \tag{8}$$

which are both trivially verified.

Armed with these four invariants, it is a simple matter to derive a bilocal equilibrium distribution describing the probability that two particles a distance $r$ apart are found to have velocities $\mathbf{v}_1$ and $\mathbf{v}_2$. Thus we find that

$$\begin{aligned} &{}^{eq}f_2^{(r)}(\mathbf{v}_1, \mathbf{v}_2) \\ &= \nu(\theta_1, \theta_2, \mathbf{\Psi}^{(r)}) \exp(\alpha(\theta_1, \mathbf{\Psi}^{(r)})(\mathbf{v}_1 - \mathbf{u}_1)^2 + \alpha(\theta_2, \mathbf{\Psi}^{(r)})(\mathbf{v}_2 - \mathbf{u}_2)^2 + (\mathbf{v}_1 - \mathbf{u}_1)^T \mathbf{\Psi}^{(r)} (\mathbf{v}_2 - \mathbf{u}_2)), \end{aligned} \tag{9}$$

which, as might have been expected, consists of a product of Maxwellian distributions multiplied by a correlating factor. The coefficients are such that $\int d\mathbf{v}_1 d\mathbf{v}_2 (\mathbf{v}_1 - \mathbf{u}_1)^2 f_2 = \theta_1$ and $\int d\mathbf{v}_1 d\mathbf{v}_2 (v_1^i - u_1^i)(v_2^j - u_2^j) f_2 = \sqrt{\theta_1 \theta_2} \varphi_{ij}^{(r)}$ (in plain words $\theta_1$ and $\theta_2$ denote the temperature at position $\mathbf{x}$ and $\mathbf{y}$ respectively, $\varphi_{ij}^{(r)}$ denotes the correlation at distance $r$ of component $i$ of $\mathbf{v}_1 - \mathbf{u}_1$ and component $j$ of $\mathbf{v}_2 - \mathbf{u}_2$), and $\nu$ denotes a normalization factor.

### 3. Balance Equations

Our aim here is to work out the balance equations associated to our bilocal invariants. The very same kind of manipulations as used on the one-particle Boltzmann equation provide us with the generic expression

$$\int d\mathbf{v}_1 d\mathbf{v}_1 \chi(\mathbf{v}_1, \mathbf{v}_2) \left( \frac{\partial}{\partial t} + \mathbf{v}_1 \cdot \nabla_{\mathbf{x}} + \mathbf{v}_2 \cdot \nabla_{\mathbf{y}} \right) f_2 = 0. \tag{10}$$

Defining

$$\langle A \rangle = \Omega^{-1} \int d\mathbf{v}_1 d\mathbf{v}_2 A f_2 \tag{11}$$

with the bilocal density $\Omega = \int d\mathbf{v}_1 d\mathbf{v}_2 f_2$ allows rewriting Equation (10) as

$$0 = \partial_t \langle \Omega \chi \rangle + \nabla_{\mathbf{x}} \cdot \langle \Omega \chi \mathbf{v}_1 \rangle - \langle \Omega \mathbf{v}_1 \cdot \nabla_{\mathbf{x}} \chi \rangle + \nabla_{\mathbf{y}} \cdot \langle \Omega \chi \mathbf{v}_2 \rangle - \langle \Omega \mathbf{v}_2 \cdot \nabla_{\mathbf{y}} \chi \rangle. \tag{12}$$

Considering now in turn the four collisional invariants introduced above, we obtain for $\chi = 1$ that

$$\partial_t \Omega + \nabla_{\mathbf{x}} \cdot \langle \Omega \mathbf{v}_1 \rangle + \nabla_{\mathbf{y}} \cdot \langle \Omega \mathbf{v}_2 \rangle = 0. \tag{13}$$

This is a bilocal continuity equation for the bilocal density $\Omega(\mathbf{x}, \mathbf{y})$, which is the exact counterpart of the standard local continuity equation.

Then, for $\chi = (v_1^i + v_2^i)$, we have for the conservation of momentum

$$\partial_t \langle \Omega (v_1^i + v_2^i) \rangle + \nabla_{\mathbf{x}} \cdot \langle \Omega (v_1^i + v_2^i) \mathbf{v}_1 \rangle + \nabla_{\mathbf{y}} \cdot \langle \Omega (v_1^i + v_2^i) \mathbf{v}_2 \rangle = 0. \tag{14}$$

Using the continuity equation given by Equation (13) above, this can be rewritten as

$$\begin{aligned} 0 = {} & \Omega(\partial_t + \mathbf{u}_1 \cdot \nabla_{\mathbf{x}}) u_1^i + \Omega(\partial_t + \mathbf{u}_2 \cdot \nabla_{\mathbf{y}}) u_2^i \\ & + \nabla_{\mathbf{x}} \cdot \langle \Omega (v_1^i - u_1^i)(\mathbf{v}_1 - \mathbf{u}_1) \rangle + \nabla_{\mathbf{x}} \cdot \langle \Omega (v_2^i - u_2^i)(\mathbf{v}_1 - \mathbf{u}_1) \rangle \\ & + \nabla_{\mathbf{y}} \cdot \langle \Omega (v_1^i - u_1^i)(\mathbf{v}_2 - \mathbf{u}_2) \rangle + \nabla_{\mathbf{y}} \cdot \langle \Omega (v_2^i - u_2^i)(\mathbf{v}_2 - \mathbf{u}_2) \rangle. \end{aligned} \tag{15}$$

We therefore obtain two copies of the pre-Euler/Navier–Stokes conservation equation for the velocity field (each acting at a different point in space), but which are coupled through a kind of bilocal pressure tensor $\langle (v_1^i - u_1^i)(v_2^j - u_2^j) \rangle$.

Next, for $\chi = (\mathbf{v}_1 - \mathbf{u}_1)^2 + (\mathbf{v}_2 - \mathbf{u}_2)^2$ we obtain in a similar way, remembering that by definition $\langle(\mathbf{v}_1 - \mathbf{u}_1)^2 + (\mathbf{v}_2 - \mathbf{u}_2)^2\rangle = \theta_1 + \theta_2$:

$$\begin{aligned}0 = &\Omega(\partial_t + \mathbf{u}_1 \cdot \nabla_{\mathbf{x}})\theta_1 + \Omega(\partial_t + \mathbf{u}_2 \cdot \nabla_{\mathbf{y}})\theta_2 \\ &+ \nabla_{\mathbf{x}} \cdot \langle\Omega(\mathbf{v}_1 - \mathbf{u}_1)^2(\mathbf{v}_1 - \mathbf{u}_1)\rangle + \nabla_{\mathbf{x}} \cdot \langle\Omega(\mathbf{v}_2 - \mathbf{v}_2)^2(\mathbf{v}_1 - \mathbf{u}_1)\rangle \\ &+ \nabla_{\mathbf{y}} \cdot \langle\Omega(\mathbf{v}_1 - \mathbf{u}_1)^2(\mathbf{v}_2 - \mathbf{u}_2)\rangle + \nabla_{\mathbf{y}} \cdot \langle\Omega(\mathbf{v}_2 - \mathbf{v}_2)^2(\mathbf{v}_2 - \mathbf{u}_2)\rangle \\ &- 2\Omega\langle(\mathbf{v}_1 - \mathbf{u}_1) \cdot (\mathbf{v}_1 - \mathbf{u}_1)\rangle\nabla_{\mathbf{x}} \cdot \mathbf{u}_1 - 2\Omega\langle(\mathbf{v}_2 - \mathbf{u}_2) \cdot (\mathbf{v}_2 - \mathbf{u}_2)\rangle\nabla_{\mathbf{y}} \cdot \mathbf{u}_2.\end{aligned} \tag{16}$$

Here, again, we obtain two copies of the local heat transport equation that are coupled through a bilocal heat flux.

We finally come to $\chi = (v_1^i - u_1^i)(v_2^j - u_2^j)$, for which we eventually obtain

$$\begin{aligned}0 = &\Omega(\partial_t + \mathbf{u}_1 \cdot \nabla_{\mathbf{x}} + \mathbf{u}_2 \cdot \nabla_{\mathbf{y}})\langle(v_1^i - u_1^i)(v_2^j - u_2^j)\rangle \\ &+ \nabla_{\mathbf{x}} \cdot \langle\Omega(v_1^i - u_1^i)(v_2^j - u_2^j)(\mathbf{v}_1 - \mathbf{u}_1)\rangle + \nabla_{\mathbf{y}} \cdot \langle\Omega(v_1^i - u_1^i)(v_2^j - u_2^j)(\mathbf{v}_2 - \mathbf{u}_2)\rangle \\ &+ \Omega\langle(\mathbf{v}_1 - \mathbf{u}_1)(v_2^j - u_2^j)\rangle \cdot \nabla_{\mathbf{x}} u_1^i + \Omega\langle(\mathbf{v}_2 - \mathbf{u}_2)(v_1^i - u_1^i)\rangle \cdot \nabla_{\mathbf{y}} u_2^j,\end{aligned} \tag{17}$$

which provides a transport equation for the bilocal pressure tensor.

## 4. Non-Viscous Hydrodynamics

Our goal now is to close the balance equations, given by expressions (13), (15)–(17), by evaluating the averages over a local equilibrium solution given by Equation (9), with $\theta_1 \to \theta_1(\mathbf{x})$, $\theta_2 \to \theta_2(\mathbf{y})$, $\mathbf{u}_1 \to \mathbf{u}_1(\mathbf{x})$, $\mathbf{u}_2 \to \mathbf{u}_2(\mathbf{y})$ and $\mathbf{\Psi} \to \mathbf{\Psi}(\mathbf{x}, \mathbf{y})$, so as to deduce the bilocal non-viscous hydrodynamical equations. (It might be argued that considering turbulent flows in the non-viscous case is somewhat vain, since viscosity plays a crucial role in the dissipation of small-scale vortices. However, the fundamental difficulty that makes the study of turbulence particularly challenging is present in the non-viscous case as well, so that from the conceptual standpoint of the present paper, considering non-viscous flows is enough for our purpose.) We have (defining at the same time the local pressure tensors $P_1(\mathbf{x})$ and $P_2(\mathbf{y})$ and their bilocal counterpart $\Phi(\mathbf{x}, \mathbf{y})$) :

$$\Omega\langle(v_1^i - u_1^i)(v_1^j - u_1^j)\rangle = \delta_{ij} P_1 = \delta_{ij}\frac{\theta_1}{3} \tag{18}$$

$$\Omega\langle(v_1^i - u_1^i)(v_2^j - u_2^j)\rangle = \sqrt{\theta_1\theta_2}\varphi_{ij} = \Phi^{ij} \tag{19}$$

$$\Omega\langle(\mathbf{v}_1 - \mathbf{u}_1)^2(\mathbf{v}_1 - \mathbf{u}_1)\rangle = 0 \tag{20}$$

$$\Omega\langle(\mathbf{v}_2 - \mathbf{v}_2)^2(\mathbf{v}_1 - \mathbf{u}_1)\rangle = 0 \tag{21}$$

$$\Omega\langle(\mathbf{v}_1 - \mathbf{u}_1) \cdot (\mathbf{v}_1 - \mathbf{u}_1)\rangle = 3P_1 = \theta_1 \tag{22}$$

$$\Omega\langle(v_1^i - u_1^i)(v_2^j - u_2^j)(\mathbf{v}_1 - \mathbf{u}_1)\rangle = 0. \tag{23}$$

Hence, our conservation equations become at zeroth order, first the bilocal continuity equation (now written in components)

$$\frac{\partial\Omega}{\partial t} + \frac{\partial(\Omega u_1^k)}{\partial x^k} + \frac{\partial(\Omega u_2^k)}{\partial y^k} = 0, \tag{24}$$

then the bilocal Euler equation

$$0 = \Omega\left(\frac{\partial}{\partial t} + u_1^k\frac{\partial}{\partial x^k}\right)u_1^i + \Omega\left(\frac{\partial}{\partial t} + u_2^k\frac{\partial}{\partial y^k}\right)u_2^i + \frac{\partial}{\partial x^i}P_1 + \frac{\partial}{\partial x^k}\Phi^{ki} + \frac{\partial}{\partial y^k}\Phi^{ik} + \frac{\partial}{\partial y^i}P_2, \tag{25}$$

the bilocal heat equation

$$0=\Omega\left(\frac{\partial}{\partial t}+u_1^k\frac{\partial}{\partial x^k}\right)\theta_1+\Omega\left(\frac{\partial}{\partial t}+u_2^k\frac{\partial}{\partial y^k}\right)\theta_2-\frac{2}{3}\left(\theta_1\frac{\partial u_1^k}{\partial x^k}+\theta_2\frac{\partial u_2^k}{\partial y^k}\right), \tag{26}$$

and the transport equation for the bilocal pressure tensor

$$0=\Omega\left(\frac{\partial}{\partial t}+u_1^k\frac{\partial}{\partial x^k}+u_2^k\frac{\partial}{\partial y^k}\right)\Phi^{ij}+\Phi^{kj}\frac{\partial u_1^i}{\partial x^k}+\Phi^{ik}\frac{\partial u_2^j}{\partial y^k}. \tag{27}$$

Finally, one might wish to obtain a transport equation for the product $u_1^i(\mathbf{x})u_2^j(\mathbf{y})$. This can be done by using Equation (25) twice to obtain

$$\begin{aligned}0=\Omega\left(\frac{\partial}{\partial t}+u_1^k\frac{\partial}{\partial x^k}+u_2^k\frac{\partial}{\partial y^k}\right)(u_1^iu_2^j)+\Omega u_1^i\left(\frac{\partial}{\partial t}+u_1^k\frac{\partial}{\partial x^k}\right)u_1^j+\Omega u_2^j\left(\frac{\partial}{\partial t}+u_2^k\frac{\partial}{\partial y^k}\right)u_2^i\\+u_2^j\frac{\partial P_1}{\partial x^i}+u_1^i\frac{\partial P_1}{\partial x^j}+u_1^i\frac{\partial P_2}{\partial y^j}+u_2^j\frac{\partial P_2}{\partial x^i}+u_1^i\frac{\partial \Phi^{kj}}{\partial x^k}+u_2^j\frac{\partial \Phi^{ki}}{\partial x^k}+u_1^i\frac{\partial \Phi^{jk}}{\partial y^k}+u_2^j\frac{\partial \Phi^{ik}}{\partial y^k}.\end{aligned} \tag{28}$$

## 5. Conclusions

It follows from our analysis that Equation (28), supplemented by expressions (25) and (27), provides a dynamical equation for the product of fluid velocities at different points in space, addressing the point raised in the introduction regarding the non-local character of complex flows. It must be emphasized that this result is deduced purely from the considerations of kinetic theory, and without resorting to any further hypotheses.

However, we considered here the full velocity field and not its fluctuating part only. Coming back to the second point regarding the non-linearity of the resulting equations, if we decompose each quantity involved as the sum of its Reynolds average plus a fluctuating component, we shall face in our bilocal Euler equation, given by Equation (25), the same problem as in the local case, with the emergence of extra stresses that are the bilocal counterparts of Reynolds stresses. Nevertheless, Equation (28) provides a dynamical equation for these stresses, so that the closure problem should not degenerate into a *hierarchical* closure problem.

It is worth reminding our assumption that the points have to be separated by a distance at least equal to the typical length characteristic of the interaction. One should therefore refrain from the temptation of taking the limit such that the points become confounded, which in the present setting would be ill-supported mathematically. That being said, this typical length is likely to be much smaller than the distances of interest in a hydrodynamical setting. It should also be recalled that the equations of hydrodynamics are notoriously robust against the breaking down of the assumptions made in first-principles derivations, so that the range of validity of the theory presented here might well turn out to be wider than expected. This will eventually be a matter for experimental confirmation or invalidation. Anyway, the theory presented here is conceived less as a fully developed scheme, and more as an invitation to explore bilocal kinetics further. We cannot but hope that we have partly reached this goal.

**Author Contributions:** G.C. and Y.T. performed the research; G.C. wrote the manuscript. All authors have read and approved the final manuscript.

**Funding:** This research received no external funding.

**Conflicts of Interest:** The authors declare no conflict of interest.

## References

1. Batchelor, G.K. *The Theory of Homogeneous Turbulence*; Cambridge University Press: Cambridge, UK, 1953.

2. Davidson, P.A. *Turbulence: An Introduction for Scientists and Engineers*; Oxford University Press: Oxford, UK, 2015.
3. De Kármán, T.; Howarth, L. On the Statistical Theory of Isotropic Turbulence. *Proc. R. Soc. A* **1938**, *164*, 192–215. [CrossRef]
4. Onsager, L. Statistical hydrodynamics. *Nuovo Cimento Suppl.* **1949**, *6*, 279–287. [CrossRef]
5. Eyink, G.L.; Sreenivasan, K.R. Onsager and the theory of hydrodynamic turbulence. *Rev. Mod. Phys.* **2006**, *78*, 87–135. [CrossRef]
6. Kraichnan, R.H. Structure of isotropic turbulence at very high Reynolds numbers. *J. Fluid Mech.* **1959**, *5*, 497–543. [CrossRef]
7. Kreuzer, H.J. *Nonequilibrium Thermodynamics and Its Statistical Foundations*; Oxford University Press: Oxford, UK, 1981.
8. Liboff, R.L. *Kinetic Theory*; Springer: New York, NY, USA, 2003.
9. Chliamovitch, G.; Malaspinas, O.; Chopard, B. A Truncation Scheme for the BBGKY2 Equation. *Entropy* **2015**, *17*, 7522–7529. [CrossRef]
10. Chliamovitch, G.; Malaspinas, O.; Chopard, B. Kinetic Theory beyond the Stosszahlansatz. *Entropy* **2017**, *19*, 381. [CrossRef]
11. Sznitman, A. Equations de type de Boltzmann, spatialement homogènes. *Wahrscheinlichkeitstheor. Geb.* **1984**, *66*, 559–592. (In French) [CrossRef]
12. Mischler, S.; Mouhot, C. Kac's program in kinetic theory. *Invent. Math.* **2013**, *193*, 1–147. [CrossRef]
13. Stephens, G.J.; Bialek, W. Statistical Mechanics of Letters in Words. *Phys. Rev. E* **2010**, *81*, 066119. [CrossRef] [PubMed]
14. Jaynes, E.T. On the rationale of maximum entropy methods. *Proc. IEEE* **1982**, *70*, 939–952. [CrossRef]
15. Sagara, K.; Tsuge, S. A bimodal Maxwellian distribution as the equilibrium solution of the two-particle regime. *Phys. Fluids* **1982**, *25*, 1970–1977. [CrossRef]
16. Huang, K. *Statistical Mechanics*; John Wiley & Sons: New York, NY, USA, 1963.
17. Harris, S. *An Introduction to the Theory of the Boltzmann Equation*; Holt, Rinehart, and Winston: New York, NY, USA, 1971.
18. Cercignani, C. *The Boltzmann Equation and Its Applications*; Springer: New York, NY, USA, 1988.
19. Pareschi, L.; Russo, G.; Toscani, G. *Modelling and Numerics of Kinetic Dissipative Systems*; Nova Science Publishers: Hauppauge, NY, USA, 2006.

*entropy*

MDPI

*Article*

# Time-Dependent Probability Density Functions and Attractor Structure in Self-Organised Shear Flows

**Quentin Jacquet [1,2], Eun-jin Kim [3,*] and Rainer Hollerbach [1]**

1 Department of Applied Mathematics, University of Leeds, Leeds LS2 9JT, UK; quentin.jacquet@ensta-paristech.fr (Q.J.); R.Hollerbach@leeds.ac.uk (R.H.)
2 ENSTA ParisTech Université Paris-Saclay, 828 Boulevard des Maréchaux, 91120 Palaiseau, France
3 School of Mathematics and Statistics, University of Sheffield, Sheffield S3 7RH, UK
* Correspondence: e.kim@sheffield.ac.uk; Tel.: +44-114-222-3876

Received: 30 July 2018; Accepted: 16 August 2018; Published: 17 August 2018

**Abstract:** We report the time-evolution of Probability Density Functions (PDFs) in a toy model of self-organised shear flows, where the formation of shear flows is induced by a finite memory time of a stochastic forcing, manifested by the emergence of a bimodal PDF with the two peaks representing non-zero mean values of a shear flow. Using theoretical analyses of limiting cases, as well as numerical solutions of the full Fokker–Planck equation, we present a thorough parameter study of PDFs for different values of the correlation time and amplitude of stochastic forcing. From time-dependent PDFs, we calculate the information length ($\mathcal{L}$), which is the total number of statistically different states that a system passes through in time and utilise it to understand the information geometry associated with the formation of bimodal or unimodal PDFs. We identify the difference between the relaxation and build-up of the shear gradient in view of information change and discuss the total information length ($\mathcal{L}_\infty = \mathcal{L}(t \to \infty)$) which maps out the underlying attractor structures, highlighting a unique property of $\mathcal{L}_\infty$ which depends on the trajectory/history of a PDF's evolution.

**Keywords:** self-organisation; shear flows; coherent structures; turbulence; stochastic processes; Langevin equation; Fokker-Planck equation; information length

## 1. Introduction

Many systems in nature and laboratories are far from equilibrium, constantly changing in time and space and exhibiting very complex behaviour. Examples include turbulence in astrophysical and laboratory plasmas, the stock market, and biological ecosystems. Despite having apparently different manifestations of complexity, these systems have much in common and are often governed by similar nonlinear dynamics. In particular, an 'ordered' collective behaviour (e.g., in the form of coherent structures) emerges on the macroscale out of complexity as a novel consequence of self-organisation. For example, in the laboratory, in geophysical and astrophysical systems, coherent structures such as large-scale shear flows (such as zonal flows and streamers in laboratory plasmas, in the atmosphere and oceans, and in giant planets) and differential rotations in the Sun and other stars emerge from small-scale turbulence. There is overwhelming evidence from laboratory experiments, observations, and computational studies that these coherent structures play an absolutely critical role in determining the level of transport in the flow.

In particular, one crucial effect of shear flows is the suppression of transport in the direction orthogonal to the flow (the shear direction) by shear-induced enhanced dissipation [1–11]. This occurs as a shear flow distorts fluid eddies, accelerates the formation of small scales, and dissipates them when molecular diffusion becomes effective on small scales. This turbulence regulation leads to the formation of a transport barrier where transport is significantly reduced locally, providing one of the crucial mechanisms for controlling the mixing and transport in a variety of systems. Important examples include (i) the low-to-high (L-H) transition (or internal transport barrier formation), during which a system undergoes a remarkable, spontaneous transition to a more ordered state, despite the increase in free energy (e.g., [3–5]); (ii) equatorial winds and polar vortices [12] (azimuthal flows in the east–west direction) which have long been known to reduce transport, acting as a transport barrier in the latitudinal direction [13]; (iii) transport barrier due to shear layers [14] in oceans which is called shear sheltering; and (iv) the solar tachocline—the boundary layer between the stable radiative interior and unstable convective layer which has a strong radial differential rotation—which can also act as a transport barrier, leading to weak anisotropic turbulence and mixing [5,7]. Our theoretical predictions of turbulent quenching in different systems have been confirmed by various numerical simulations (e.g., refs. [15,16]).

The foregoing statements underscore the importance of self-regulation between small-scale fluctuations and large-scale shear flows. We proposed a one-dimensional (1D) continuous model of self-organised shear flow [17] by extending a prototypical sand-pile model which evolves in discrete time. Specifically, we considered the formation of a shear flow driven by a short-correlated (white-noise) random forcing, where the shear gradient increases until it becomes unstable according to the stability criterion. For instance, in a strongly stratified medium, the stability is determined by the Richardson criterion: fluctuations on small scales (or internal gravity waves) amplify a shear gradient and thus, act as a forcing until the gradient exceeds the critical value given by the Richardson criterion, $R = (A/N)^2 > R_c = (A_c/N)^2 = 1/4$. Here, $N$ is the buoyancy frequency due to the restoring force (buoyancy) in a stably stratified medium, and $A$ is the shear gradient with the critical value $A_c$. When unstable, the shear flow then relaxes its gradient and generates small-scale fluctuations, and this relaxation was modelled by nonlinear (cubic) diffusion; the shear gradient then grows again when small-scale turbulence becomes sufficiently strong to drive a shear flow. The same cycle repeats itself, exhibiting continuous growth and damping. This highlights that a self-organised state is never stationary in time, but involves persistent fluctuations.

The extension of refs. [17,18] solved a stochastic differential equation with a fourth-order stochastic Runga–Kutta method for Gaussian coloured noise in 1D and showed the transition from an unimodal stationary Probability Density Function (PDF) to a bimodal stationary PDF when the correlation time of a random forcing exceeds a critical value. The mean shear gradient is zero for a unimodal PDF, while its non-zero value represents the critical shear gradient around which a shear gradient continuously grows and damps through the interaction with fluctuations. The transition from a unimodal to bimodal PDF represents the formation of a non-zero mean shear gradient, or the formation of jets. Interestingly, In ref. [18], we found similar results in a 0D model and 2D hydrodynamic turbulence. In particular, the 2D results showed that a shear flow evolves through the competition between its growth and damping due to a localized instability, maintaining a stationary PDF, and that the bimodal PDF results from a self-organising shear flow with a linear profile.

The purpose of this paper is to investigate the evolution of a time-dependent PDF to understand how a given initial (global) shear gradient modelled by a narrow PDF relaxes into a bimodal or unimodal stationary PDF. We are particularly interested in understanding the information geometry associated with this process. Our information geometry theory is based on the Fisher metric [19] extended to time-dependent problems. (Note that we use information about statistically different states, refraining from the debate on the exact definition of information [19,20]). We recall that for a Gaussian PDF whose evolution is described by the movement of a peak and the change in its width, the uncertainty measuring the mean value of $x$ is set by the standard deviation. Two PDFs with the same standard deviation would differ by one statistical state when their mean values differ by the standard deviation (e.g., see ref. [21]). To formalise this idea to quantify the information change associated with the time evolution of PDFs [22–32], we define an infinitesimal distance at any time by comparing two PDFs at adjacent times and sum these distances. The total distance gives us the number of statistically different states that a system passes through in time and is called the information length ($\mathcal{L}$). While the detailed derivation of $\mathcal{L}$ and its applications are given in refs. [22–32], it is useful to highlight that $\mathcal{L}$ is a measure of the total elapsed time in units of a dynamical timescale for information change. To show this, we define the dynamical time ($\tau(t)$) [22–30] as follows:

$$\mathcal{E} \equiv \frac{1}{[\tau(t)]^2} = \int \frac{1}{p(x,t)} \left[\frac{\partial p(x,t)}{\partial t}\right]^2 dx. \tag{1}$$

Here, $\tau(t)$ is the characteristic timescale over which the information changes. Having units of time, $\tau(t)$ quantifies the correlation time of a PDF. Alternatively, $1/\tau$ quantifies the (average) rate of change of information in time. $\mathcal{L}(t)$ is then defined by measuring the total elapsed time ($t$) in units of $\tau$ as

$$\mathcal{L}(t) = \int_0^t \frac{dt_1}{\tau(t_1)} = \int_0^t \sqrt{\int dx \frac{1}{p(x,t_1)} \left[\frac{\partial p(x,t_1)}{\partial t_1}\right]^2}\, dt_1. \tag{2}$$

$\mathcal{L}(t)$ measures the cumulative change in $p(x,t)$, and depends on the intermediate states that a system evolves through between times 0 and $t$. Thus, it is a Lagrangian quantity (unlike entropy or relative entropy) which depends on the time history of $p(x,t)$, uniquely defined as a function of time $t$ for a given initial PDF. $\mathcal{L}$ represents the total number of statistically distinguishable states that a system evolves through, providing a very convenient methodology for measuring the distance between $p(x,t)$ and $p(x,0)$ continuously in time for a given $p(x,0)$. References [22–32] showed that $\mathcal{L}_\infty$ is a new diagnostic for understanding a dynamical system and for mapping out an attractor structure. In particular, $\mathcal{L}_\infty$ captures the effect of different deterministic forces through the scaling of $\mathcal{L}_\infty$ against the peak position of a narrow initial PDF. For a stable equilibrium, the minimum value of $\mathcal{L}_\infty$ occurs at the equilibrium point. In comparison, in the case of a chaotic attractor, $\mathcal{L}_\infty$ exhibits a sensitive dependence on initial conditions like a Lyapunov exponent.

In this paper, we investigate the evolution of a shear gradient ($x$) starting from a relatively narrow PDF ($p(x,0)$) with an initial mean value of $x_0$ which represents the mean value of an initial shear gradient. For a unimodal stationary PDF, the mean shear gradient decreases to zero in the long time limit, while for a bimodal stationary PDF with a peak of $\pm x_*$, the case of $x_0 > x_*$ models the relaxation of an initial super-critical gradient ($x_0$) to the critical value ($x_*$), and the case of $x_0 < x_*$ models the build-up of the gradient from a subcritical initial value to the critical value ($x_*$). We are interested in the information changes in these processes and in identifying the differences between the relaxation and build-up of the shear gradient in view of these information changes and in mapping out an attractor structure by using $\mathcal{L}$.

The remainder of this paper is organised as follows. We introduce our model and provide analytical solutions of time-dependent PDFs in limiting cases in Section 2. In order to systematically undertake a numerical study, in Section 3, we first provide a detailed discussion on stationary PDFs for different parameter values to determine the parameter space for unimodal versus bimodal PDFs. Section 4 provides numerical solutions for time-dependent PDFs and $\mathcal{L}$. The discussion and conclusions are found in Section 5.

## 2. Model

In this section, we introduce our model and provide analytical solutions for time-dependent PDFs in limiting cases. As noted in Section 1, given the universality of self-organisation in 0D, 1D, and 2D models and the challenge of the computation of time-dependent PDFs, we utilised a 0D model to facilitate the calculation of PDFs. Our 0D model is based on the cubic process for a stochastic variable ($x$) (e.g., representing a shear gradient). Specifically, we considered $x$ driven by a finite correlated forcing ($f$), governed by the following Langevin equations

$$\partial_t x = -(ax + bx^3) + f \equiv -g(x) + f, \tag{3}$$

$$\partial_t f = -\gamma f + \xi. \tag{4}$$

Here, $g(x) = ax + bx^3$; $a, b \geq 0$ are constants; $\xi$ is a stochastic noise with a short correlation time with the correlation function

$$\langle \xi(t)\xi(t') \rangle = 2D\delta(t - t'). \tag{5}$$

The highest cubic nonlinearity in our 0D model mimics a nonlinear cubic diffusion in the 1D model in refs. [17,18]. Equation (3) is the Ornstein–Uhlenbeck process [33] with the solution

$$f(t) = f(0)e^{-\gamma t} + \int_0^t dt_1 e^{-\gamma(t-t_1)}\xi(t_1). \tag{6}$$

For $f(0) = 0$, the correlation time of $f(t)$ is approximately $1/\gamma$, as follows:

$$\begin{aligned} \langle f(t)f(t') \rangle &= \int_0^t dt_1 \int_0^{t'} dt_2 e^{-\gamma(t-t_1)} e^{-\gamma(t'-t_2)} \langle \xi(t_1)\xi(t_2) \rangle \\ &= \frac{D}{\gamma}\left[e^{-\gamma(t'-t)} - e^{-\gamma(t+t')}\right] \approx \frac{D}{\gamma} e^{-\gamma|t'-t|}, \end{aligned} \tag{7}$$

where we assumed $t' > t$ and used Equation (5). Thus, $x$ in Equation (3) is driven by the Gaussian noise with the correlation time $\gamma^{-1}$. While the set of Equations (3) and (4) give a PDF in two dimensions $(x, f)$, it is useful to obtain an approximate PDF in the $x$ dimension only. To this end, we combine Equations (3) and (4) to obtain the equation for $x$ as

$$\partial_{tt}x + (\gamma + \partial_x g)\partial_t x = -\gamma g + \xi, \tag{8}$$

and consider the overdamped limit where $\partial_{tt}x$ is negligible compared with the damping term. This is the so-called unified-colored noise approximation [34], and turns Equation (8) into

$$(\gamma + \partial_x g)\partial_t x \simeq -\gamma g + \xi. \tag{9}$$

We observe that for sufficiently small $\gamma$, to $O(\gamma)$ Equation (9) is, again, an Ornstein–Uhlenbeck process [33] for $Q = g + \gamma x$:

$$\partial_t Q = -\gamma Q + \gamma^2 x + \xi \approx -\gamma Q + \xi. \tag{10}$$

Thus, the mean value of $\langle Q(t) \rangle = Q_0 e^{-\gamma t}$, where $Q_0 = \langle Q(t = 0) \rangle$, decays exponentially in time while the variance, $\langle (Q - \langle Q \rangle)^2 \rangle = \frac{1}{2\beta}$, evolves according to

$$\frac{1}{2\beta} = \frac{e^{-2\gamma t}}{2\beta_0} + \frac{D(1 - e^{-2\gamma t})}{\gamma}, \tag{11}$$

where $\beta$ and $\beta_0 = \beta(t = 0)$ are the inverse temperatures of $p(Q, t)$ and its initial value, respectively. Therefore, the time-dependent PDF of $Q$ is a Gaussian process and is given by

$$p(Q, t) = \sqrt{\frac{\beta}{\pi}} e^{-\beta(Q - \langle Q \rangle)^2}, \tag{12}$$

where $\beta$ is the inverse temperature that satisfies Equation (11).

Since $\mathcal{E}$ in Equation (1) and $\mathcal{L}$ in Equation (2) are invariant under the change of variables, the Gaussian PDF of $Q$ in Equation (12) provides us with a convenient way of calculating them by utilising the property of the Gaussian PDF. Specifically, for the Gaussian PDF of $Q$, $\mathcal{E}$ is given by

$$\mathcal{E} = \frac{(\partial_t \beta)^2}{2\beta^2} + 2\beta(\partial_t \langle Q \rangle)^2, \tag{13}$$

where the first and second terms on the right-hand side are due to the temporal changes in the width and peak position of the PDF. For sufficiently small $D$ (large $\beta$) and/or large $\langle Q \rangle$, $\mathcal{E}$ in Equation (13) is dominated by the second term. Furthermore, with a small $D$, Equation (11) becomes $2\beta \sim 2\beta_0 e^{2\gamma t}$. Thus, by substituting $2\beta \sim 2\beta_0 e^{2\gamma t}$, $\partial_t \langle Q \rangle = -\gamma Q_0 e^{-\gamma t}$ into Equation (13), we obtain

$$\mathcal{E} \sim 2\gamma^2 \beta_0 Q_0^2, \tag{14}$$

where $Q_0 = (a+\gamma)x_0 + bx_0^3$, and $x_0 = \langle x(t=0) \rangle$ is the mean position of $x$ at $t = 0$. To relate Equation (14) to what is observed in the PDF of $x$, we need to find the initial inverse temperature, $\beta_0^x = 1/2\langle (x(0) - x_0)^2 \rangle$, for $p(x, t=0)$ that corresponds to $\beta_0 = 1/2\langle (Q(0) - Q_0)^2 \rangle$ (which is the inverse temperature of the PDF of $Q$ at $t = 0$). To this end, we use $Q - \langle Q \rangle = (a+\gamma)x + bx^3 - \langle (a + \gamma)x + bx^3 \rangle \sim (x - \langle x \rangle)(a + \gamma + 3b\langle x \rangle^2)$ to leading order for $|\langle x \rangle| \gg |x - \langle x \rangle|$ and obtain

$$\langle (Q - \langle Q \rangle)^2 \rangle \sim \langle (x - \langle x \rangle)^2 \rangle (a + \gamma + 3b\langle x \rangle^2)^2. \tag{15}$$

For $x_0 \gg \gamma, a$, Equation (15) evaluated at $t = 0$ gives us

$$\beta_0 \sim \frac{\beta_0^x}{9b^2 x_0^4}. \tag{16}$$

Equations (14) and (16) give us

$$\mathcal{E} \sim \frac{2\beta_0^x \gamma^2 x_0^2}{9b^2}, \quad \mathcal{L}(t) \sim \sqrt{\frac{2\beta_0^x}{9b^2}}\, \gamma x_0 t. \tag{17}$$

Thus, $\mathcal{L}(t)$ increases linearly with time with a slope that is proportional to $\gamma$ and $x_0$ (for small time, small $D$, small $\gamma$, and large $x_0$). The numerical simulations in Section 4 examine this behaviour in more detail.

Then, by using the conservation of the probability, the time-dependent PDF of $x$ is obtained as

$$p(x,t) = \left| \frac{dQ}{dx} \right| p(Q,t) = \sqrt{\frac{\beta}{\pi}} |\partial_x g + \gamma| \exp\left( -\beta (Q - \langle Q \rangle)^2 \right). \tag{18}$$

It is interesting to note that $p(x, t \to \infty)$ in Equation (18) can be either unimodal or bimodal depending on the values of the parameters. This is discussed in detail in Section 3.

Having gained some insight into the leading order behaviour of $p(x,t)$ for small $\gamma$, we investigate a more general case of Equation (9). To this end, it is convenient to recast Equation (9) as

$$\partial_t x = -\frac{\gamma g}{G} + \frac{1}{G}\xi, \tag{19}$$

where $G = \partial_x g + \gamma$. The corresponding Fokker–Planck equation for $p(x,t)$ is

$$\frac{\partial}{\partial t} p(x,t) = \frac{\partial}{\partial x}\left[ \frac{\gamma g}{G} p(x,t) \right] + D \frac{\partial}{\partial x}\left[ \frac{1}{G} \frac{\partial}{\partial x} \left( \frac{1}{G} p(x,t) \right) \right]. \tag{20}$$

In Equation (20), we used the Stratonovich calculus [33,35–37], which recovers the limit of a short correlated forcing from the finite correlated forcing [37]. Although a time-dependent solution to Equation (20) is not easily obtained analytically, a stationary solution can be found and is discussed in detail in Section 3.

## 3. Stationary PDFs

In order to undertake a systematic numerical study in Section 4, we here provide a detailed discussion of stationary PDFs for different parameter values, and determine the parameter space for unimodal versus bimodal PDFs. A stationary PDF found from Equation (18) is

$$p(x) \propto |G(x)| \exp\Big(-\frac{\gamma}{D}\int^x g(x_1)G(x_1)dx_1\Big) = |\partial_x g + \gamma| \exp\Big(-\frac{\gamma}{2D}[g(x)^2 + 2\gamma\int^x g(x_1)dx_1]\Big). \quad (21)$$

To $O(\gamma)$, Equation (21) reproduces Equation (18). To determine the location of the local maxima and minima of $p(x)$ in Equation (21), we calculate

$$\partial_x p(x) = 0 \qquad \Longrightarrow \qquad -\frac{\gamma}{D}(\partial_x g + \gamma)^2 g + \partial_{xx} g = 0. \quad (22)$$

For $g = ax + bx^3$, Equation (22) can be rewritten as

$$x\left[-\frac{\gamma}{D}(a+\gamma+3bx^2)^2(a+bx^2) + 6b\right] = 0. \quad (23)$$

Equation (23) gives the solution $x = 0$ and $x \neq 0$, indicating the possibility of the bimodal PDF. We then find the non-zero solution by solving

$$-\frac{\gamma}{D}(a+\gamma+3bx^2)^2(a+bx^2) + 6b = 0. \quad (24)$$

To this end, it is convenient to make the following three successive changes in variables:

$$\begin{cases} X = a + bx^2, \\ \alpha = (\Omega + 3X)^2 X, \end{cases} \rightarrow \begin{cases} Y = 1 + \frac{3X}{\Omega}, \\ \frac{3\alpha}{\Omega^3} = Y^2(Y-1), \end{cases} \rightarrow \begin{cases} Z = 1/Y, \\ Z^3 + \delta Z - \delta = 0, \end{cases} \quad (25)$$

with $\Omega$, $\alpha$, $\delta$ defined as

$$\Omega = \gamma - 2a, \quad \alpha = \frac{6Db}{\gamma}, \quad \delta = \frac{\gamma(\gamma-2a)^3}{18Db}. \quad (26)$$

In order to solve the equation for $Z$ in Equation (25), we use the Cardano formula and find the following three roots:

$$Z = \begin{cases} \sqrt[3]{\frac{\delta}{2}}(S+T), \\ \sqrt[3]{\frac{\delta}{2}}(jS + j^2T), \\ \sqrt[3]{\frac{\delta}{2}}(j^2S + jT). \end{cases} \quad (27)$$

Here, $j = -\frac{1}{2} + i\sqrt{\frac{3}{2}}$ and

$$S = \sqrt[3]{1 + \sqrt{1 + \frac{4\delta}{27}}}, \quad T = \sqrt[3]{1 - \sqrt{1 + \frac{4\delta}{27}}}. \quad (28)$$

Equation (27) gives the non-zero solutions of Equation (24):

$$x_*^2 = \sqrt[3]{\frac{4D}{3\gamma b^2}}\Psi - \frac{\gamma + a}{3b}, \quad (29)$$

where

$$\frac{1}{\Psi} = \begin{cases} S + T, \\ jS + j^2T, \\ j^2S + jT. \end{cases} \quad (30)$$

To find real solutions, we check the discriminant ($\Delta$) of the last equation of Equation (25),

$$\Delta = -27(-\delta)^2 - 4(\delta)^3 = -4\delta^2 \left[\frac{27}{4} + \delta\right], \tag{31}$$

as the sign of $\Delta$ determines the number of the real root as follows:

- If $\Delta < 0$, then one root is real, and two are complex conjugates,
- If $\Delta = 0$, then all roots are real, and at least two are equal,
- If $\Delta > 0$, then all roots are real and unequal.

From a detailed analysis of different cases provided in Appendix A, we conclude that the existence of a bimodal PDF requires $\Delta \leq 0$ in Equation (31), and that the peak position of a bimodal PDF is given by

$$x_* = \pm\sqrt{\sqrt[3]{\frac{4D}{3\gamma b^2}}\frac{1}{S+T} - \frac{\gamma + a}{3b}}, \tag{32}$$

where

$$\delta = \frac{\gamma(\gamma - 2a)^3}{18Db}, \quad S = \sqrt[3]{1 + \sqrt{1 + \frac{4\delta}{27}}}, \quad T = \sqrt[3]{1 - \sqrt{1 + \frac{4\delta}{27}}}.$$

Finally, a convenient method of identifying parameter values for unimodal versus bimodal PDFs is to check the sign of $\partial_{xx}p(x)$ at $x = 0$:

$$\partial_{xx}p\Big|_{x=0} = \left[6b - \frac{\gamma}{D}(a+\gamma)^2 a\right]. \tag{33}$$

Since a unimodal PDF takes a local maximum at $x = 0$ when $\partial_{xx}p < 0$ and a local minimum at $x = 0$ when $\partial_{xx}p > 0$, we can see from Equation (33) that a unimodal PDF with $\partial_{xx}p(x = 0) < 0$ is more likely for larger $\gamma$ and smaller $D$. Alternatively, a finite correlation time of $f$ (small $\gamma$) and a large diffusion ($D$) facilitate the formation of a bimodal PDF.

To illustrate these results, Figures 1 and 2 show how the peak position $x_*$ and peak amplitude $p(x_*)$, respectively, vary with $\gamma$ for a range of $D$ values. Figure 3 shows the boundary between the unimodal and bimodal PDFs in the $\{\gamma, D\}$ parameter space. These results are for $a = b = 1$, but other values yield the same general boundary shapes, and in particular, the same agreement occurs between the two different evaluation methods, $R = 0$ and (33). The condition $\partial_{xx}p(x = 0) > 0$ is therefore a necessary and sufficient condition to have a bimodal PDF. Figure 4 shows what the PDFs look like and how the transition between unimodal and bimodal PDFs comes about.

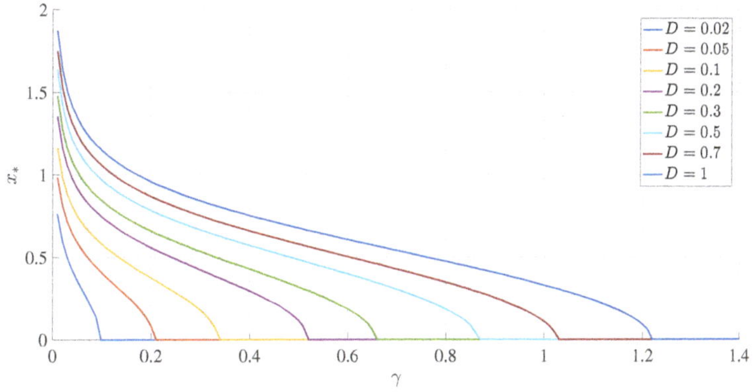


**Figure 1.** The peak positions ($x_*$) as functions of $\gamma$, for different values of $D$, as indicated, and $a = b = 1$.

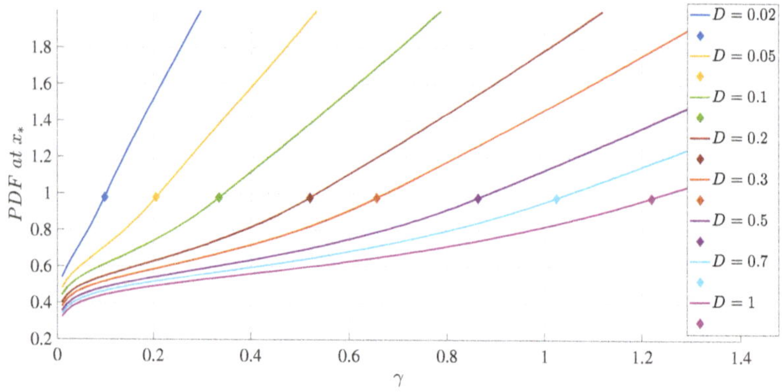


**Figure 2.** The peak amplitudes ($p(x_*)$) as functions of $\gamma$, for different values of $D$, as indicated, and $a = b = 1$. The small diamonds indicate the transition points between unimodal and bimodal Probability Density Functions (PDFs).

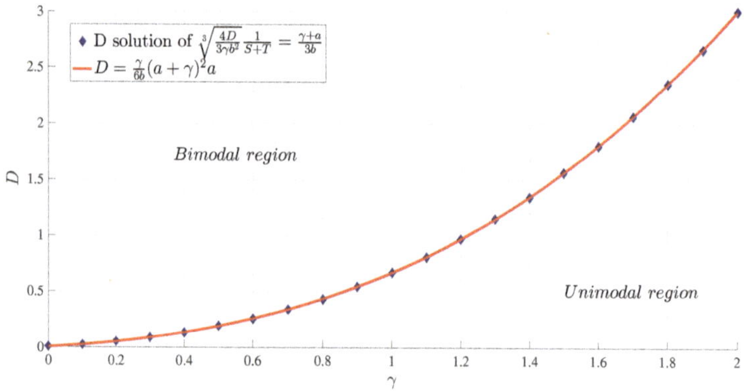


**Figure 3.** The boundary between unimodal and bimodal PDFs in the parameter space $\{\gamma, D\}$, for $a = b = 1$. The red curve is the solution of $\partial_{xx}p(x = 0) = 0$, whereas the blue diamonds are the result of setting $R = 0$.

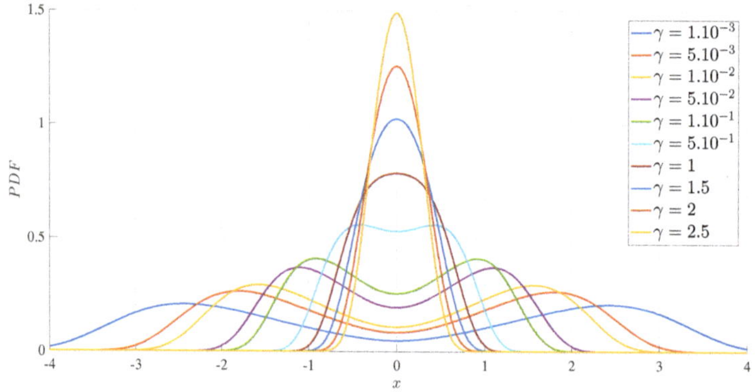


**Figure 4.** The stationary PDFs for $D = 0.7$, $a = b = 1$, and $\gamma$, as indicated. Note the transition between unimodal PDFs for large $\gamma$ and bimodal PDFs for small $\gamma$, in agreement with the boundary shown in Figure 3.

## 4. Numerical Results

We provided analytical solutions for a time-dependent PDF in certain limiting cases, such as small $\gamma$ (e.g., Equation (12)), large $x_0$ and small time (e.g., Equation (17)) in Section 2, and in the limit of large time, where the PDF settles into a stationary solution, in Section 3. To obtain exact time-dependent solutions to the Fokker–Planck equation (22) for any parameter values, we now use numerical methods in this section and utilise results from Section 3 to perform our numerical simulation systematically. As shown in Appendix B, we can set $a = b = 1$ without any real loss of generality by rescaling the other quantities appropriately. The effective parameter space is therefore reduced to $\{\gamma, D\}$, together with whatever parameters define the initial condition, which we take to be $p(x, t = 0) \propto \exp\left[-(x - x_0)^2/10^{-3}\right]$. That is, $\beta_0^x = 10^3$ remains fixed, corresponding to a relatively narrow PDF, and the initial peak position ($x_0$) is the one additional parameter. The initial condition ($p(x, t = 0)$) represents the PDF for an initial shear gradient. When the final stationary PDF is unimodal, the mean shear will decrease to zero in the long time limit; when the stationary PDF is bimodal with a peak of $\pm x_*$, $x_0 > x_*$ models the relaxation of an initial super-critical gradient ($x_0$) to the critical value ($x_*$) while $x_0 < x_*$ models the build-up of the gradient from an initial subcritical value to the critical value ($x_*$). We are interested in the information change in this relaxation problem and in identifying the difference between the relaxation and build-up of the shear gradient in view of the information change. The numerical implementation of Equation (22) is based on second-order accurate finite-differencing in both $x$ and $t$, with up to $10^4$ grid points in $x$, and timesteps as small as $10^{-4}$. The domain in $x$ is truncated to the interval $[-10, 10]$ rather than the original unbounded interval for which the analytic theory applies. As seen in Figure 4, for example, for the parameter values of interest here, the PDFs are well-confined to the interval $|x| \leq 10$, making a numerical solution of (22) with boundary conditions of $p = 0$ at $x = \pm 10$ an excellent equivalent to an infinite interval.

### 4.1. *Time Evolution of PDFs*

Figure 5 shows examples of how different values of $x_0$ ultimately all relax to the same final PDF. Panels (a–d) correspond to $x_0 = 0$, 0.32, 0.6, 1, respectively. $\gamma = D = 1$, according to Figure 3, is slightly in the bimodal regime, consistent with the final PDF seen here. Figure 6 focuses specifically on how the positions of the peaks evolve in time. Important observations that we can make from Figures 5 and 6 are as follows:

(a) An initial PDF with a peak at $x_0 = 0$ remains unimodal before becoming a bimodal PDF;

(b) An initial PDF with a peak at $x_0 = x_*$ (0.32 for this case) does not maintain the same peak position at $x_*$, but moves outward first to $x > x_*$ and then inwards to $x_*$. This initial outward movement explains why the minimum $\mathcal{L}_\infty = \mathcal{L}(t \to \infty)$ does not occur for $x_0 = x_*$ in Section 4.2;

(c) An initial PDF with a peak at $x = x_{\mathcal{L}}$ (where $x_{\mathcal{L}}$ is the $x_0$ value which minimises $\mathcal{L}_\infty$, as defined in Section 4.2) constitutes the border line between different PDF evolutions (an initial PDF with a peak at $x_0 < x_{\mathcal{L}}$ goes outwards and then inwards, while an initial PDF with a peak at $x_0 > x_{\mathcal{L}}$ monotonically moves inwards to $x_*$);

(d) An initial PDF with a peak at $x_0 = 1 > x_{\mathcal{L}}$ monotonically moves inwards.

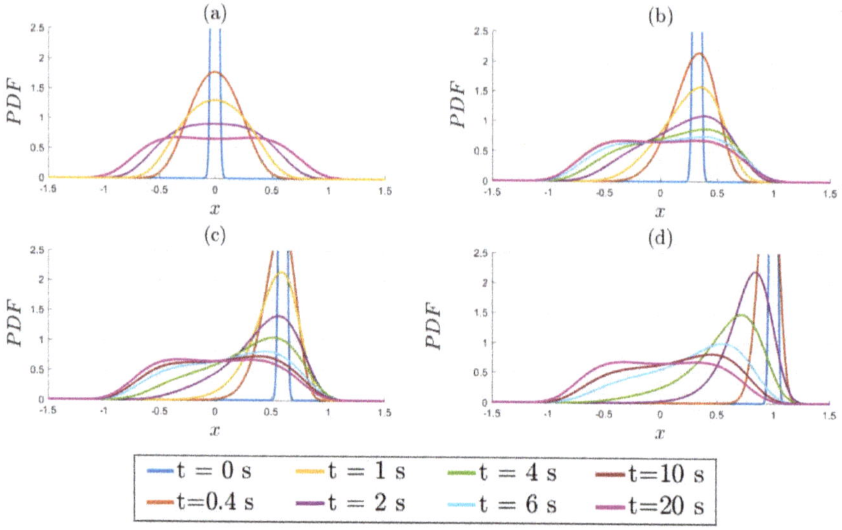


**Figure 5.** Time evolution of the PDFs for the following initial conditions: (**a**) $x_0 = 0$; (**b**) $x_0 = 0.32$; (**c**) $x_0 = 0.6$; (**d**) $x_0 = 1$. $\gamma = D = 1$ for all four.

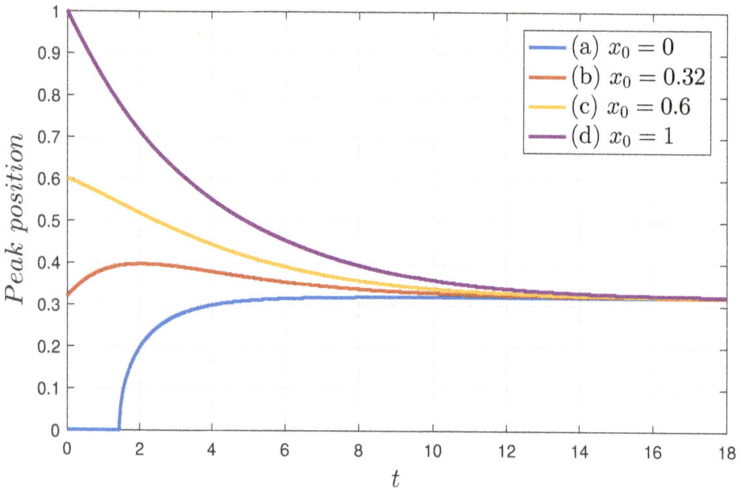


**Figure 6.** The peak positions of the solutions in Figure 5 as functions of time.

## 4.2. *Information Length: Attractor Structure*

Since $\mathcal{L}(t)$ represents the cumulative change in information, it is zero at $t = 0$ and increases with time. As a PDF settles into a stationary PDF in the limits of a large time, the temporal change in PDFs becomes smaller and then becomes zero, $\mathcal{L}(t)$ settling to a constant value of $\mathcal{L}_\infty(x_0, D, \gamma)$. A typical evolution of $\mathcal{L}(t)$ is shown in Figure 7 for $D = 0.5$, $x_0 = 3$, and 4, and a range of $\gamma$ values. The logarithmic scale on the right makes it especially clear that for small times, $\mathcal{L}$ grows linearly in time, before eventually equilibrating to its final value, $\mathcal{L}_\infty$. In order to make more precise comparisons with the analytic prediction (17), Figure 8 shows the results of extracting a numerically computed slope, call it $\mu = \frac{d}{dt}\mathcal{L}(t)$, and compares with the analytic expectation $\sqrt{2\beta_0^x/9b^2}\,\gamma x_0$ in Equation (17). That $\mu$ is expected to scale linearly with $\gamma$ and $x_0$ and be independent of $D$, is reasonably well reproduced by the numerical data with less than a 10% difference between the theoretical prediction and simulation results (note the small range of the $y$-axis).

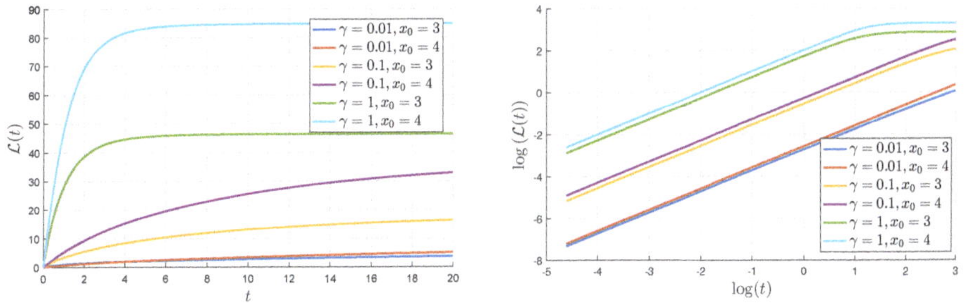


**Figure 7.** $\mathcal{L}(t)$ as a function of $t$, with a linear scale on the left and a logarithmic scale on the right.

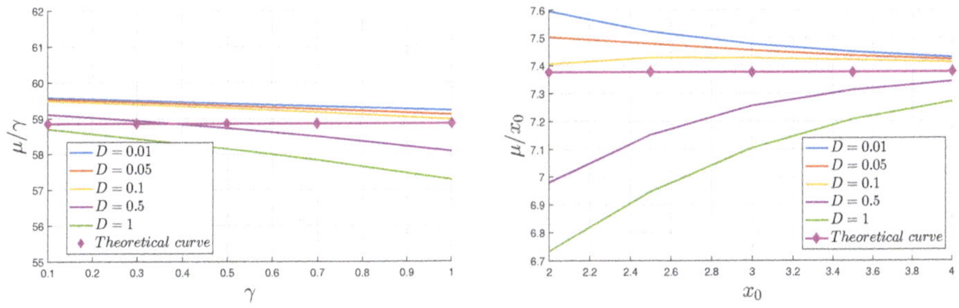


**Figure 8.** Letting $\mu$ denote the numerically computed slope $\mathcal{L}(t)/t$ (for small $t$), the left panel shows $\mu/\gamma$ as a function of $\gamma$, for $x_0 = 4$, and the right panel shows $\mu/x_0$ as a function of $x_0$ for $\gamma = 0.5$. The agreement with the expectations from Equation (17) is seen to be reasonably good.

$\mathcal{L}_\infty(x_0, D, \gamma)$ is a unique representation of the total number of statistically different states that a PDF evolves through to reach a final unimodal or bimodal PDF. The smaller $\mathcal{L}_\infty$ is, the smaller the number of states that the initial PDF passes through to reach the final equilibrium. Therefore, $\mathcal{L}_\infty$ provides us with a path-dependent Lagrangian measure of the distance between a given initial and final PDF. Thus, by choosing a narrow initial PDF at different peak positions ($x_0$), we can map out the attractor structure (the proximity of $x_0$ to an equilibrium) by measuring $\mathcal{L}_\infty$ as a function of $x_0$. We were particularly interested in how differently $\mathcal{L}_\infty$ would behave for the final unimodal and bimodal PDFs, which have different stable equilibrium points: $x = 0$ and $x = x_* \neq 0$, respectively. To this end, Figure 9 shows $\mathcal{L}_\infty$ as a function ($x_0$) for a range of $D$ values. For final bimodal PDFs, the location of the final peak position ($x_*$) is shown by a little vertical line.

We note first in Figure 9 that the overall shapes of the curves are drastically different depending on whether the final PDF is unimodal or bimodal. For a unimodal final PDF, the minimum value of $\mathcal{L}_\infty$ occurs for $x_0 = 0$. This is because $x_0 = 0$ is a stable equilibrium for a unimodal PDF and thus, an initial PDF with the peak ($x_0$) closer to $x = 0$ undergoes less change during the evolution of time and is more similar to the final PDF. Therefore, the absolute minimum of $\mathcal{L}_\infty$ occurs at $x_{\mathcal{L}} = \text{argmin}_{x_0}\mathcal{L}_\infty(x_0) = 0$, as can be seen in the orange and yellow curves in Figure 9.

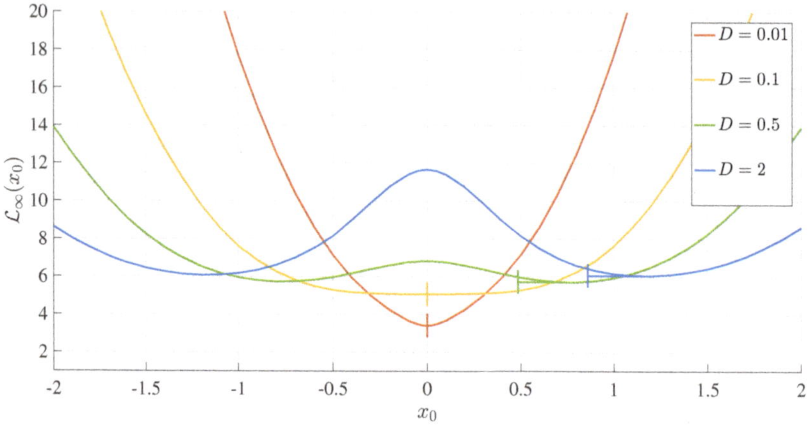


**Figure 9.** $\mathcal{L}_\infty$ as a function of $x_0$ for $D$, as indicated, and $\gamma = 0.5$.

In comparison, $x = 0$ is an unstable equilibrium point for a final bimodal PDF, while $x_* \neq 0$, given by Equation (32), is a stable equilibrium point. Therefore, $\mathcal{L}_\infty$ has a local maximum around $x_0 = 0$ (unstable point). Naively, the minimum value of $\mathcal{L}_\infty$ would be expected to occur for an initial PDF with $x_0 = x_*$, that is, when the peak position of an initial PDF ($x_0$) coincides with that of the final PDF ($x_*$). However, the blue and green curves in Figure 9 reveal the very interesting fact that $\mathcal{L}_\infty$ is actually minimised for $x_0 = x_{\mathcal{L}} > x_*$. As noted from Figures 5 and 6, this is because the initial peaks that are sufficiently far away move inwards monotonically, but the initial peaks near $x_*$ actually have a more complicated evolution (moving outwards and then inwards).

These observations confirm that $\mathcal{L}_\infty$ is a good Lagrangian measure that captures the attractor structure and dynamics. It is, thus, of particular interest to compare $\mathcal{L}_\infty$ with the Kullback–Leibler divergence [19] (that is commonly used in comparing PDFs), defined as

$$D(p||q) = \int p(x) \ln \left[\frac{p(x)}{q(x)}\right] dx, \tag{34}$$

where $p(x)$ is the initial PDF and $q(x)$ is the final one. Obviously, unlike $\mathcal{L}_\infty$, $D(p||q)$ depends only on the initial and final PDFs, and thus, does not provide any information on dynamics (e.g., what different states an initial PDF passes through in the time evolution, or how the locations and the shapes of the PDFs evolve in time between initial and final PDFs). Since we have an analytic expression for the stationary PDFs, we computed $D(p||q)$ by numerical integration with the initial PDF used above. Figure 10 shows these results, where the little vertical lines represent the positions of $x_*$.

We can see that the absolute minimum relative entropy always occurs when $x_0 = 0$ or $x_*$ for unimodal and bimodal PDFs, respectively, unlike $\mathcal{L}_\infty$. In retrospect, this is not particularly surprising, since the relative entropy only measures the difference between the two PDFs, and an initial PDF located at the final peak position is most similar to the final PDF. Specifically, for a bimodal PDF, the initial PDF at the peak position of the final PDF has the strongest resemblance to the final PDF, with the minimum $D(p||q)$ occurring for $x_0 = x_*$.

For completeness, we also show $D(q||p)$ in Figure 11. Unlike Figure 10, the absolute minimum value occurs at $x_0 = 0$, even when the final PDF is bimodal, failing to capture the attractor structure associated with a bimodal PDF. Furthermore, the values of $D(q||p)$ are much larger than those of $D(p||q)$, and thus, a symmetric version ($[D(p||q) + D(q||p)]/2$) would be dominated by $D(q||p)$. This drastic difference between $D(p||q)$ and $D(q||p)$ calls for care in using symmetric versions.

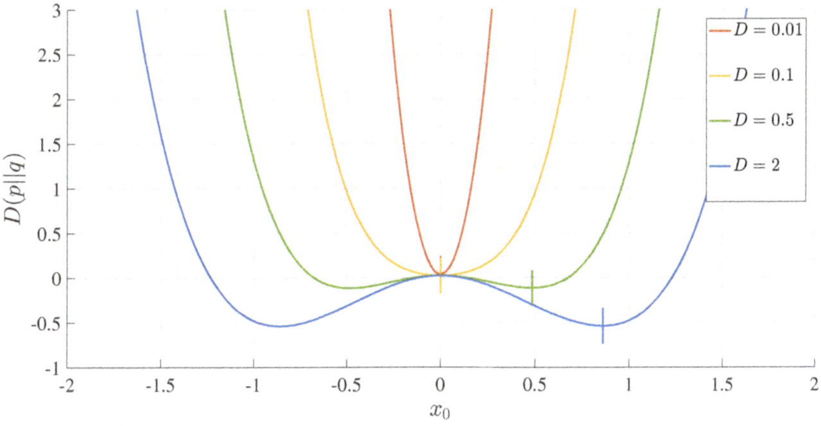


**Figure 10.** Relative entropy ($D(p||q)$) as a function of $x_0$ for $D$, as indicated, and $\gamma = 0.5$.

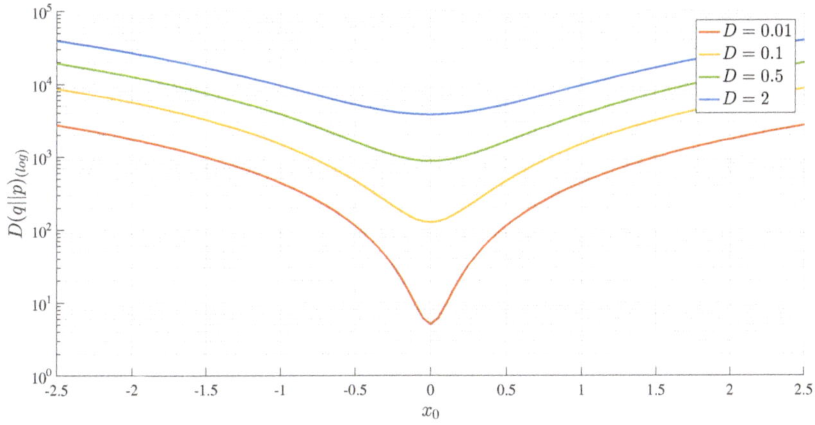


**Figure 11.** $D(q||p)$ as a function of $x_0$ for $D$, as indicated, and $\gamma = 0.5$.

## 5. Discussion and Conclusions

We investigated the time evolution of PDFs in a toy model of self-organised shear flows using a unified coloured approximation, and utilised the information length to understand information changes and attractor structures. In our model, the formation of shear flows was induced by a finite memory time of a stochastic forcing and was manifested by the emergence of a bimodal PDF, with the two peaks representing non-zero mean values of a shear flow (gradient). We presented a thorough study of PDFs for different correlation time and amplitude values for the stochastic forcing. By solving the Fokker–Planck equation numerically, we investigated the time evolution of PDFs starting with a narrow PDF at different peak positions ($x_0$) at time $t = 0$. The cumulative change in information ($\mathcal{L}_\infty$) beautifully maps out the underlying attractor structures. Specifically, for a unimodal PDF, the minimum value of $\mathcal{L}_\infty$ occurs for $x_0 = 0$, since $x_0 = 0$ is a stable equilibrium for a unimodal PDF and thus, an initial PDF with a peak ($x_0$) closer to $x = 0$ undergoes less change during the time evolution and is more similar to the final PDF; for a bimodal PDF, $\mathcal{L}$ is minimised for $x_0 = x_{\mathcal{L}} > x_*$, where $x_*$ is the peak position of a bimodal PDF. Recalling that $x_0$ represents the mean shear gradient at $t = 0$ while $x_*$ is a critical shear gradient, $x_0 = x_{\mathcal{L}} > x_*$ implies that an initial narrow PDF with a super-critical shear gradient is, in fact, more similar to a final stationary state, while an initial narrow PDF with a mean critical shear gradient undergoes a complicated evolution through the interaction with fluctuations. This is likely to be due to the rapid relaxation of instability at the super-critical state, similar to what was observed in the forward process in the phase transition in [27] (e.g., compare Figures 6b and 7b). That is, a process triggered by instability involves a smaller change in information and thus, a larger change in entropy (as might be expected as a consequence of instability). This reflects a unique property of $\mathcal{L}_\infty$ which depends on a trajectory/history of a PDF evolution. In comparison, the relative entropy, which only measures the difference between the initial and final PDFs, does not provide any information on the dynamics between the initial and final times. In summary, we demonstrated the importance of studying the dynamics and the merit of the information length in understanding the dynamics and the evolution of PDFs in a toy model of self-organised shear flow. Further work will include the extension of this work to the analysis of our model without unified colored-noise approximation and to other turbulence models, in particular, to quantify the information change associated with intermittency and self-organisation.

**Author Contributions:** Original research idea, E.K.; Investigation, Q.J., E.K. and R.H.; Software, R.H.; Visualization, Q.J.; Supervision, E.K and R.H; Writing—original draft, E.K, Q.J. and R.H.; Writing—review and editing, E.K., Q.J. and R.H.

**Funding:** This research received no external funding.

**Conflicts of Interest:** The authors declare no conflict of interest.

## Appendix A. Derivation of Equation (32)

### *Appendix A.1. Case $\delta = -\frac{27}{4} \iff \Delta = 0$*

According to the definitions of $S$ and $T$ in Equation (28), $S = T = 1$. So, by using $1 + j + j^2 = 0$, we calculated Equation (30):

$$\frac{1}{\Psi} = \begin{cases} S + T = 2, \\ jS + j^2 T = j + j^2 = -1, \\ j^2 S + jT = j^2 + j = -1. \end{cases} \tag{A1}$$

Consistent with the statement above, we obtained three real solutions. The last two solutions with the same value of $\frac{1}{\Psi} = -1$ in Equation (A1) make Equation (29):

$$x_*^2 = -\left[\sqrt[3]{\frac{4D}{3\gamma b^2}} + \frac{\gamma + a}{3b}\right] < 0, \tag{A2}$$

which is inconsistent, since $x_*$ is a real number. On the other hand, the solution $\frac{1}{\Psi} = S + T = 2$ can give a consistent solution if $x_*^2$ in Equation (29) is positive, that is

$$R \equiv \sqrt[3]{\frac{4D}{3\gamma b^2}}\frac{1}{S+T} - \frac{\gamma + a}{3b} > 0. \tag{A3}$$

If not, the PDF is unimodal.

### *Appendix A.2. Case $\delta > -\frac{27}{4} \iff \Delta < 0$*

$\Delta < 0$ gives a unique real solution and two complex solutions which are complex conjugates. It is easy to see that the real solution of our interest corresponds to $\frac{1}{\Psi} = S + T$ because $\sqrt{1 + \frac{4\delta}{27}}$ is real. Therefore, as long as Equation (A3) holds, we have a bimodal PDF.

*Appendix A.3. Case $\delta < -\frac{27}{4} \iff \Delta > 0$*

We can take $\gamma < 2a$, because if $\gamma > 2a$, then $\delta > 0$ (see the last equation in Equation (26)). We thus take a root ($Z_*$) of the last equation of Equation (25) with $\delta < -\frac{27}{4}$.

Appendix A.3.1. Subcase $Z_* > 0$

Obviously, in this case,

$$Y_* = \frac{1}{Z_*} > 0. \tag{A4}$$

We recall that

$$X_* = \frac{\gamma - 2a}{3}(Y_* - 1). \tag{A5}$$

Using $\gamma - 2a < 0$ and Equation (A4), we have

$$X_* = \frac{2a - \gamma}{3}(1 - Y_*) < \frac{2a - \gamma}{3} < \frac{2}{3}a, \tag{A6}$$

which is in contradiction to $X_* = a + bx_0^2 > a$. Thus, there is no consistent non-zero solution in this case.

Appendix A.3.2. Subcase $Z_* < 0$

Because $\delta < -\frac{27}{4} < 0$, $\delta Z_* > 0$. Using the last equation in Equation (25), we then have

$$Z_*^3 - \delta = -\delta Z_* < 0, \tag{A7}$$

and thus,

$$Z_*^3 < \delta < \frac{-27}{4} \Rightarrow Z_* < -\frac{3}{\sqrt[3]{4}}. \tag{A8}$$

Using Equation (A7) in Equation (A8) then gives us

$$Z_*^3 - \delta = \delta(-Z_*) < \frac{27}{4}Z_* < -\frac{27}{4}\frac{3}{\sqrt[3]{4}}. \tag{A9}$$

Therefore,

$$Z_*^3 < \delta - \frac{27}{4}\frac{3}{\sqrt[3]{4}} < -\frac{27}{4}(1 + \frac{3}{\sqrt[3]{4}}), \tag{A10}$$

and

$$Y_* = \frac{1}{Z_*} > -\frac{1}{\sqrt[3]{\frac{27}{4}(1 + \frac{3}{\sqrt[3]{4}})}}. \tag{A11}$$

So, using Equation (A11) in Equation (A5) gives us

$$X_* = \frac{2a - \gamma}{3}(1 - Y_*) < \frac{2a - \gamma}{3}\left[1 + \frac{1}{\sqrt[3]{\frac{27}{4}(1 + \frac{3}{\sqrt[3]{4}})}}\right] < \frac{2}{3}\left[1 + \frac{1}{\sqrt[3]{\frac{27}{4}(1 + \frac{3}{\sqrt[3]{4}})}}\right] a \simeq 0.91a. \tag{A12}$$

Equation (A12) is in contradiction with $X_* = a + bx_*^2 > a$. Therefore, there is no consistent solution in this case. This proves that in the case $\Delta > 0$, there is no consistent non-zero solution ($x_*$) for a bimodal PDF.

*Appendix A.4. Summary*

From the analyses above, we can conclude that the existence of a bimodal PDF requires $\Delta \leq 0$ and

$$R \equiv \sqrt[3]{\frac{4D}{3\gamma b^2}\frac{1}{S+T}} - \frac{\gamma + a}{3b} > 0. \tag{A13}$$

To make sure that this solution in Equation (29) corresponds to a local maximum, we need to show $\partial_{xx} p(x = x_*) < 0$. To this end, we recall that $x_*$ satisfies Equation (23)

$$-\frac{\gamma}{D}(\partial_x g(x_*) + \gamma)^2 \frac{g(x_*)}{x_*} + 6b = 0. \tag{A14}$$

We then find the second derivative of the stationary PDF at $x_*$, as follows:

$$\begin{aligned} \partial_{xx} p(x_*) \exp\Big(\frac{\gamma}{2D}\big[g(x_*)^2 + 2\gamma \int^{x_*} g(x_1) dx_1\big]\Big) &= \Big[-\frac{2\gamma}{D}(\partial_x g + \gamma) g \partial_{xx} g \\ -\frac{\gamma}{D}(\partial_x g + \gamma)^2 \partial_x g + 6b + \Big[-\frac{\gamma}{D} g \partial_x g\Big]\Big[-\frac{\gamma}{D}(\partial_x g + \gamma)^2 g + \partial_{xx} g\Big]\Big]_{x=x_*}&. \end{aligned} \tag{A15}$$

By using Equation (23), we see that the last term in Equation (A15) is identically zero. Furthermore, using Equation (A14), we simplify Equation (A15) as

$$\begin{aligned} &\partial_{xx} p(x_*) \exp\Big(\frac{\gamma}{2D}\big[g(x_*)^2 + 2\gamma \int^{x_*} g(x_1) dx_1\big]\Big) \\ &= \Big[\frac{\gamma}{D}(\partial_x g + \gamma)\Big[-2g\partial_{xx} g + (\partial_x g + \gamma)\Big(\frac{g}{x} - \partial_x g\Big)\Big]\Big]_{|x=x_*} \\ &= -\frac{\gamma}{D}(a + \gamma + 3bx_*^2)\Big[12b(ax_*^2 + bx_*^4) + 2x_*^2(a + \gamma + 3bx_*^2)\Big]. \end{aligned} \tag{A16}$$

Since the exponential is positive, $\partial_{xx} p(x_*) < 0$ in Equation (A16), confirming a local maximum of a PDF at $x = x_*$.

## Appendix B. Rescaling

We show, in detail, how we rescale Equations (3) and (4) to make $a = b = 1$. We first rescale $x$ by using $x = \sqrt{\frac{a}{b}}\tilde{x}$ in Equations (3) and (4) to recast them as

$$\begin{cases} \partial_t \tilde{x} = \sqrt{\frac{b}{a}}\Big[-\frac{a\sqrt{a}\tilde{x}}{\sqrt{b}} - \frac{a\sqrt{a}\tilde{x}^3}{\sqrt{b}} + f\Big], \\ \partial_t f = -\gamma f + \xi, \end{cases} \tag{A17}$$

and thus,

$$\begin{cases} \partial_t \tilde{x} = -a(\tilde{x} + \tilde{x}^3) + \sqrt{\frac{b}{a}} f, \\ \partial_t f = -\gamma f + \xi. \end{cases} \tag{A18}$$

Next, we rescale $t$ by using $t = \frac{\tilde{t}}{a}$

$$\begin{cases} \partial_{\tilde{t}} \tilde{x} = -\tilde{x} - \tilde{x}^3 + \frac{1}{a}\sqrt{\frac{b}{a}} f, \\ \partial_{\tilde{t}} f = -\frac{\gamma}{a} f + \frac{\xi}{a}. \end{cases} \tag{A19}$$

We let $\tilde{f} = \frac{1}{a}\sqrt{\frac{b}{a}}f$

$$\begin{cases} \partial_{\tilde{t}}\tilde{x} = -\tilde{x} - \tilde{x}^3 + \tilde{f}, \\ \partial_{\tilde{t}}\tilde{f} = -\frac{1}{a}\gamma\tilde{f} + \frac{1}{a^2}\sqrt{\frac{b}{a}}\xi. \end{cases} \tag{A20}$$

We then let $\tilde{\gamma} = \frac{1}{a}\gamma$ and $\tilde{\xi} = \frac{1}{a^2}\sqrt{\frac{b}{a}}\xi$

$$\begin{cases} \partial_{\tilde{t}}\tilde{x} = -\tilde{x} - \tilde{x}^3 + \tilde{f}, \\ \partial_{\tilde{t}}\tilde{f} = -\tilde{\gamma}\tilde{f} + \tilde{\xi}. \end{cases} \tag{A21}$$

Finally, we rescale $\langle\tilde{\xi}(t)\tilde{\xi}(t')\rangle = \frac{b}{a^5}\langle\xi(t)\xi(t')\rangle = \frac{b}{a^5}D\delta(t'-t) = \frac{b}{a^4}D\delta(\tilde{t}'-\tilde{t}) = \tilde{D}\delta(\tilde{t}'-\tilde{t})$.

## References


1. Dam, M.; Brons, M.; Rasmussen, J.J.; Naulin, V.; Hesthaven, J.S. Identification of a predator-prey system from simulation data of a convection model. *Phys. Plasmas* **2017**, *24*, 022310. [CrossRef]
2. Chang, C.S.; Ku, S.; Tynan, G.R.; Hager, R.; Churchill, R.M.; Cziegler, I.; Greenwald, M.; Hubbard, A.E.; Hughes, J.W. Fast Low-to-High confinement mode bifurcation dynamics in a tokamak edge plasma gyrokinetic simulation. *Phys. Rev. Lett.* **2017**, *118*, 175001. [CrossRef] [PubMed]
3. Kim, E. Consistent theory of turbulent transport in two dimensional magnetohydrodynamics. *Phys. Rev. Lett.* **2006**, *96*, 084504. [CrossRef] [PubMed]
4. Kim, E.; Dubrulle, B. Turbulent transport and equilibrium profile in 2D MHD with background shear. *Phys. Plasmas* **2001**, *8*, 813. [CrossRef]
5. Kim, E. Self-consistent theory of turbulent transport in the solar tachocline I. Anisotropic turbulence. *Astron. Astrophys.* **2005**, *441*, 763.
6. Leprovost, N.; Kim, E. Dynamo quenching due to shear. *Phys. Rev. Lett.* **2008**, *100*, 144502. [CrossRef] [PubMed]
7. Kim, E.; Diamond, P.H. Zonal flows and transient dynamics of the L-H transition. *Phys. Rev. Lett.* **2003**, *91*, 075001. [CrossRef] [PubMed]
8. Li, J.; Kishimoto, Y. Numerical study of zonal flow dynamics and electron transport in electron temperature gradient driven turbulence. *Phys. Plasmas* **2004**, *11*, 1493. [CrossRef]
9. Idomura, Y.; Tokuda, S.; Kishimoto, Y. Global profile effects and structure formations in toroidal electron temperature gradient driven turbulence. *Nucl. Fusion* **2005**, *45*, 1571. [CrossRef]
10. Xu, G.S.; Wu, X.Q. Understanding L-H transition in tokamak fusion plasmas. *Plasma Sci. Technol.* **2017**, *19*, 033001. [CrossRef]
11. Itoh, K. Physics of zonal flows. *Phys. Plasmas* **2006**, *13*, 055502. [CrossRef]
12. Piani, C.; Norton, W.A.; Iwi, A.M.; Ray, E.A.; Elkins, J.W. Transport of ozone-depleted air on the breakup of the stratospheric polar vortex in spring/summer 2000. *J. Geophys. Res. Atmos.* **2002**, *107*, 8270. [CrossRef]
13. Shepherd, T.G. Rossby waves and two-dimensional turbulence in a large-scale zonal jet. *J. Fluid Mech.* **1987**, *183*, 467. [CrossRef]
14. Hunt, J.C.R.; Durbin, P.A. Perturbed vortical layers and shear sheltering. *Fluid Dyn. Res.* **1999**, *23*, 375. [CrossRef]
15. Sood, A.; Kim, E.; Hollerbach, R. Suppression of a laminar kinematic dynamo by a prescribed large-scale shear. *J. Phys. A Math. Theor.* **2016**, *49*, 425501. [CrossRef]
16. Newton, A.P.; Kim, E. A generic model for transport in turbulent shear flows. *Phys. Plasmas* **2011**, *18*, 052305. [CrossRef]
17. Kim, E.; Liu, H.; Anderson, J. Probability distribution function for self-organization of shear flows. *Phys. Plasmas* **2009**, *16*, 052304. [CrossRef]
18. Newton, A.P.; Kim, E. On the self-organizing process of large scale shear flows. *Phys. Plasmas* **2013**, *20*, 092306. [CrossRef]
19. Frieden, B.R. *Science from Fisher Information*; Cambridge University Press: Cambridge, UK, 2004.
20. Wilde, M.M. *Quantum Information Theory*; Cambridge University Press: Cambridge, UK, 2017.

21. Wootters, W.K. Statistical distance and Hilbert space. *Phys. Rev. D* **1981**, *23*, 357. [CrossRef]
22. Nicholson, S.B.; Kim, E. Investigation of the statistical distance to reach stationary distributions. *Phys. Lett. A* **2015**, *379*, 83–88. [CrossRef]
23. Nicholson, S.B.; Kim, E. Structures in sound: Analysis of classical music using the information length. *Entropy* **2016**, *18*, 258. [CrossRef]
24. Heseltine, J.; Kim, E. Novel mapping in non-equilibrium stochastic processes. *J. Phys. A* **2016**, *49*, 175002. [CrossRef]
25. Kim, E.; Lee, U.; Heseltine, J.; Hollerbach, R. Geometric structure and geodesic in a solvable model of nonequilibrium process. *Phys. Rev. E* **2016**, *93*, 062127. [CrossRef] [PubMed]
26. Kim, E.; Hollerbach, R. Signature of nonlinear damping in geometric structure of a nonequilibrium process. *Phys. Rev. E* **2017**, *95*, 022137. [CrossRef] [PubMed]
27. Hollerbach, R.; Kim, E. Information geometry of non-equilibrium processes in a bistable system with a cubic damping. *Entropy* **2017**, *19*, 268. [CrossRef]
28. Kim, E.; Tenkès, L.-M.; Hollerbach, R.; Radulescu, O. Far-from-equilibrium time evolution between two gamma distributions. *Entropy* **2017**, *19*, 511. [CrossRef]
29. Tenkès, L.-M.; Hollerbach, R.; Kim, E. Time-dependent probability density functions and information geometry in stochastic logistic and Gompertz models. *J. Stat. Mech. Theor. Exp.* **2017**, *123201*. [CrossRef]
30. Kim, E.; Lewis, P. Information length in quantum systems. *J. Stat. Mech. Theor. Exp.* **2018**, *043106*. [CrossRef]
31. Kim, E. Investigating information geometry in classical and quantum systems through information length. *Entropy* **2018**, *20*, 574. [CrossRef]
32. Hollerbach, R.; Dimanche, D.; Kim, E. Information geometry of nonlinear stochastic systems. *Entropy* **2018**, *20*, 550. [CrossRef]
33. Risken, H. *The Fokker-Planck Equation: Methods of Solution and Applications*; Springer: Berlin, Germany, 1996.
34. Jung, P.; Hanggi, P. Dynamical systems: A unified colored-noise approximation. *Phys. Rev. A* **1987**, *35*, 4464. [CrossRef]
35. Klebaner, F. *Introduction to Stochastic Calculus with Applications*; Imperial College Press: London, UK, 2012.
36. Gardiner, C. *Stochastic Methods*, 4th ed.; Springer: Berlin, Germany, 2008.
37. Wong, E.; Zakai, M. On the convergence of ordinary integrals to stochastic integrals. *Ann. Math. Stat.* **1960**, *36*, 1560. [CrossRef]

*entropy*

MDPI

*Article*

# Trailing-Edge Flap Control for Mitigating Rotor Power Fluctuations of a Large-Scale Offshore Floating Wind Turbine under the Turbulent Wind Condition

**Bofeng Xu [1,*], Junheng Feng [1], Tongguang Wang [2], Yue Yuan [1], Zhenzhou Zhao [1] and Wei Zhong [2]**

1 College of Energy and Electrical Engineering, Hohai University, Nanjing 211100, China; fengjunheng0@163.com (J.F.); yyuan@hhu.edu.cn (Y.Y.); zhaozhzh_2008@hhu.edu.cn (Z.Z.)

2 Jiangsu Key Laboratory of Hi-Tech Research for Wind Turbine Design, Nanjing University of Aeronautics and Astronautics, Nanjing 210016, China; tgwang@nuaa.edu.cn (T.W.); zhongwei@nuaa.edu.cn (W.Z.)

* Correspondence: bfxu1985@hhu.edu.cn

Received: 13 August 2018; Accepted: 4 September 2018; Published: 6 September 2018

**Abstract:** A trailing-edge flap control strategy for mitigating rotor power fluctuations of a 5 MW offshore floating wind turbine is developed under turbulent wind inflow. The wind shear must be considered because of the large rotor diameter. The trailing-edge flap control strategy is based on the turbulent wind speed, the blade azimuth angle, and the platform motions. The rotor power is predicted using the free vortex wake method, coupled with the control strategy. The effect of the trailing-edge flap control on the rotor power is determined by a comparison with the rotor power of a turbine without a trailing-edge flap control. The optimal values of the three control factors are obtained. The results show that the trailing-edge flap control strategy is effective for improving the stability of the output rotor power of the floating wind turbine under the turbulent wind condition.

**Keywords:** trailing-edge flap; control strategy; floating wind turbine; turbulence; free vortex wake

---

## 1. Introduction

Wind power has been developing rapidly worldwide due to fossil fuel energy depletion and environmental pollution. Offshore wind power is characterized by high wind energy density, high annual utilization hours, and close proximity to a power load center, and these advantages provide an important direction for future wind power development [1]. Although offshore wind power technology is relatively mature, the economy and reliability of wind turbines with fixed bases have decreased significantly with the expansion of wind turbine installation in deep-water areas and the associated advantages of a floating base. In 2009, the first spar-type full-scale floating offshore wind turbine, Hywind, was successfully installed and commissioned [2]. Subsequently, additional full-scale floating offshore wind turbines have been installed worldwide [3].

Ocean waves, ocean currents, strong winds, and the complex marine environment pose a series of problems to the normal operation of floating wind turbines. First, turbulent wind affects the inflow velocity and angle of the wind turbine blades. The inertia of large-scale wind blades is large and the individual pitch control is difficult to achieve, due to the rapidly changing aerodynamic loads under turbulent wind conditions [4]. Second, although the wind shear coefficient is smaller offshore than on land, the influence of the wind shear on the rotor power of the wind turbine cannot be ignored because the large hub height and long blades of offshore wind turbines cause a significant difference in the wind speed at the highest and lowest points of the wheel. Third, the floating platform swings periodically in a certain direction under the influence of the waves and currents. This phenomenon produces a large negative effect on the fatigue loads of the relevant components of the wind turbine [5]. The periodic motion of the floating foundation also causes periodic fluctuations in the rotor power, and affects the quality of the electric power output [1]. The 5 MW floating offshore wind turbine platform built by the US National Renewable Energy Laboratory (NREL) provides reliable basic data for offshore wind energy research [6]. Since then, research on load fluctuations and optimal control of floating wind turbines has also been conducted extensively in the industry [7]. Research on the use of blade flaps in wind turbines [8,9] has indicated that the trailing-edge flap structure of the wind turbine blade has technical advantages on the floating wind turbine. The wind turbine blade flaps do not only mitigate the load fluctuation effectively, but also supplement the traditional pitch control, which results in a more flexible and robust wind turbine control system.

Most of the aerodynamic analysis software packages such as Bladed and FAST are based on the Blade Element Momentum (BEM) theory. However, the wake-induced velocity in the BEM theory is the average induced velocity, and large errors can occur in the calculations, requiring many corrections. The vortex theory provides more accurate results for the wake flow field of the wind turbine when calculating the aerodynamic performance, and the induced velocity of the flow field is obtained directly from the wake flow [10]. Therefore, the vortex theory is suitable for calculating the unsteady aerodynamic performance of wind turbines. The free vortex wake (FVW) method is based on the vortex theory, and has been successfully applied to the aerodynamic performance calculation of wind turbines.

The main objective of this study is to propose a trailing-edge flap control strategy for a large-scale offshore floating wind turbine to mitigate the rotor power fluctuations in the turbulent wind condition. The NREL 5 MW floating wind turbine is used as an example for the calculations. The previously developed FVW model [10,11] is used to calculate the rotor power of the wind turbine with the proposed trailing-edge flap control.

In Section 2, the platform motions of the floating wind turbine are described. Section 3 describes the turbulent wind condition. Section 4 describes the blade structure with the trailing-edge flap and control strategy for mitigating the rotor power fluctuations. Section 5 briefly introduces the FVW method. The results are presented in Section 6, and they include the control effect on the rotor power under different unsteady conditions. The conclusions are drawn in Section 7.

## 2. Platform Motions of the Floating Wind Turbine

As shown in Figure 1, the motion of the floating wind turbine platform can be described using a global coordinate system ($X_F$, $Y_F$, $Z_F$) and a shaft coordinate system ($x$, $y$, $z$) originating at the shaft of the turbine. Assuming that the wind turbine is a rigid structure, the floating wind turbine platform has six degree of freedom, in terms of translational and rotational motions in the global coordinate system; these are surge, sway, heave, pitch, roll, and yaw.

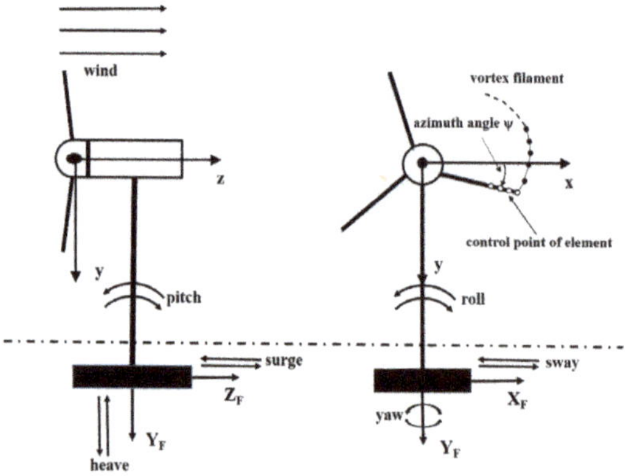


**Figure 1.** Coordinate system and the six degree of freedom motions.

The NREL 5 MW wind turbine is located on a floating tension leg platform (TLP). The rated operating condition is maintained at the rated speed of 11.4 m/s and the rotor speed is 12.1 rpm. The resulting FAST-simulated platform kinematics of the TLP [12] for the rated operating condition for a 300 s simulation was used in this study. Figure 2a shows the three translational motions, and Figure 2b shows the three rotational motions. It was observed that the mean and amplitude of the yaw, heave, and roll motions were small. The yaw motions had little effect on the wind turbine for a small angle, as the analysis of the influence of each degree-of-freedom motion on the aerodynamic performance showed. However, the pitch and surge motions had a large influence on the blade inflow of the wind turbine [12]. Therefore, in this study, we only analyzed the influence of the pitch and surge motions on the aerodynamic performance of the floating wind turbine.

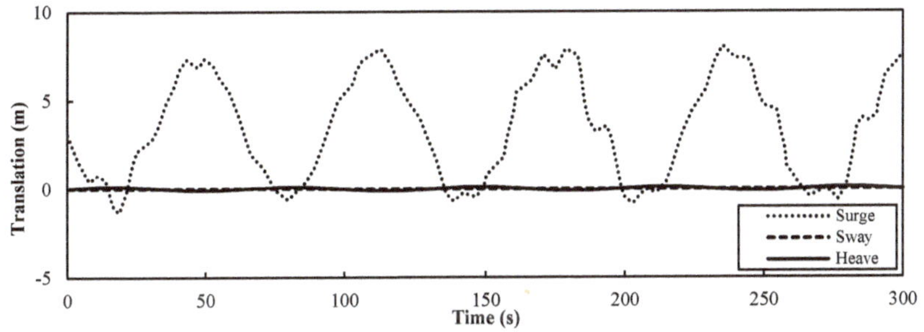


(**a**) Translational motions

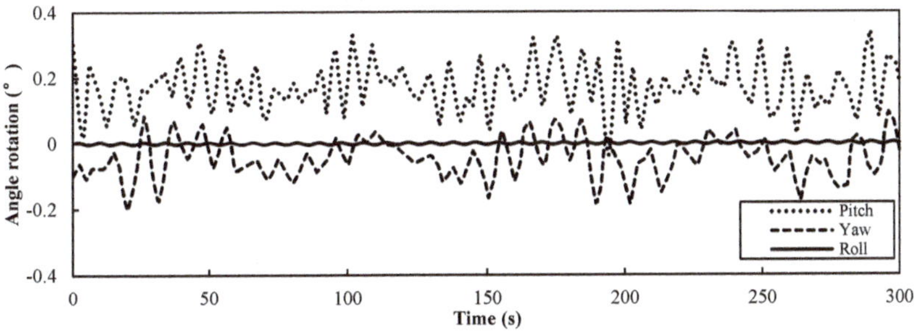


(**b**) Rotational motions

**Figure 2.** Time history of the tension leg platform (TLP) motions of the US National Renewable Energy Laboratory (NREL) 5 MW turbine for the rated operating condition.

## 3. Turbulence Wind Condition

In land-based wind farms or offshore wind farms, the main factor affecting the load fluctuations of wind turbines, is turbulent wind. Due to the lack of relevant measured data of offshore turbulent wind, we simulated the turbulence of the offshore wind field using a turbulence model. The wavelet transform is an appropriate method to detect the local similarity in time-series data of turbulence [13]. In this study, a one-dimensional velocity change (axial component) was considered at the hub center height of 90 m under the turbulent condition. The wavelet inverse transformation method [14] was used to calculate the turbulence wind field according to the advanced von Karman power density spectrum. The axial velocity varied around the rated wind speed of 11.4 m/s from 0 to 300 s, as shown in Figure 3. The roughness of the ocean surface was 0.001 mm, and the turbulence intensity was 0.0933.

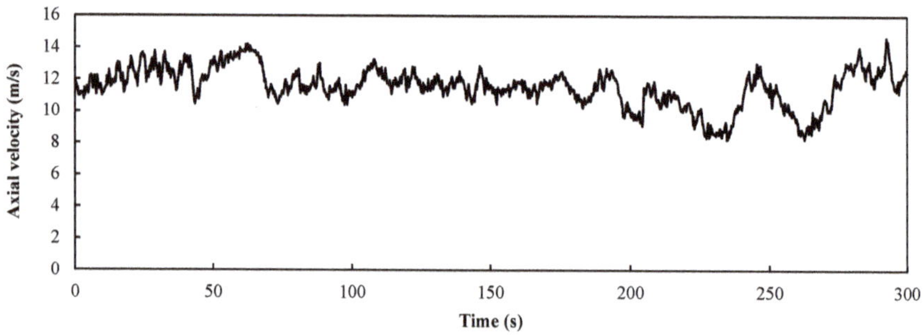


**Figure 3.** Axial velocity for the rated speed of 11.4 m/s in the turbulent condition.

Wind shear exists in the atmosphere near the ground and sea surface because of the topography and the sea surface roughness. The oncoming boundary layer wind velocity profile is described as:

$$U(h) = U_{ref} \cdot \left(\frac{h}{h_{ref}}\right)^{\alpha} \tag{1}$$

where $h_{ref}$ is the reference height (hub center height) and $U_{ref}$ is the wind speed at the reference height. The power law exponent $\alpha$ is associated with the local terrain roughness. Figure 4 shows the wind shear distribution near the ground, in which $\alpha$ = 0.1 is the value for the offshore sea and $\alpha$ = 0.2 is the value for land. Although the power law exponent is smaller offshore, the wind turbine output force changes due to the wind shear, because the large-scale offshore wind turbine has a diameter of more than 100 m. The hub center height of the NREL 5 MW wind turbine was 90 m, and the diameter of the rotor is 126 m. At a hub wind speed of 11.4 m/s, the maximum wind speed of the wheel reached 12.02 m/s, and the lowest wind speed is 10.11 m/s. Figure 5 shows the aerodynamic torque of a single blade of the NREL 5 MW wind turbine during a rotating period under wind shear at the rated wind speed. The range of the aerodynamic torque during the period is 1110–1482 k·Nm. The amplitude is 28% of the average. Therefore, the wind shear needs to be taken into account.

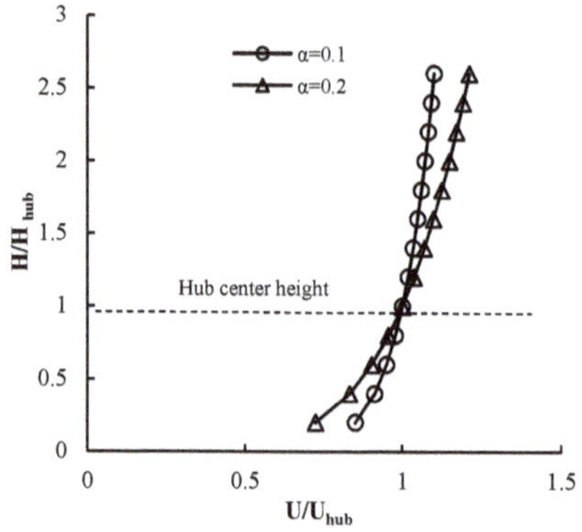


**Figure 4.** Atmospheric boundary layer profiles.

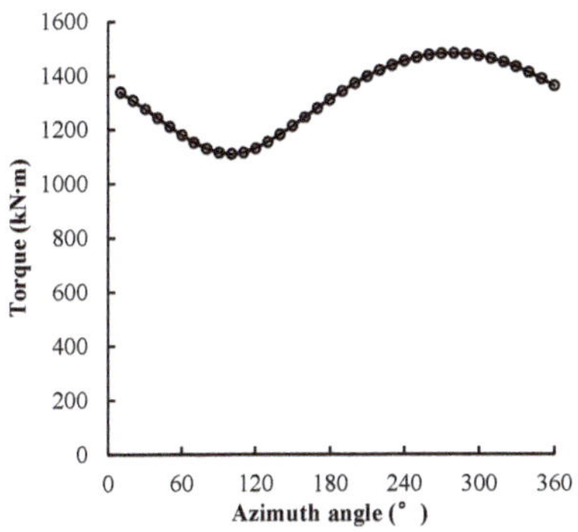


**Figure 5.** Aerodynamic torque of a single blade of the NREL 5 MW wind turbine during a rotation period under wind shear at the rated wind speed.

## 4. The Trailing-Edge Flap Control Strategy

### 4.1. The Blade Structure with the Trailing-Edge Flap

In this study, a trailing-edge flap structure for the wind turbine was proposed. The trailing-edge flap structure used in the NREL 5 MW wind turbine blade was based on the results of a study by Zhang et al. [15] on the optimization of the structural parameters of the trailing-edge flap of wind turbines. As shown in Figure 6a, the blade length was 61.5 m and the radial flap length of the red portion was 14 m along the axis direction of the blade; the flap was located at 1.2 m from the tip of the blade, and extended to a distance of 15.2 m from the tip. The relative thickness of the airfoil with the flaps was changed to 18%, which was convenient for installation and calculation purposes. The flap length comprised 20% of the length of the chord, as shown in Figure 6b.

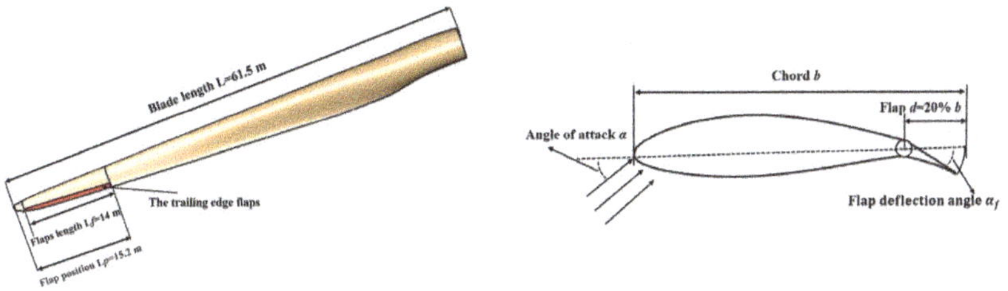


(**a**) Blade structure with trailing-edge flap (**b**) Airfoil of the flap

**Figure 6.** NREL 5 MW wind turbine blade with trailing-edge flap.

### 4.2. The Aerodynamic Performance of the Trailing-Edge Flap

The control system adjusts the torque coefficient and the thrust coefficient of the blades, to mitigate the load fluctuations of the blade through the deflection control of the flap. Therefore, the aerodynamic performance of the airfoil with the flaps is vital. Figure 7 shows the lift and drag coefficient data of the airfoil with different flap deflection angles. In this study, the NREL 5 MW wind turbine was simulated under the condition of the rated wind speed. When the NREL 5 MW wind turbine ran at the rated wind speed of 11.4 m/s, the blade pitch angle was 0° and the rotational speed was 12.1 rpm. The angle of attack at the blade tip was about 8°, and the lift and drag coefficient data, in the range of 4–14°, are given in Figure 7. It can be seen that the lift coefficient increased with the increase in the flap deflection angle for the same attack angle, and the lift-to-drag ratio increased first and then decreased. It is worth noting that the smaller the attack angle, the faster the lift-to-drag ratio increased, and the larger the adjustable range became. Therefore, a smaller attack angle of the blade is preferred. In the range of the flap deflection angle of 10~15°, the lift-to-drag ratio decreased, whereas the lift-to-drag ratio was unstable in the range of 15~20°. A comprehensive analysis indicated that $\alpha_f = -5°$ was optimal for the origin of the flap, and −20~10° was the optimum range of the flap deflection angle.

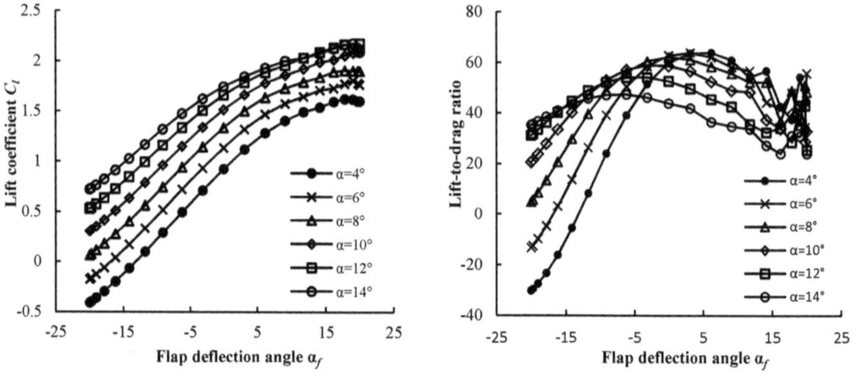


(**a**) Lift coefficient vs the flap deflection angle (**b**) Lift-to-drag ratio vs the flap deflection angle

**Figure 7.** Aerodynamic performance of the airfoil with a trailing-edge flap.

### 4.3. Control Strategy

The flap control method is also important and determines the effectiveness of the flap control. A simple flap control method based on wind speed, blade azimuth angle, and platform motion was proposed to verify the effectiveness of the flap control. More efficient and practical control methods require further study.

The control strategy under the turbulent wind was based on the two average wind speeds ($U$ and $U_t$). It was assumed that the instantaneous wind speed was measured four times per second by the nacelle anemometer. The control scheme is defined as:

$$\alpha_{ft}(t) = a \cdot (U - U_t) \quad (2)$$

where $a$ is the control factor for the turbulence and $U$ is the average wind speed over 60 s. In order to avoid the frequent change of the flap deflection angle, $U_t$ is defined as the average wind speed over one second, and is expressed in Equation (3):

$$U_t = \frac{u_t + u_{t-0.25} + u_{t-0.5} + u_{t-0.75} + u_{t-1}}{5} \quad (3)$$

The effect of the wind shear on the wind turbine is related to the position of the blade. The maximum load is obtained at the highest point of the turbine wheel and the minimum load is obtained at the lowest point. The aerodynamic loads of the turbine are determined by the superposition of three blades and theoretically, there are three peaks and three lows in one cycle. The proposed control scheme controls the three blades to mitigate the load fluctuations of the wind turbine according to the azimuth angle of the blade, which can be described as:

$$\alpha_{fs}(t) = b \cdot \left(\frac{U_{hub} - U_{tip}}{|U_{hub} - U_{\max}|}\right) \cdot \left|\frac{U_{hub} - U_{tip}}{U_{hub} - U_{\max}}\right| \quad (4)$$

where $b$ is the flap control factor of the wind shear. $U_{hub}$ is the wind speed at the hub and $U_{tip}$ is the wind speed at the blade tip, which is calculated by Equation (1) with $\alpha$ = 0.1 and it can described as:

$$U_{tip} = U_{hub}\left(\frac{h_{hub} - R\sin\psi}{h_{hub}}\right)^{\alpha} \tag{5}$$

where $h_{hub}$ is the height of the hub, $R$ is the radius of the turbine wheel, and $\psi$ is the blade azimuth angle. $U_{\max}$ is the maximum value of $U_{tip}$.

For the control of the pitch and surge motions, the deflection angle $\alpha_p(t)$, and the angular velocity $\omega_p(t)$ of the pitch motion, the velocity $v(t)$ of the surge motion, and the azimuth angle of the blade $\psi$ have to be measured. Equation (6) describes the flap control method for the platform motion:

$$\alpha_{fp}(t) = c \cdot \left(\omega_p(t) \cdot (h_{hub} - R\sin\psi) + v(t) \cdot \cos\alpha_p(t)\right) + (1 - \cos\alpha_p(t)) \tag{6}$$

where $c$ is the control factor for the platform motion. In this scheme, the position of the flaps were determined by calculating the wind velocity component of the blade inflow, caused by the platform motion.

$\alpha_{ft}$, $\alpha_{fs}$, and $\alpha_{fp}$ are the flap deflection angles due to wind speed change, blade azimuth angle, and platform motion. A linear superposition method was used to determine the position of the flaps under multi-input conditions. The flap deflection angle is defined as:

$$\alpha_f = \begin{cases} -20 & , \quad \alpha_{ft}(t) + \alpha_{fs}(t) + \alpha_{fp}(t) < -15 \\ \alpha_{ft}(t) + \alpha_{fs}(t) + \alpha_{fp}(t) - 5 & , \quad -15 \le \alpha_{ft}(t) + \alpha_{fs}(t) + \alpha_{fp}(t) \le 15 \\ 10 & , \quad \alpha_{ft}(t) + \alpha_{fs}(t) + \alpha_{fp}(t) > 15 \end{cases} \tag{7}$$

## 5. Description of the FVW Model

The FVW model assumes that the flow field is incompressible and potential. The blade is modeled by a Weissinger-L model [16] as a series of straight constant strength vortex segments lying along the blade quarter chord line. The control points are located at a 3/4-chord at the center of each panel. The wake vortices extend downstream from the 1/4-chord, forming a series of horseshoe filaments. The trailing filaments cut off at a wake age angle of 60° in the near-wake and roll up and form a single tip vortex filament in the far-wake. The strength of the tip vortex equals the global maximum bound vorticity over the span of the blade. The release point of the tip vortex is the tip of the blade. The detailed calculation process of the FVW model can be found in [10]. The validation of the FVW model on blade airload predictions of offshore floating wind turbines is also presented in Ref. [10].

## 6. Results and Discussion

Appropriate flap control parameters are required to achieve a good control effect. In the following section, we discussed the influence of the control parameters on the control performance under the unsteady conditions comprised of turbulent wind, wind shear, and platform motion.

We set $a$ as the ratio of the flap control angle $\Delta\alpha_{ft}$ for the turbulence to the maximum deviation value $\Delta u_t$ of the turbulent wind ($a = \Delta\alpha_{ft}/\Delta u_t$). It can be seen from Figure 3 that the maximum deviation $\Delta u_t$ of the turbulent wind was 2.6 m/s, and the power curves of the four control factors are obtained by setting $\Delta\alpha_{ft}$ as 0°, 8°, 12°, and 15° respectively. The power response of the wind turbine from 0 to 200 s is shown in Figure 8 and was based on the turbulent wind condition shown in Figure 3. The input of the control group is a constant wind and the rotor power of the wind turbine is 4.7 MW, slightly less than the rated power. This is because the origin of the flap was −5°. It was obvious that the turbulent wind had a large influence on the stability of the wind turbine and the maximum deviation of the power fluctuation reaches 60% of the stable power when $a$ = 0. We present four curves with different control factors. The amplitude of the curve for the turbine with the controlled flap ($a$ > 0) was clearly smaller than that of the turbine with the fixed flap ($a$ = 0). In Table 1, the statistical results of the rotor power with the four different control factor values of $a$ are summarized. It is evident that the average value varied little, but the larger the value $a$, the smaller the standard deviation was, and the closer the maximum and minimum were to the mean value. This shows that the larger the value of $a$, the better the flap control system worked. Therefore, the maximum value of $a$ (5.77) was the appropriate value for this turbulent condition.

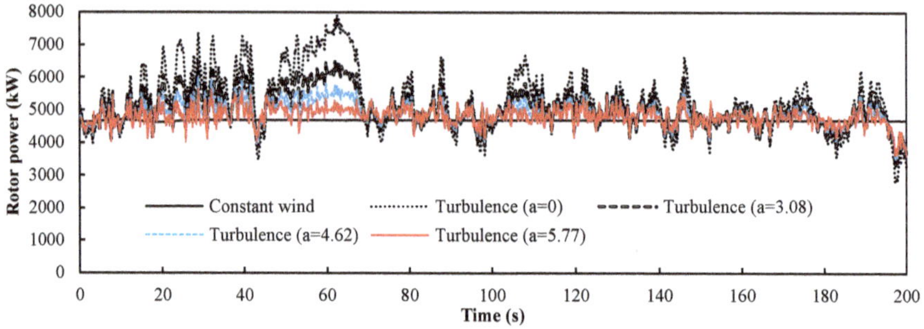


**Figure 8.** Power response of the floating wind turbine under turbulent wind from 0 to 200 s.

**Table 1.** The statistical metrics of the rotor power with different control factor values $a$.

| $a$ | 0 | 3.08 | 4.62 | 5.77 |
|---|---|---|---|---|
| Mean value (kW) | 5295 | 5050 | 4913 | 4807 |
| Standard deviation (kW) | 905 | 533 | 365 | 279 |
| Maximum (kW) | 7921 | 6493 | 6007 | 5640 |
| Minimum (kW) | 2763 | 3290 | 3501 | 3615 |

The control factor $b$ for the wind shear was equal to the maximum flap deflection angle $\Delta\alpha_{fs}$, which was set as 0°, 0.4°, 0.8°, and 1.2°. It can be seen from Figure 9 that the wind shear could cause obvious aerodynamic bending moment fluctuations at the root of a single blade. The control effect on the aerodynamic bending moment at the 90° azimuth angle was more obvious than that at the 270° azimuth angle, because of the different wind speed gradient. However, due to the superposition of the three blades, the total aerodynamic power of the turbine did not fluctuate as much, as shown in Figure 10. Under the wind shear condition, the rotor power curve exhibits low-frequency fluctuations in the first 30 s. This occurs because the flow field was calculated by the constant flow field prior to time 0, and the wind shear field was calculated after time 0, and the induction of the new wake had a continuous effect on the blade load. In the four curves, the amplitude decreased as the value of $b$ increased, but when $b = 1.2$, the amplitude of the power fluctuation reversed. The statistical metrics of the four values of the control factor $b$ are shown in Table 2. The mean value of each group of data was very close, and the maximum and minimum values were very similar. The change values between the maximum and minimum values were about 70 kW. At $b = 0.8$, the standard deviation reached the minimum value; therefore, 0.8 was the appropriate value of the control factor $b$.

**Table 2.** The statistical metrics of the rotor power curves with different control factor values $b$.

| $b$ | 0 | 0.4 | 0.8 | 1.2 |
|---|---|---|---|---|
| Mean value (kW) | 4592 | 4598 | 4602 | 4609 |
| Standard deviation (kW) | 16.1 | 13.6 | 12.8 | 13.7 |
| Maximum (kW) | 4639 | 4639 | 4640 | 4651 |
| Minimum (kW) | 4553 | 4565 | 4571 | 4576 |

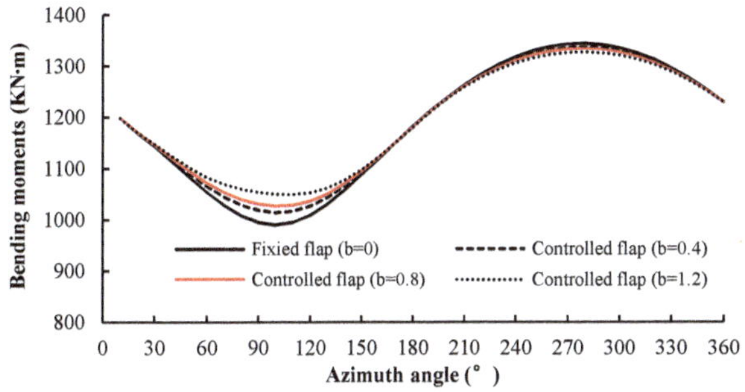


**Figure 9.** Blade root aerodynamic bending moments of a single blade during a rotation period under shear wind.

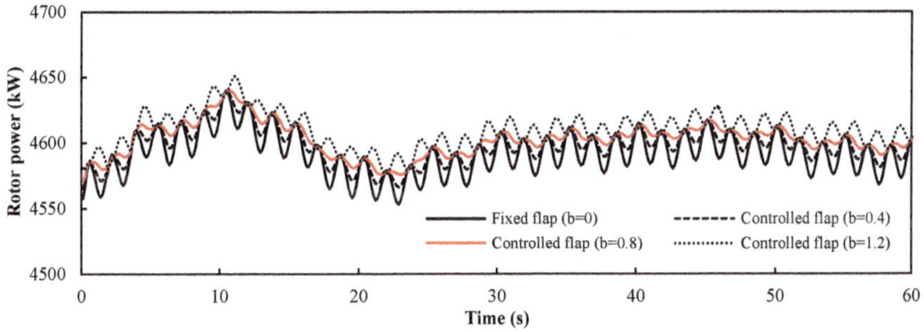


**Figure 10.** Power response of the floating wind turbine under wind shear from 0 to 60 s.

The analysis of the platform motions indicates that the influence on the wind turbine load mainly consists of pitch and surge motion. The control factor $c$ is set as the ratio of the maximum flap deflection angle $\Delta\alpha_{fp}$ to the maximum deviation $\Delta u_p$ of the axial wind velocity caused by two kinds of platform motions ($c = \Delta\alpha_{fp}/\Delta u_p$). Here $\Delta u_p = 2$ m/s can be obtained from the platform motion data. Four values were selected to study the influence of the control factor on the control performance. Figure 11 is the load response of the pitch motions, and Figure 12 is the load response of the surge motions. It was evident that the power fluctuation caused by the surge motions was larger than that caused by the pitch motions for the same flap deflection angle. The frequency of the load fluctuation caused by the platform was smaller than that caused by the turbulent wind and the greater the control factor, the better the control effect is. Figure 13 shows the load response of the wind turbine under the combined pitch and surge motions. The result is similar to the load response of the surge motion, which indicates that the surge motion is the dominant motion type. There are no apparent fluctuations when the control factor $c$ is equal to the maximum value 7.5, which shows that the trailing-edge flaps have a positive effect on the low-frequency fluctuations, such as the platform motion. The statistical metrics of the curves with different control factor value $c$ are shown in Table 3. As $c$ increases, the mean value decreases only slightly but the standard deviation decreases markedly. When $c$ reaches the maximum value, the standard deviation reaches the minimum value. As a result, 7.5 is the optimal value of the control factor $c$.

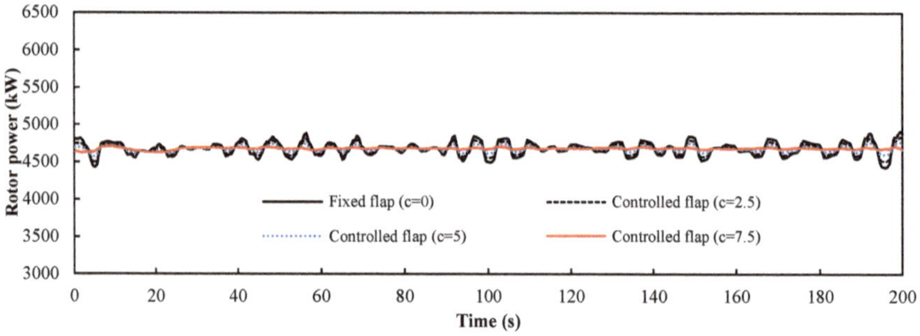


**Figure 11.** The power response of the floating wind turbine under pitch motions from 0 to 200 s.

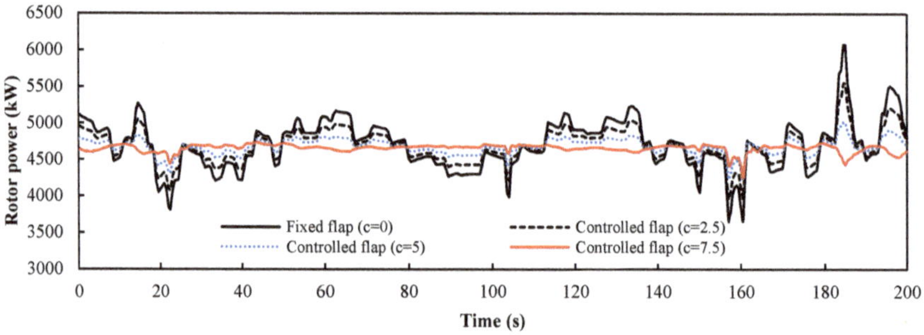


**Figure 12.** Power response of the floating wind turbine under surge motions from 0 to 200 s.

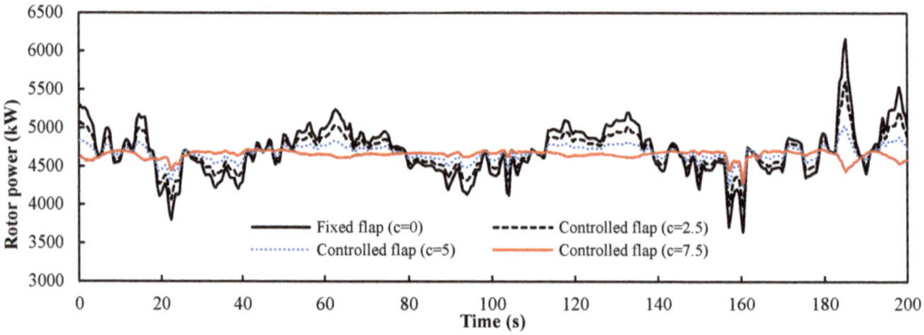


**Figure 13.** Power response of the floating wind turbine under pitch and surge motions from 0 to 200 s.

**Table 3.** The statistical metrics of the curves with different control factors for the pitch and surge motions.

| *c* | 0 | 2.5 | 5 | 7.5 |
|---|---|---|---|---|
| Mean value (kW) | 4702 | 4691 | 4675 | 4650 |
| Standard deviation (kW) | 333 | 218 | 105 | 52 |
| Maximum (kW) | 6166 | 5609 | 5025 | 4721 |
| Minimum (kW) | 3646 | 3967 | 4220 | 4272 |

When the input wind condition is the turbulent wind, turbulent wind, wind shear and platform motions will affect the rotor power together. The above results indicated that the trailing-edge flap had a good mitigation effect on the power fluctuations caused by the three kinds of dynamic inputs when $a = 5.77$, $b = 0.8$, and $c = 7.5$. We used these values for the three parameters to achieve the linear control of the flap. Figure 14 shows the power response of the turbine for all three dynamic inputs. Table 4 shows the statistical metrics of the two curves in Figure 14. It was observed that the controlled trailing-edge flap mitigates the rotor power fluctuations of the wind turbine under these conditions. Under the combined action of the three kinds of dynamic inputs, the controlled flap reduced the standard deviation of the power fluctuations from 1095 to 404.

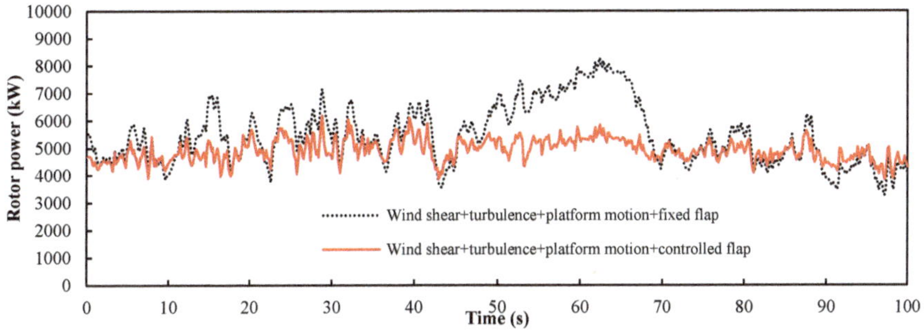


**Figure 14.** Power response of the NREL 5 MW floating wind turbine under the turbulent wind condition.

**Table 4.** Statistical metrics of the two curves in Figure 14.

| Input Conditions | Wind Shear, Turbulence, Platform Motion, Fixed Flap | Wind Shear, Turbulence, Platform Motion, Controlled Flap |
|---|---|---|
| Mean value (kW) | 5477 | 4902 |
| Standard deviation (kW) | 1095 | 404 |
| Maximum (kW) | 8283 | 6176 |
| Minimum (kW) | 3266 | 3857 |

## 7. Conclusions

In this study, a trailing-edge flap control strategy is integrated into the blade of the NREL 5 MW wind turbine to cope with the turbulent wind, wind shear, and the motions of the floating platform. The rotor power of the wind turbine is calculated using the FVW method. A simple flap control method based on wind speed, blade azimuth angle, and platform motion is proposed to achieve effective flap control. The results show that turbulent wind has the largest impact on the stability of the floating wind turbine. The optimal values of three control factors for turbulent wind, wind shear, and platform motions are obtained. The controlled trailing-edge flap with appropriate control factors effectively mitigates the power fluctuations caused by turbulent wind, wind shear, and platform motion. The proposed trailing-edge flap control strategy need to be validated by some experimental studies in the future, and then it can be applied in the large-scale offshore floating wind turbine.

**Author Contributions:** B.X. and J.F. run the FVW codes and prepared this manuscript under the guidance of T.W. and Y.Y., Z.Z., and W.Z. supervised the work and contributed in the interpretation of the results. All authors carried out data analysis, discussed the results, and contributed to writing the paper.

**Funding:** This work was supported by the National Natural Science Foundation of China (grant numbers 51607058, 51506088); the National Basic Research Program of China (973 Program) (grant number 2014CB046200); and the Fundamental Research Funds for the Central Universities (grant number 2016B01514).

**Conflicts of Interest:** The authors declare no conflict of interest.

## References

1. Ma, Y.; Hu, Z.Q.; Xiao, L.F. Wind-wave induced dynamic response analysis for motions and mooring loads of a spar-type offshore floating wind turbine. *J. Hydrodyn.* **2014**, *26*, 865–874. [CrossRef]

2. Driscoll, F.; Jonkman, J.; Robertson, A.; Sirnivas, S.; Skaare, B.; Nielsen, F.G. Validation of a FAST model of the Statoil-Hywind Demo floating wind turbine. *Energy Procedia* **2016**, *94*, 3–19. [CrossRef]
3. Rodrigues, S.; Restrepo, C.; Kontos, E.; Pinto, R.T.; Bauer, P. Trends of offshore wind projects. *Renew. Sustain. Energy Rev.* **2015**, *49*, 1114–1135. [CrossRef]
4. Lobitz, D.W.; Veers, P.S. Load mitigation with bending/twist-coupled blades on rotors using modern control strategies. *Wind Energy* **2003**, *6*, 105–117. [CrossRef]
5. Wu, C.K.; Nguyen, V. Aerodynamic simulations of offshore floating wind turbine in platform-induced pitching motion. *Wind Energy* **2016**, *20*, 835–858. [CrossRef]
6. Jonkman, J.; Butterfield, S.; Musial, W.; Scott, G. *Definition of a 5-MW Reference Wind Turbine for Offshore System Development*; Technical Report, NREL/TP-500-38060; National Renewable Energy Laboratory: Golden, CO, USA, 2009.
7. Jiang, Z.Y.; Karimirad, M.; Moan, T. Dynamic response analysis of wind turbines under blade pitch system fault, grid loss, and shutdown events. *Wind Energy* **2014**, *17*, 1385–1409. [CrossRef]
8. Bergami, L.; Riziotis, V.; Gaunaa, M. Aerodynamic response of an airfoil section undergoing pitch motion and trailing edge flap deflection: A comparison of simulation methods. *Wind Energy* **2014**, *17*, 389–406. [CrossRef]
9. Barlas, T.K.; van Kuik, G.A.M. Review of state of the art in smart rotor control research for wind turbines. *Prog. Aerosp. Sci.* **2010**, *46*, 1–27. [CrossRef]
10. Xu, B.F.; Wang, T.G.; Yuan, Y.; Cao, J.F. Unsteady aerodynamic analysis for offshore floating wind turbines under different wind conditions. *Philos. Trans. A Math Phys. Eng. Sci.* **2015**, *28*, 373. [CrossRef] [PubMed]
11. Xu, B.F.; Wang, T.G.; Yuan, Y.; Zhao, Z.Z.; Liu, H.M. A simplified free vortex wake model of wind turbines for axial steady conditions. *Appl. Sci.* **2018**, *8*, 866. [CrossRef]
12. Sebastian, T.; Lackner, M.A. Characterization of the unsteady aerodynamics of offshore floating wind turbines. *Wind Energy* **2013**, *16*, 339–352. [CrossRef]
13. Yamada, M.; Ohkitani, K. Orthonormal wavelet analysis of turbulence. *Fluid Dyn. Res.* **1991**, *8*, 101–204. [CrossRef]
14. Kitagawa, T.; Nomura, T. A wavelet-based method to generate artificial wind fluctuation data. *J. Wind Eng. Ind. Aerodyn.* **2003**, *91*, 943–964. [CrossRef]
15. Zhang, W.G.; Bai, X.J.; Han, Y. Trailing edge flap structure parameters optimization of a wind turbine and analysis of the control performance. *J. Chin. Soc. Power Eng.* **2017**, *37*, 1023–1030.
16. Weissinger, J. *The Lift Distribution of Swept-Back Wings*; Technical Report, NACA-TM-1120; National Advisory Committee for Aeronautics: Washington, DC, USA, 1947.

entropy

MDPI

Article

# Non-Linear Langevin and Fractional Fokker–Planck Equations for Anomalous Diffusion by Lévy Stable Processes

**Johan Anderson** [1,*], **Sara Moradi** [2] **and Tariq Rafiq** [3]

[1] Department of Space, Earth and Environment, Chalmers University of Technology, SE-412 96 Göteborg, Sweden
[2] Laboratory for Plasma Physics—LPP-ERM/KMS, Royal Military Academy, 1000 Brussels, Belgium; sara.moradi@ukaea.uk
[3] Department of Mechanical Engineering and Mechanics, Lehigh University, Bethlehem, PA 18015, USA; rafiq@lehigh.edu
* Correspondence: anderson.johan@gmail.com; Tel.: +46-10-516-5926

Received: 28 July 2018; Accepted: 29 September 2018; Published: 3 October 2018

**Abstract:** The numerical solutions to a non-linear Fractional Fokker–Planck (FFP) equation are studied estimating the generalized diffusion coefficients. The aim is to model anomalous diffusion using an FFP description with fractional velocity derivatives and Langevin dynamics where Lévy fluctuations are introduced to model the effect of non-local transport due to fractional diffusion in velocity space. Distribution functions are found using numerical means for varying degrees of fractionality of the stable Lévy distribution as solutions to the FFP equation. The statistical properties of the distribution functions are assessed by a generalized normalized expectation measure and entropy and modified transport coefficient. The transport coefficient significantly increases with decreasing fractality which is corroborated by analysis of experimental data.

**Keywords:** non-local theory; Lévy noise; Tsallis entropy; fractional Fokker–Plank equation; anomalous diffusion

**PACS:** 05.40Fb; 02.50Ey; 05.40-a

## 1. Introduction

In magnetically confined (MC) plasma devices transport driven by turbulent fluctuations often severely limit the confinement time and thus impede the development of fusion as an alternative for electricity production. It is pertinent to understand and mitigate the effects of the turbulently driven transport where simplified models often are employed in order to elucidate the main features of the plasma turbulence. In magnetised plasmas, it is commonly accepted that turbulence is the primary cause of anomalous (i.e., elevated compared to collisional) transport [1,2]. It has been recognized that the nature of the anomalous transport processes is dominated by a significant ballistic or non-local component where a diffusive description is improper. The turbulence in MC tokamak plasmas is anisotropic in the parallel and perpendicular length scales to the magnetic field and taps free energy from the pressure gradient that can drive fluctuations in electrostatic potential and density [1,2]. The super-diffusive properties are often ubiquitously found in plasmas, such as the thermal and particle fluxes in the gradient region or in the Scrape-Off Layer (SOL) where the transport is dominated by the coherent structures (blobs) [3–9] and inherently possess a non-local character [10–16]. Moreover, there is a large quantity of experimental evidence that density and potential fluctuations measured by Langmuir probes at different fusion devices support the idea that these fluctuations are distributed according to Lévy statistics. This was illustrated for example in [4], where probability density functions (PDFs) of the turbulence induced fluxes at the edge of the W7-AS stellarator were shown to exhibit power law characteristics in contrast to exponential decay expected from Gaussian statistics. Furthermore, the experimental evidence of the wave-number spectrum characterized by power laws over a wide range of wave-numbers can be directly linked to the values of Lévy index $\alpha$ of the PDFs of the underlying turbulent processes. One widely used simplified model of a plasma is the Hasegawa–Wakatani model which was recently analysed by statistical methods in Reference [17]. It was concluded that even simplified models may have components of fractionality stemming from the non-linear interactions and the generation of large scale modes such as zonal and shear flows. The Hasegawa–Wakatani model allows for the electrons to dynamically and self-consistently determine the relationship between the density and the electrostatic potential through the turbulence. Moreover, fractal features in transport have been observed in many experiments in many different fields of research. In particular it has been found that there is strong evidence of non-local heat transport in JET plasmas [18]. In this paper, fractal features is synonymous to a system where power law statistics is found. Here it is important to keep in mind that, although a simplified fractional transport model is used, it indicates that there is a lack of physics in the current transport models based on the mean field theory, namely the super-diffusive character of heat transport. Finding a proper kinetic description of dynamical systems with chaotic behaviour is one of the main problems in classical physics [19–30]. Over the past two decades it has become obvious that behaviour much more complex than standard diffusion can occur in dynamical Hamiltonian chaotic systems. In principle, the orbits in dynamical systems are always theoretically predictable since they arise as solutions to simple system of equations such as Newton's equations; however, these orbits are sensitive to initial conditions and thus very small changes in initial conditions may yield widely different outcomes. From the macroscopic point of view, the rapid mixing of orbits has been used as a motivation for assumptions of randomness of the motion and the random walk models [19]. In characterizing the diffusion processes in plasmas, the starting point is often Brownian motion where the mean value vanishes, whereas the second moment or variance grows linearly in time according to $\langle \delta x^2 \rangle = 2Dt$. However, taking into account the experimental data found in plasma experiments, it is evident that many phenomena exhibit anomalous diffusion where variance grows non-linearly in time such that $\langle \delta x^2 \rangle = 2Dt^\alpha$. The reason an anomalous diffusion approach is needed is due to the restrictive assumptions of locality in space and time, and lack of long-range correlations that is the basis of Brownian motion. There are two limits of interest for $\alpha$ where the first is super-diffusion with $\alpha > 1$ and the second is sub-diffusion with $\alpha < 1$. A super-diffusive description is most often appropriate for fusion plasmas. Lévy statistics describing fractal processes (Lévy index $\alpha$ where $0 < \alpha < 2$) lie at the heart of complex processes such as anomalous diffusion. Lévy statistics can be generated by random processes that are scale-invariant. This means that a trajectory will possess many scales, but no single scale will be characteristic and dominate the process. Geometrically, this implies the fractal property that a trajectory, viewed at different resolutions, will look self-similar. Such strange kinetics [19,24] may be generated by accelerated or sticky motions along the trajectory of the random walk [2]. Super-diffusivity may also occur as a result of variation in the step length of the motion, which breaks the assumption that a unique step length may, e.g., give rise to long-range correlations in the dynamics generated by the presence of anomalously large particle displacements connecting otherwise physically disjoint domains.

We note that, although sub-diffusive processes are beyond the scope of the present work, its properties have been studied in many different contexts where transport is often inhibited by sticky motion. Among sub-diffusive phenomena are holes in amorphous semiconductors, where a waiting time distribution with long tails has been introduced [31]. The sub-diffusive processes within a single protein molecule have been described by generalized Langevin equation with fractional Gaussian noise [32]. Turbulence and related anomalous diffusion phenomena have been observed in a wide variety of complex systems such as high energy plasmas, semiconductors, glassy materials, nanopores, biological cells, and epidemic proliferation.

The objective of the present paper is to explore the non-linear character of the fractional Fokker–Planck (FFP) equation resulting from a Langevin description driven by Lévy stochastic force with a non-linear interaction in the velocity. The present work is based on previous efforts reported in Reference [29] and may provide new insights on the recent developments in the modelling of the anomalous transport of charged particles in magnetised plasmas, such as the non-local heat transport found in JET plasmas.

The paper is organized in the following way: in Section 2, the model is presented, and the numerical results are shown and discussed in Section 3. The final section presents a discussion and conclusions.

## 2. The Fokker–Planck and Langevin Equations

Fractional kinetics is a powerful framework in describing anomalous transport processes exhibiting Lévy statistics. It is able to reproduce key aspects of anomalous transport including the non-Gaussian self-similar nature of the PDFs of particle displacement, and the anomalous scaling of the moments. It has been shown that the chaotic dynamics can be described by using the FFP equation with coordinate fractional derivatives as a possible tool for the description of anomalous diffusion [33]. Previous papers on plasma transport have used models including a fractional derivatives on phenomenological premises [6,34,35]. Additionally, the integro-differential nature of the fractional derivative operators allows the description of spatiotemporal nonlocal transport processes. In particular, in fractional diffusion, the local Fourier–Fick's law is replaced by an integral operator in which the flux at a given point in space depends globally on the spatial distribution of the transported scalar and on the time history of the transport process. Using fractional generalizations of the Liouville equation, kinetic descriptions have been developed [36–38]. The currently applied model is based on the Langevin equation with a Lévy-stable noise term, where the applied noise exhibits a power law tail [39,40]. The generalized Central Limit Theorem for Lévy-stable processes is a particular weak-convergence theorem in probability theory. It expresses the fact that a sum of many independent and identically distributed (i.i.d.) random elements, or alternatively, random elements with specific types of dependence, will tend to be distributed according to one of a small set of attractor distributions. There are here two cases of special interest: the first is when the variance of the i.i.d. variables is finite and the attractor distribution is then a normal distribution, and the second is where the sum of a number of i.i.d. random elements with power law tail distributions decreasing as $|x|^{-\alpha-1}$ where $0 < \alpha < 2$ (therefore having infinite variance) will tend to a Lévy-stable distribution with a fractality index of $\alpha$ as the number of elements in the set increases.

The motion of a colloidal particle can be described by the Langevin equation in the case of Brownian motion and it will take the form

$$\frac{d}{dt}v = -\nu v + A(t), \tag{1}$$

where $v$ is the speed of the particle, $-\nu v$ is the friction, and $A(t)$ is the white stochastic force such that $\langle A(t)A(t')\rangle = 2D\delta(t-t')$. Moreover, by assuming that $A(t)$ is a Gaussian stochastic force, a Maxwellian velocity distribution may be obtained and would lead to the standard Fokker–Planck (FP) equation describing the evolution of the distribution function:

$$\frac{\partial}{\partial t}P + v\frac{\partial P}{\partial r} + \frac{F}{m}\frac{\partial P}{\partial v} = \nu\frac{\partial}{\partial v}(vP) + D\frac{\partial^2 P}{\partial^2 v}. \tag{2}$$

Here $P$ is the distribution function, $v$ is the velocity, $F$ is an external force, e.g., the electromagnetic force, $m$ is the mass, $\nu$ is the friction, and $D$ is the diffusion coefficient. The corresponding reduced FP equation, where the Lorentz force is neglected, to the Langevin equation is

$$\frac{\partial}{\partial t}P = \nu\frac{\partial}{\partial v}(vP) + D\frac{\partial^2 P}{\partial^2 v}, \tag{3}$$

which yields to the stationary state Maxwellian velocity distribution for $P(v)$ [41,42]. However, if $A(t)$ is a stochastic noise with the properties of a Lévy-stable process, the FP equation has to be modified in order to accommodate for power law tails of the form $P(v) \propto v^{-\alpha-1}$ for a Lévy stable with fractional index $\alpha$. The FFP equation becomes

$$\frac{\partial}{\partial t}P(v,t) = \nu\frac{\partial}{\partial v}(vP(v,y)) + D\frac{\partial^{\alpha}P(v,t)}{\partial^{\alpha}|v|} \tag{4}$$

where $0 < \alpha \leq 2$ and $|v| < \infty$. The time-dependent solution is readily found in the Fourier space where the fractional Riesz operator in 1 + 1D can be transformed to

$$\frac{\partial}{\partial t}\hat{P}(k,t) = -\nu k\frac{\partial}{\partial k}(\hat{P}(k,t)) - D|k|^{\alpha}\hat{P}(k,t) \tag{5}$$

where the Fourier transformed distribution function can be determined to be

$$\hat{P}(k,t) = \exp(-\frac{D|k|^{\alpha}}{\nu\alpha}(1-\exp(-\nu\alpha t))). \tag{6}$$

The fractional Riesz derivative is defined through its Fourier transform ${}_{-\infty}\hat{D}^{\mu}_{x}f(x) = \frac{\partial^{\mu}f(x)}{\partial^{\mu}|x|} = -|k|^{\mu}f(k)$, see, e.g., [22] for more information. Here it should be noted that the time derivative only introduces a relaxation time dependent on the friction and the fractionality $\alpha$, where a smaller $\alpha$ yield a longer relaxation time.

In Figure 1, the exponentially fast relaxation of the velocity PDFs with time is displayed. The PDFs of a Gaussian ($\alpha = 2.0$) and for a PDF with fractional index $\alpha = 1.5$ for times $t = 0.1, 0.5, 1.0$, and 10.0 are computed numerically. We note that, at $t = 10.0$, the PDFs are close to the stationary state PDF, whereas the time evolution of the PDF depends on the fractional index $\alpha$ such that the relaxation process is slower for a PDF with a lower fractional index. In general, the distributions found for the $\alpha = 1.5$ have more pronounced tails and sharper peaks, whereas, in the $\alpha = 2.0$ case, the system has a shorter relaxation time.

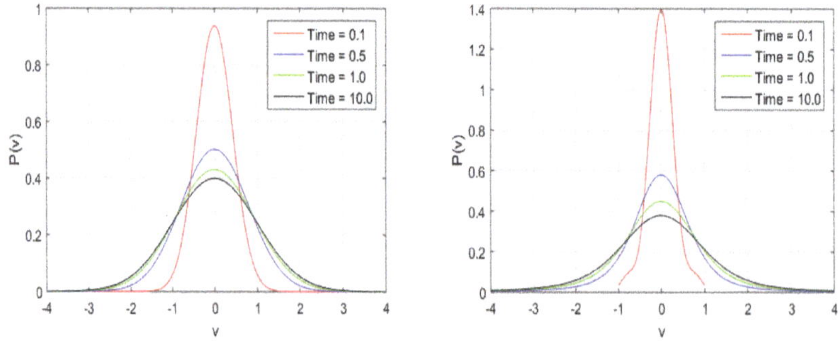


**Figure 1.** The probability density function (PDF) of velocity computed by the inverse Fourier transform of Equation (6) with $\alpha = 2.0$ (**left**) and $\alpha = 1.5$ (**right**) for $t = 0.1, 0.5, 1.0, 10.0$.

## 3. Results

The aim of the present paper is to look into the effects of a non-linear interaction in the Langevin equation, but it is here assumed that we can neglect the time dependence, i.e., the stationary state PDF ($\frac{dP}{dt} = 0$), and the FFP Equation can be written as

$$0 = \nu \frac{\partial}{\partial v} \left( (v + \beta v^3) P \right) + D \frac{\partial^\alpha}{\partial^\alpha |v|} P. \tag{7}$$

Here $\nu$, $\beta$, and $D$ are constants. The equation is obtained by inclusion of the quartic potential, leading to the addition of a term of third order of the form $\beta v^3$. The main effect of retaining the temporal dynamics is to introduce a relaxation time. In the current model, square and quartic terms will be retained. The properties of the current non-linear terms are analogous to a potential well with square and quartic terms. Note that even terms in the potential provide proper stable equilibria, whereas odd terms yield an unstable equilibrium; thus, the third and fifth order terms are neglected. The Equation (7) is directly integrated by using a predictor—corrector method according to Adams-Bashforth-Moulton [43].

To find an analytical solution of the original Langevin equation, the Fourier transform can be used;

$$\nu \left[ \frac{\partial}{\partial k} - \beta \frac{\partial^3}{\partial k^3} \right] \tilde{P} + D|k|^{\alpha-1} \tilde{P} = 0. \tag{8}$$

The found equation is a third order ordinary differential equation with variable coefficients. The general solutions to Equation (3) can only be determined by numerical means however a similar system was investigated in Reference [26] suggesting a PDF proportional to $exp(-a * v^4)$, where $a$ is a constant. Furthermore, it is also possible to find an analytical solution for the tail of the PDF to leading order by using the Wentzel–Kramers–Brillouin (WKB) approximation for small values of $\beta$. The WKB anzats is to assume a series solution to the Fourier transformed equation (3), of the form $\tilde{P}(k) = exp(\frac{1}{\epsilon} \sum_{n=0}^{\infty} \epsilon^n S_n(k))$, here $\epsilon$ will be taken small and to be determined in terms of $\beta$. It is then found that, the leading order tail contribution corroborates the findings in [26] for $\alpha = 2.0$. We note that the real space distribution function is convergent for $\beta > 0$ and can only in general be obtained by numerical integration, and is here solved by using method described above. We note that there are three different interesting regimes: the first is where the diffusion is much larger than the quartic potential strength $D/\nu >> \beta v^2$, the second is where the diffusion is comparable to the quartic potrential strength $D/\nu \sim \beta v^2$, and the third is where the diffusion is negligible to the quartic potential strength $D/\nu << \beta v^2$. In the third regime, the PDFs become may be expected to have similarities to the results found in [26] for $\alpha = 2.0$ where $P(v) \sim \exp(-av^4)$ for some constant $a$. The values used in this study are chosen to illustrate these three regimes of interest. Note that the non-linear interaction, i.e., the $\beta v^3$ term introduces three different regimes with richer dynamics which is in contrast to what was found in Reference [29]. In any linear fractal model based on the Lévy statistics the power law tails of the velocity PDF will be $P(v) \propto |v|^{-\alpha-1}$. Even more interestingly, in non-linear models the precise scaling of the PDF tails are still open.

In Figures 2–4, the numerically found PDFs, by solving Equation (7) by the Adams-Bashforth-Moulton method [43], are shown for the three different regimes: $D/\nu = 1.0$ and $\beta = 0.1$, $D/\nu = 1.0$ and $\beta = 1.0$, and $D/\nu = 0.1$ and $\beta = 1.0$, respectively. The resolution in $v$ is $2^{-12}$ except in the case of $\alpha = 1.25$ for $D/\nu = 1.0$ and $\beta = 1.0$ where the resolution is increased to $2^{-20}$, however increased resolution would not change the $P(v)$ in any significant way for smaller $|v|$. As expected, in Figure 2 almost independently of $\alpha$ where $D/\nu > \beta v^2$, the PDFs exhibit power law tails, although in the case of $\alpha = 2$ some exponential behaviour is observed at the tail of the distribution (for large modulus of velocity $|v|$). In Figures 3 and 4, we find more pronounced tails in particular in the low-$\alpha$ case. In the low-$\alpha$ case, the fractal term dominates the dynamics. We note that the PDFs displayed in Figure 3 exhibit a hybrid between fractal and Gaussian behaviour ($D/\nu \sim \beta v^2$), where in some cases the PDF is retains some Gaussian behaviour, which is particularly visible for small $|v|$. In the regime $D/\nu << \beta v^2$ the non-Gaussian effects of the PDFs are clearly visible in Figure 4. The PDFs are used to evaluate the dynamics of the system in terms of Tsallis' statistical mechanics where $q$-entropy and $q$-energy determines the properties of the system for the three different regimes: $D/\nu = 1.0$ and $\beta = 0.1$, $D/\nu = 1.0$ and $\beta = 1.0$, and $D/\nu = 0.1$ and $\beta = 1.0$ in Figures 5 and 6, respectively. The $q$, $q$-entropy, and $q$-energy values are determined by the following relations (see References [29,44–54]):

$$q = \frac{3+\alpha}{1+\alpha} \tag{9}$$

$$S_q = \frac{1-\int dv(F(v))^q}{q-1} \tag{10}$$

$$\langle v^2 \rangle_q = \frac{\int dv(F(v))^q v^2}{\int dv(F(v))^q}. \tag{11}$$

The entropy and $q$-entropy are displayed in Figure 5. A maximum in the entropy, computed according to $S_B = \int dvP(v)lnP(v)$, is found, whereas almost constant $q$-entropy, computed according to Equation (10), is found with increasing $q$, where $q$ is determined according to Equation (9). An increasing trend with importance of non-local effects are also visible where the entropy and $q$-entropy is increased in the cases with $\beta = 1.0$, which is in the regime where the fractal nature is more prominent. In Figure 6, a decreased energy and increased $q$-energy with increasing $q$ for $\beta = 1.0$ and small $D/\nu$ case is found.

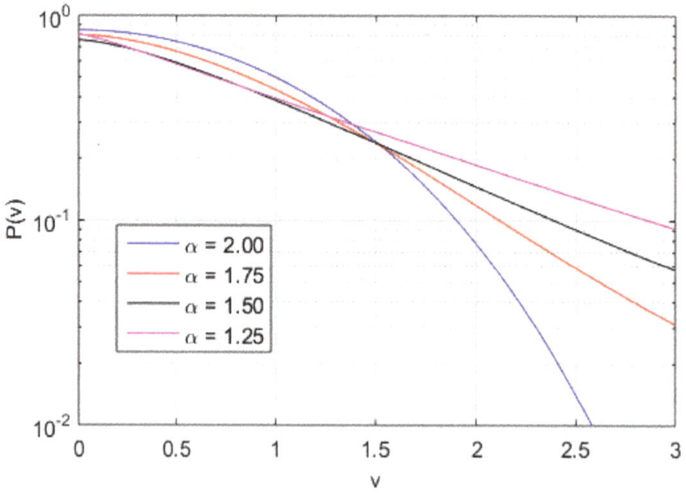


**Figure 2.** The PDF of velocity computed by integration of Equation (7) with with $\alpha = 1.25$ (magenta line), $\alpha = 1.5$ (black line), $\alpha = 1.75$ (red line), and $\alpha = 2.0$ (blue line) for $D/\nu = 1.0$ and $\beta = 0.1$.

The interpretation of this strange kinetics has to be based on the results from experimental data since there is no first principle method to compute the value of $\alpha$ and thus $q$ is indeterminable. However, recent findings suggest that JET plasmas have a significant degree of super-diffusive transport with an $\alpha < 2$, and it was found that this super-diffusive transport is slightly different for the ion and electron channels [18]. The analysis is based on a power balance where a large set of JET shots are used whereby a distribution in $\alpha$ can be obtained with a mean value of approximately 1, suggesting that a convective model would be more appropriate with $q \approx 2$. The diffusion coefficient can be estimated by the velocity autocorrelation functions according to the Kubo formula, but such an estimate looks at the ratio of the generalized diffusion coefficient ($D_\alpha$) and the Brownian diffusion coefficient $D_0 = D(\alpha = 2.0)$, using the tempered $q$-velocity correlators, computed by Equation (11).

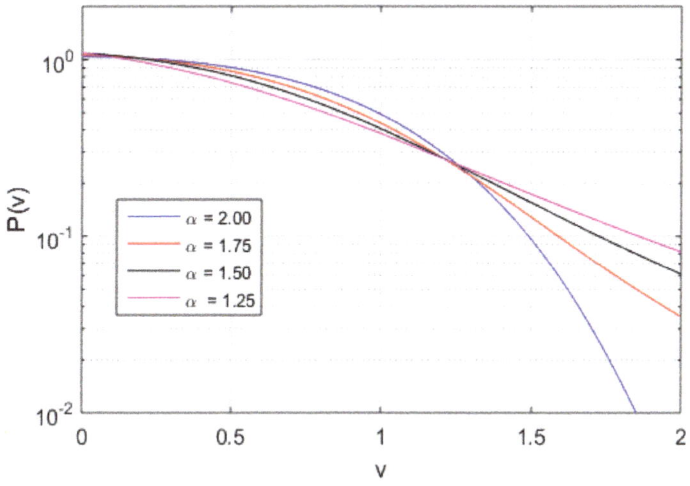


**Figure 3.** The PDF of velocity computed by integration of Equation (7) with $\alpha = 1.25$ (magenta line), $\alpha = 1.5$ (black line), $\alpha = 1.75$ (red line), and $\alpha = 2.0$ (blue line) for $D/\nu = 1.0$ and $\beta = 1.0$.

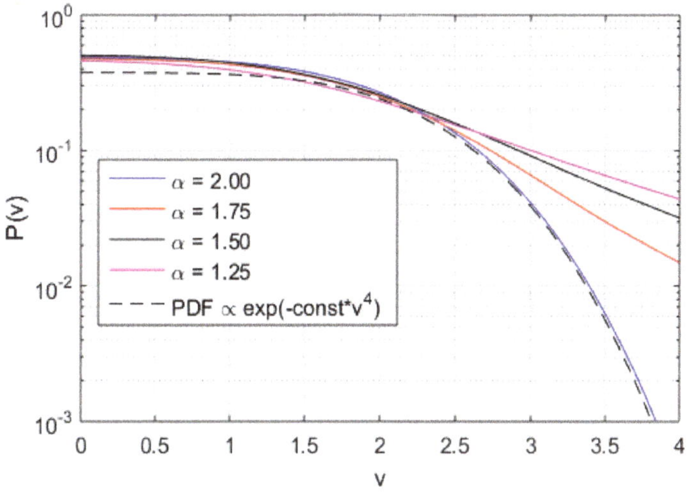


**Figure 4.** The PDF of velocity computed by integration of Equation (7) with with $\alpha = 1.25$ (magenta line), $\alpha = 1.5$ (red line), $\alpha = 1.75$ (red line), and $\alpha = 2.0$ (blue line) for $D/\nu = 0.1$ and $\beta = 1.0$.

We find that the ratio of the diffusion coefficients increases with smaller $\alpha$ and significantly increases in the regime where fractality is pronounced, as shown in Figure 7. Interestingly enough, in the analysis presented in [18] it is evident that in the cases with increased transport a lower value of $\alpha$ is obtained, indicating a strong non-diffusive component or equivalently, a significantly increased transport where processes following Lévy statistics dominate the transport. The qualitative increase in the generalized transport coefficient $D_q$ is thus qualitatively corroborated by what is seen experimentally using the power balance analysis.

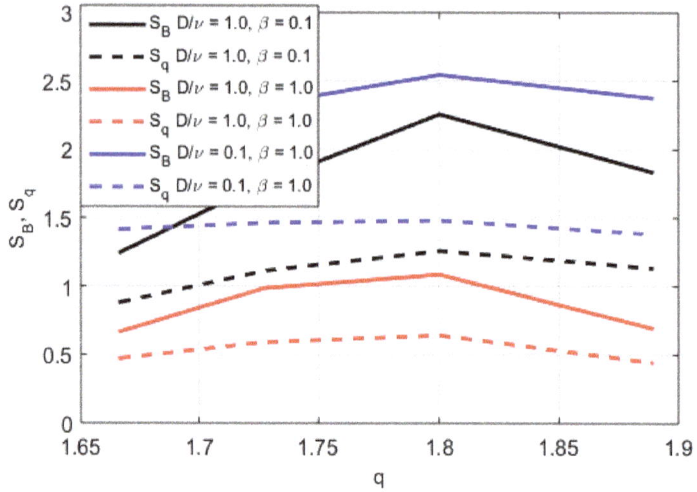


**Figure 5.** The Boltzmann–Gibbs entropy and the Tsallis' entropy as functions of the fractality index $q$ for $D/\nu = 1.0$ and $\beta = 0.1$ (solid black line and dashed black line, respectively), $D/\nu = 1.0$ and $\beta = 1.0$ (solid red line and dashed red, respectively), $D/\nu = 0.1$ and $\beta = 1.0$ (solid blue line and dashed blue line, respectively).

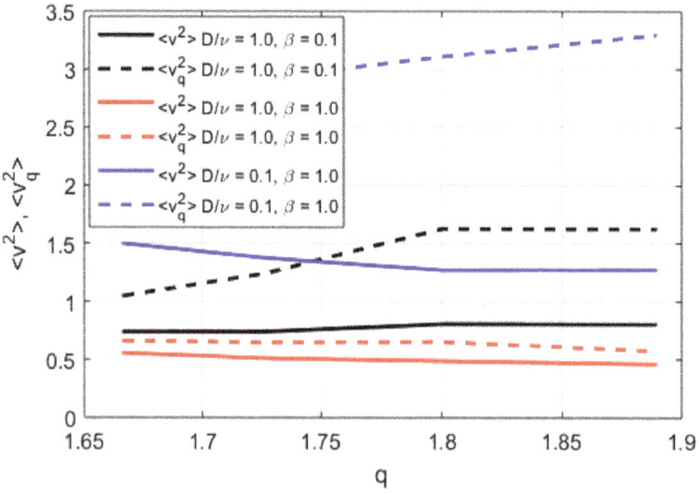


**Figure 6.** The energy and the generalized $q$-energy as functions of the fractality index $q$ for $D/\nu = 1.0$ and $\beta = 0.1$ (solid black line and dashed black line, respectively), $D/\nu = 1.0$ and $\beta = 1.0$ (solid red line and dashed red, respectively), $D/\nu = 0.1$ and $\beta = 1.0$ (solid blue line and dashed blue line, respectively).

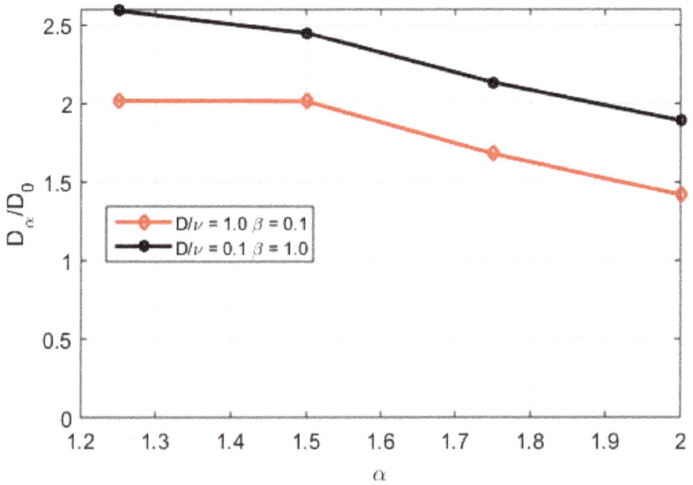


**Figure 7.** The ratio of the generalized diffusion coefficient ($D_\alpha$) and the Brownian diffusion coefficient as functions of the fractality index $\alpha$ for $D/\nu = 1.0$ and $\beta = 0.1$ (red line) and $D/\nu = 0.1$ and $\beta = 1.0$ (black line).

## 4. Discussion and Conclusions

Understanding anomalous transport in MC plasmas is an outstanding issue in controlled fusion research. It is commonly accepted that, in these plasmas, turbulence is the primary cause of anomalous (i.e., elevated compared to collisional) transport. It has also been recognized that the nature of the anomalous transport processes is dominated by a significant ballistic or non-local component where a diffusive description is improper. A satisfactorily understanding of the non-local features as well as the non-Gaussian PDFs found in experimental measurements of particle and heat fluxes is still lacking [25,26], but there has been some recent progress in this direction. Fractional kinetics has been put forward for building a more physically relevant kinetic description for such dynamics. In these situations, kinetic descriptions, which arise as a consequence of averaging over the well-known Gaussian and Poissonian statistics (for diffusion in space and temporal measures, respectively), seem to fall short in describing the apparent randomness of dynamical chaotic systems [19]. This is due to the restrictive assumptions of locality in space and time, and the lack of long-range correlations that is the basis of these descriptions.

In magnetised plasma experiments, a predator–prey system exists with avalanches (strong driver of transport) and zonal flows (sheared flows that decorrelate turbulent eddies reducing transport). It has been suggested that an Fractional Fokker–Planck Equation on a comb-like potential background can be applied where meso-scale transport events (avalanching) occurs in between regions of strong zonal flow activity (see Milovanov and Rasmussen [55]). This method is straightforward for applications in this setting; by assuming the used potential in between the zonal flow regions, it is suggested that the potential should be of degree 4 (or higher), as has been used here.

Although there has previously been some criticism on the appropriateness of using the Tsallis method in describing processes with Lévy statistics, this is mainly concerning descriptions based on fractality in coordinate space not in velocity space. However, the aim of the present work was to shed light on the non-extensive properties of the velocity space statistics and characterization of the fractal processes by estimating the generalized diffusion coefficients of the FFP equation in terms of Tsallis statistics. Jespersen et al. [56] showed an example of the Langevin equation with a harmonic potential, and the Tsallis $q$-statistics had limited usefulness. The reason for this is that, using the variational calculus of Equation (10) with the appropriate constraints, the relation between $\alpha$ and $q$ is $\alpha = \frac{4-2q}{q-1}$, which is different from Equation (9) and thus cannot reproduce the correct scaling. They then concluded that the Tsallis entropy was not the appropriate framework for Lévy flights in a harmonic potential described by the generalized FP equation. However, this limitation seems not to impede the usefulness of the application of Tsallis entropy on this Langevin equation where the correct scaling is obtained.

In summary, we have employed an FFP equation to find the PDFs and studied the $q$-entropy and $q$-energies in this system with a non-linear interaction in the FFP equation. We found a significantly elevated diffusion coefficient, which is qualitatively similar to what was expected in light of the analysis of the experimental data.

**Author Contributions:** Conceptualization, J.A. and S.M.; Draft Preparation, J.A., S.M. and T.R.

**Funding:** This research was partly funded by US DOE grant numbers DE-SC0013977 and DE-FG02-92ER54141.

**Acknowledgments:** The author would like to acknowledge contributions and fruitful discussions with Kim, E. and del Castillo Negrete, D. One of the co-authors, Rafiq, T., would like to acknowledge support by US DOE Grants DE-SC0013977 and DE-FG02-92ER54141.

**Conflicts of Interest:** The authors declare no conflict of interest.

## References

1. Horton, W. *Turbulent Transport in Magnetized Plasmas*; World Scientific: Danvers, MA, USA, 2017; ISBN 978-981-3225-88-6.
2. Krommes, J.A. Fundamental statistical descriptions of plasma turbulence in magnetic fields. *Phys. Rep.* **2002**, *360*, 1–352. [CrossRef]
3. Carreras, B.A.; Hidalgo, C.; Sanchez, E.; Pedrosa, M.A.; Balbin, R.; Garcia, C.I.; van Milligen, B.; Newman, D.E.; Lynch, V.E. Fluctuation-induced flux at the plasma edge in toroidal devices. *Phys. Plasmas* **1996**, *3*, 2664–2672. [CrossRef]
4. Carreras, B.A.; van Milligen, B.; Hidalgo, C.; Balbin, R.; Sanchez, E.; Cortes, I.G.; Pedrosa, M.A.; Bleuel, J.; Endler, M. Self-similarity properties of the probability distribution function of turbulence-induced particle fluxes at the plasma edge. *Phys. Rev. Lett.* **1999**, *83*, 3653–3656. [CrossRef]
5. Van Milligen, B.P.; Sanchez, R.; Carreras, B.A.; Lynch, V.E.; LaBombard, B.; Pedrosa, M.A.; Hidalgo, C.; Gonçalves, B.; Balbín, R.; The W7-AS Team. Additional evidence for the universality of the probability distribution of turbulent fluctuations and fluxes in the scrape-off layer region of fusion plasmas. *Phys. Plasmas* **2005**, *12*, 52501–52507. [CrossRef]
6. Sanchez, R.; Newman, D.E.; Leboeuf, J.N.; Decyk, V.K.; Carreras, B.A. Nature of Transport across Sheared Zonal Flows in Electrostatic Ion-Temperature-Gradient Gyrokinetic Plasma Turbulence. *Phys. Rev. Lett.* **2008**, *101*, 205002–205004. [CrossRef] [PubMed]
7. del-Castillo-Negrete, D.; Carreras, B.A.; Lynch, V.E. Front Dynamics in Reaction-Diffusion Systems with Levy Flights: A Fractional Diffusion Approach. *Phys. Rev. Lett.* **2005**, *94*, 18302–18304. [CrossRef] [PubMed]
8. Sanchez, R.; Carreras, B.A.; Newman, D.E.; Lynch, V.E.; van Milligen, B.P. Renormalization of tracer turbulence leading to fractional differential equations. *Phys. Rev. E* **2006**, *74*, 16305–16311. [CrossRef] [PubMed]
9. Hahm, T.S. Nonlinear gyrokinetic equations for tokamak microturbulence. *Phys. Fluids* **1988**, *31*, 2670–2673. [CrossRef]
10. Zweben, S.J. Search for coherent structure within tokamak plasma turbulence. *Phys. Fluids* **2007**, *28*, 974–982. [CrossRef]
11. Naulin, V. Turbulent transport and the plasma edge. *J. Nucl. Mater.* **2007**, *363-365*, 24–31. [CrossRef]
12. Kaye, S.M.; Barnes, C.W.; Bell, M.G.; DeBoo, J.C.; Greenwald, M.; Riedel, K.; Sigmar, D.; Uckan, N.; Waltz, N. Status of global energy confinement studies. *Phys. Plasmas* **1990**, **2**, 2926–2940.
13. Cardozo, N.J.L. Perturbative transport studies in fusion plasmas. *Plasma Phys. Control. Fusion* **1995**, *37*, 799–852. [CrossRef]

14. Gentle, K.W.; Bravenec, R.V.; Cima, G.; Gasquet, H.; Hallock, G.A.; Phillips, P.E.; Ross, D.W.; Rowan, W.L.; Wootton, A.J. An experimental counter-example to the local transport paradigm. *Phys. Plasmas* **1995**, *2*, 2292–2298. [CrossRef]
15. Mantica, P.; Galli, P.; Gorini, G.; Hogeweij, G.M.D.; de Kloe, J.; Cardozo, N.J.L.; RTP Team. Nonlocal transient transport and thermal barriers in rijnhuizen tokamak project plasmas. *Phys. Rev. Lett.* **1999**, *82*, 5048–5051. [CrossRef]
16. Van-Milligen, B.P.; de la Luna, E.; Tabars, F.L.; Ascasíbar, E.; Estrada, T.; Castejón, F.; Castellano, J.; Cortés, I.G.; Herranz, J.; Hidalgo, C.; et al. Ballistic transport phenomena in TJ-II. *Nucl. Fusion* **2002**, *42*, 787–795. [CrossRef]
17. Anderson, J.; Hnat, B. Statistical analysis of Hasegawa-Wakatani turbulence. *Phys. Plasmas* **2017**, *24*, 62301–62308. [CrossRef]
18. Moradi, S.; Anderson, J.; Romanelli, M. Evidence of non-local heat transport model in JET plasmas. Presented at EU-US Transport Task Force Meeting, Seville, Spain, 11–14 September 2018.
19. Schlesinger, M.F.; Zaslavsky, G.M.; Klafter, J. Strange kinetics. *Nature* **1993**, *363*, 31–37. [CrossRef]
20. Sokolov, I.M.; Klafter, J.; Blumen, A. Fractional kinetics. *Phys. Today* **2002**, *55*, 48–54. [CrossRef]
21. Klafter, J.; Sokolov, I.M. Anomalous diffusion spreads its wings. *Phys. World* **2005** *18*, 29–32. [CrossRef]
22. Metzler, R.; Klafter, J. The random walk's guide to anomalous diffusion: A fractional dynamics approach. *Phys. Rep.* **2000**, *339*, 1–77. [CrossRef]
23. Metzler, R.; Klafter, J. The restaurant at the end of the random walk: recent developments in the description of anomalous transport by fractional dynamics. *J. Phys. A* **2004**, *37*, 161–208. [CrossRef]
24. Mandelbrot, B.B. *Fractals and Geometry of Nature;* W. H. Freeman and Company: San Francisco, CA, USA, 1982; pp. 170–180.
25. Anderson, J.; Xanthopoulos, P. Signature of a universal statistical description for drift-wave plasma turbulence. *Phys. Plasmas* **2010**, *17*, 110702–110704. [CrossRef]
26. Kim, E.; Liu, H.L.; Anderson, J. Probability distribution function for self-organization of shear flows. *Phys. Plasmas* **2009** *16*, 52301–52304. [CrossRef]
27. Moradi, S.; Anderson, J.; Weyssow, B. A theory of non-local linear drift wave transport. *Phys. Plasmas* **2011**, *18*, 062101–062106. [CrossRef]
28. Moradi, S.; Anderson, J. Non-local gyrokinetic model of linear ion-temperature-gradient modes. *Phys. Plasmas* **2012**, *19*, 82301–82307. [CrossRef]
29. Anderson, J.; Kim, E.; Moradi, S. A fractional Fokker–Planck model for anomalous diffusion. *Phys. Plasmas* **2014**, *21*, 122101–122108. [CrossRef]
30. Moradi, S.; del Castillo, N.D.; Anderson, J. Charged particle dynamics in the presence of non-Gaussian Lévy electrostatic fluctuations. *Phys. Plasmas* **2016**, *23*, 907041–907045. [CrossRef]
31. Montroll, E.W.; Scher, H. Random walks on lattices. IV. Continuous-time walks and influence of absorbing boundaries. *J. Stat. Phys.* **1973**, *9*, 101–135. [CrossRef]
32. Kou, S.C.; Sunney, X. Generalized langevin equation with fractional Gaussian noise: Subdiffusion within a single protein molecule. *Phys. Rev. Lett.* **2004**, *93*, 1806031–1806034. [CrossRef] [PubMed]
33. Combescure, M. *Hamiltonian Chaos and Fractional Dynamics;* Oxford University Press: Oxford, UK, 2005; pp. 0305–4470.
34. del Castillo, N.D.; Carreras, B.A.; Lynch, V.E. Fractional diffusion in plasma turbulence. *Phys. Plasmas* **2004**, *11*, 3854–3864. [CrossRef]
35. Del Castillo, N.D. Non-diffusive, non-local transport in fluids and plasmas. *Nonlinear Proc. Geophys.* **2010**, *17*, 795–807. [CrossRef]
36. Zaslavsky, G.M. Chaos, fractional kinetics, and anomalous transport. *Phys. Rep.* **2002**, *371*, 461–580. [CrossRef]
37. Tarasov, V.E. Fractional Liouville and BBGKI equations. *J. Phys.* **2005**, *7*, 17–33. [CrossRef]
38. Tarasov, V.E. Fractional statistical mechanics. *Chaos* **2006**, *16*, 331081–331087. [CrossRef] [PubMed]
39. Lévy, P. *Théorie de L'addition des Variables Aléatoires;* Gauthier-Villiers: Paris, France, 1937.
40. West, B.J.; Seshadri, V. Linear systems with Lévy fluctuations. *Physical A* **1982**, *113*, 203–216. [CrossRef]
41. Fogedby, H.C. Langevin equations for continuous time Lévy flights. *Phys. Rev. E* **1994**, *50*, 1657–1660. [CrossRef]

42. Fogedby, H.C. Lévy Flights in Random Environments. *Phys. Rev. Lett.* **1994**, *73*, 2517–2520. [CrossRef] [PubMed]
43. Diethelm, K.; Freed, A.D. The Fractional PECE Subroutine for the numerical solution of differential equations of fractional order. In *Forschung und Wissenschaftliches Rechnen*; Heinzel, S., Plesser, T., Eds.; Gessellschaft fur Wissenschaftliche Datenverarbeitung: Gottingen, Germany, 1999; pp. 57–71.
44. Tsallis, C.; de Souza, A.M.C.; Maynard, R. Derivation of Lévy-type anomalous superdiffusion from generalized statistical mechanics. In *Lévy Flights and Related Topics in Physics;* Springer: New York, NY, USA, 1995; Volume 450, pp. 269–289.
45. Tsallis, C.; Bukman, D.J. Anomalous diffusion in the presence of external forces: Exact time-dependent solutions and their thermostatistical basis. *Phys. Rev. E* **1996**, *54*, 2197–2200. [CrossRef]
46. Tsallis, C.; Mendes, R.S.; Plastino, A.R. The role of constraints within generalized nonextensive statistics. *Physical A* **1998**, *261*, 534–554. [CrossRef]
47. Hamza, A.B.; Krim, H. Jensen-Rényi divergence measure: Theoretical and Computational Perspectives. In Proceedings of the 2003 IEEE International Symposium on Information Theory, Yokohama, Japan, 29 June–4 July 2003.
48. Barkai, E. Stable equilibrium based on Lévy statistics: Stochastic collision models approach. *Phys. Rev. E.* **2003**, *68*, 551041–551044. [CrossRef] [PubMed]
49. Angulo, J.M.; Esquivel, F.J. Multifractal dimensional dependence assessment based on Tsallis mutual information. *Entropy* **2015**, *17*, 5382–5401. [CrossRef]
50. Balasis, G.; Daglis, I.A.; Anastasiadis, A.; Papadimitriou, C.; Mandea, M.; Eftaxiasb, K. Universality in solar flare, magnetic storm and earthquake dynamics using Tsallis statistical mechanics. *Physical A* **2011**, *390*, 341–346. [CrossRef]
51. Pavlos, G.P.; Karkatsanis, L.P.; Xenakis, M.N.; Sarafopoulos D.; Pavlos, E.G. Tsallis statistics and magnetospheric self-organization. *Physical A* **2012**, *391*, 3069–3080. [CrossRef]
52. Pavlos, G.P.; Karkatsanis, L.P.; Xenakis, M.N. Tsallis non-extensive statistics, intermittent turbulence, SOC and chaos in the solar plasma, Part one: Sunspot dynamics. *Physical A* **2012**, *391*, 6287–6319. [CrossRef]
53. Tsallis, C.; Lévy, S.V.F.; Souza, A.M.C.; Maynard, R. Statistical-mechanical foundation of the ubiquity of Lévy distributions in nature. *Phys. Rev. Lett.* **1995**, *75*, 3589–3593. [CrossRef] [PubMed]
54. Prato, D.; Tsallis, C. Nonextensive foundation of Lévy distributions. *Phys. Rev. E* **1999**, *60*, 2398–2401. [CrossRef]
55. Milovanov, A.V.; Rasmussen, J.J. Lévy flights on a comb and the plasma staircase. *Phys. Rev. E* **2018**, *98*, 222081–222092. [CrossRef] [PubMed]
56. Jespersen, S.; Metzler, R.; Fogedby, H.C. Lévy flights in external force fields: Langevin and fractional Fokker–Planck equations and their solutions. *Phys. Rev. E* **1999**, *59*, 2736–2745. [CrossRef]

*entropy*

MDPI

*Article*

# Effects of Near Wall Modeling in the Improved-Delayed-Detached-Eddy-Simulation (IDDES) Methodology

**Rohit Saini [1], Nader Karimi [1,2], Lian Duan [3], Amsini Sadiki [4] and Amirfarhang Mehdizadeh [1,*]**

1 Civil and Mechanical Engineering Department, School of Computing and Engineering, University of Missouri-Kansas City, Kansas City, MO 64110, USA; rohitsaini@mail.umkc.edu
2 School of Engineering, University of Glasgow, Glasgow G12 8QQ, UK; Nader.Karimi@glasgow.ac.uk
3 Department of Mechanical and Aerospace Engineering, Missouri University of Science and Technology, Rolla, MO 65409, USA; duanl@mst.edu
4 Department of Mechanical Engineering, Institute of Energy and Power Plant Technology, Technische Universität Darmstadt, 64289 Darmstadt, Germany; Sadiki@ekt.tu-darmstadt.de
* Correspondence: mehdizadeha@umkc.edu

Received: 6 September 2018; Accepted: 29 September 2018; Published: 8 October 2018

**Abstract:** The present study aims to assess the effects of two different underlying RANS models on overall behavior of the IDDES methodology when applied to different flow configurations ranging from fully attached (plane channel flow) to separated flows (periodic hill flow). This includes investigating prediction accuracy of first and second order statistics, response to grid refinement, grey area dynamics and triggering mechanism. Further, several criteria have been investigated to assess reliability and quality of the methodology when operating in scale resolving mode. It turns out that irrespective of the near wall modeling strategy, the IDDES methodology does not satisfy all criteria required to make this methodology reliable when applied to various flow configurations at different Reynolds numbers with different grid resolutions. Further, it is found that using more advanced underlying RANS model to improve prediction accuracy of the near wall dynamics results in extension of the grey area, which may delay the transition to scale resolving mode. This systematic study for attached and separated flows suggests that the shortcomings of IDDES methodology mostly lie in inaccurate prediction of the dynamics inside the grey area and demands further investigation in this direction to make this methodology capable of dealing with different flow situations reliably.

**Keywords:** hybrid (U)RANS-LES; IDDES methodology; attached and separated flows

## 1. Introduction

High Reynolds number flows are a classical research theme that retains its vitality at several levels from real-world applications, through physical and computational modeling, up to rigorous mathematical analysis. The main reason for the sustained relevance of this topic is in the ubiquity of such flows in practical situations, such as blood flow in large caliber vessels, various energy systems, aerodynamics, combustion systems, to name only a few. Numerical simulation of high Reynolds number flows is supposed to serve the purpose of providing necessary data for design and optimization. However, modeling high Reynolds number flows is immensely challenging due to the complex interaction among disparate turbulent length scales associated with different regimes in these flows. Advanced modeling strategies are needed to describe the interaction between different flow regions, e.g., surface viscous layers and outer turbulent flow regions in wall-bounded turbulent flows.

Reynolds-Averaged Navier–Stokes (RANS) models are generally used to simulate stationary high Reynolds number turbulent flows with industrial applications. However, it is becoming increasingly clear that there is a need to capture unsteady dynamics of the complex turbulent flows where classical RANS models either cannot provide necessary data (e.g., acoustics simulations where the turbulence generates noise sources, which cannot be extracted accurately from RANS simulations) or they are not accurate enough even in the first order statistics (e.g., strongly separated flows such as flow past a building and a re-entry vehicle). Unsteady extensions of RANS models (denoted as (U)RANS) attempt to capture some level of unsteady dynamics. However, because (U)RANS methods are not designed to capture integral-scale dynamics, Large Eddy Simulation (LES) is sometimes needed to capture essential energetic unsteady dynamics in complex flows. Unfortunately, LES is not feasible for many engineering applications due to the high computational cost associated with grid refinement to resolve the energy-containing eddies appropriately. This limitation becomes immense to capture near-wall dynamics. In the vicinity of the walls, the LES philosophy of resolving energy-containing vortical structures requires grid refinement probably close to the Direct Numerical Simulation (DNS) level, which makes LES prohibitively expensive to apply to wall-bounded flows at high Reynolds number.

Recognizing the limitations of the classical RANS/(U)RANS and LES and in search for more efficient solution methods for practical applications, the CFD community has turned its attention to hybrid (U)RANS-LES and wall-modeled LES approaches as alternative strategies for complex turbulent flow with high Reynolds numbers. The primary goal of a hybrid (U)RANS-LES/wall-modeled LES approach is to achieve time-dependent and three-dimensional space-resolved simulation of large-scale structures, which describe the turbulence dynamics at an affordable cost, while near-wall dynamics are accurately modeled. Several hybrid (U)RANS-LES approaches have been proposed within the last two decades. These included detached eddy simulation [1], scale-adaptive simulation [2], partially-averaged Navier–Stokes [3], etc. Among them, the Detached Eddy Simulation (DES) developed originally by Spalart [1], including its variants Delayed Detached Eddy Simulation (DDES) [4] and Improved Delayed Detached Eddy Simulation (IDDES) [5], has attracted the most attention due to its simplicity in implementation, and it is widely used to simulate high Reynolds number flows relevant for industrial applications [6].

The first version of DES was based on a modified transport equation for turbulent eddy viscosity ($\nu_t$) that uses distance from the wall as the RANS length scale. Local grid refinement is used to alter the length scale away from the wall to drive the model into a scale-resolving mode. However, this approach faced several practical issues, in particular, Grid-Induced Separation (GIS), Model Stress Depletion (MSD) and Log-layer Mismatch (LMM), which were discussed in [7]. DDES [4] and IDDES [5] have consequently been proposed to mitigate these issues. IDDES features Wall-Modeled-LES (WMLES) capabilities, depending on inflow condition and, therefore, includes more empiricism. The IDDES methodology can basically be combined with various RANS models to form a hybrid approach [6]. In the present work, focus will be on Spalart–Allmaras IDDES (uses distance from the wall to provide RANS length scale [5]) and $k$-$\omega$-SSTIDDES (uses the two-equation model to provide the RANS length scale [8]). In particular, the effect of the underlying RANS model in overall model behavior will be investigated. This includes response to grid refinement, prediction accuracy, grey-area dynamics and the triggering mechanism. Toward this end, Spalart–Allmaras (S-A) IDDES and $k$-$\omega$-SST IDDES will be applied to different configurations ranging from fully-attached to complex separated flows.

The paper is organized as follows: In the next section, the IDDES formulations will be briefly presented and discussed. In Section 3, an overview of the test cases is provided. Sections 4 and 5 are dedicated to present quality assessment criteria and the numerical approach. Section 6 will present and discuss the results obtained from the IDDES methodology using different near-wall modeling strategies. Section 7 concludes the paper with a summary, conclusion and outlook.

## 2. Improved Delayed Detached Eddy Simulation Methodology

In this section, a brief description of the governing transport equations of S-A IDDES and $k$-$\omega$-SST IDDES models along with the triggering mechanism involved in the IDDES methodology will be presented.

### 2.1. Spalart–Allmaras IDDES

S-A IDDES is defined based on the transport equation for modified eddy viscosity ($\tilde{\nu}$) and is given as follows:

$$\frac{\partial \tilde{\nu}}{\partial t} + U_i \frac{\partial \tilde{\nu}}{\partial x_j} = c_{b1} \tilde{S} \tilde{\nu} + \frac{1}{\sigma} [\nabla . (\tilde{\nu} \nabla \tilde{\nu}) + c_{b2} (\nabla \tilde{\nu})^2] - c_{w1} f_w (\tilde{r} (\frac{\tilde{\nu}}{l_{IDDES}})^2) \quad (1)$$

where the turbulent eddy viscosity is defined as $\nu_t = f_{v1} \tilde{\nu}$. Functions $f_{v1}$ and $f_w$ are introduced for near-wall corrections in the case of finite and high Reynolds number flows, respectively. $\tilde{S}$ is the strain rate tensor, and $\tilde{r}$ is the non-dimensional term defined as $\nu_t / (\tilde{S} \kappa^2 {d_w}^2)$, where $\kappa$ and $d_w$ are the von-Karman constant and distance from the wall. $\sigma$, $c_{b1}$, $c_{b2}$ and $c_w$ are the model constants imported from the original Spalart–Allmaras (S-A) model [9]. A complete description of the model is provided in Shur et al. [5]. The $l_{IDDES}$ term is a modified length scale responsible for triggering to a scale-resolving mode and will be discussed in Section 2.3.

### 2.2. k-ω-SST IDDES

$k$-$\omega$-SST IDDES employs a modified version of $k$-$\omega$-SST model to improve near-wall prediction and is defined as below:

$$\frac{\partial k}{\partial t} + \nabla . (\tilde{U} k) = \nabla . [(\nu + \sigma_k \nu_t) \nabla k] + P_k - \sqrt{k^3} / l_{IDDES}, \quad (2)$$

$$\frac{\partial \omega}{\partial t} + \nabla . (\tilde{U} \omega) = \nabla . [(\nu + \sigma_\omega \nu_t) \nabla \omega] + 2(1 - F_1) \sigma_{\omega 2} \frac{\nabla k . \nabla \omega}{\omega} + \alpha \frac{1}{\nu_t} P_k - \beta \omega^2, \quad (3)$$

where blending function $F_1$ and model constants ($\alpha$, $\sigma_k$, $\sigma_\omega$, $\sigma_{\omega 2}$ and $\beta$) are imported from the original $k$-$\omega$-SST model [10]. It should be noted that within $k$-$\omega$-SST IDDES, only the destruction term in the $k$-equation is modified by introducing the $l_{IDDES}$ term, whereas the $\omega$ equation remains unchanged. Similar to S-A IDDES, $l_{IDDES}$ is responsible for triggering a transition from (U)RANS mode into a scale-resolving mode.

### 2.3. Triggering Mechanism

The goal in the IDDES methodology is to trigger a transition from (U)RANS to a scale-resolving mode, depending on a criterion based on the turbulent length scale. In this context, the $l_{IDDES}$ term is applied to the destruction term in the modified eddy viscosity, $\tilde{\nu}$ (Equation (1)) and turbulent kinetic energy, $k$ (Equation (2)), transport equations. The intention is to increase dissipation (reduce the level of turbulent eddy viscosity) as we transverse away from the wall to trigger a transition to scale-resolving mode. The $l_{IDDES}$ term is defined as follows:

$$l_{IDDES} = \tilde{f_d} (1 + f_e) l_{RANS} + (1 - \tilde{f_d}) l_{LES}, \quad (4)$$

where $l_{RANS}$ for S-A IDDES is simply distance from the wall ($d_w$) and for $k$-$\omega$-SST IDDES corresponds to $k^2 / (C_\mu \omega)$, with $C_\mu = 0.09$. In scale-resolving mode, $l_{LES}$ for S-A IDDES is defined as $C_{DES} \psi \triangle$, where $C_{DES} = 0.65$ and $\psi$ is the low Reynolds number correction, which accommodates near-wall corrections, as discussed in [4]. In case of $k$-$\omega$-SST IDDES, $\psi$ equals one, and $C_{DES}$ is calculated algebraically as below:

$$C_{DES} = C_{DES1} \cdot F_1 + C_{DES2} \cdot (1 - F_1), \quad (5)$$

with $C_{DES1} = 0.78$, $C_{DES2} = 0.61$ and $F_1$ is calculated as per the original $k$-$\omega$-SST turbulence model [10]. Filter width or the characteristic cut-off length scale ($\triangle$), used in calculating LES length scale, is a piece-wise function containing wall-distance dependency and local cell dimension information:

$$\triangle = min(max[C_w d_w, C_w h_{max}, h_{wn}], h_{max}), \quad (6)$$

where $C_w$ is an empirical constant and is 0.15, $h_{max}$ is the maximum of the local cell size in streamwise, wall-normal and lateral directions and $h_{wn}$ is the wall-normal grid spacing. The $\widetilde{f_d}$ function in Equation (4) includes a set of blending functions responsible for switching from (U)RANS mode (defined by $\widetilde{f_d}$ = 1) to scale-resolving mode (defined by $\widetilde{f_d}$ = 0). However, transition from (U)RANS to scale-resolving mode does occur through an intermediate area called the grey area, where $0 < \widetilde{f_d} < 1$. The function $\widetilde{f_d}$ is defined as follows:

$$\widetilde{f_d} = max(1 - f_d, f_B). \tag{7}$$

$f_d$ is called the delaying function and is defined as below:

$$f_d = 1 - \tanh[8(r_d{}^3)], \tag{8}$$

where the $r_d$ term is borrowed from the original S-A model [9].

To provide a remedy to the log layer mismatch at the interface of (U)RANS and scale-resolving region, the IDDES methodology includes wall modeling capability called Wall-Modeled LES (WMLES). The WMLES branch of IDDES is intended to be active if turbulent inflow content is provided and the grid is fine enough to resolve the dominant vortical structures in the boundary layer. Under appropriate conditions for WMLES operation, the $l_{IDDES}$ defined in Equation (4) is modified to $l_{WMLES}$ as follows:

$$l_{WMLES} = f_B(1 + f_e)l_{RANS} + (1 - f_B)l_{LES}, \tag{9}$$

where the blending function $f_B$ is purely grid dependent and is based on the distance from the wall and the local maximum cell edge length. $f_B$ is described as follows:

$$f_B = min[2exp(-9\alpha^2), 1.0], \tag{10}$$

where the grid-dependent parameter $\alpha$ is calculated as $\alpha$ = 0.25 − $(d_w/h_{max})$. $f_B$ varies from zero to one and should provide a rapid transition from (U)RANS mode to scale-resolving mode within the range of wall distance $0.5h_{max} < d_w < h_{max}$. Another empirical function called elevating function $f_e$, included in Equation (9), helps in preventing the excessive reduction of the Reynolds stresses in the near-wall region ((U)RANS region). $f_e$ is described as follows:

$$f_e = max[(f_{e1} - 1), 0]\psi f_{e2}. \tag{11}$$

$f_{e1}$ solely is grid dependent, whereas $f_{e2}$ is a function of the flow field quantities. Further details regarding this methodology and related functions can be found in Shur et al. [5] and Gritskevich et al. [8].

As discussed, S-A IDDES and $k$-$\omega$-SST IDDES employ the same triggering mechanism by introducing a modified length scale ($l_{IDDES}$) into the destruction term of $\widetilde{\nu}$ (S-A IDDES) and $k$ ($k$-$\omega$-SST IDDES) transport equations. Therefore, the main focus of the present study is to investigate the effect of the underlying RANS model on overall model behavior when applied to different configurations ranging from fully-attached to separated flows.

## 3. Overview of the Test Cases

S-A IDDES and $k$-$\omega$-SST IDDES are applied to two benchmark test cases, with increasing geometrical complexities. These include fully-developed turbulent channel flow and flow over a periodic hill. Various criteria/functions will be assessed on different grid resolutions to demonstrate the effect of near-wall modeling on model performance/behavior.

### 3.1. Turbulent Channel Flow

This test case will demonstrate the effect of underlying RANS model under stable and attached flow conditions. The size of the computational domain used for simulation is $L_x$ = $2\pi h$, $L_y$ = $2h$

and $L_z = \pi h$, where $x$, $y$ and $z$ denote the streamwise, wall-normal and the spanwise directions, respectively. Two friction Reynolds numbers have been considered; $Re_\tau = 395$ and $Re_\tau = 4200$, where $Re_\tau = (u_\tau * h)/\nu$ is based on friction velocity $u_\tau$ and channel half height $h$. Corresponding resulting mean Reynolds numbers based on the bulk mean velocity and channel half height are $Re_b \approx 13{,}000$ and $Re_b \approx 200{,}000$, respectively. Note that the computational domain is kept the same for all the considered Reynolds numbers. A constant pressure gradient is applied via the source term in the momentum equation to derive the flow at the required Reynolds number. Periodic boundary conditions are imposed in the streamwise and the spanwise direction, and the no-slip boundary condition is used for top and bottom walls. Three different grid resolutions (coarse, medium and fine) are used to simulate channel flow at $Re_\tau = 395$. Constant geometric stretching of approximately 11% is used in the wall-normal direction. Please note that the resolution of the fine grid for $Re_\tau = 395$ is equivalent to the DNS grid used by Moser et al. [11]. For $Re_\tau = 4200$, two grid resolutions have been used. The coarse grid corresponds to the medium resolution used for $Re_\tau = 395$ simulation. The fine grid is designed based on criteria to support transition to LES mode after the buffer layer and the beginning of the surface layer (i.e., $y^+ \geq 40$–$50$), proposed by Brasseur and Wei [12]. This guarantees that the appropriate grid is available for transition from (U)RANS to LES mode at this Reynolds number, if the model functions properly. More details about the different grid resolutions used in this test case are summarized in Table 1. The results are compared with the DNS results of Moser et al. [11] (for $Re_\tau = 395$) and Duran and Jiménez [13] (for $Re_\tau = 4200$).

**Table 1.** Details of the grid resolution for turbulent developed channel flow.

| $Re_\tau$ | Grids | $\Delta x^+$ | $\Delta y^+_w$ | $\Delta z^+$ | $N_x$ | $N_y$ | $N_z$ |
|---|---|---|---|---|---|---|---|
| 395 | Coarse | 41.60 | 0.1 | 27.7 | 64 | 192 | 48 |
| | Medium | 20.84 | 0.1 | 13.90 | 128 | 192 | 96 |
| | Fine | 10.04 | 0.09 | 6.56 | 256 | 192 | 196 |
| 4200 | Coarse | 212.2 | 1.03 | 140.2 | 128 | 192 | 96 |
| | Fine | 117.6 | 1.09 | 57.9 | 234 | 146 | 234 |

### 3.2. Periodic Hill Flow

This is a typical test case to study separation and reattachment dynamics over a smooth curved surface. The size of the computational domain is 9H, 3H and 4.5H in the streamwise, wall-normal and spanwise directions, respectively, where H is the hill Height at the crest. The schematic of the flow domain is shown in Figure 1. Two bulk Reynolds numbers are investigated in the present study; $Re_b = 10{,}595$ and $Re_b = 37{,}000$, based on the hill Height (H) and the bulk velocity ($U_b$) at the crest. Two different grid-resolutions, similar to what is used in Razi et al. [14] to evaluate the PANS hybrid method, are used in this investigation (summarized in Table 2). Similar to the turbulent channel flow, the flow is driven by a constant pressure gradient, which is added as a source term in the streamwise momentum equation. Periodic inlet/outlet and spanwise boundary conditions were chosen, and therefore, the mean flow properties are also averaged in the spanwise direction. Results are compared with available experimental measurements [15] and high fidelity numerical simulation [16].

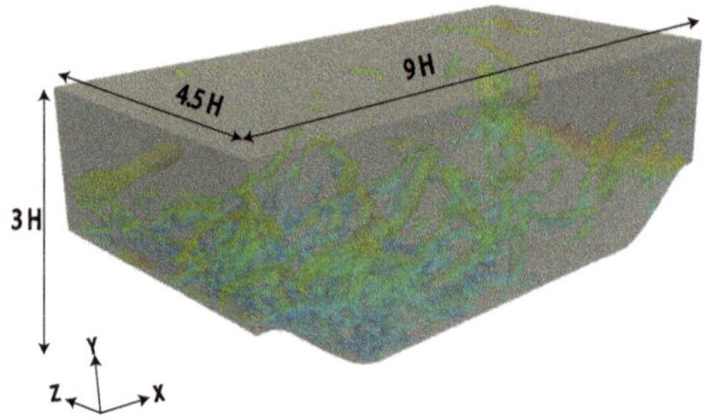


**Figure 1.** A three-dimensional geometry considered for periodic hill flow, with dimensions and the coordinate system employed in the present study. H, Height.

**Table 2.** Details of the grid resolution for periodic hill flow.

| $Re_b$ | Grid | $N_x$ | $N_y$ | $N_z$ | $N_{total}$ |
|---|---|---|---|---|---|
| 10,595 and 37,000 | Coarse | 100 | 100 | 30 | 300,000 |
| | Fine | 150 | 100 | 60 | 900,000 |
| LES for 10,595 [16] | - | - | - | - | 11,300,000 |

## 4. Quality Assessment

Hybrid (U)RANS-LES methods are increasingly applied to complex high Reynolds number flows relevant for industrial applications. In this context, assessment of the quality and reliability of hybrid model when operating in scale-resolving mode is essential. Several grid-based criteria have been proposed for LES quality assessment. These include Meyers et al. [17], Klein [18] and Celik et al. [19], to name a few.

In the present investigation, the scale-resolving region is assessed through various criteria summarized in Table 3. These will help in assessing the various capabilities of the IDDES methodology.

**Table 3.** List of grid assessment criteria for the scale-resolving region in the present study.

| Equation | Criterion | Description |
|---|---|---|
| (1) | $k_{resolved}/(k_{resolved}+k_{sgs})$ | ratio of the resolved turbulent kinetic energy to the total turbulent kinetic energy, where resolved turbulent kinetic energy is defined as $k = \frac{1}{2}\langle u'_i u'_i\rangle$ and modeled turbulent kinetic energy is defined as Equation (12). |
| (2) | $1/(1+0.05\left(\frac{\nu+\nu_t}{\nu}\right)^{0.53})$ | relative sub-grid scale viscosity ratio |
| (3) | $\Delta/\eta$ | ratio of the characteristic cut-off length scale to the relative Kolmogorov length scale |
| (4) | $L_{sgs.}/\Delta$ | the ratio of the sub-grid length scale and the characteristic cut-off length scale |

According to Pope [20], a well-resolved LES region can be defined when 80% of the total turbulent kinetic energy is resolved. Therefore, the ratio of modeled turbulent kinetic energy to total turbulent kinetic energy ($k_{modeled}/(k_{modeled}+k_{resolved})$, Criterion 1 in Table 3), should attain its maximum (theoretically one) in the wall vicinity, where the model is supposed to operate in (U)RANS mode and should decrease to about 0.2 in the scale-resolving region of the simulation away from the wall. This criterion will also be used to evaluate the amount of intrusion of the scale-resolving region into the (U)RANS region. The intrusion could negatively affect the (U)RANS dynamics and ultimately near-wall prediction quality. Note that the modeled turbulent kinetic energy for S-A IDDES is calculated by relying on the Smagorinsky algebraic relation and is expressed as:

$$k_{sgs.} = \left(\frac{\nu_t}{C_s\Delta}\right)^2. \tag{12}$$

where $C_s \approx 0.16$ is obtained from Gong and Tanner [21].

The second criterion, related to the relative sub-grid scale viscosity, provides similar results as the first criterion in the scale-resolving region, therefore not shown in the present study.

The third criterion used in assessing grid resolution compares the characteristic cut-off length scale ($\triangle$, as defined in Equation (6)) to an estimated Kolmogorov length scale, ($\eta$), characterizing the length scale of the dissipative motion. In this criterion, $\eta$ is obtained from the dissipation rate ($\epsilon$) by the following relation:

$$\eta = \left(\frac{\nu^3}{\epsilon}\right)^{1/4}. \tag{13}$$

where $\epsilon = \frac{(k_{sgs.})^{3/2}}{L_{sgs.}}$. It should be noted that Equation (13) is merely a scale relation and provides a very conservative estimate of the finest scale in the turbulent flow. Considering a carefully-devised energy spectrum according to the Kolmogorov hypothesis, this criterion is particularly applicable to high Reynolds number flows [20]. However, Fröhlich et al. [22] applied this criterion to fairly low Reynolds number ($Re_b$ = 10,000) in wall-bounded flows and defined that the ratio should be around eight to ten in the scale-resolving region to resemble well-resolved LES. Furthermore, it demonstrates the significance of the sub-grid scale model to assess grid resolution requirement. In the present study, we assess this criterion only on channel flow with $Re_\tau = 4200$ and periodic hill flow with $Re_b$ = 37,000.

The last criterion is based on the ratio of the sub-grid ($L_{sgs}$) and characteristic cut-off length scales. In $k$-$\omega$-SST IDDES, the two-equation model is used to determine the sub-grid length scale ($k^2/(C_\mu\omega)$), whereas for S-A IDDES, the sub-grid length scale is a modified distance from the wall, $d_w\ \psi$. The ratio of the sub-grid length scale and grid length scale should be of the same order in the scale-resolving region.

These criteria will be evaluated on the test cases discussed in the previous section. Importantly, these criteria are expected to respond appropriately to the grid refinement such that the scale-resolving mode will reflect the characteristics of the systematic eddy-resolving approach.

## 5. Numerical Procedure

All computations are performed using the open-source CFD code, Open-FOAM [23]. Two test cases, as described in Section 3, are investigated in the present study. In all cases, second-order central differencing for velocity, turbulent kinetic energy $k$ and specific dissipation rate $\omega$ is used. The second-order time discretization method is used for all the simulations. Unsteady SIMPLE and PISO algorithms are used for momentum advancement and to solve the Poisson equation, respectively.

## 6. Results and Discussion

In this section, results obtained from $k$-$\omega$-SST IDDES and S-A IDDES, when applied to channel flow and periodic hill flow will be presented and discussed. As mentioned before, the effect of near-wall modeling will be assessed on overall model prediction capability.

### *6.1. Turbulent Channel Flow*

#### 6.1.1. $Re_\tau$ = 395

Figure 2a,b shows non-dimensionalized velocity profiles obtained from three different grids under the two-equation ($k$-$\omega$-SST IDDES) and one-equation (S-A IDDES) models, respectively. The log layer mismatch (LLM) in the outer log-layer is marginal in the case of the two-equation model, whereas it is observable for the one-equation model. It is expected that the LLM will become less and ultimately vanish at the DNS level mesh. However, both models and particularly S-A IDDES showed an inconsistent response to mesh refinement. To assess the LMM issue more accurately, the log-law indicator defined by Brasseur and Wei [12] was used. It is defined as the gradient of the mean streamwise velocity normalized by the inertial Law-Of-The-Wall (LOTW) surface-layer velocity and length scales:

$$\phi(y) = \frac{y}{u_\tau}\frac{\partial \overline{U}}{\partial y}, \tag{14}$$

where $y$ and $u_\tau$ are the wall-normal distance and friction velocity at the wall. Figure 3 shows the variation of $\phi_m$ ($\phi_m \equiv \kappa\,\phi(y)$) plotted against wall-normal distance, where $\kappa$ is the von-Karman constant, assumed to be 0.41 in the present study. According to the LOTW scaling, the log-law indicator should be constant and equal to one in the plateau region, which denotes the logarithmic region. However, it should be noted here that a true log-layer is not expected to appear, as $Re_\tau$ = 395 is too low [24]. Therefore, the main reason for accessing this quantity at $Re_\tau$ = 395 is to determine the discrepancy between models and DNS data in a more accurate manner. It is clearly shown that both models indicated an inconsistent behavior as the grid was further refined from a medium to a fine (DNS level) grid, i.e., deviation from DNS data became more noticeable.

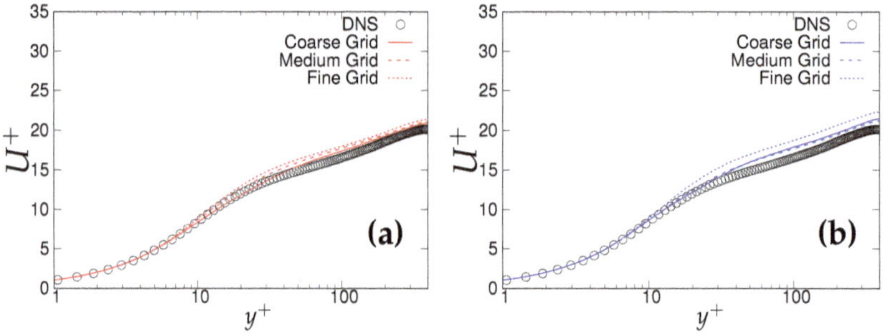


**Figure 2.** Channel flow at $Re_\tau = 395$: wall normal variation of the non-dimensional velocity profile; (**a**) left column = *k*-$\omega$-SSTIDDES; (**b**) right column = Spalart–Allmaras (S-A) IDDES.

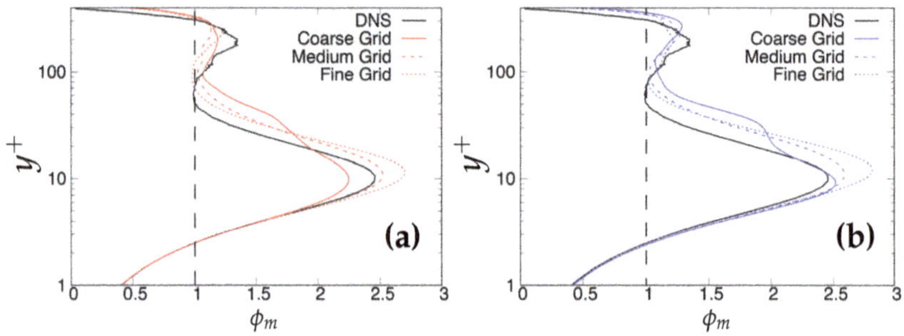


**Figure 3.** Channel flow at $Re_\tau = 395$: variation of the normalized mean shear ($\phi_m$) as a diagnostic quantity for a log law. (**a**) left column = *k*-$\omega$-SST IDDES; (**b**) right column = Spalart–Allmaras (S-A) IDDES.

Further, we compare the resolved streamwise, wall-normal and the spanwise turbulent fluctuations in Figure 4a,b. In the vicinity of the wall, the fluctuations are well captured by the *k*-$\omega$-SST IDDES model, and predictions improved in response to the grid refinement. In the S-A

IDDES model, wall-normal and spanwise fluctuations were captured well in the near-wall region, whereas streamwise velocity fluctuation was over-predicted while progressing from the coarser to finer grid resolution. However, prediction of the resolved velocity fluctuations within both models was improved in the core region with advancement in grid resolution, signifying that the scale-resolving region responds appropriately to grid refinement at the present Reynolds number. Figure 4c,d shows the variation of total turbulent kinetic energy (resolved + modeled) along the wall normal direction. It can be observed that the k-$\omega$-SST IDDES model responded more appropriately to grid refinement and the peak of total turbulent kinetic energy was well captured on the medium and fine (DNS level) grid. The inconsistent response of S-A IDDES might have been due to the existing uncertainties in calculating modeled turbulent kinetic energy. This is definitely a short-coming of S-A IDDES when sub-grid (modeled) quantities are important.

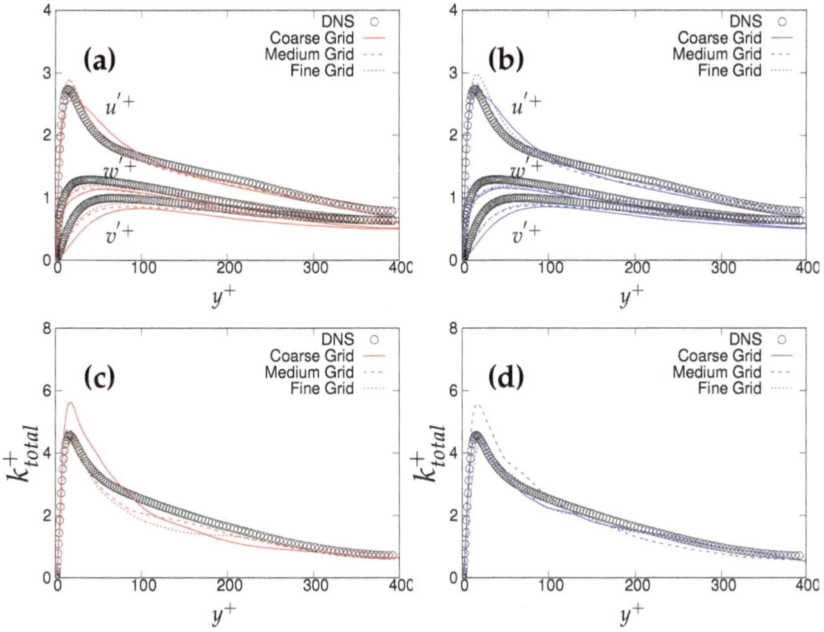


**Figure 4.** Channel flow at $Re_\tau = 395$: wall-normal variation of: (**a**,**b**) resolved turbulence fluctuations; (**c**,**d**) non-dimensionalized total turbulent kinetic energy; symbols represent DNS. Left column = *k*-$\omega$-SST IDDES, right column = Spalart–Allmaras (S-A) IDDES.

Later, we systematically investigate the effect of grid resolution on the main function ($\widetilde{f}_d$) involved in the triggering mechanism and responsible for switching from (U)RANS to the scale-resolving mode. As mentioned previously, $\widetilde{f}_d$ (also equals $max(1 - f_d, f_B)$) defines the (U)RANS and scale-resolving region at $\widetilde{f}_d = 1$ and $\widetilde{f}_d = 0$, respectively. The grey region/area is defined in the range $0 < \widetilde{f}_d < 1$ and is shown as the shaded region between (U)RANS and the Scale-Resolving Region (SRR) in the subsequent figures. The grid refinement should result in shrinkage of the grey region, as well as the (U)RANS region and ultimately vanishing on the finest (DNS level) grid. Figure 5a,c,e depicts the behavior of $\widetilde{f}_d$ for the *k*-$\omega$-SST IDDES model. As mentioned before, this function will help mainly to understand and evaluate the transition dynamics from (U)RANS to scale-resolving mode. Figure 5a,c,e shows that the grey area becomes larger while shifting towards the wall in response to grid refinement. This makes the (U)RANS region smaller, which is expected, as it allows the model to trigger to scale-resolving mode since the appropriate grid was provided. However, it is expected that similar to the (U)RANS region, the grey area would become smaller and allow the model to operate in scale-resolving mode in most part of the simulation on the finest grid.

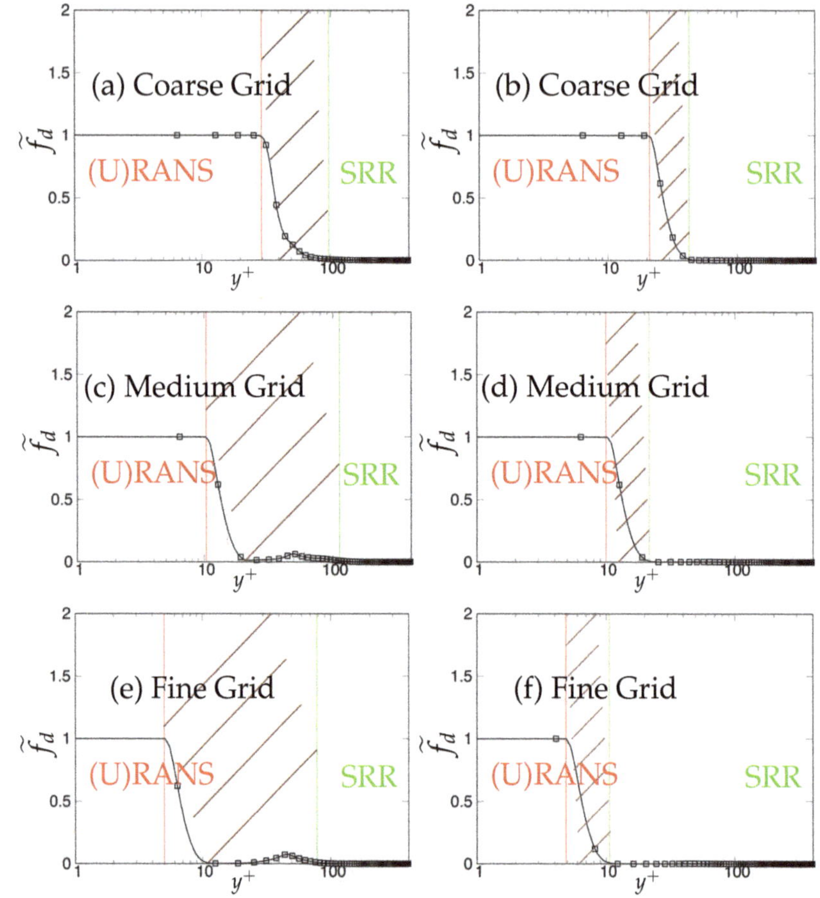


**Figure 5.** Channel flow at $Re_\tau = 395$: response of $\widetilde{f}_d$ to the grid refinement in $k$-$\omega$-SST IDDES and S-A IDDES framework; (U)RANS = Unsteady RANS, SRR = Scale-Resolving Region and shaded region = grey area; left column = $k$-$\omega$-SST IDDES, right column = Spalart–Allmaras (S-A) IDDES.

In contrast to the two-equation model, the thickness of the grey area in S-A IDDES was not responsive to grid refinement and only was shifted towards the wall when the grid became finer. This behavior can be seen in Figure 5b,d,f. Comparing the results obtained from $k$-$\omega$-SST IDDES and S-A IDDES may lead to the conclusion that improving the underlying RANS model does not necessarily improve the triggering mechanism and in fact might negatively affect it and lead to prolonged transition to scale-resolving mode due to the thicker grey region.

As the next step, the criteria stated in Table 3 will be assessed for determining the quality and reliability of the scale-resolving region. The arrows (shown in Figure 6) on the corresponding vertical lines towards increasing wall normal distance describe the Scale-Resolving Region (SRR). The ratio of the modeled to total turbulent kinetic energy (shown in Figure 6a,b) provides the extent of the modeled velocity scales, which should be around one in the wall vicinity in the case of coarse and medium grid. This ratio should become fairly negligible when the DNS level grid is used [20]. This is expected from both models and resembles the first criterion stated in Table 3. For both models, the ratio was significantly less than the one under the coarse and medium grid and reduced to nearly zero with the finest grid. This clearly shows a significant amount of intrusion from the scale-resolving region into (U)RANS part of the simulation that could be detrimental when near-wall effects need to be accurately modeled. The intrusion was due to the weak shielding provided to the underlying (U)RANS model and was confirmed by the variation of the function $f_e$, shown in Figure 7. As discussed in Section 2, function $f_e$ should provide necessary shielding to the near-wall (U)RANS region by preventing excessive reduction of the Reynolds stresses. Therefore, the behavior of $f_e$ under the $k$-$\omega$-SST IDDES and S-A IDDES models on the coarse grid is compared in Figure 7. This function plays an integral role particularly when the IDDES methodology is applied to simulate high Reynolds number flows by preventing transition to scale-resolving mode when appropriate grid support is not available. Although shielding is stronger for the $k$-$\omega$-SST IDDES model, which could be the reason for more amount of intrusion in S-A IDDES, it is not enough to prevent the intrusion. The $f_e$ went to zero under the medium and fine grid resolution.

Lastly, Figure 6c,d shows the variation of the ratio of the sub-grid length scale (obtained from underlying RANS model) and the characteristic cut-off length scale. The ratio should correspond to the same order in the scale-resolving region to represent the correct spectral dynamics on the energy spectrum. Both models satisfy the criterion in the core region. In the S-A IDDES model, the sub-grid length scale (distance from the wall) increases while traversing from the wall to the core of channel, and therefore, the ratio increases linearly and reaches its maximum at the channel center line while still preserving the correct spectral information.

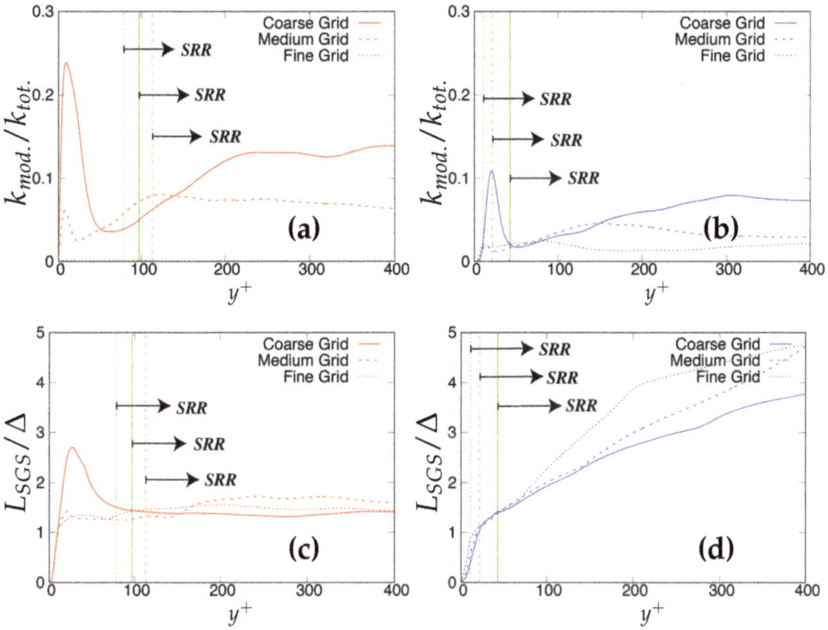


**Figure 6.** Channel flow at $Re_\tau = 395$: variation of the ratio of (**a**,**b**) modeled to total turbulent kinetic energy and (**c**,**d**) integral length scale to characteristic cut-off length scale, along the wall normal direction; (U)RANS = Unsteady RANS, SRR = Scale-Resolving Region and shaded region = grey area; left column = $k$-$\omega$-SST IDDES, right column = Spalart–Allmaras (S-A) IDDES.

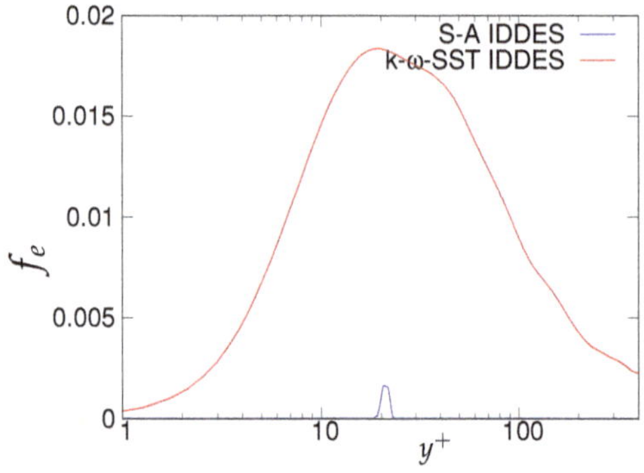


**Figure 7.** Channel flow at $Re_\tau = 395$: response of elevating function ($f_e$) to the coarse grid resolution ($64 \times 192 \times 48$) in the $k$-$\omega$-SST IDDES and Spalart–Allmaras (S-A) IDDES framework.

The anisotropic behavior of the turbulence was analyzed through the "Lumley triangle [25]". The Reynolds stress anisotropy tensor is defined by:

$$b_{ij} = \frac{\langle u_i u_j \rangle}{\langle u_k u_k \rangle} - \frac{1}{3}\delta_{ij}. \tag{15}$$

where the trace of $b_{ij}$ is zero and departure from isotropy is defined between the two bounding lines, often represented in the $\xi$-$\eta$ plane [20], where,

$$\xi = \left(\frac{b_{ij}b_{jk}b_{kl}}{6}\right)^{\frac{1}{3}}, \quad \eta = \left(\frac{b_{ij}b_{ij}}{3}\right)^{\frac{1}{2}}.$$

All physically realistic states of the turbulence should lie inside the triangle. The upper curve corresponds to the two-component turbulence, the left-hand curve to "axisymmetric contraction" and the right-hand curve to the "axisymmetric expansion". The (0,0) point on the $\xi$-$\eta$ plane corresponds to the isotropy point. Details can be found in Pope [20] and Sagaut [26].

Figure 8 shows the anisotropy invariant map and compares the Reynolds stress structure obtained from S-A IDDES and $k$-$\omega$-SST IDDES on the medium and fine grid. Furthermore, the behavior is compared with the DNS as it is expected that the fine (DNS level) grid should closely resemble the DNS profile obtained from Moser et al. [11]. Walking along the DNS profile, starting from the origin, we begin with an isotropic state for Reynolds stresses in the core region of the channel, moving forward to the small kink at $\Delta y^+ \approx 100$, and then, the two-component turbulence state is achieved as we move close to the wall. The arrows shown in the Figure 8 point towards the core of the channel. The discrepancies in the Reynolds stress structure can be seen in the medium and fine grids, especially near the core of the channel for both models, which clearly demonstrate that it is independent of the underlying RANS model, and the IDDES methodology does not respond appropriately to grid refinement and may not be considered as a systematic eddy-resolving method.

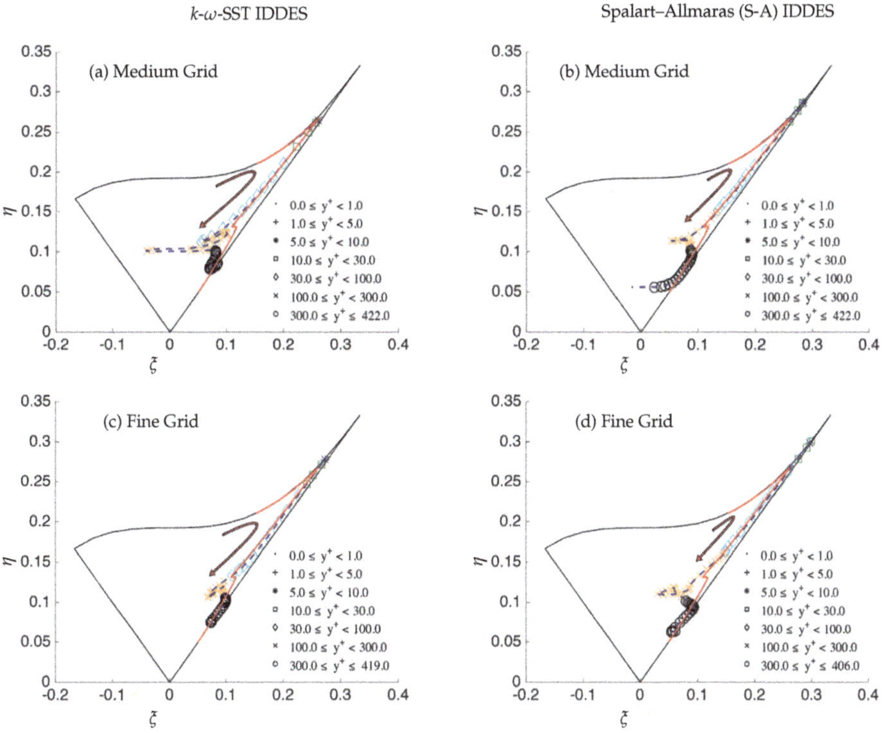


**Figure 8.** Channel flow at $Re_\tau = 395$: anisotropy invariant map for three different mesh resolutions along the wall-normal direction; solid red line = DNS and points dash line = IDDES.

### 6.1.2. $Re_\tau = 4200$

In this sub-section, effects of underlying RANS model on the overall behavior of IDDES methodology at higher Reynolds number are presented using a similar analysis approach as $Re_\tau$ = 395. Figure 9a,b demonstrates the mean velocity profiles obtained from the $k$-$\omega$-SST IDDES and S-A IDDES models on coarse and fine grids. As can be seen, there was a clear overshoot in predictions of both models, which was more severe for the S-A IDDES model. Further, mesh refinement improved the situation only marginally. In order to show this more clearly, the log-law indicator ($\phi_m$) is plotted in Figure 10a,b. Deviation from the law of the wall (also called LMM) can distinctly be seen between starting from the upper part of the surface layer ($y^+ \approx 70$) up to $y^+ \approx 1000$, where $\phi_m$ should be close to unity. In addition, neither of the models were responsive to grid refinement, and non-significant improvement was observed when the grid became much finer. It may be deduced that the triggering mechanism of the IDDES methodology inappropriately responds to grid refinement at high Reynolds numbers.

The wall-normal variation of non-dimensionalized total turbulent kinetic energy (modeled + resolved) and its response to grid refinement, for $k$-$\omega$-SST IDDES and S-A IDDES model, is shown in Figure 11a,b, respectively. The peak of turbulent kinetic energy was pretty well captured by $k$-$\omega$-SST IDDES in the wall vicinity, while S-A IDDES significantly over-predicted the peak magnitude. This was mainly due to the lack of an appropriate near-wall model in S-A IDDES that could lead to uncertainty in calculating the modeled turbulent kinetic energy. More importantly, the amount of uncertainty was considerably higher for a higher Reynolds number. In the core region, both models predicted the total turbulent kinetic energy fairly well, even on the coarse grid.

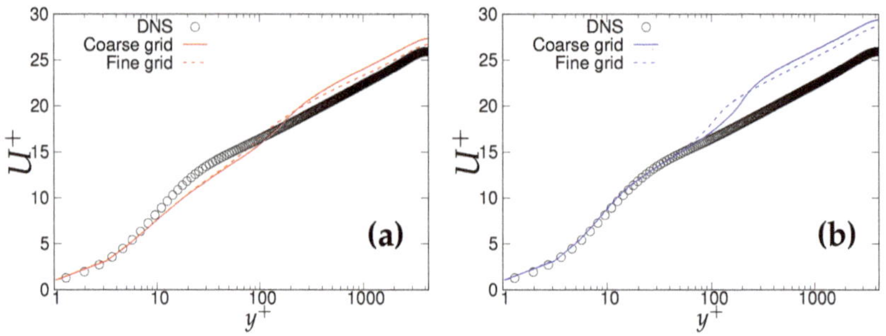


**Figure 9.** Channel flow at $Re_\tau = 4200$: variation of the non-dimensionalized velocity profile along wall-normal direction: (**a**) left column = $k$-$\omega$-SST IDDES, (**b**) right column = S-A IDDES.

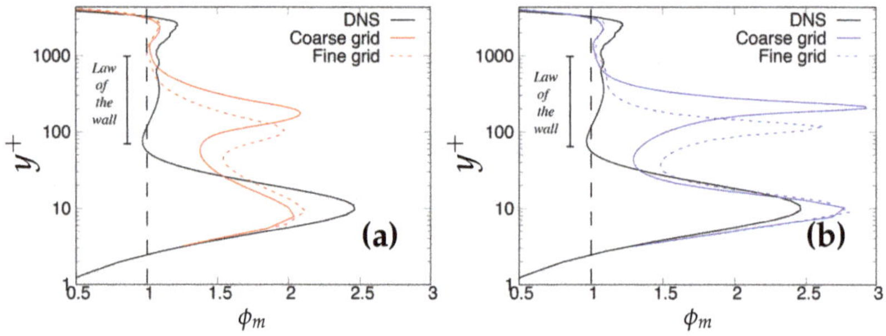


**Figure 10.** Channel flow at $Re_\tau = 4200$: variation of the normalized mean shear ($\phi_m$) as a diagnostic quantity for a log law: (**a**) left column = $k$-$\omega$-SST IDDES; (**b**) right column = S-A IDDES.

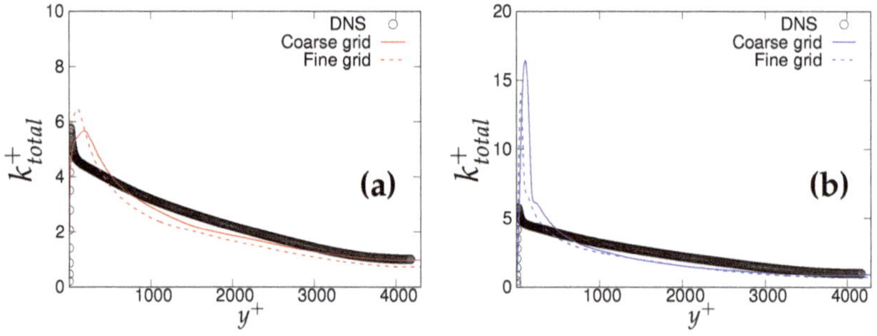


**Figure 11.** Channel flow at $Re_\tau = 4200$: variation of non-dimensionalized total turbulent kinetic energy: (**a**) left column = $k$-$\omega$-SST IDDES; (**b**) right column = S-A IDDES.

The effect of underlying (U)RANS models and grid resolution on the dynamics of the grey area, depicted by the function $\widetilde{f}_d$, is shown in Figure 12. For the coarse grid, shown in Figure 12a,b, the grey area ($0 < \widetilde{f}_d < 1$) was significantly thinner in the case of the S-A IDDES model, compared to the one of the $k$-$\omega$-SST IDDES model, indicating a delay in transition to scale-resolving mode, which might have been due to the more diffusive nature of the underlying RANS model in the $k$-$\omega$-SST IDDES model. Further, the grey area in both models showed only slight sensitivity to grid refinement, as shown in Figure 12c,d. This may explain why grid refinement did not help to rectify the overshoot problem shown in Figure 9.

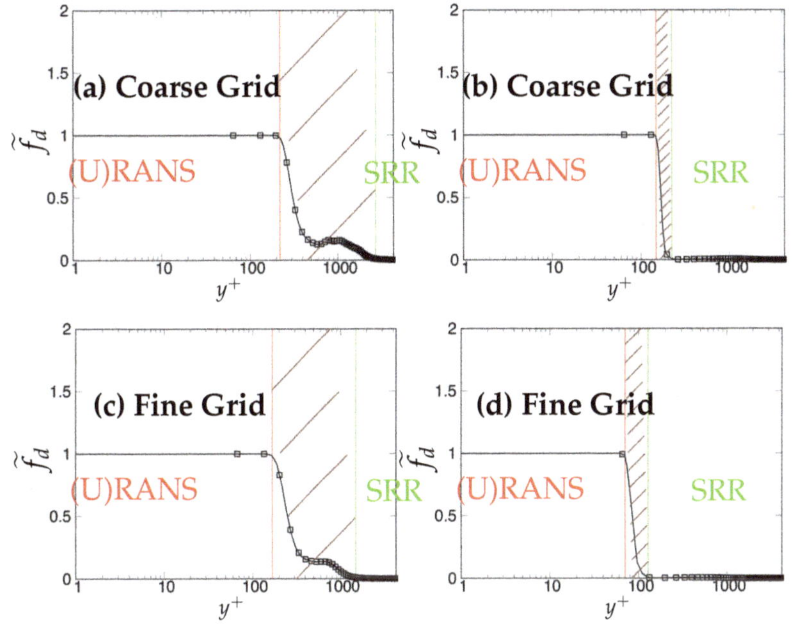


**Figure 12.** Channel flow at $Re_\tau = 4200$: response of $\widetilde{f_d}$ to the grid refinement in the k$\omega$SST IDDES and S-A IDDES framework; (U)RANS = Unsteady RANS, SRR = Scale-Resolving Region and shaded region = grey area; left column = $k$-$\omega$-SST IDDES, right column = Spalart–Allmaras (S-A) IDDES.

The criteria stated in Table 3, for assessment of the scale-resolving region, are shown in Figure 13. The modeled the total turbulent kinetic energy ratio should attain a value of one (theoretically) in the wall vicinity, to prevent intrusion of the resolved scales from scale-resolving simulation to the (U)RANS solving simulation. The intrusion of resolved velocity scales in the (U)RANS region remained persistent under both models and could be observed from Figure 13a, showing around 25% and 50% of intrusion in the case of S-A IDDES and $k$-$\omega$-SST IDDES, respectively. Unlike the $Re_\tau = 395$ case, intrusion is less severe under both models. This can be attributed to the enhanced role of elevating function $f_e$, shown in Figure 14, which seems to be more responsive at a higher Reynolds number.

Figure 13b shows the variation of the ratio of the characteristic cut-off length scale and the Kolmogorov length scale for two different grid resolutions. For coarser grid resolution, the $k$-$\omega$-SST IDDES model demonstrated that there was not appropriate grid support for LES, as the ratio in the core region was approximately one order of magnitude higher. In contrast, S-A IDDES satisfied the criterion, i.e., inaccurately confirmed appropriate grid support for LES. This clearly shows the relevance of sub-grid scale modeling to have an accurate assessment for grid resolution. By grid refinement, both models showed appropriate behavior in the core region and fell in the acceptable values to satisfy the criterion.

The last criterion, i.e., the ratio of the length scale provided by the underlying RANS model when operating in scale-resolving mode to the characteristic cut-off length scale, is shown in Figure 13c and should be of the same order in the scale-resolving region to satisfy the criterion. For S-A the IDDES model, an order increment in the fraction was seen in the core region. This was probably due to the length scale associated with the S-A IDDES model (distance from the wall), which increased while traversing from the wall to the core of the channel. It can be inferred that the S-A IDDES model did not satisfy this criterion at high Reynolds number, as the core region possessed the wrong spectral information. Therefore, it is hard to conclude that S-A IDDES switched to a true LES mode in the core region and could negatively impact the model prediction in the case of complex wall-bounded flows. In contrast, the $k$-$\omega$-SST IDDES model satisfied this criterion as the ratio was of the same order in the core of the channel.

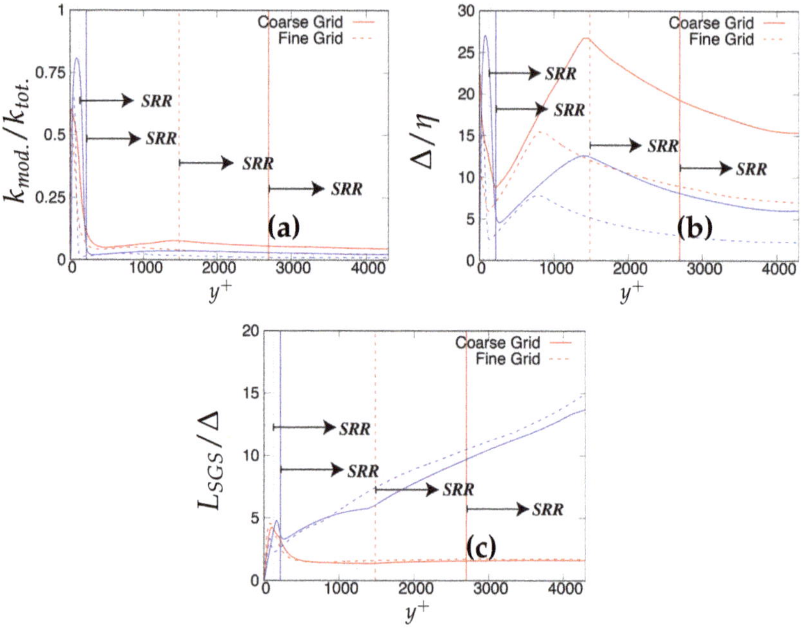


**Figure 13.** Channel flow at $Re_\tau = 4200$: variation of the ratio of (**a**) modeled to total turbulent kinetic energy; (**b**) characteristic cut-off length scale to Kolmogorov length scale and (**c**) integral length scale to characteristic cut-off length scale, along the wall normal direction; SRR = Scale-Resolving Region; red = $k$-$\omega$-SST IDDES, blue = Spalart–Allmaras (S-A) IDDES.

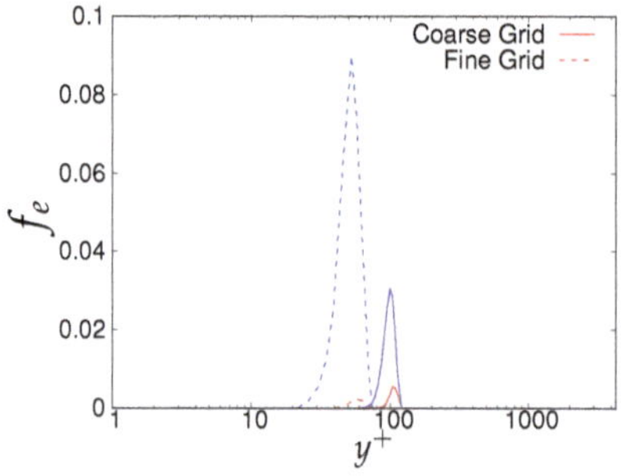


**Figure 14.** Channel flow at $Re_\tau = 4200$: response of elevating function ($f_e$) to the different grid resolution in the $k$-$\omega$-SST IDDES and S-A IDDES framework; red = $k$-$\omega$-SST IDDES, blue = S-A IDDES.

The anisotropy behavior of the $k$-$\omega$-SST IDDES and S-A IDDES models was analyzed qualitatively through the Lumley triangle, under fine grid resolution, and is shown in Figure 15a,b, where red colored arrows signify the starting point on the $\xi$ and $\eta$ plane. It can be seen that all Reynolds stress tensor invariants lied inside the boundaries of the Lumley triangle, which shows that the realizability constraint was well satisfied under both IDDES models. The effect of near-wall modeling (underlying RANS model) can distinctly be seen in the starting point in the Lumley triangle. In the case of the

$k$-$\omega$-SST IDDES model, the starting point resided in the three-component isotropic state, which is compatible with the RANS modeling assumption. However, due to the intrusion model, it was not able to retain it within the whole (U)RANS region, which was about $y^+ \approx 100$. In contrast, S-A IDDES did not show any indication of RANS-like behavior, even very close to wall. Further, for both models, a tendency to reach the two-component turbulence state (completely opposite to DNS) was observed from $y^+ \approx 70$ to $y^+ \approx 1000$ and is shown in the zoomed view in Figure 15a,b. This region overlaps with the grey area for both models, which may indicate that inaccurate prediction of the dynamics of the grey area was contributing to this behavior. Moreover, this behavior confirms again that the near-wall model did not have much effect on overall model performance.

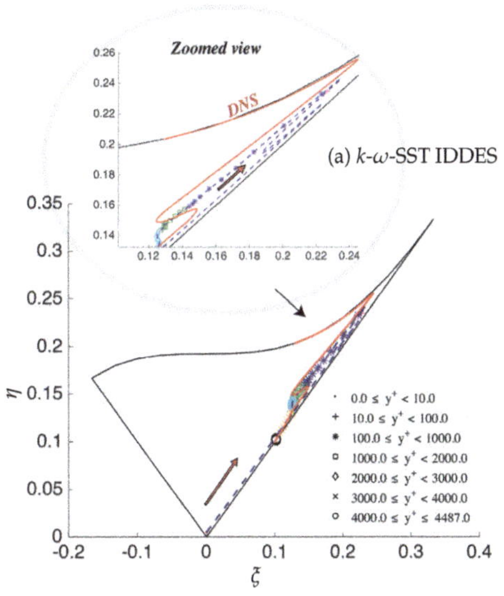


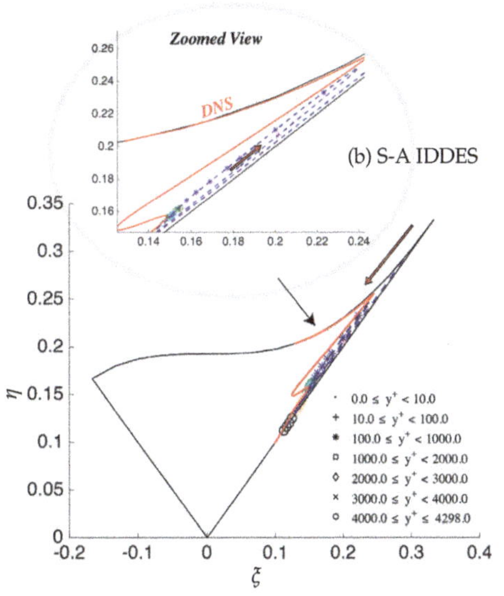


**Figure 15.** Channel flow at $Re_\tau = 4200$: anisotropy invariant map of fine grid resolution along the wall-normal direction; solid red line = DNS and points and dashed line = IDDES.

### 6.2. Periodic Hill Flow

The numerical modeling of flow separation around smoothly curved surfaces is challenging as compared to sharp-edge separation. This is mainly due to the prediction of a separation point or line, which is not fixed in space and is very sensitive to parameters like external flow properties, turbulence level, and development of a streamwise pressure gradient. The IDDES methodology, as one of the widely-used hybrid RANS-LES methods, is applied to periodic hill flow at two different Reynolds numbers. The main focus is to assess the capability of the method only to predict this flow with respect to the underlying RANS model.

#### 6.2.1. $Re_b = 10{,}595$

Figure 16 shows different streamwise locations, i.e., $x/H = 0.05, 2, 6,$ and 8, which are chosen for analyzing first and second order statistics, since these locations dictate the most critical physics in this flow configuration as reported in the experimental investigation of Rapp and Manhart [15] and the LES studies of Breuer et al. [16]. Note that the grid resolution used in LES prediction [16] is fine enough to resolve the near-wall dynamics. Figure 17 depicts the mean velocity profiles at the above-mentioned four streamwise locations. At $x/H = 0.05$, the flow acceleration in the lower wall vicinity was fairly well predicted by both models with marginal differences between fine and coarse grids. At the next location, $x/H = 2$, where the interaction of free shear layer separating from the crest of the hill and the reverse flow exists, mean streamwise velocity was well captured with moderate differences between fine and coarse grids. At $x/H = 6$, after reattachment, both models were able to capture the flow recovery from the low-energy separated region accurately on the fine grid. The flow started accelerating on the windward side of the hill at $x/H = 8$, and again, the behavior of both models lied between LES predictions and experimental measurements, with no significant sensitivity to grid refinement observed.

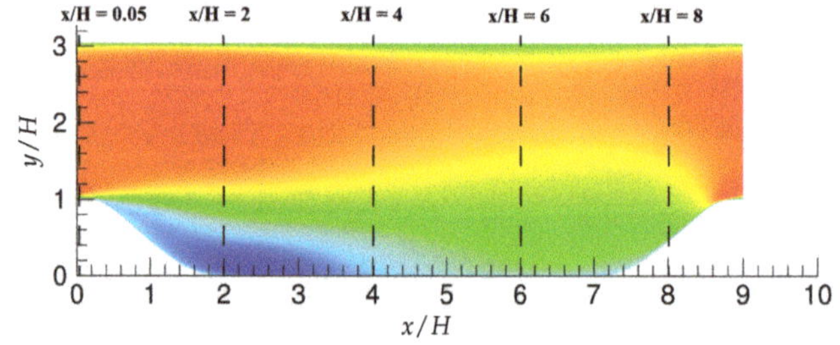


**Figure 16.** Periodic hill flow: considered streamwise locations, $x/H$ = 0.05, 2, 6, and 8 for $Re_b$ = 10,000 and $x/H$ = 0.05, 2, 4 and 8 for $Re_b$ = 37,000.

Figures 18–20 show the streamwise, wall normal and shear stresses at the same locations. The total stress was computed as the sum of modeled and resolved stresses. Inside the recirculation zone, none of the models were able to capture the peak streamwise stress distribution, shown in Figure 18b. However, predictions were improved after flow reattachment, and results from both models were in fair agreement with LES predictions and experimental measurements. Further, a strong grid dependency on streamwise stresses could be seen after the reattachment location. The wall-normal stress profiles are shown in Figure 19. It can be seen that overall (except for $x/H = 0.05$), the $k$-$\omega$-SST IDDES model delivered more accurate results, which showed more grid sensitivity compared to the streamwise stress. Discrepancies at the first location might have been due to the interference of the periodic boundary condition with model performance. Furthermore, a significant difference in the experiment [15] confirms the interference. Figure 20 depicts shear stress profiles at all four streamwise locations. Overall, good agreement with LES predictions and experimental measurements was observed for both models. Another important observation was that the shear stress in the vicinity of the lower wall remained the same within the S-A IDDES and $k$-$\omega$-SST IDDES models, denoting insensitivity towards the near-wall model.

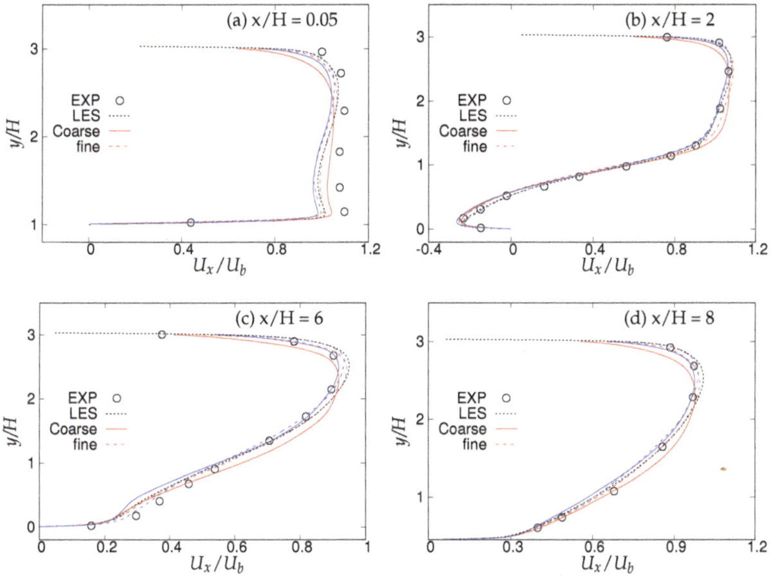


**Figure 17.** Periodic hill flow at $Re_b = 10{,}590$: profiles of mean streamwise velocity at four different axial locations; red = $k$-$\omega$-SST IDDES, blue = Spalart–Allmaras (S-A) IDDES.

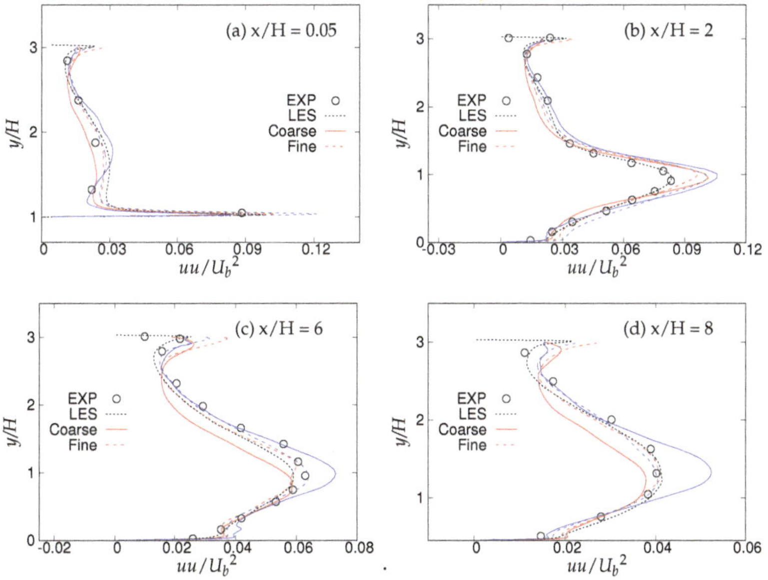


**Figure 18.** Periodic hill flow at $Re_b = 10{,}590$: profiles of streamwise stress at four different axial locations; red = $k$-$\omega$-SST IDDES, blue = Spalart–Allmaras (S-A) IDDES.

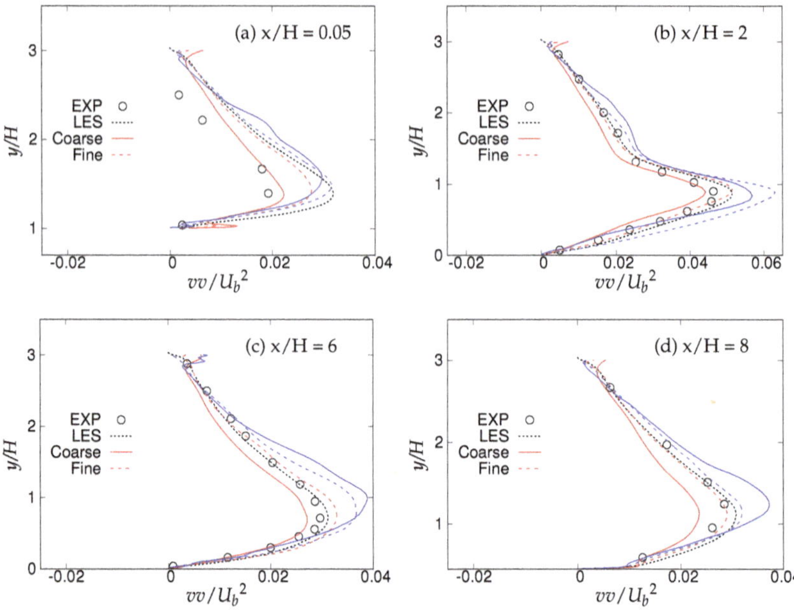


**Figure 19.** Periodic hill flow at $Re_b = 10{,}590$: profiles of wall-normal stress at four different axial locations; red = $k$-$\omega$-SST IDDES, blue = Spalart–Allmaras (S-A) IDDES.

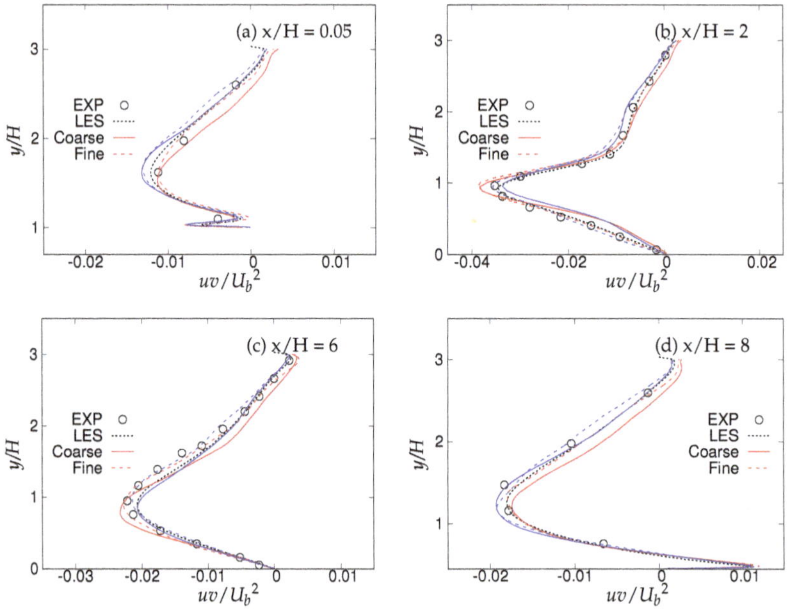


**Figure 20.** Periodic hill flow at $Re_b = 10{,}590$: profiles of shear stress at four different axial locations; red = $k$-$\omega$-SST IDDES, blue = Spalart–Allmaras (S-A) IDDES.

Figure 21 shows the variation of skin friction coefficient over the lower wall-region for fine grid resolution. It can be seen that re-attachment locations predicted by both models were in a good agreement with both reference data, as shown in Table 4.

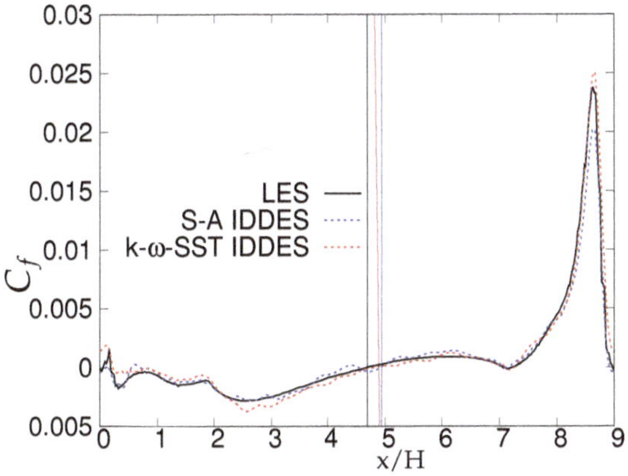


**Figure 21.** Periodic hill flow at $Re_b = 10{,}590$: distribution of averaged skin friction coefficient for the fine grid; vertical lines denote the reattachment point.

**Table 4.** Dependency of grid-resolution on the reattachment location.

| Cases | Modeling Framework | Grid-Resolution | $(x/h)_{Reattach.}$ |
|---|---|---|---|
| $Re_b = 10{,}000$ | One-equation model | $100 \times 100 \times 30$ | 4.8172 |
| | | $150 \times 100 \times 60$ | 5.021 |
| | Two-equation model | $100 \times 100 \times 30$ | 4.6072 |
| | | $150 \times 100 \times 60$ | 4.9166 |
| Fröhlich et al. (2005) | LES | | 4.6 |
| Rapp and Manhar (2011) | Experimental location | | 4.21 |
| $Re_b = 37{,}000$ | One-equation model | $100 \times 100 \times 30$ | 3.9578 |
| | | $150 \times 100 \times 60$ | 4.2922 |
| | Two-equation model | $100 \times 100 \times 30$ | 3.9578 |
| | | $150 \times 100 \times 60$ | 4.7078 |
| Rapp and Manhar (2011) | Experimental location | | 3.76 |

The behavior of the $\widetilde{f_d}$ function responsible for transition from (U)RANS to scale-resolving mode under coarse and fine grids is shown in Figures 22 and 23, respectively. In spite of shear layer instabilities emanating from the crest of the hill, the grey area predicted by the coarse grid resolution of $k$-$\omega$-SST IDDES model was significantly larger than S-A IDDES. This might be an indication of the diffusive nature of the underlying RANS model in $k$-$\omega$-SST IDDES. However, the grey area in S-A IDDES remained minimal at x/H = 0.05 and 2, signifying the strong influence of shear layer instabilities, and further downstream, broadened marginally at x/H = 6 and 8. For fine grid resolution, (U)RANS, as well as the grey region reduced significantly in the $k$-$\omega$-SST IDDES model, but it was still considerably wider at locations where flow was attached, i.e., x/H = 0.05 and 8. However, for S-A IDDES, the grey area remained nearly constant at all four streamwise locations after grid refinement and only was shifted towards the lower wall. The insensitivity towards oncoming flow instabilities in the $k$-$\omega$-SST IDDES model, under the fine grid resolution, could be seen from the extended attached shear layer convecting downstream from the crest of the hill, as shown in Figure 24a, whereas in the S-A IDDES model, the shear layer was highly unstable and may have been the reason for the swift transition to the scale-resolving mode.

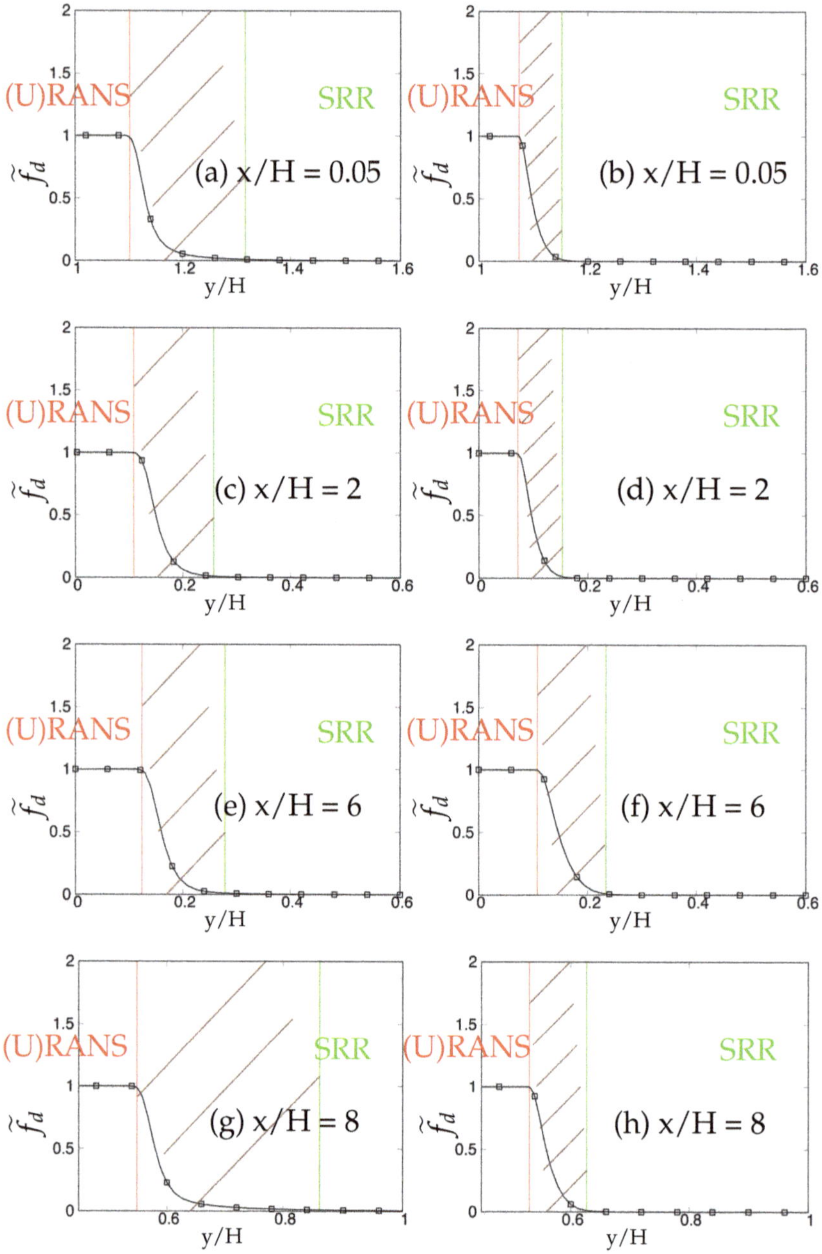


**Figure 22.** Periodic hill flow at $Re_b = 10{,}590$: Response of $\widetilde{f}_d$ to coarse grid resolution; (U)RANS = Unsteady RANS and SRR = Scale-Resolving Region; left column = $k$-$\omega$-SST IDDES, right column = Spalart–Allmaras (S-A) IDDES.

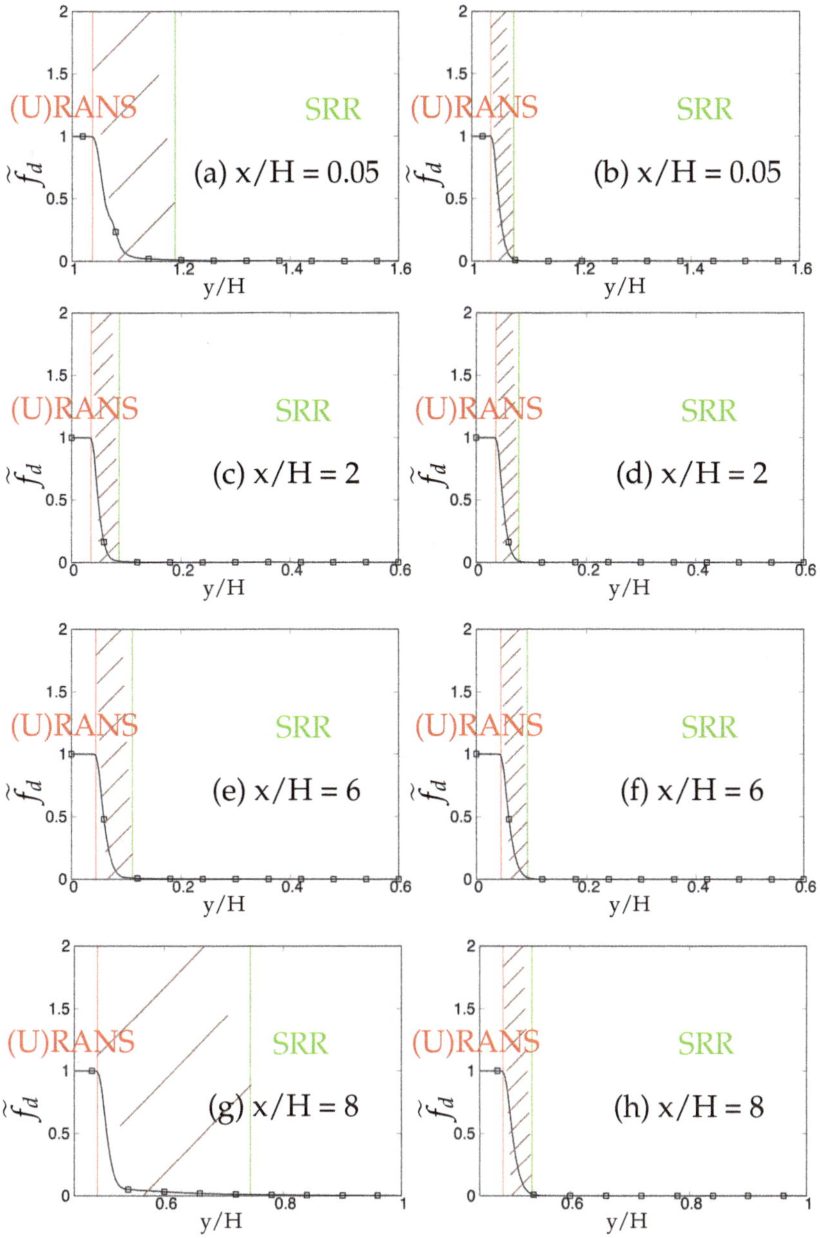


**Figure 23.** Periodic hill flow at $Re_b$ = 10,590: response of $\widetilde{f}_d$ to fine grid resolution; (U)RANS = Unsteady RANS and SRR = Scale-Resolving Region; left column = *k*-*ω*-SST IDDES, right column = Spalart–Allmaras (S-A) IDDES.

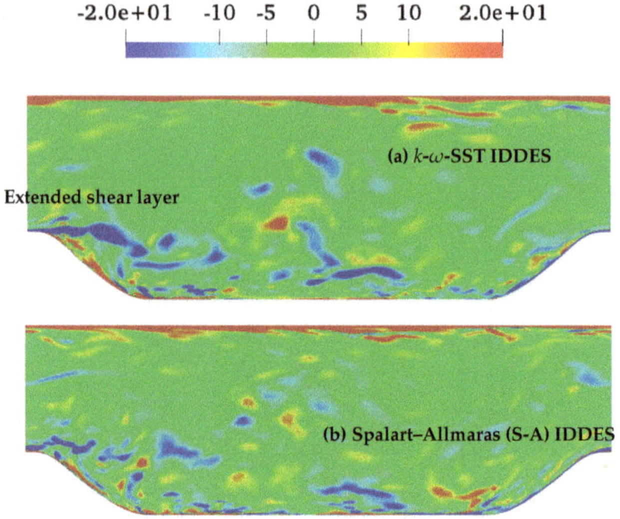


**Figure 24.** Periodic hill flow at $Re_b = 10{,}590$: instantaneous field of spanwise vorticity for (**a**) $k$-$\omega$-SST IDDES and (**b**) Spalart–Allmaras (S-A) IDDES, with fine grid resolution.

Now, the criteria listed in Table 3 for assessing the quality and reliability of the scale-resolving region under different underlying (U)RANS models will be analyzed. Firstly, these criteria were applied to the coarse grid resolution and are shown in Figure 25. It can be seen from Figure 25a,b, where the ratio of modeled to total turbulent kinetic energy close to lower wall region signifies the intrusion from the scale-resolving region to the (U)RANS region is persistent among both models at all four streamwise locations. Similar to turbulent channel flow, the intrusion of the scale-resolved simulation could be attributed to the weak shielding of the (U)RANS region provided by the elevating function $f_e$. Further, from near-wall peak ratio, it is observed that the amount of intrusion for S-A IDDES was more severe as compared to $k$-$\omega$-SST IDDES. As discussed in Section 4, the current Reynolds number is fairly low, and therefore, Criterion 3 in Table 3 was not applied in this section. The ratio of the sub-grid to characteristic cut-off length scale is shown in Figure 25c,d and is expected to be of the same order in the scale-resolving region to retain the spectral consistency. This criterion was satisfied by both models, depicting that the correct length scale is used in the scale-resolving mode by both models.

When applied to the fine grid resolution (shown in Figure 26), these criteria responded with different sensitivity under the given underlying (U)RANS model. Overall, intrusion of the scale resolving simulation into the (U)RANS region estimated from the ratio of modeled to total turbulent kinetic energy increased further with grid refinement. Similar to the coarse grid, it was more severe under the S-A IDDES model as compared to the $k$-$\omega$-SST IDDES model.

The ratio of the sub-grid to characteristic cut-off length scale increased to one order magnitude higher at $x/H$ = 2 and 6 for the S-A IDDES model and can be seen in Figure 26d. This outcome may allows us to conclude that in the S-A IDDES model, the characteristic length scale did not correspond to the appropriate sub-grid scale when the model was operating as a sub-grid scale model. In contrast, in the $k$-$\omega$-SST IDDES model, the ratio remained at the same order at all four streamwise locations, confirming the appropriate length scale in scale-resolving mode.

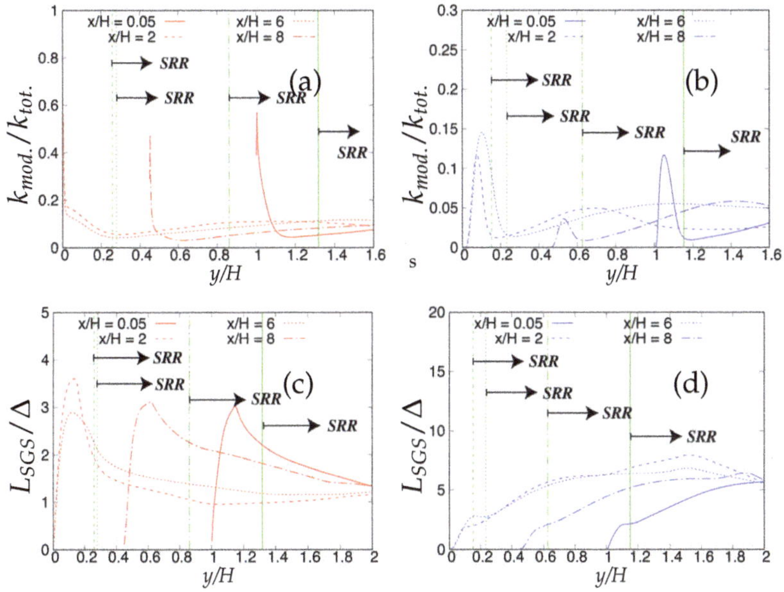


**Figure 25.** Periodic hill flow at $Re_b = 10{,}590$: variation of the ratio of (**a**,**b**) modeled to total turbulent kinetic energy and (**c**,**d**) sub-grid length scale to characteristic cut-off length scale, along the wall normal direction under coarse grid resolution; SRR = Scale-Resolving Region; left column = $k$-$\omega$-SST IDDES, right column = Spalart–Allmaras (S-A) IDDES.

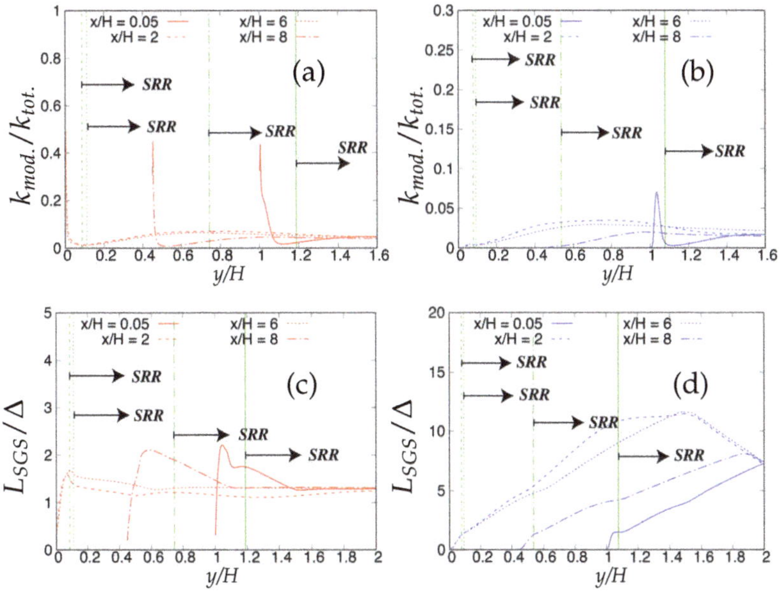


**Figure 26.** Periodic hill flow at $Re_b = 10{,}590$: variation of the ratio of (**a**,**b**) modeled to total turbulent kinetic energy and (**c**,**d**) sub-grid length scale to characteristic cut-off length scale, along the wall normal direction under fine grid resolution; SRR = Scale-Resolving Region; left column = $k$-$\omega$-SST IDDES, right column = Spalart–Allmaras (S-A) IDDES.

Figure 27 shows the anisotropy invariant map at four streamwise locations in the flow direction for fine grid resolution. Important information about the anisotropic/isotropic states can be inferred from these plots when compared with highly resolved LES predictions [22]. First, all the data for the Reynolds stress tensor invariants lied inside the Lumley triangle, therefore satisfying the realizability constraint. Under S-A IDDES, the starting point was located at two-component turbulence state at all four streamwise locations. Up to $x/H \leq 6$, the two-component turbulence state was achieved through the axisymmetric contraction line, while this approach was along the axisymmetric expansion line for $x/H = 8$. This behavior of the S-A IDDES model is consistent with the highly resolved LES findings of Fröhlich et al. [22]. However, the two-component turbulence state was never achieved within the $k$-$\omega$-SST IDDES model while traveling towards the lower wall region from the core of the channel. At attached flow regions ($x/H = 6$ and 8), the Reynolds stress invariant map commenced from the near-isotropic line, i.e., the line crossing the point where $\xi$ and $\eta$ equal zero, indicating the underlying (U)RANS assumption of isotropic turbulence being well satisfied in $k$-$\omega$-SST IDDES. The Lumley triangle provides qualitative assessment of flow anisotropy, while the quantitative measure of the Reynolds stress invariants can be assessed through the flatness parameter.

The flatness parameter ($A$) is shown in Figure 28, also proposed by Lumley, combining Reynolds stress invariants using the following expression:

$$A = 1 + 9\left(\frac{b_{ij}b_{jk}b_{ki}}{3} - \frac{b_{ij}b_{ij}}{2}\right) \tag{16}$$

The value of $A$ went to one for isotropic flow and defined the two-component turbulence state at $A = 0$. As shown in Figure 28, both models indicated more isotropic behavior away from solid walls compared to LES [22]. The flatness parameter ($A$) profiles for S-A IDDES depicted that Reynolds stress invariants have the tendency to reach the two-component turbulence state at the lower wall region. However, for the $k$-$\omega$-SST IDDES model, the flow near the lower wall region closely resembled the isotropic state, which is consistent with assumption of isotropic turbulence in two-equation (U)RANS model concept. This clearly explains the effect of near-wall RANS model, which was shown to be more appropriate in the case of $k$-$\omega$-SST IDDES, as the near-wall dynamics was supposed to be captured using the (U)RANS method.

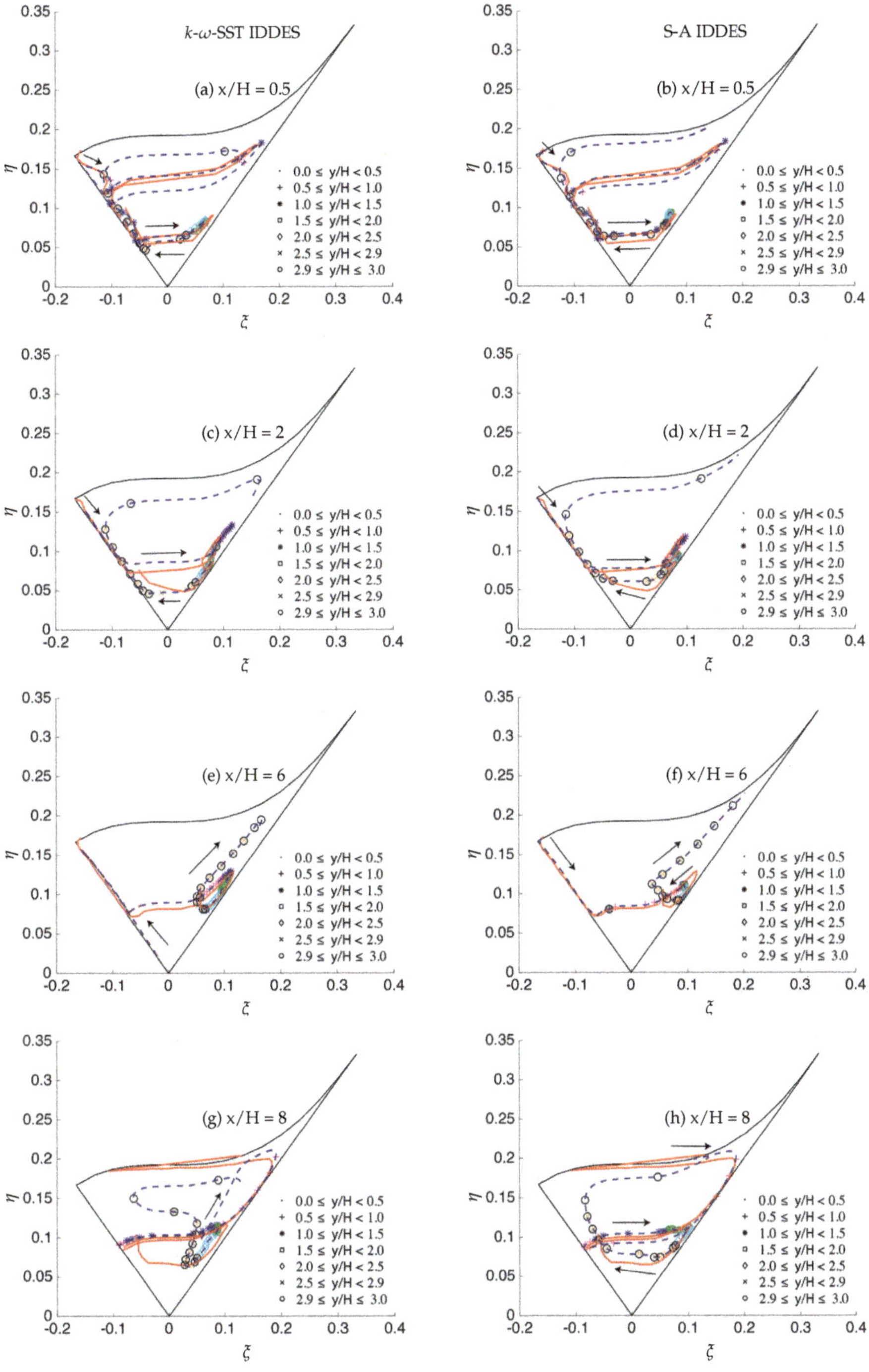


**Figure 27.** Periodic hill flow at $Re_b = 10{,}590$: anisotropy invariant map for fine grid resolution along the wall-normal direction; solid red = well-resolved LES, points and dashed lines = IDDES.

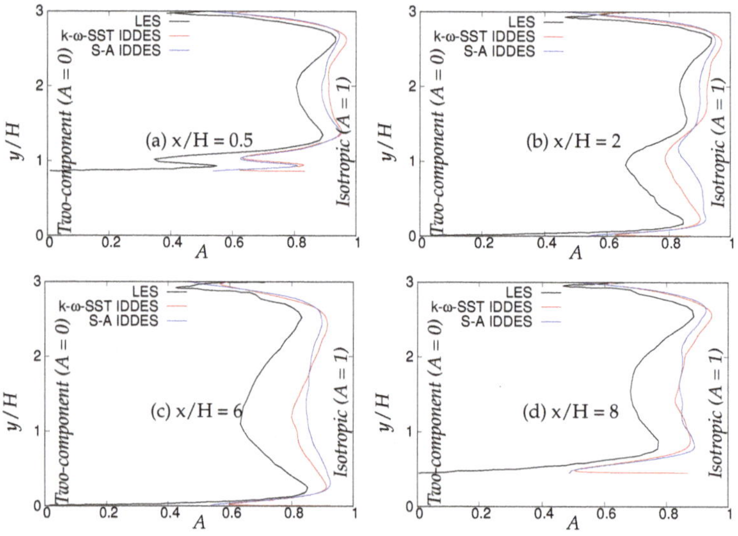


**Figure 28.** Periodic hill flow at $Re_b = 10{,}590$: variation of the flatness parameter, $A$, at four streamwise locations.

6.2.2. $Re_b = 37{,}000$

We now proceed with examining the overall performance of the underlying (U)RANS model in response to the grey area and assessment of grid refinement for higher a bulk Reynolds number of 37,000. Indeed, at this Reynolds number, LES computations are prohibitively expensive and have not been performed in the literature. Therefore, the IDDES methodology will be evaluated using the experimental measurement provided by Rapp and Manhart [15].

We first investigate the effects of the underlying (U)RANS models and grid resolution on first-order flow statistics. The recirculation bubble became smaller with increasing Reynolds number [15]; therefore, we compared predictions with experimental measurements at streamwise locations (x/H) equal to 0.05, 2, 4 and 8, for this Reynolds number. Figure 29 shows the streamwise velocity profiles at four locations of x/H = 0.05, 2, 4 and 8 using coarse and fine grid resolutions. The flow acceleration close to the lower wall at x/H = 0.05 was under-predicted by both models, and prediction showed poor performance with increasing grid resolution. However, part of the disagreement might be due to the interference of the periodic boundary condition, as discussed before. The velocity profile in the wall vicinity was well predicted inside the recirculation region, at x/H = 2, by both models, with marginal discrepancies. According to experimental reference [15], the flow was expected to reattach at x/H = 4, but the velocity profile in the vicinity of lower wall at x/H = 4 (shown in Figure 29c) was under-predicted, which may have caused slow recovery from the upstream negative energy flow. x/H = 8 corresponds to the post-reattachment region, wherein the flow recovered from the upstream separated flow and the velocity profile was fairly well predicted by both models. Overall streamwise velocity predictions were in good agreement with experimental measurements with minor discrepancies.

Figures 30–32 show the components of the stress tensor at the four different streamwise locations. The effect of the grid sensitivity can be easily seen in the second-order statistics, where major discrepancies in the core region are shown under coarse grid resolution. Predictions obtained from the fine grid resolution showed the tendency to follow experimental measurements in the core region, however without an acceptable level of accuracy. The oscillating behavior of the results obtained on coarse grid, particularly in the near-wall region, may be indicative of detrimental effect of intrusion from the scale-resolving region into the (U)RANS region, allowing the model to resolve structures on a grid that is too coarse. This was more severe in the case of S-A IDDES, as the shielding function $f_e$ was weaker, as confirmed previously. This behavior strongly implies the negative effect of grid resolution on the IDDES methodology. The slow flow-reattachment process discussed earlier was confirmed by the plot of the friction coefficient shown in Figure 33, indicating a larger recirculation region predicted under the IDDES methodology. Unfortunately, at a higher Reynolds number of 37,000, the LES results were not available for quantitative comparison. Another interesting observation of this plot is the flow behavior right after the reattachment region. After reattachment and partial recovery from the negative energy flow, the flow appeared to be prone to separation at around $x/H \approx 7.5$, where flow decelerated while moving towards the downstream hill, resulting in the local minimum in the friction coefficient plot. However, a further sharp rise in the friction coefficient was observed at the end of periodic hill.

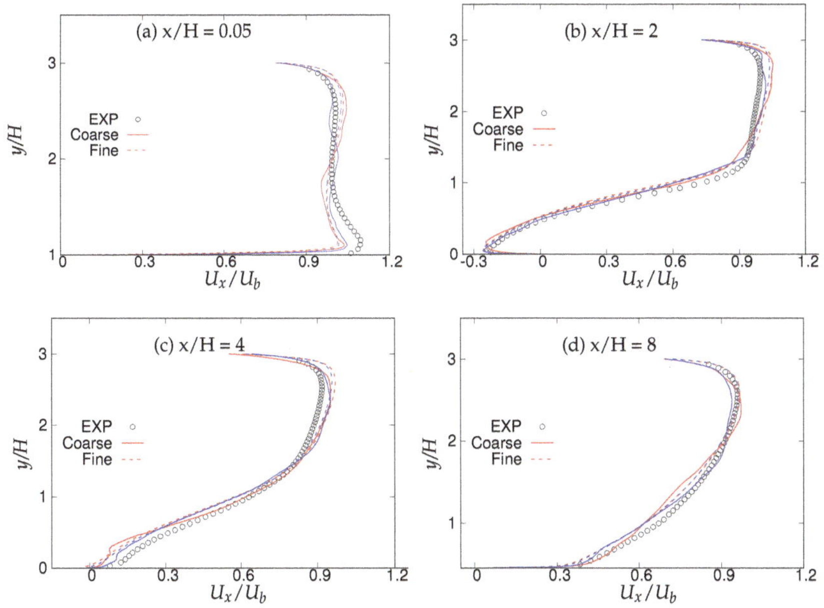


**Figure 29.** Periodic hill flow at $Re_b = 37{,}000$: profiles of mean streamwise velocity at four different axial locations; red = $k$-$\omega$-SST IDDES, blue = Spalart–Allmaras (S-A) IDDES.

The dependency of the $\tilde{f}_d$ function on two different underlying (U)RANS models and grid refinement is shown in Figures 34 and 35. At $Re_b = 37{,}000$, the shear layer emanating from the crest of the hill was expected to be highly unstable and should have provided a substantial amount of flow instabilities to allow a swift transition from (U)RANS to the scale-resolving mode with a minimal grey area, especially in the recirculation region at $x/H = 2$. Within coarse grid resolution (shown in Figure 34), the grey region obtained at all four streamwise locations in the $k$-$\omega$-SST IDDES model was significantly larger than the S-A IDDES model, indicating the prolonged transition from (U)RANS to scale-resolving mode in the $k$-$\omega$-SST IDDES model. After grid refinement (shown in Figure 35), the grey area was extended within the $k$-$\omega$-SST IDDES model at $x/H = 2$ and 4, demonstrating the inconsistent behavior of the model. It is concluded that the grey area obtained from the IDDES methodology using the advanced underlying (U)RANS model did not respond appropriately to grid refinement and in fact further grew with increasing Reynolds number under separated flows. On the other hand, the grey area in the S-A IDDES model after grid refinement stayed nearly constant or insensitive and only shifted spatially towards the decreasing wall-normal distance, which was not an expected outcome in the case of systematic-eddy-resolving simulation.

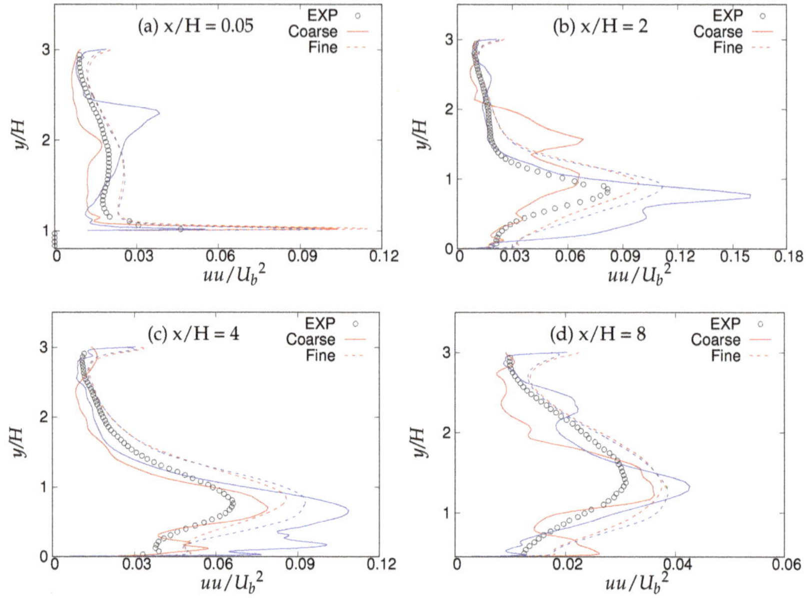


**Figure 30.** Periodic hill flow at $Re_b$ = 37,000: Profiles of streamwise stress at four different axial locations; red = $k$-$\omega$-SST IDDES, blue = Spalart–Allmaras (S-A) IDDES.

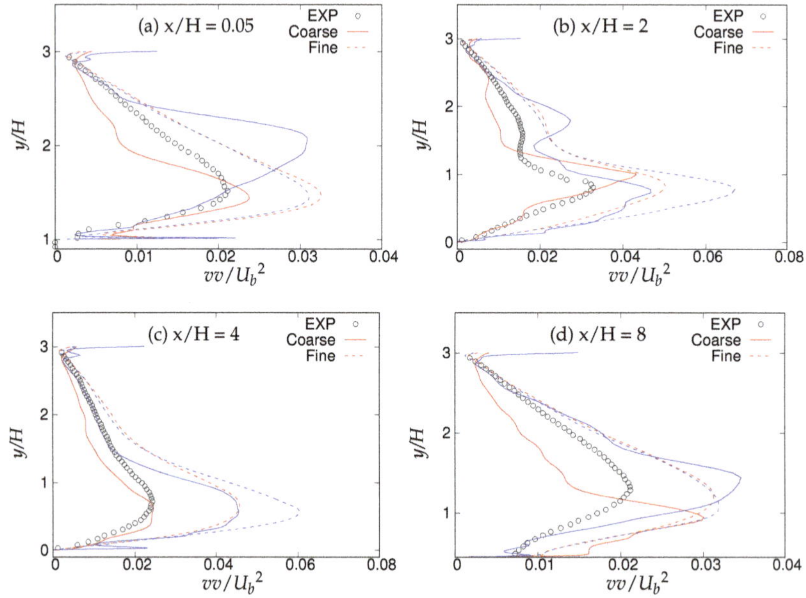


**Figure 31.** Periodic hill flow at $Re_b$ = 37,000: profiles of wall-normal stress at four different axial locations; red = $k$-$\omega$-SST IDDES, blue = Spalart–Allmaras (S-A) IDDES.

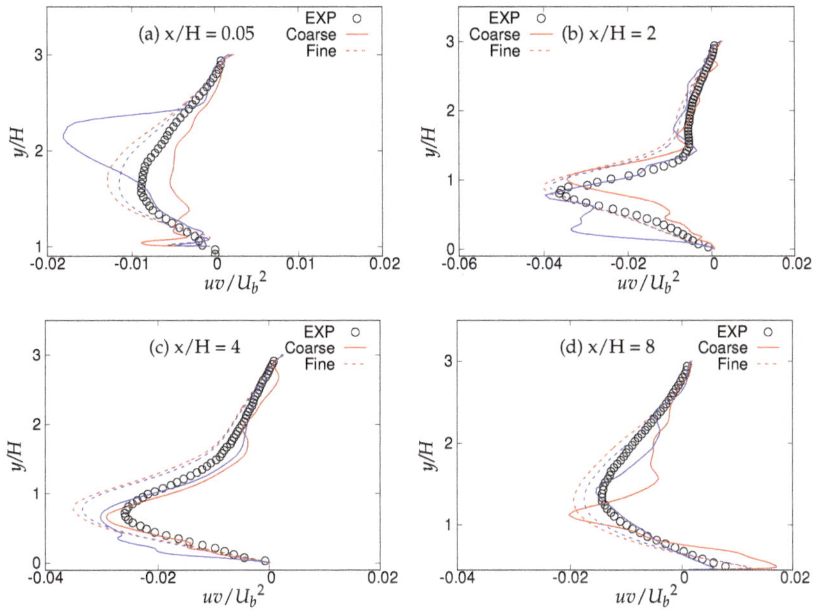


**Figure 32.** Periodic hill flow at $Re_b = 37{,}000$: profiles of shear stress at four different axial locations; red = $k$-$\omega$-SST IDDES, blue = Spalart–Allmaras (S-A) IDDES.

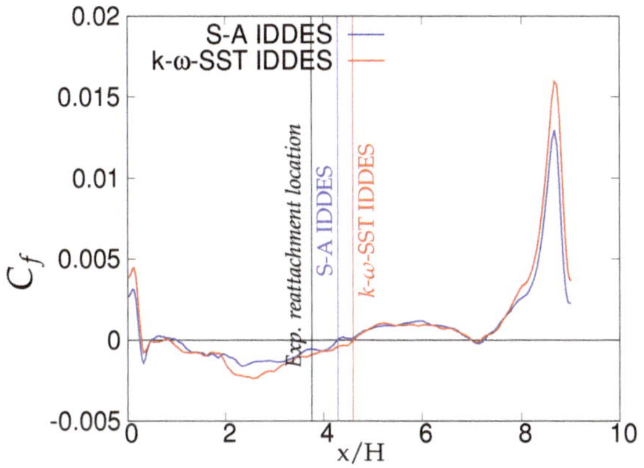


**Figure 33.** Periodic hill flow at $Re_b = 37{,}000$: distribution of the averaged skin-friction coefficient for fine grid resolution; red = $k$-$\omega$-SST IDDES, blue = Spalart–Allmaras (S-A) IDDES.

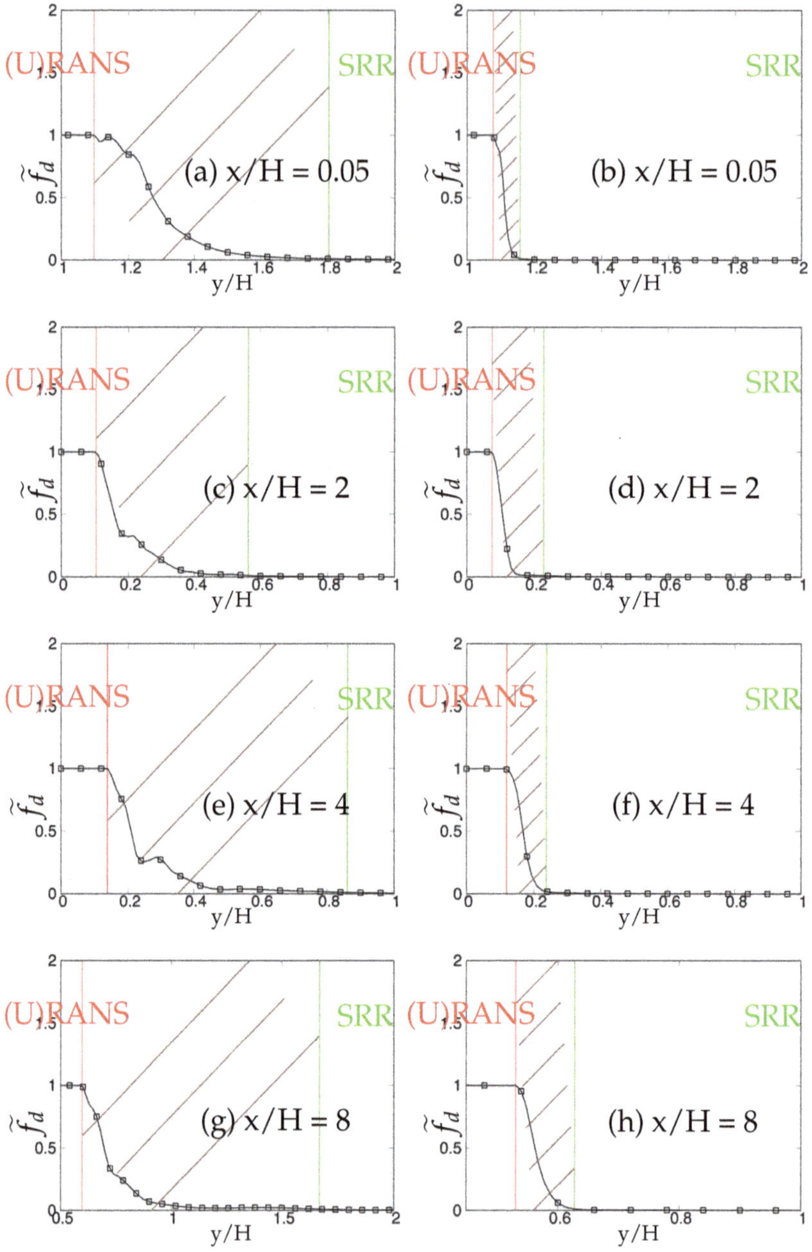


**Figure 34.** Periodic hill flow at $Re_b = 37{,}000$: response of the $\widetilde{f}_d$ function to coarse grid resolution; left column = $k$-$\omega$-SST IDDES, right column = Spalart–Allmaras (S-A) IDDES.

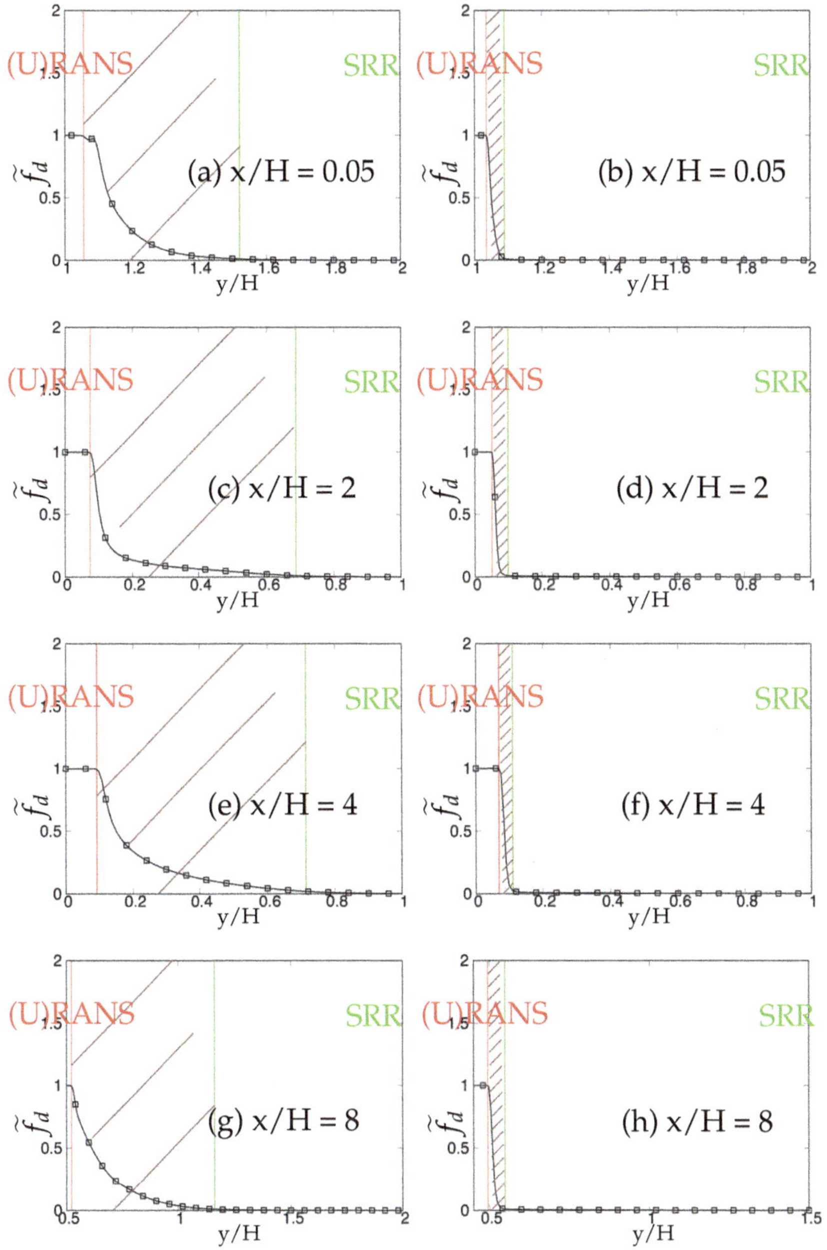


**Figure 35.** Periodic hill flow at $Re_b = 37{,}000$: response of the $\widetilde{f_d}$ function to fine grid resolution; left column = $k$-$\omega$-SST IDDES, right column = Spalart–Allmaras (S-A) IDDES.

In the next step, the criteria listed in Table 3 were investigated on coarse and fine grid resolutions, subsequently. As one of the major issues we have seen so far, the intrusion of scale-resolving simulation into the (U)RANS simulation could be seen from the ratio of modeled to total turbulent kinetic energy in coarse grid resolution under both models, shown in Figure 36a,b. The average amount of intrusion was around 60% and 75% in the $k$-$\omega$-SST IDDES and S-A IDDES models, respectively. While comparing with the periodic hill flow case at $Re_b = 10,000$, it is observed that the shielding effect seemed more responsive for S-A IDDES at higher Reynolds number, resulting in a lesser amount of intrusion, whereas the opposite trend was seen under the $k$-$\omega$-SST IDDES model at a higher Reynolds number. This behavior under different underlying (U)RANS models was consistent with the channel flow case at higher Reynolds number.

Further, the ratio of the characteristic cut-off length scale and Kolmogorov length scale for coarse grid is shown in Figure 36c,d. First of all, the maximum peak ratios signify that the most dissipation occurred inside the scale-resolving region, and next, the level of ratios in the core region compares the accuracy of sub-grid scale models used in each model. However, calculation states that the current grid resolution was coarse for LES/scale-resolving simulation in the scale-resolving region. During the sub-grid scale operation mode under the coarse grid, both models satisfied the third criterion in Table 3, i.e., the ratio of sub-grid length scale and the characteristic cut-off length scale, shown in Figure 36e,f, was of the same order at all four streamwise locations.

Under fine grid resolution, the severity of the intrusion of scale-resolved simulation into (U)RANS simulation could be seen from the ratio of modeled to total turbulent kinetic energy in Figure 37a,b, especially in the S-A IDDES model. The variation of the ratio of the characteristic cut-off length scale and Kolmogorov length scale is shown in Figure 37c,d. It is important to note here that the current grid resolution was extremely coarse to address the well-resolved LES region at such a high Reynolds number. For $k$-$\omega$-SST IDDES, the ratio in the scale-resolving region indicated that grid support was not sufficient for LES simulation at all four streamwise locations, whereas for S-A IDDES, the level of the ratio falsely satisfied the criterion while traversing to the core region, as the current grid resolution is not fine enough for well-resolved LES simulation. Comparing the ratios in the $k$-$\omega$-SST IDDES and S-A IDDES models demonstrates the importance of sub-grid scale modeling to have an accurate assessment for grid resolution. An inappropriate sub-grid length scale to characteristic length scale was obtained in the S-A IDDES model, as shown in Figure 37f. Like the turbulent channel flow discussed in the previous section of channel flow at $Re_\tau = 4200$, a higher Reynolds number flow defies the systematic-eddy-resolving approach in this case, as well, mostly due to the length scale associated with the S-A IDDES model, which is simply the distance from the wall.

Based on the results obtained from the channel flow and periodic hill flow configuration, we have found that choosing the more advanced underlying (U)RANS model for accurate modeling of near-wall dynamics was insensitive to the flow instabilities under the IDDES methodology. Therefore, a significant amount of grey area was obtained using the advanced underlying (U)RANS model, while the grey area obtained from the simple (U)RANS model, i.e., one-equation model, remained minimal and of nearly constant thickness at different locations in the flow. Overall, it can be concluded that irrespective of grid resolution, neither models showed the characteristics of being the systematic-eddy-resolving approach in the scale-resolving region.

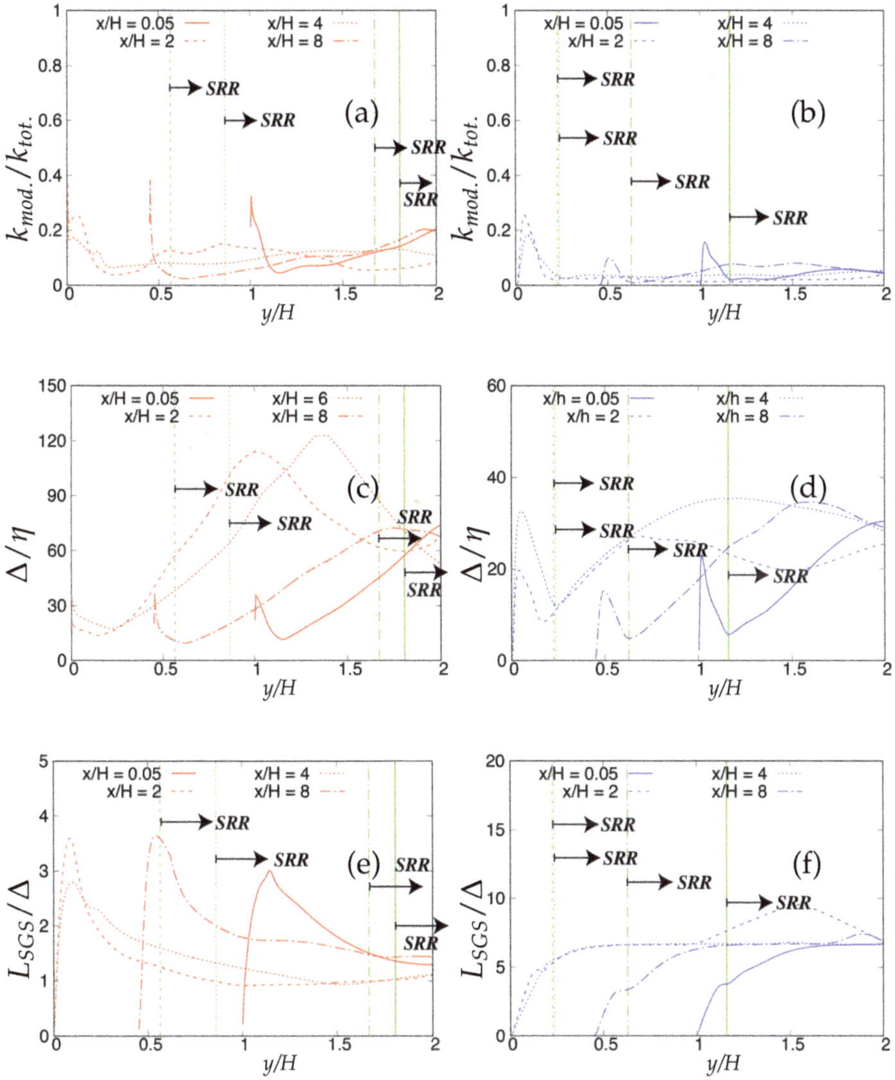


**Figure 36.** Periodic hill flow at $Re_b = 37{,}000$: variation of the ratio of (**a**,**b**) modeled to total turbulent kinetic energy; (**c**,**d**) the characteristic cut-off length scale to the Kolmogorov length scale and (**e**,**f**) the sub-grid length scale to characteristic the cut-off length scale, at four different streamwise locations under coarse grid resolution; SRR = Scale-Resolving Region; left column = $k$-$\omega$-SST IDDES, right column = Spalart–Allmaras (S-A) IDDES.

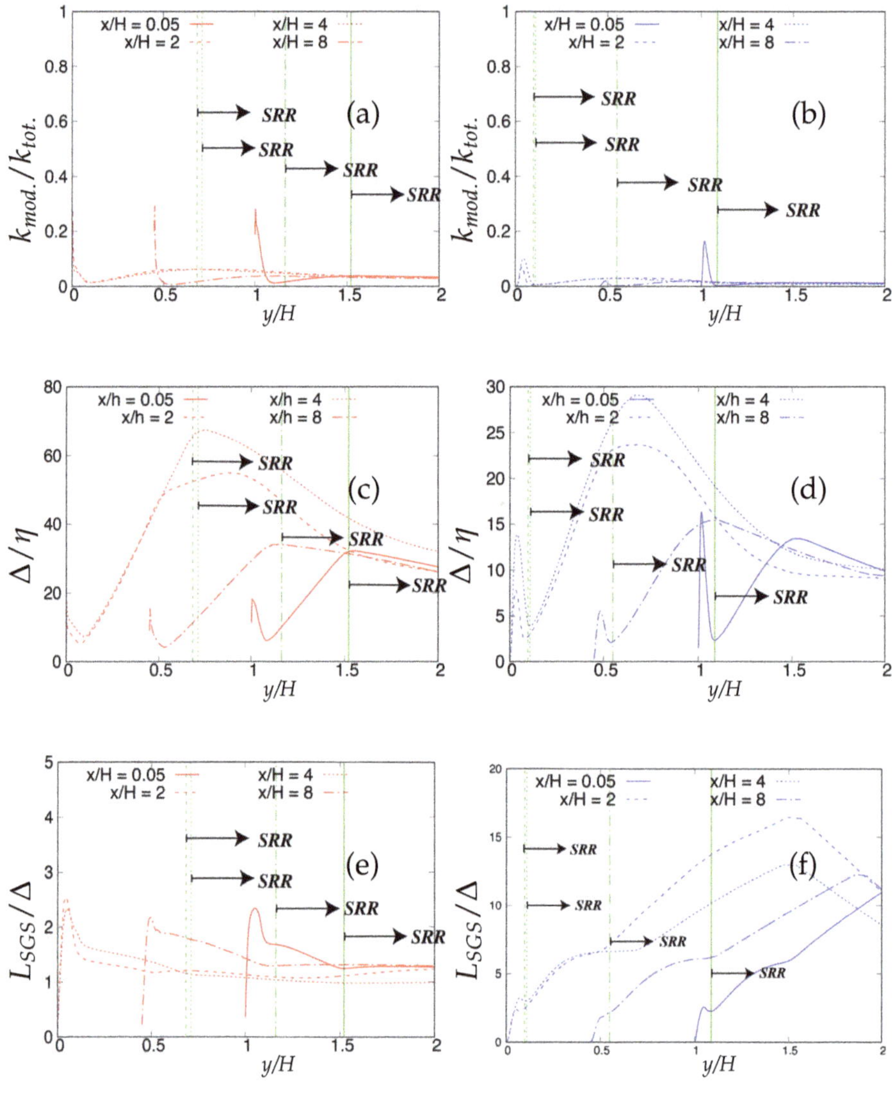


**Figure 37.** Periodic hill flow at $Re_b = 37{,}000$: variation of the ratio of (**a**,**b**) modeled to total turbulent kinetic energy; (**c**,**d**) the characteristic cut-off length scale to the Kolmogorov length scale and (**e**,**f**) the sub-grid length scale to the characteristic cut-off length scale, at four different streamwise locations under fine grid resolution; SRR = Scale-Resolving region; left column = $k$-$\omega$-SST IDDES, right column = Spalart–Allmaras (S-A) IDDES.

## 7. Summary and Conclusions

The IDDES methodology using two different underlying (U)RANS models to capture near-wall dynamics has been applied to two distinct benchmark test cases: channel flow and periodic hill flow, at two different Reynolds numbers. The main focus was to investigate the effect of the near-wall model on overall model prediction capability, the dynamics of the grey area and the response to grid refinement.

It turns out that near-wall model does not have any significant effects on the prediction of first and second order statistics. Further, it was shown that using an advanced underlying (U)RANS model ($k$-$\omega$-SST IDDES) provides an extended grey region compared to the one-equation model (S-A IDDES), resulting in a delayed transition to the scale-resolving mode, which might be due to the diffusive nature of the two-equation model ($k$-$\omega$-SST IDDES).

Moreover, inconsistent responses of the grey area to grid refinement were observed for both models, such as not vanishing when the DNS level grid was used within both models or getting extended (particularly for $k$-$\omega$-SST IDDES). Furthermore, it was observed that there is a Reynolds number dependency in the response of the grey area to grid refinement, generally with more inconsistency at higher Reynolds numbers in the case of $k$-$\omega$-SST IDDES. In contrast, the grey area in S-A IDDES is thinner, indicates slight sensitivity regarding grid refinement and, generally, shifted towards near-wall when the grid becomes finer.

Inconsistent behavior may suggest that the dynamics of the grey area (responsible for allowing transition from (U)RANS to scale-resolving mode) cannot be captured using empirical blending functions mostly dominated by geometrical parameters rather than flow field quantities.

Three different criteria have been applied to assess the reliability and quality of the scale-resolving region within the IDDES methodology. First, intrusion of the scale-resolving simulation into (U)RANS simulation has been observed within both models, which is inappropriate, as the grid design in the near-wall region is not viable to support scale-resolving simulation and, therefore, may result in inaccurate modeling of near-wall dynamics. This can be seen in the oscillatory behavior of the statistics for periodic hill flow at $Re_b$ = 37,000 when the coarse grid is used. Secondly, S-A IDDES falsely satisfies the third criterion (stated in Table 3) and reports well-resolved LES simulation residing in the core region even in the case when the grid is too coarse. Conversely, the $k$-$\omega$-SST IDDES model correctly satisfies the criterion under both benchmark test cases. This clearly shows the relevance of sub-grid scale modeling to have an accurate assessment for grid resolution. Regarding the last criterion, at higher Reynolds number flows, the S-A IDDES model when operating in sub-grid scale operation mode cannot be considered as true LES.

D'Alessandro et al. [27] compared the S-A IDDES model to a DES methodology built on a non-linear (U)RANS model and a $k - \epsilon - v^2$ based model [28] when applied to separated flow (there is no significant difference between the IDDES and DES methodology in separated flows [5]). Results did not show any significant difference among the above-mentioned models.

Results presented in this study along with other observations mentioned before may lead to the conclusion that improving the underlying (U)RANS model alone would not significantly improve the capabilities of the IDDES. Further, they suggest that the main reason for the observed shortcomings in the IDDES methodology mostly likely are due to inaccurate predictions of the grey area. Further progress will require additional focus on capturing the dynamics of the grey area accurately to make this methodology a reliable tool that can be applied to various flow configurations at different Reynolds numbers and grid resolutions.

**Author Contributions:** R.S. performed numerical simulations, which were designed along with Amirfarhang Mehdizadeh. Amirfarhang Mehdizadeh and R.S. analyzed the results and prepared the manuscript. N.K., L.D. A.S. and A.M. further supported improving the manuscript.

**Funding:** The authors gratefully acknowledge the financial support provided by the University of Missouri Research Board (UMRB).

**Acknowledgments:** The corresponding author gratefully acknowledges fruitful discussions with Prof. Sadiki's research group within the CRC/Transregio 150 "Turbulent, chemially reactive, multi-phase flows near wall" program.

**Conflicts of Interest:** The authors declare no conflict of interest.

## References


1. Spalart, P.R. Comments on the Feasibility of LES for Wings, and on Hybrid RANS/LES Approach. In Proceedings of the First AFOSR International Conference on DNS/LES, Ruston, LA, USA, 4–8 August 1997.
2. Menter, F.; Egorov, Y. The scale-adaptive simulation method for unsteady turbulent flow predictions. Part 1: Theory and model description. *Flow Turbul. Combust.* **2010**, *85*, 113–138. [CrossRef]
3. Girimaji, S.; Abdol-Hamid, K. Partially-averaged navier stokes model for turbulence: Implementation and validation. In Proceedings of the 43rd AIAA Aerospace Sciences Meeting and Exhibit, Reno, NV, USA, 10–13 January 2005; p. 502.
4. Spalart, P.R.; Deck, S.; Shur, M.L.; Squires, K.D.; Strelets, M.K.; Travin, A. A new version of detached-eddy simulation, resistant to ambiguous grid densities. *Theor. Comput. Fluid Dyn.* **2006**, *20*, 181. [CrossRef]
5. Shur, M.L.; Spalart, P.R.; Strelets, M.K.; Travin, A.K. A hybrid RANS-LES approach with delayed-DES and wall-modelled LES capabilities. *Int. J. Heat Fluid Flow* **2008**, *29*, 1638–1649. [CrossRef]
6. Chaouat, B. The State of the Art of Hybrid RANS/LES Modeling for the Simulation of Turbulent Flows. *Flow Turbul. Combust.* **2017**, *99*, 279–327. [CrossRef] [PubMed]
7. Spalart, P.R. Detached-eddy simulation. *Ann. Rev. Fluid Mech.* **2009**, *41*, 181–202. [CrossRef]
8. Gritskevich, M.S.; Garbaruk, A.V.; Schütze, J.; Menter, F.R. Development of DDES and IDDES formulations for the k-$\omega$ shear stress transport model. *Flow Turbul. Combust.* **2012**, *88*, 431–449. [CrossRef]
9. Spalart, P.; Allmaras, S. A one-equation turbulence model for aerodynamic flows. In Proceedings of the 30th Aerospace Sciences Meeting and Exhibit, Reno, NV, USA, 6–9 January 1992; p. 439.
10. Menter, F.R.; Kuntz, M.; Langtry, R. Ten years of industrial experience with the SST turbulence model. *Turbul. Heat Mass Transf.* **2003**, *4*, 625–632.
11. Moser, R.D.; Kim, J.; Mansour, N.N. Direct numerical simulation of turbulent channel flow up to $Re_\tau = 590$. *Phys. Fluids* **1999**, *11*, 943–945. [CrossRef]
12. Brasseur, J.G.; Wei, T. Designing large-eddy simulation of the turbulent boundary layer to capture law-of-the-wall scaling. *Phys. Fluids* **2010**, *22*, 021303. [CrossRef]
13. Lozano-Durán, A.; Jiménez, J. Effect of the computational domain on direct simulations of turbulent channels up to $Re_\tau = 4200$. *Phys. Fluids* **2014**, *26*, 011702. [CrossRef]
14. Razi, P.; Tazraei, P.; Girimaji, S. Partially-averaged Navier–Stokes (PANS) simulations of flow separation over smooth curved surfaces. *Int. J. Heat Fluid Flow* **2017**, *66*, 157–171. [CrossRef]
15. Rapp, C.; Manhart, M. Flow over periodic hills: An experimental study. *Exp. Fluids* **2011**, *51*, 247–269. [CrossRef]
16. Breuer, M.; Peller, N.; Rapp, C.; Manhart, M. Flow over periodic hills–numerical and experimental study in a wide range of Reynolds numbers. *Comput. Fluids* **2009**, *38*, 433–457. [CrossRef]
17. Meyers, J.; Geurts, B.J.; Baelmans, M. Database analysis of errors in large-eddy simulation. *Phys. Fluids* **2003**, *15*, 2740–2755. [CrossRef]
18. Klein, M. An attempt to assess the quality of large eddy simulations in the context of implicit filtering. *Flow Turbul. Combust.* **2005**, *75*, 131–147. [CrossRef]
19. Celik, I.; Klein, M.; Janicka, J. Assessment measures for engineering LES applications. *J. Fluids Eng.* **2009**, *131*, 031102. [CrossRef]
20. Pope, S.B. *Turbulent Flows*; Cambridge University Press: Cambridge, UK, 2001.
21. Gong, Y.; Tanner, F.X. Comparison of RANS and LES models in the laminar limit for a flow over a backward-facing step using OpenFOAM. In Proceedings of the Nineteenth International Multidimensional Engine Modeling Meeting at the SAE Congress, Detroit, MI, USA, 19 April 2009; pp. 1–6.
22. Fröhlich, J.; Mellen, C.P.; Rodi, W.; Temmerman, L.; Leschziner, M.A. Highly resolved large-eddy simulation of separated flow in a channel with streamwise periodic constrictions. *J. Fluid Mech.* **2005**, *526*, 19–66. [CrossRef]
23. Weller, H.G.; Tabor, G.; Jasak, H.; Fureby, C. A tensorial approach to computational continuum mechanics using object-oriented techniques. *Comput. Phys.* **1998**, *12*, 620–631. [CrossRef]
24. George, W.; Castillo, L.; Wosnik, M. *A Theory for Turbulent Pipe and Channel Flow at High Reynolds Numbers*; Technical Report, TAM Report; Department of Theoretical and Applied Mechanics, University of Illinois at Urbana: Champaign, IL, USA, 1997.

25. Lumley, J.L.; Newman, G.R. The return to isotropy of homogeneous turbulence. *J. Fluid Mech.* **1977**, *82*, 161–178. [CrossRef]
26. Sagaut, P. *Large Eddy Simulation for Incompressible Flows: An Introduction;* Springer Science and Business Media: Berlin, Germany, 2006.
27. D'Alessandro, V.; Montelpare, S.; Ricci, R. Detached–eddy simulations of the flow over a cylinder at Re = 3900 using OpenFOAM. *Comput. Fluids* **2016**, *136*, 152–169. [CrossRef]
28. Durbin, P.A. Separated flow computations with the k-epsilon-v-squared model. *AIAA J.* **1995**, *33*, 659–664. [CrossRef]

entropy

MDPI

Article

# Alternation of Defects and Phase Turbulence Induces Extreme Events in an Extended Microcavity Laser

**Sylvain Barbay** [1,*], **Saliya Coulibaly** [2] **and Marcel G. Clerc** [3]

1 Centre de Nanosciences et de Nanotechnologies, CNRS, Université Paris-Sud, Université Paris-Saclay, Avenue de la Vauve, 91120 Palaiseau, France
2 Université de Lille, CNRS, UMR 8523-PhLAM—Physique des Lasers Atomes et Molécules, F-59000 Lille, France; saliya.coulibaly@univ-lille.fr
3 Departamento de Física and Millennium Institute for Research in Optics, FCFM, Universidad de Chile, Casilla 487-3, 8370456 Santiago, Chile; marcel@dfi.uchile.cl
* Correspondence: sylvain.barbay@c2n.upsaclay.fr

Received: 10 August 2018; Accepted: 9 October 2018; Published: 15 October 2018

**Abstract:** Out-of-equilibrium systems exhibit complex spatiotemporal behaviors when they present a secondary bifurcation to an oscillatory instability. Here, we investigate the complex dynamics shown by a pulsing regime in an extended, one-dimensional semiconductor microcavity laser whose cavity is composed by integrated gain and saturable absorber media. This system is known to give rise experimentally and theoretically to extreme events characterized by rare and high amplitude optical pulses following the onset of spatiotemporal chaos. Based on a theoretical model, we reveal a dynamical behavior characterized by the chaotic alternation of phase and amplitude turbulence. The highest amplitude pulses, i.e., the extreme events, are observed in the phase turbulence zones. This chaotic alternation behavior between different turbulent regimes is at contrast to what is usually observed in a generic amplitude equation model such as the Ginzburg–Landau model. Hence, these regimes provide some insight into the poorly known properties of the complex spatiotemporal dynamics exhibited by secondary instabilities of an Andronov–Hopf bifurcation.

**Keywords:** complex dynamics; microcavity laser; spatiotemporal chaos

## 1. Introduction

Out-of-equilibrium systems exhibit permanent complex dynamical behaviors as a consequence of the balance between the injection and dissipation of energy, momentum, and particles [1–3]. In particular, nonequilibrium processes often lead in nature to the formation of patterns—dissipative structures [1]—developed from a uniform state thanks to the spontaneous breaking of symmetries present in the system under study [1–5]. Close to this spatial instability, one generically observes the emergence of spatial structures such as stripes and hexagons. As one increases the strength of the control parameter, these patterns exhibit bifurcations that, for example, generate the emergence of more complex stationary patterns such as superlattice and quasi-crystals [5]. One strategy that has allowed a unified description of all these bifurcations and the dynamics of these stationary patterns is based on the amplitude or envelope equations [5–7]. As the stationary patterns develop more complex textures, these are described analytically by the inclusion of additional critical amplitudes.

The previous scenario changes radically when the patterns exhibit an oscillatory instability [8], that is, an Andronov–Hopf bifurcation between a stationary pattern to one of an oscillatory nature. The oscillatory patterns are characterized by oscillations in a synchronized manner over a wide range of parameters. By increasing the control parameter, they exhibit a quasi-periodic behavior through a secondary instability [9–11]. As a consequence, the Fourier transform of the amplitude shows multiple peaks with incommensurate frequencies. As the control parameter is further increased, this quasi-periodic behavior is replaced by spatiotemporal chaotic behavior. The previous route is known as extended quasi-periodicity [9]. Hence, the pattern exhibits a complex spatiotemporal behavior characterized by a continuous Lyapunov spectrum. Indeed, small modifications or disturbances in the initial conditions generate unpredictability. A simple physical system that presents the former scenario is an extended semiconductor microcavity laser with saturable gain and absorber layers [10,12]. In this system, it has been shown theoretically that spatiotemporal chaos emerges through the mechanism of quasiperiodic, extended spatiotemporal intermittency [10]. The onset of spatiotemporal chaos also gives rise almost simultaneously to extreme events in the form of rare and high amplitude optical pulses. A straightforward correspondence between the proportion of extreme events and the dimension of the strange attractor was established in [12] by comparing experimental and numerical results. The universal envelope model, the Ginzburg–Landau equation [13], which generically describes the dynamics close to an Andronov–Hopf bifurcation, does not adequately account for the dynamics previously described, even though this equation exhibits complex and appealing behaviors such as phase turbulence, amplitude turbulence, and spatiotemporal intermittency [13,14]. Phase turbulence is characterized by a complex dynamics of modes described by a field phase that exhibits a decaying power law in its power spectrum [15]. The corresponding dynamics is of spatiotemporal chaos-type, in which the magnitude of the field is never zero, that is, the real and imaginary parts of the field never cross the zero axis simultaneously. Hence, the field is said to be free of phase singularity or defects in its magnitude. Amplitude turbulence is also characterized by a complex dynamics of modes that exhibit a power law in the field energy power spectrum. However, its main feature is the permanent nucleation of amplitude defects, where the phase is undeterminate [14]. This dynamics requires a strong coupling between the phase and the module of the field envelope. Hence, amplitude turbulence exhibits a dynamical behavior of greater complexity than phase turbulence. The aperiodic emergence of phase singularities characterizes spatiotemporal intermittence, but unlike the dynamics observed in amplitude turbulence, the disappearance of defects is governed by self-organization that engenders transitions between coherent and incoherent regions in the spatiotemporal evolution [14]. Despite the rich dynamics contained in the Ginzburg–Landau equation, this model fails in the adequate physical description of the microcavity laser due to the assumption that the envelope is a slow spatiotemporal variable compared to the wavelength of the underlying pattern. As a consequence of this type of scale mismatch, amplitude equations do not describe several physical phenomena, such as the pinning effect of fronts [16], noise-induced front propagation [17], and the homoclinic snaking bifurcation of localized patterns [18,19].

The characterization of the complex spatiotemporal dynamics exhibited by secondary instabilities of an Andronov–Hopf bifurcation is an open problem in nonlinear science. This paper aims to investigate the complex dynamics shown by the patterns in an extended, one-dimensional semiconductor microcavity laser with an intracavity saturable absorber that displays such secondary instability. Based on a theoretical model, we reveal a dynamic behavior characterized by the chaotic alternation of phase and amplitude turbulence. We stress that this type of dynamics is not contained in the Ginzburg–Landau equation. Interaction and superposition between wave packets characterize phase and defect turbulence [14]. Phase turbulence is distinguished by exhibiting a cascade of the power law for energy versus wavenumber [15]. In the case of defects turbulence, it is characterized by the wave interaction, which permanently gives rise to phase singularities [14]. In the following, we identify the different turbulent behaviors and give new insights into the physical origin of extreme events in our system. Moreover, we find that extreme events occur during the phase turbulence zones.

The manuscript is organized as follows: In Section 2, we review the emergence of extreme events and spatiotemporal chaos in a spatially extended microcavity laser with saturable gain and absorption media. The theoretical model that describes the laser microcavity is presented and analyzed in Section 3. Sections 2 and 3 constitute a review of our previous results [10,12]. Alternation of defects and phase turbulence in an extended microcavity laser is analyzed in Section 4. Section 5 shows how the alternation of defects and phase turbulence induces extreme events. Finally, conclusions are presented in Section 6.

## 2. Extreme Events in a Microcavity Laser

Extreme events have attracted a great deal of attention lately, in particular in optical systems where reliable statistics can be obtained and where many different and controlled physical situations can be explored [20,21]. In dissipative optical systems, extreme events have been found in the intensity dynamics of fibre lasers [22], semiconductor lasers with injected signal [23], and solid-state lasers with a saturable absorber [24]. Vertical-cavity surface emitting lasers with an integrated saturable absorber (VCSEL-SAs) [25,26] are good candidates for studying complex dynamical phenomena and extreme events in self-pulsing spatially extended systems thanks to their small footprint and high aspect ratio. Moreover, the fast timescales associated to semiconductor materials allow for gathering a large amount of information in a short amount of time, which is interesting for statistical analyses and tracking rare phenomena such as extreme events. Broad-area VCSEL-SAs may also have interesting applications, e.g., high-power lasers with vertical cavity emission. These laser devices are composed of two multilayer mirrors, which optimize optical pumping, and of an active zone. This active zone is made up of two InGaAs quantum wells for the gain section and one InGaAs quantum well for the saturable absorber section, forming a $2\lambda$ optical length cavity ($\lambda = 980$ nm). By contrast to a standard laser composed solely of a gain section, the laser with a saturable absorber can sustain self-pulsing at the laser threshold [25]. In the limit of a single transverse mode laser (i.e., with a low aspect ratio cavity), the dynamics is always regular with typical experimental parameters [27]. However, in an extended cavity laser, a more complex dynamics can set in thanks to the interplay between the system nonlinearity and spatial coupling through the light diffraction inside the cavity. In addition, while the typical timescale for the intracavity electromagnetic field is of the order of several picoseconds, the material excitation timescale is much longer (typically the non-radiative recombination of semiconductor carriers is of the order of 1 ns or less). It is thus not possible to reduce the dynamics to the one of the optical intensity. The experimental setup is shown in Figure 1a. The microcavity laser is coated with a thin gold layer with a rectangular opening to define the pumped region. The rectangular mask has an 80 µm length and a 10 µm width, thus forming a quasi one-dimensional line laser. The microcavity laser is optically pumped through a dichroic mirror at 800 nm and emits around 980 nm. Laser emission is imaged on a screen provided with one or two holes. These holes allow for selecting the detection area, which correspond to a disk of a 5 µm diameter on the sample surface. The line VCSEL-SA emission intensity is monitored and recorded with a fast avalanche photodiode (>5 GHz bandwidth). Likewise, the temporal signal is amplified in a low noise, high bandwidth amplifier (3 kHz–18 GHz bandwidth) and acquired with a 6 GHz bandwidth oscilloscope at a sampling rate of 20 GS/s. This allows for easy statistical analysis of the recorded data since very large time traces can be collected in a short amount of time. Figure 1b shows the near field of the laser above threshold with a camera placed at the screen position.

Excerpts of time traces of the laser intensity recorded at the center of the laser are shown in Figure 2 for different pumping intensities. With the full time traces recorded, the histogram of the heights $H$ can be constructed. The height $H$ is defined by the average of the left and right pulse heights, as in hydrodynamics. From these analyses one can conclude that the system exhibits a complex dynamics of extreme events [10,12]. Figure 2 depicts heights histograms for different values of the pump parameter $P$. Let us introduce $P_{th}$ as the laser threshold pump. At normalized pump power $P/P_{th} = 1.02$, the histogram in a semi-log plot is characterized by a quadratic decay in the tails. Figure 2a shows the probability density function (PDF), which resembles a Rayleigh distribution for a positive valued Gaussian process. Increasing the pump parameter, the PDF develops long tails with an initial exponential decay (cf. Figure 2b). Increasing further the pump values, the PDF becomes an exponential distribution ($P/P_{th} = 1.20$). For a still higher pump value ($P/P_{th} = 1.25$) the PDF redisplays a Gaussian tail. To determine the threshold amplitude for extreme events, we consider the standard hydrodynamical criterion, that is, an extreme event corresponds to an event having a height $H$ twice the significant height $H_s$, where $H_s$ stands for the mean of the highest tertile of the PDF. Namely, extreme events are characterized by an abnormality index $AI \equiv H/H_s > 2$ [28]. To ignore a large number of small peaks due to detector noise to the left of the PDF, one can determine the relevant or significant height $H_s$ by considering events whose altitude is higher than the observed maximum peak dark noise amplitude. On Figure 2, extreme events are in orange in the PDF. When the PDF presents a non-Gaussian tail, we observe that the system exhibits a large number of extreme events (a normalized pump of 1.17). When increasing the pump parameter, a complicated dynamical behavior characterized by intermittent pulsations of the recorded intensity is observed. Indeed, the dynamics shows irregular oscillations of the intensity characterized by sharp peaks that appear irregularly in the temporal domain; that is, the peaks exhibit an aperiodic behavior, which is a typical signature of chaos [10]. Hence, the dynamics of the microcavity laser is characterized by a supercritical intermittency route to chaos [29], and has thus been called extended spatiotemporal intermittency [10]. The experimental results discussed so far are well reproduced by a theoretical model of an extended microcavity laser with a saturable absorber, which we present in Section 3.

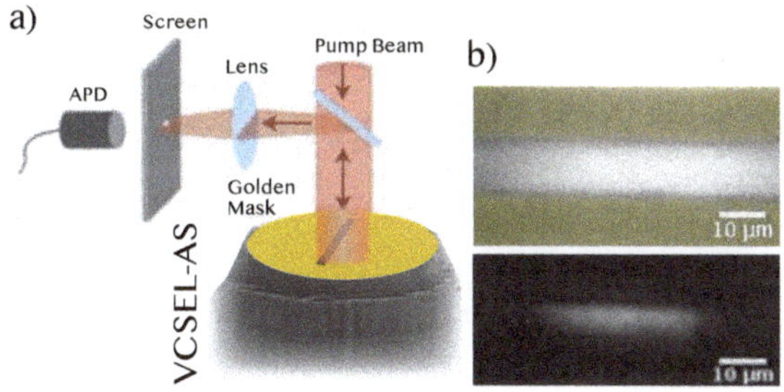


**Figure 1.** Experimental set up. (**a**) Schematic representation of an extended planar vertical cavity surface emitting laser with an integrated saturable absorber medium (VCSEL-SA). (**b**) Right panels account for the top-view camera snapshots of the one-dimensional line VCSEL-SA surface below (upper image, with the mask visible) and above laser threshold (lower image).

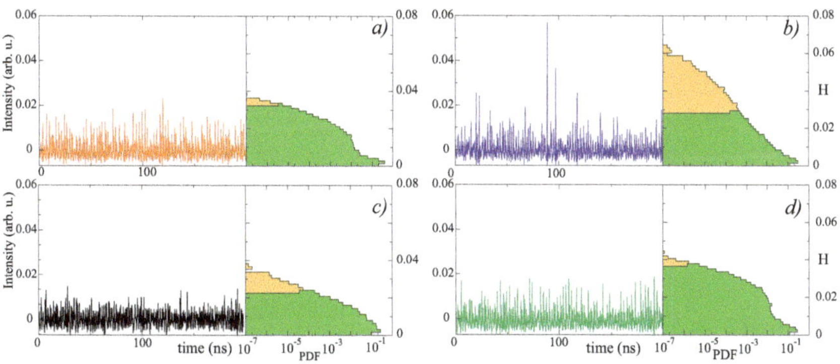


**Figure 2.** Typical temporal evolution of the experimentally recorded intensity and semi-log graph of the associated probability density distribution of the intensity height $H$ for different normalized pump values (adapted from [12]): (**a**) $P/P_{th} = 1.02$; (**b**) $P/P_{th} = 1.17$; (**c**) $P/P_{th} = 1.20$; and (**d**) $P/P_{th} = 1.25$. Normal and extreme events are shown in orange and green, respectively ($AI > 2$).

The emergence of extreme events is related to the onset of spatiotemporal chaos, or at the beginning of the transition from a complex dynamical behavior to another [10,12]. The total intensity

$I_{tot}(t) \equiv \int |E(x,t)|^2 dx$ and local intensity $I_{loc} \equiv |E(x,t)|^2$, where $E(x,t)$ is the intracavity electric-field envelope, are two relevant physical quantities to characterize the dynamics of the extended microcavity laser. The latter quantity, in particular, is only accessible through numerics because it is not possible to record the full spatiotemporal evolution in the experiment, due to the very short timescales at stake. This justifies the numerical approach that we present hereafter. Figure 3a,b show the proportion of extreme events in all the numerically observed events ($p_{EE}$), and the deviations from the Gaussian distribution of the numerical PDF (excess kurtosis $\gamma_2$) as a function of the pump parameter $\mu$. The same analysis is done for the two observables, namely the total intensity emitted by the laser $I_{tot}$ (cf. Figure 3a,b) and the intensity of the spatiotemporal peaks $I_{loc}$ (cf. Figure 3e,f). Note that $p_{EE}$ and $\gamma_2$ are correlated in both cases. However, they follow different trends with $\mu$: in the case of the observables associated with the intensity, both extreme events indicators tend to grow as a function of the pumping parameter. However, extreme events indicators linked to spatiotemporal intensity peaks tend to increase near the bifurcation of the spatiotemporal chaos and subsequently decay strongly.

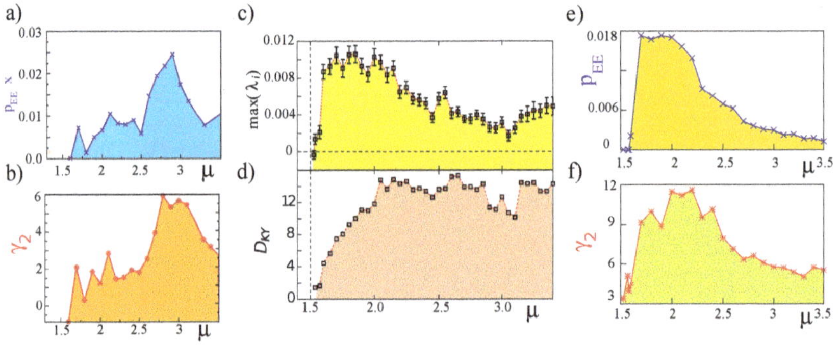


**Figure 3.** Numerical characterization of the emergence of extreme events in an extended, planar vertical-cavity surface-emitting laser with an integrated saturable absorber medium obtained from Equation (1). Graph of the proportion of extreme events $p_{EE}$ ($\times$) (**a**) and excess kurtosis $\gamma_2$ ($*$) (**b**) as a function of pump parameter $\mu = P/P_{th}$ considering the height $H$ of the laser intensity. Graph of the largest Lyapunov exponent max($\lambda_i$) (squares) (**c**) and Kaplan–Yorke dimension $D_{KY}$ (**d**) as a function of pump parameter $\mu$. Graph of the proportion of extreme events $p_{EE}$ (**e**) and excess kurtosis $\gamma_2$ (**f**) as a function of pump parameter $\mu$ considering the local intensity spatiotemporal maxima (adapted from [10]).

## 3. Theoretical Description of a One-Dimensional Spatially Extended Laser

A planar vertical-cavity surface-emitting line laser with a saturable absorber medium can be described to a good approximation by a one-dimensional spatially extended laser with a saturable absorber layer [30]. This model has been shown to successfully describe different phenomena in the system under study, such as pattern and localized structure formation [31] and spatiotemporal chaos. In this latter case, we have shown that the model captures very well the evolution with the pump parameter of the intensity statistics and of the intensity cross-correlation computed at two different locations, as well as the evlution of the power spectrum of the intensity and of the extreme event indicators [10,12]. The dimensionless model reads

$$\begin{aligned}
\frac{\partial E(x,t)}{\partial t} &= \left[(1-i\alpha)G + (1-i\beta)Q - 1\right]E + i\frac{\partial^2 E}{\partial x^2} \\
\frac{\partial G(x,t)}{\partial t} &= \gamma_g\left[\mu - G(1+|E|^2)\right] \\
\frac{\partial Q(x,t)}{\partial t} &= \gamma_q\left[-\gamma - Q(1+s|E|^2)\right]
\end{aligned} \tag{1}$$

where the fields $E(x,t)$, $G(x,t)$, and $Q(x,t)$, respectively, account for the intracavity electric-field envelope, the carrier density in the gain, and the saturable absorber medium. $x$ and $t$ stand for the spatial coordinate and time. The non-radiative carrier recombination rates are $\gamma_g$ and $\gamma_q$ . The parameters $\mu$ and $\gamma$ are the pumping and linear absorption processes, respectively. The parameters $\alpha$ and $\beta$ account for the Henry enhancement factors in both the gain and absorber regions, respectively. These parameters are related to phase-amplitude coupling in semiconductor media. The Laplacian term stands for the diffraction process. Diffusion processes of carriers are smaller than diffraction ones and are ignored in the first approximation. The time and spatial variables have been rescaled to the field lifetime and the diffraction length $w_d$ in the cavity, respectively. Considering the parameters of the cavity, the time and spatial scales correspond to 8.0 ps and 7.4 μm. Since the pumped region has a length $w_p \sim 80$ μm, we obtain $w_p/w_d \sim 11$ as a direct estimate of the Fresnel number of the line microlaser. Considering parameters compatible with our semiconductor system, we obtain $\alpha = 2$, $\beta = 0$, $s = 10$, $\gamma_g = \gamma_q = 0.005$, and $\gamma = 0.5$. The Henry enhancement factors are chosen with usual values [32]. Assuming that the carriers recombinations times are of the order of 800 ps, one can determine the other physical parameters straighforwardly.

The bifurcation diagram of Equation (1) has been studied in detail (see [30] and references therein). For small pumping, the system is in the no-lasing state. When increasing the pumping parameter above $\mu_{th} = 1 + \gamma$, the (plane-wave) lasing threshold is reached. Further increasing the pumping parameter, Equation (1) exhibits an Andronov–Hopf bifurcation for plane waves $\mu(I) < \mu_H(I) \equiv r(2rsI\gamma - \gamma_g(1+I)(1+sI)(1+I+r+rsI))/2I$ with $r = \gamma_q/\gamma_g$ [30]. Due to the complex dynamics presented by the system, analytical studies are inaccessible. To figure out the dynamics exhibited by the microcavity extended laser with a saturable absorber medium, we have numerically studied model (1). Our strategy has been to consider only one parameter in the analysis, for better comparison with the experiment where this parameter is easily accessible, namely the power pump parameter $\mu$. For pumping power values such that $\mu > \mu_{th}$, the laser turns on through a transcritical bifurcation. When increasing the pump power value ($\mu/\mu_{th} \sim 1.047$), the total intensity $I_{\text{tot}}$ exhibits a quasi-periodic dynamical behavior. Indeed, the temporal evolution of the total intensity of the electric field envelope is aperiodic and presents fluctuations around its average value [10]. Note that extreme events are not observed in this parameter regime. Unexpectedly, increasing the value of the pumping power parameter ($\mu/\mu_{th} \sim 1.333$), the system presents a bifurcation. In this parameter regime, the total intensity exhibits intermittent pulsations in its temporal evolution characterized by aperiodic fluctuations, in which sharp peaks randomly appear. This dynamical behavior is compatible with the experimental observations as shown in Figure 2.

To understand the complex dynamics observed, we can determine its sensitivity to perturbations by means of the Lyapunov spectrum (with Lyapunov exponents $\lambda_i$). One of the main characteristics of this spectrum is that the system presents a temporal or low dimensional chaotic behavior if and only if the largest Lyapunov exponent $\max(\lambda_i)$ is positive. However, to conclude a spatiotemporal or high dimensional chaos, the latter condition is necessary but not sufficient. Spatiotemporal chaos is a permanent, aperiodic spatiotemporal dynamical behavior. In addition, this dynamical behavior is characterized by being of an extensive nature [33]. The Lyapunov spectrum is composed by the set of the Lyapunov exponents arranged in decreasing order considering their real parts. This spectrum allows the distinction between chaos and spatiotemporal chaos. Indeed, a Lyapunov spectrum with a continuous set of positive values characterizes spatiotemporal chaos. In contrast, a Lyapunov spectrum with a discrete set of positive values characterizes chaos of low dimensions. The Kaplan–Yorke dimension $D_{KY}$ [34] can be determined from the Lyapunov spectrum. This dimension accounts for the dimension of the strange attractor under study. The largest Lyapunov exponent $\max(\lambda_i)$ and the Kaplan–Yorke dimension are right quantities to characterize complex dynamical behaviors and transitions between them [35]. For instance, steady-state solutions are characterized by a negative and zero largest Lyapunov exponent and Kaplan–Yorke dimension, respectively. Periodic or quasi-periodic solutions have a zero largest Lyapunov exponent and Kaplan–Yorke dimension. When both the largest Lyapunov exponent and Kaplan–Yorke dimension are strictly positive, this corresponds to a chaotic dynamical behavior. In the region of aperiodic intermittent pulsations, Equation (1) shows a characteristic Lyapunov spectrum of a spatiotemporal nature [10,12]. Figure 3c,d show $\max(\lambda_i)$ and $D_{KY}$ as a function of the pumping parameter $\mu$, obtained numerically. We observe that the emergence of extreme events in the microcavity laser is correlated to the appearance of spatiotemporal chaos. Indeed, extreme events are observed only when the largest Lyapunov exponent and the Kaplan–Yorke dimension are both strictly positive.

In addition, when increasing the pump power parameter, the spatiotemporal complexity increases (see the onset of spatiotemporal chaos in Figure 3). Note that $\max(\lambda_i)$ and $D_{KY}$ both consistently increase with the pumping value $\mu$. The microcavity laser with a saturable absorber medium exhibits extreme events when the system is in a regime of spatiotemporal chaos. However, the kind of spatiotemporal chaos displayed by Equation (1) is not determined by this analysis and will be the subject of the next section.

## 4. Characterization of Spatiotemporal Dynamics of an Extended Laser with a Saturable Absorber: Alternation of Defects and Phase Turbulence

To figure out the complicated dynamical behaviors presented by the microcavity laser model with a saturable absorber, we simulated numerically the set of Equation (1). We used a split-operator method to accurately compute the Laplacian term, while the nonlinear temporal evolution is taken care of in real space. The non-zero pump is restricted to a finite domain ($[-5,5]$ interval) and is zero otherwise (not shown), thus giving absorbing boundaries. Figure 4 displays the space–time evolution of the laser intensity together with spatiotemporal positions of defects and of extreme events computed for different pumping parameters. Defects correspond to zeros of the envelope of the electric field $E(x,t)$; that is, in these points, the phase is not defined: they correspond to phase singularities [13]. From this figure, we observe that the system presents interchange between a region of phase turbulence and defects turbulence. The region of phase turbulence is characterized by a complex dynamics of wave interaction. In this region, the phase is always well defined; that is, the amplitude of the waves is never zero. Note that, in this region, the spatial modes of the system exhibit complex spatiotemporal dynamics (cf. Figure 4). We monitored and determined the spatiotemporal positions of the amplitude defects in the temporal progression of the envelope of the electric field (see blue dots in Figure 4). Note that amplitude defects tend to gather for low pumping and generally display a complex spatiotemporal distribution. The regions of phase turbulence are separated by areas with low intensities that exhibit amplitude defects (phase singularities). Likewise, we monitored and determined the spatiotemporal position of extreme events in the electric field envelope $E$ (see red dots in Figure 4 and corresponding dash signs). One expects complex behaviors such as phase or defects turbulence to exhibit extreme events due to the strong temporal correlation of deterministic dynamics. Unexpectedly, extreme events are mostly observed in the regions of phase turbulence. We can therefore conclude that the spatiotemporal dynamics of the system is characterized by the chaotic alternation of phase singularities (amplitude defects) and the observation of large amplitude pulsations (extreme events). This type of complex spatiotemporal dynamics is not contained in universal models, such as the Kuramoto–Sivashinsky [15] and the Ginzburg–Landau equations [13], which account for the dynamics around an Andronov–Hopf instability. Hence, the dynamics observed in Equation (1) goes beyond the dynamics contained arround the Andronov–Hopf bifurcation, and the alternation between phase and defects turbulence is a new kind of complex dynamics.

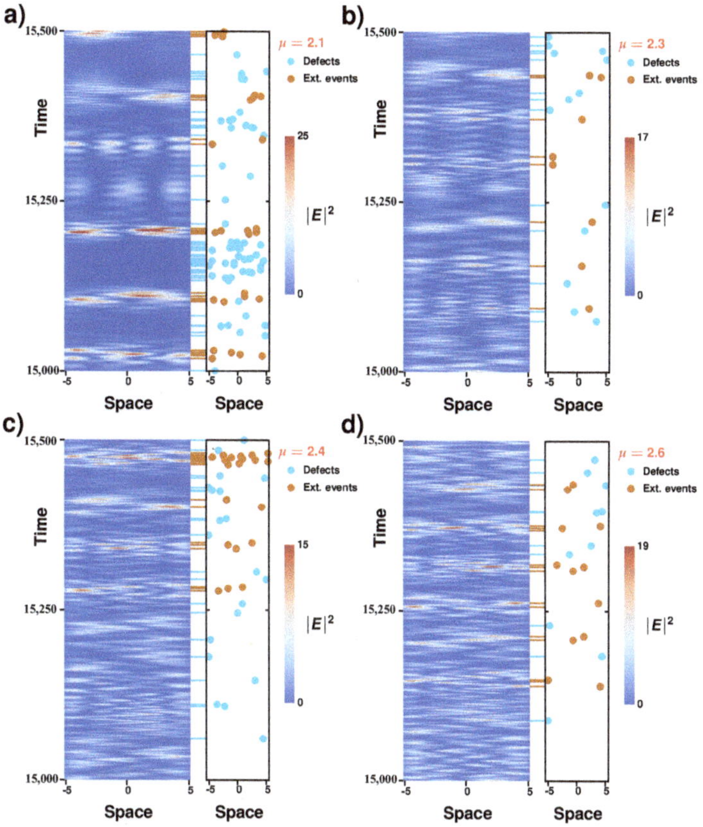


**Figure 4.** Alternation of defects and phase turbulence in the laser with saturable absorber model expressed by Equation (1). Spatiotemporal evolution of the electric field intensity, together with the spatiotemporal positions of phase singularities of the electric field envelope $E(x,t)$ and of the extreme events (blue and red dots, respectively; temporal location of respective events are highlighted by dash signs) in the spatiotemporal complex regime with $\alpha = 2$, $\beta = 0$, $\gamma_g = 0.005$, $\gamma_q = 0.005$, $\gamma = 0.5$, $s = 10$, and the following $\mu$ values: (**a**) $\mu = 2.1$; (**b**) $\mu = 2.3$; (**c**) $\mu = 2.4$; (**d**) $\mu = 2.6$.

To characterize more accurately the dynamics exhibited by the system, we calculate the phase associated with the envelope

$$\varphi(x,t) \equiv \frac{\Im[E(x,t)]}{\Re[E(x,t)]} \tag{2}$$

and analyze its spatiotemporal evolution. Close to the Andronov–Hopf bifurcation, the equations governing the phase and envelope amplitude can be decoupled. Notably, around the Benjamin–Feir instability [2], the phase satisfies the Kuramoto–Sivashinsky equation. This model has been an angular footing in the study of complex spatiotemporal dynamics, since it corresponds to the simplest scalar model that describes the dynamics of coupled oscillators and exhibits turbulence dynamics [15]. Likewise, this is one of the first models to be used to rigorously unveil spatiotemporal chaos and display a continuous Lyapunov spectrum [36]. However, the dynamics displayed in the spatiotemporal diagrams of the amplitude (cf. Figure 4) shows a regular appearance of phase singularities, which is a prohibitive condition for the separation of dynamics from the phase and the magnitude of the envelope. This rules out a mechanism similar to the one found in the Kuramoto–Sivashinsky equation. We investigated the spatiotemporal evolution of the phase as defined by Equation (2) for different values of the pumping parameter and plotted the result in Figure 5. These diagrams illustrate a complex wave dynamics since no visible structure emerge. To characterize this dynamic from a statistical point of view, we calculate the average spectrum of phase fluctuations defined by [15]

$$\langle \bar{\varphi}(k) \rangle \equiv \lim_{T\to\infty} \frac{1}{T} \int_0^T \left[ \frac{1}{L} \int_{-L/2}^{L/2} \varphi(x,t) e^{ikx} dx \right]^2 dt \tag{3}$$

$L$ accounts for the system size, $T$ is a long enough time, to perform an average on the statistics, and $k$ is a wavenumber. This quantity allows one to characterize the transport of energy between the different scales of the coupled oscillators [15]. Figure 5 shows the average spectrum of phase fluctuations $\langle \bar{\varphi}(k) \rangle$ for different pumping values in semi-log and log-log plot. It is clearly visible there that the averaged phase spectrum exhibits a power-law behavior in a specific range of wave numbers. From this observation, one can conclude that the dynamics presented by the microcavity laser with a saturable absorber medium is of a turbulent nature. Hence, the dynamical behavior characterized by alternation of defects and phase spatiotemporal complexity is of a turbulent nature.

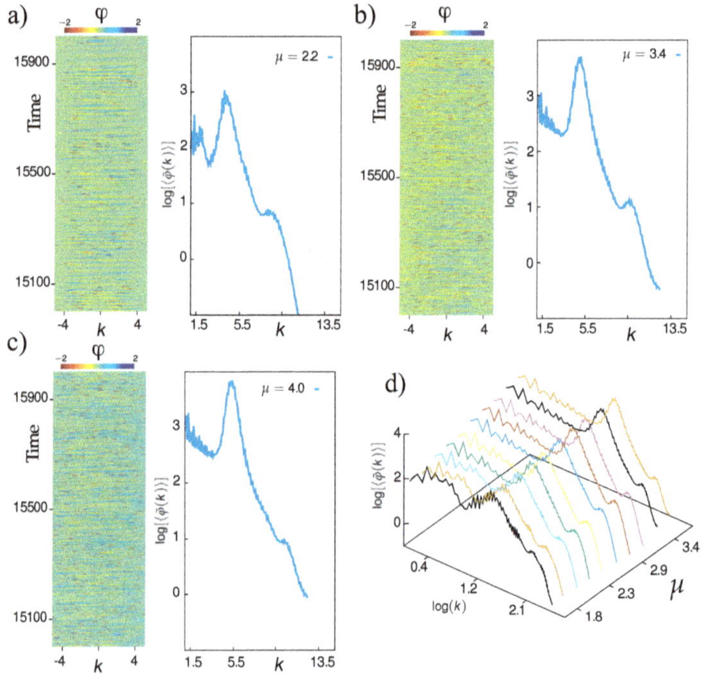


**Figure 5.** Turbulence dynamics of the one-dimensional microcavity laser with a saturable absorber medium. Spatiotemporal diagram and the average spectrum $\bar{\varphi}_k$ of the phase of the electric field envelope of Equation (1) by $\alpha = 2$, $\beta = 0$, $\gamma_g = 0.005$, $\gamma_q = 0.005$, $\gamma = 0.5$, $s = 10$, and the following $\mu$ values: (**a**) $\mu = 2.2$; (**b**) $\mu = 3.4$; (**c**) $\mu = 4.0$. (**d**) The average spectrum $\bar{\varphi}_k$ of the phase of the electric field envelope for different pumping parameters.

## 5. Alternation of Defects and Phase Turbulence Induces Extreme Events

In order to emphasize the relationship between the alternation dynamics from phase turbulence to defects turbulence and the appearance of extreme events, we analyzed the spatiotemporal diagrams in a larger simulation time window in Figure 6, and for different pumping parameters. Near the lasing bifurcation, there are globally many defects and those have a tendency to bunch in the low laser intensity zones to give clear alternations with the zones of phase turbulence where, by contrast, extremes events can be found. The chaotic pulsation (and the alternation dynamics between the turbulent regimes) consists of large areas of defect turbulence (low intensity zones) and small areas of phase turbulence (higher intensities), which in turn is consistent with the observation of a large number of extreme events (i.e., rare and high intensity peaks). However, as one moves away from the bifurcation point, the number of defects is much smaller and amplitude defects tend to spread all over the spatiotemporal diagram. This is consistent with a faster alternation of the turbulent regimes (defects and phase mediated) and with the fact that the proportion of extreme events globally decreases (see Figure 3).

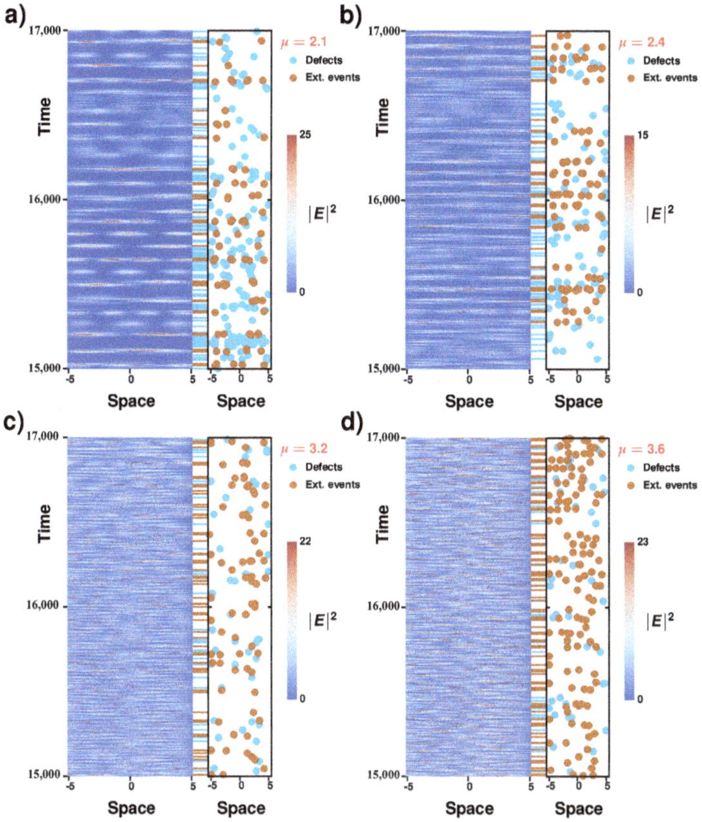


**Figure 6.** Complex dynamics exhibited by the laser with saturable absorber model computed with Equation (1) in a large time window. Spatiotemporal progression of the electric field magnitude and spatiotemporal positions of the defects of the electric field envelope $E(x,t)$ and of the extreme events (blue and red dots, respectively; temporal location of respective events are highlighted by dash signs). Parameters are identical to those in Figure 4, with pumping: (**a**) $\mu = 2.1$; (**b**) $\mu = 2.4$; (**c**) $\mu = 3.2$; and (**d**) $\mu = 3.6$.

As illustrated in Figure 3, extreme events appear almost simultaneously with the emergence of spatiotemporal chaos. One can understand this phenomenon because the observed spatiotemporal chaotic dynamics is of an intermittent nature, namely, the system moves between different dynamical behaviors. However, the occurrence of spatiotemporal chaos does not necessarily mean in general that the system will display extreme events. Nonetheless, the aperiodic alternation between different (complex) dynamical behaviors can generate extreme events. That is, chaotic behaviors that are characterized by the variation between different dynamical behaviors is a natural context where one can observe extreme events. The above argument explains why in the laser with a saturable absorber, one verifies the simultaneous emergence of extreme events and spatiotemporal chaos. It can also allow one to establish a parallel with the context of (temporal) chaos, where it has been shown that deterministic extreme events are linked to multistability and to the occurrence of crises [37,38].

## 6. Conclusions

Out-of-equilibrium extended systems exhibit complex dynamical spatiotemporal behaviors. One strategy for understanding this type of dynamical behavior is to investigate its bifurcations and routes to complexity. Nevertheless, the greatest successes have been achieved in understanding primary instabilities, thanks to the use of amplitude equations, perturbation singular, and normal forms theory. The characterization and classification of complex behaviors in extended systems are one of the fundamental problems of nonlinear science. We investigated the complex dynamics shown by oscillatory patterns in a spatially extended semiconductor microcavity laser with an intracavity saturable absorber. Based on a theoretical model of the microcavity laser, which has proven to be qualitatively accurate in the experimental system's description, a numerical analysis has revealed a complex spatiotemporal dynamical behavior characterized by the alternation of phase and amplitude turbulence. To our knowledge, this is the first time that this intriguing dynamical behavior has been reported since the two turbulent regimes are usually not observed in current models within the same parameter regions. It is also remarkable to note that this kind of dynamics is beyond the Ginzburg–Landau world [13]. Likewise, the alternation between turbulent behaviors is characterized by the occurrence of the highest amplitude optical pulses, which are observed in the phase turbulence zones. Indeed, it was already known that the appearance of spatiotemporal chaos generates extreme events, but we give here a much finer account of the kind of dynamical mechanism that is responsible for the observation of extremes. At last, the complex spatiotemporal dynamics observed here is believed to be observable in other systems that exhibit an Andronov–Hopf bifurcation. Work in this direction is in progress.

**Author Contributions:** All authors contributed to the conceptualization of the problem. S.C. performed the numerical simulations and analysis with inputs from M.G.C. and S.B.; S.B. and S.C. performed the statistical analysis of the numerical and experimental data. All the authors discussed the results. All authors participated to the elaboration of the manuscript.

**Funding:** This research was funded by the Millennium Institute for Research in Optics (MIRO) and FONDECYT projects 1180903. This research was partially funded by the CNRS Renatech Network for the fabrication of the microlasers.

**Acknowledgments:** S.C. acknowledge the LABEX CEMPI (ANR-11-LABX-0007) as well as the Ministry of Higher Education and Research, Hauts de France council and European Regional Development Fund (ERDF) through the Contrat de Projets Etat-Region (CPER Photonics for Society P4S).

**Conflicts of Interest:** The authors declare no conflict of interest.

## References

1. Nicolis, G.; Prigogine, I. *Self-Organization in Nonequilibrium Systems*; Wiley: New York, NY, USA, 1977.
2. Cross, M.C.; Hohenberg, P.C. Pattern formation outside of equilibrium. *Rev. Mod. Phys.* **1993**, *65*, 851–1112. [CrossRef]
3. Pismen, L.M. *Patterns and Interfaces in Dissipative Dynamics*; Springer: Berlin, Germany, 2006.
4. Cross, M.; Greenside, H. *Pattern Formation and Dynamics in Nonequilibrium Systems*; Cambridge University Press: New York, NY, USA, 2009.
5. Hoyle, R.B. *Pattern Formation: An Introduction to Methods*; Cambridge University Press: New York, NY, USA, 2006.
6. Newell, A.C. Envelope equations. *Lect. Appl. Math.* **1974**, *15*, 157–163.
7. Newell, A.C.; Passot, T.; Lega, J. Order parameter equations for patterns. *Annu. Rev. Fluid Mech.* **1993**, *25*, 399–453. [CrossRef]

8. Coullet, P.; Iooss, G. Instabilities of one-dimensional cellular patterns. *Phys. Rev. Lett.* **1990**, **64**, 866–869. [CrossRef] [PubMed]
9. Clerc, M.G.; Verschueren, N. Quasiperiodicity route to spatiotemporal chaos in one-dimensional pattern-forming systems. *Phys. Rev. E* **2013**, *88*, 052916. [CrossRef] [PubMed]
10. Coulibaly, S.; Clerc, M.G.; Selmi, F.; Barbay, S. Extreme events following bifurcation to spatiotemporal chaos in a spatially extended microcavity laser. *Phys. Rev. A* **2017**, *95*, 023816. [CrossRef]
11. Panajotov, K.; Clerc, M.G.; Tlidi, M. Spatiotemporal chaos and two-dimensional dissipative rogue waves in Lugiato-Lefever model. *Eur. Phys. J. D* **2017**, *71*, 176. [CrossRef]
12. Selmi, F.; Coulibaly, S.; Loghmari, Z.; Sagnes, I.; Beaudoin, G.; Clerc, M.G.; Barbay, S. Spatiotemporal chaos induces extreme events in an extended microcavity laser. *Phys. Rev. Lett.* **2016**, *116*, 013901. [CrossRef] [PubMed]
13. Aranson, I.S.; Kramer, L. The world of the complex Ginzburg–Landau equation. *Rev. Mod. Phys.* **2002**, *74*, 99–143. [CrossRef]
14. Chate, H. Spatiotemporal intermittency regimes of the one-dimensional complex Ginzburg–Landau equation. *Nonlinearity* **1994**, *7*, 185–204. [CrossRef]
15. Kuramoto, Y. *Chemical Oscillations, Waves, and Turbulence*; Springer: Berlin/Heidelberg, Germany, 2012.
16. Bensimon, D.; Shraiman, B.I.; Croquette, V. Nonadiabatic effects in convection. *Phys. Rev. A* **1988**, *38*, 5461–5464. [CrossRef]
17. Clerc, M.G.; Falcon, C.; Tirapegui, E. Additive noise induces front propagation. *Phys. Rev. Lett.* **2005**, *94*, 148302. [CrossRef] [PubMed]
18. Clerc, M.G.; Falcon, C. Localized patterns and hole solutions in one-dimensional extended systems. *Physica A* **2005**, *356*, 48–53. [CrossRef]
19. Barbay, S.; Hachair, X.; Elsass, T.; Sagnes, I.; Kuszelewicz, R. Homoclinic Snaking in a Semiconductor-Based Optical System. *Phys. Rev. Lett.* **2008**, *101*, 253902. [CrossRef] [PubMed]
20. Solli, D.R.; Ropers, C.; Koonath, P.; Jalali, B. Optical rogue waves. *Nature* **2007**, *450*, 1054–1057. [CrossRef] [PubMed]
21. Akhmediev, N.; Kibler, B.; Baronio, F.; Belić, M.; Zhong, W.-P.; Zhang, Y.; Chang, W.; Soto-Crespo, J.-M.; Vouzas, P.; Grelu, P.; et al. Roadmap on optical rogue waves and extreme events. *J. Opt.* **2016**, *18*, 063001. [CrossRef]
22. Lecaplain, C.; Grelu, P.H.; Soto-Crespo, J.M.; Akhmediev, N. Dissipative Rogue Waves Generated by Chaotic Pulse Bunching in a Mode-Locked Laser. *Phys. Rev. Lett.* **2012**, *108*, 233901. [CrossRef] [PubMed]
23. Bonatto, C.; Feyereisen, M.; Barland, S.; Giudici, M.; Masoller, C.; Leite, J.R.R.; Tredicce, J.R. Deterministic Optical Rogue Waves. *Phys. Rev. Lett.* **2011**, *107*, 053901. [CrossRef] [PubMed]
24. Bonazzola, C.R.; Hnilo, A.A.; Kovalsky, M.G.; Tredicce, J.R. Features of the extreme events observed in an all-solid-state laser with a saturable absorber. *Phys. Rev. A* **2015**, *92*, 053816. [CrossRef]
25. Elsass, T.; Gauthron, K.; Beaudoin, G.; Sagnes, I.; Kuszelewicz, R.; Barbay, S. Control of cavity solitons and dynamical states in a monolithic vertical cavity laser with saturable absorber. *Eur. Phys. J. D* **2010**, *59*, 91–96. [CrossRef]
26. Barbay, S.; Ménesguen, Y. ; Sagnes, I.; Kuszelewicz, R. Cavity optimization of optically pumped broad-area microcavity lasers. *Appl. Phys. Lett.* **2005**, *86*, 151119. [CrossRef]
27. Dubbeldam, J.L.; Krauskopf, B. Self-pulsations of lasers with saturable absorber: Dynamics and bifurcations. *Opt. Commun.* **1999**, *159*, 325–338. [CrossRef]
28. Kharif, C.; Pelinovsky, E.; Slunyaev, A. *Rogue Waves in the Ocean*; Springer: Berlin/Heidelberg, Germany, 2009.
29. Pomeau, Y.; Manneville, P. Intermittent transition to turbulence in dissipative dynamical systems. *Commun. Math. Phys.* **1980**, *74*, 189–197. [CrossRef]
30. Bache, M.; Prati, F.; Tissoni, G.; Kheradmand, R.; Lugiato, L.A.; Protsenko, I.; Brambilla, M. Cavity soliton laser based on VCSEL with saturable absorber. *Appl. Phys. B* **2005**, *81*, 913–920. [CrossRef]
31. Elsass, T.; Gauthron, K.; Beaudoin, G.; Sagnes, I.; Kuszelewicz, R.; Barbay, S. Fast manipulation of laser localized structures in a monolithic vertical cavity with saturable absorber. *Appl. Phys. B* **2010**, *98*, 327–331. [CrossRef]
32. Chow, W.; Koch, S.; Sargent, M. *Semiconductor-Laser Physics*; Springer: Berlin, Germany, 1994.
33. Nicolis, G. *Introduction to Nonlinear Science*; Cambridge University Press: Cambridge, UK, 1995.
34. Ott, E. *Chaos in Dynamical Systems*; Cambridge University Press: Cambridge, UK, 2002.

35. Liu, Z.; Ouali, M.; Coulibaly, S.; Clerc, M.G.; Taki, M.; Tlidi, M. Characterization of spatiotemporal chaos in a Kerr optical frequency comb and in all fiber cavities. *Opt. Lett.* **2017**, *42*, 1063–1066. [CrossRef] [PubMed]
36. Manneville, P. Liapunov exponents for the Kuramoto-Sivashinsky model. In *Macroscopic Modelling of Turbulent Flows*; Springer: Berlin/Heidelberg, Germany, 1985; pp. 319–326.
37. Metayer, C.; Serres, A.; Rosero, E.J.; Barbosa, W.A.S.; de Aguiar, F.M.; Rios Leite, J.R.; Tredicce, J.R. Extreme events in chaotic lasers with modulated parameter. *Opt. Express* **2014**, *22*, 19850–19859. [CrossRef] [PubMed]
38. Granese, N.M.; Lacapmesure, A.; Agüero, M.B.; Kovalsky, M.G.; Hnilo, A.A.; Tredicce, J.R. Extreme events and crises observed in an all-solid-state laser with modulation of losses. *Opt. Lett.* **2016**, *41*, 3010–3012. [CrossRef] [PubMed]

*entropy*

MDPI

*Article*

# Wall-Normal Variation of Spanwise Streak Spacing in Turbulent Boundary Layer With Low-to-Moderate Reynolds Number

**Wenkang Wang, Chong Pan * and Jinjun Wang**

Institute of Fluid Mechanics, Beihang University, Beijing 100191, China; wwk@buaa.edu.cn (W.W.); jjwang@buaa.edu.cn (J.W.)

* Correspondence: panchong@buaa.edu.cn; Tel.: +86-010-8233-8069

Received: 19 November 2018; Accepted: 27 December 2018; Published: 31 December 2018

**Abstract:** Low-speed streaks in wall-bounded turbulence are the dominant structures in the near-wall turbulent self-sustaining cycle. Existing studies have well characterized their spanwise spacing in the buffer layer and below. Recent studies suggested the existence of these small-scale structures in the higher layer where large-scale structures usually receive more attention. The present study is thus devoted to extending the understanding of the streak spacing to the log layer. An analysis is taken on two-dimensional (2D) wall-parallel velocity fields in a smooth-wall turbulent boundary layer with $Re_\tau = 440\sim2400$, obtained via either 2D Particle Image Velocimetry (PIV) measurement taken here or public Direct Numerical Simulation (DNS). Morphological-based streak identification analysis yields a *Re*-independent log-normal distribution of the streak spacing till the upper bound of the log layer, based on which an empirical model is proposed to account for its wall-normal growth. The small-scale part of the spanwise spectra of the streamwise fluctuating velocity below $y^+ = 100$ is reasonably restored by a synthetic simulation that distributes elementary streak units based on the proposed empirical streak spacing model, which highlights the physical significance of streaks in shaping the small-scale part of the velocity spectra beyond the buffer layer.

**Keywords:** turbulent boundary layer; low speed streaks

## 1. Introduction

Low-speed streaks in wall-bounded turbulence, which were first observed by Hama and Nutant [1], Ferrell et al. [2], refer to narrow strips of low-momentum coherent motions extending lengthwise in the streamwise direction. These structures populate in the near-wall region, present quasi-regular distribution along the spanwise direction, and are always accompanied by trains of quasi-streamwise vortices with shorter length located in higher layer [3,4]. The origin of these streaks was attributed to the lift-up of low-momentum fluids from the wall under the induction of streamwise vortices [3,5–9], which transfers the energy from mean shear to turbulent fluctuations [10–12] and can be mathematically explained by a transient growth of three-dimensional (3D) disturbances due to the non-orthogonal eigenmodes in the linearized Navier-Stokes operator [13–16].

Low-speed streaks have been widely accepted as the building block of the inner-layer turbulent self-sustaining cycle [4,17–20]. The so-called bursting process, which usually denotes the whole dynamic process of the streak lift-up, oscillation and breakdown [19,21–23], was found to contribute to all the turbulent production and a large portion of the Reynolds stress generation in the buffer layer and below [17,19,22,24]. The generation and self-sustaining of near-wall streamwise vortices can be well explained by a streak transient growth mechanism [4], which was supported by observations that streak breakdown leads to the generation of either streamwise vortices or hairpin vortices dependent on the symmetry of the streak perturbation [25–27]. Hwang and Bengana [23] and Hwang et al. [28]

recently reported a self-sustaining process of attached eddies in the log layer and above, in which streaks and streamwise vortices with various length scales evolve in a way similar to those in the near-wall region.

One of the 'old' issues related to low-speed streaks is their spanwise length scales. Note that in early wall-parallel flow visualizations [29,30], streamwise vortices were not differentiated from streaks. These two tightly-related structures contribute equivalently to the spanwise spectra of $u$ component fluctuating velocity, as has been evidenced in Hwang [8] and Hwang and Bengana [23]. The spanwise spacing of neighboring streaks $\lambda$ (abbreviated as streak spacing in the following) thus serves as a typical measure of the lateral length scale of near-wall structures. It is well known that in the buffer layer and below, the mean streak spacing scaled by inner variables is $\overline{\lambda}^+ = \overline{\lambda} u_\tau / \nu \sim O(10^2)$ ($u_\tau$ is the friction velocity and $\nu$ the kinematic viscosity), and grows with respect to the wall-normal height $y^+$ ($y^+ = y u_\tau / \nu$) [29,31–35]. An asymptotic linear scaling $\overline{\lambda}^+ \sim 2y^+$ beyond $y^+$=10 was reported by Nakagawa and Nezu [36], who attributed it to the streak pairing process. Smith and Metzler [30] further suggested that merging and intermittency of streaks are responsible for the increase of $\overline{\lambda}^+$ in the region of $10 < y^+ < 30$.

Previous studies examining the streak spacing using different methods are summarized in Table 1. As can be seen, most of them focused on the streak spacing in the near-wall region and suggested a *Re*-independency of the wall-normal growth of $\overline{\lambda}^+$ below $y^+ = 30$. This idea is consistent with the traditional viewpoint that near-wall energetic dynamics are independent of outer region, which is supported by both a minimum turbulent channel DNS [19,37] and a turbulent channel DNS with large-scale motions being artificially removed [8,28,38]. However, the existence of large- and very large-scale motions (LSMs and VLSMs) in outer region, which significantly affect the production of Reynolds shear stress (RSS), turbulent kinetic energy (TKE) and skin friction [39–49], forms a high-*Re* effect [50,51] by the so-called outer-layer influence. Rao et al. [21] was one of the first to experimentally show that the bursting frequency of near-wall cycle scales on the boundary layer thickness $\delta$, which implies that large scales exert influence in the near-wall region. Bradshaw and Langer [52] reported a *Re*-dependency of the strength of near-wall streaks, which was recently deemed as an amplitude modulation of small-scale fluctuations by LSMs or VLSMs [50,53–59]. To our regards, the amplitude modulation does not conflict to the invariance of the length scale of small-scale coherent motions. Nevertheless, spectral analysis by Hoyas and Jiménez [60], Jiménez and Hoyas [61] and Hwang [8] all suggest a *Re*-dependency of the energetic small scales in spectral domain.

The value of studying the wall-normal variation of the lateral scale of small-scale streaks lies in the following considerations. First, the attached-eddy hypothesis [62–66] implies a linear growth of the spanwise length scale of energy-containing motions. Various scalings, i.e., $y \approx 1\lambda_z$ in Tomkins and Adrian [41], $y \approx 1/3\lambda_z$ in Del Álamo and Jiménez [11] and $y = 0.1\lambda_z$ in Hwang [67], have been proposed to characterize the wall-normal growth of the lateral length scale $\lambda_z$ of certain kind of large-scale structures. In addition, Baars et al. [68] recently identified a self-similar wall-attached structure whose streamwise/wall-normal aspect ratio is $\lambda_x/y \approx 14$. None of these scalings seems to be suitable for small-scale ones. Indeed, whether or not these small-scale structures present an attached-eddy behavior is still unclear. Study on this issue might promote the understanding of how small energetic scales originated from the near-wall region behave in higher flow layers and what kind of influence large-scale structures exert on them. Secondly, to inhibit streak-centered near-wall dynamics via riblet [69], discrete roughness elements [70] or active wall actuator [71], the streak spacing is a key parameter to be known in advance. Moreover, large-eddy simulation (LES) might get improved if the spanwise distribution of streaks in the near-wall region can be modeled correctly.

**Table 1.** Summary of previous studies on the spanwise spacing of low-speed streaks.

| Studies | Flow Type | $Re_\tau$ | $y^+$ | $\overline{\lambda}^+$ | $\Delta Z^+$ | Distribution | Method |
|---|---|---|---|---|---|---|---|
| Coantic [32] | Pipe flow | 2500 ($Re_\theta$) | $y^+ < 5$ | 110–130 | – | – | Hot-wire with correlation analysis |
| Schraub and Kline [72] | Boundary layer | 501 | $y^+ < 5$ | $100 \pm 20$ | – | – | Dye and $H_2$ bubble visualization |
| Kline et al. [29] | Boundary layer | 431, 501 | $y^+ \approx 2$ | 91, 106 | 500 | – | Dye and $H_2$ bubble visualization |
| Bakewell Jr and Lumley [73] | Boundary layer | ∼239 | $y^+ = 0$–$7$ | 80–100 | – | – | Hot-wire with space-time correlation |
| Gupta et al. [33] | Boundary layer | 870–2160 | $y^+ = 3.4$–$10.8$ | 97.5–151.2 | 373 | – | Hot-wire with short duration correlation |
| Lee et al. [34] | Pipe flow | 1735–2045 ($Re_\theta$) | $y^+ < 0.5$ | 105-107 | 250 | Lognormal | Electrochemical measurement with spatial correlation |
| Nakagawa and Nezu [36] | Channel flow | 318, 696 | $y^+ = 10$–$100$ | 100–1000 | 3000 | Lognormal | Hot-wire with conditional correlation |
| Smith and Metzler [30] | Boundary layer | 1040 | $y^+ = 1$–$30$ | 93–146 | 1000 | Lognormal | Hydrogen bubbles visualization |
| Kim et al. [74] | Channel flow | 180 | $y^+ = 1$–$23$ | 100–125 | 1150 | – | Averaged correlation |
| Klewicki et al. [35] | Atmospheric surface layer | $3 \times 10^5$ | $y^+ = 3.4$ | 93.1 | – | – | Fog visualization |
| Lagraa et al. [75] | Boundary layer | 1170 | $y^+ = 0$–$50$ | 100–180 | 216 | – | Electrochemical measurement with space-time correlation |
| Lin et al. [76] | Boundary layer | 7800 ($Re_\theta$) | $y^+ = 15$–$50$ | 110–120 | 320 | Rayleigh | Stereo-PIV with morphological analysis |

Based on these reasons, the present work is devoted to studying the streak spacing from the buffer layer to the upper bound of the log layer at low-to-moderate *Re*. One may argue that streaks only populate in the buffer layer and below, as has been stated by Smith and Metzler [30]: Due to the streak merging and coalescence event, 'for $y^+ \geqslant 30$ streak identification becomes very uncertain, such that a process of systematic visual streak counting becomes too subjective'. To our knowledge, this statement only stresses the difficulty in detecting streaks in higher layers. Ganapathisubramani et al. [42] identified eddy packets from PIV measured wall-parallel velocity fields via feature extraction algorithm, the most probable length and width of these structures were found to follow inner scaling even in the log layer, with magnitudes comparable to those of near-wall streaks (see Figure 3 in Ganapathisubramani et al. [42] for illustration).

Hwang [8] took a numerical experiment to show that near-wall streaks and streamwise vortices can survive in outer layer if large-scale motions in the log layer and the wake region are removed. The velocity-vorticity correlation structure in turbulent channel flow at $Re_\tau \approx 180$, as recently identified by Chen et al. [77], well captures the geometrical feature of near-wall streaks and streamwise vortices, its spanwise width follows a scaling of ${\lambda_z}^+ = 0.31y^+ + 30.3$ till $y^+ \approx 140$. Moreover, Lee et al. [48] attributed the primary source of the generation of outer-layer LSMs as the growing and merging of low-speed streaks which seem to lift from the near-wall region, which was supported by a DNS study of a minimum turbulent channel flow at low *Re* [78]. These studies imply the existence of streaks in higher layers. Here, we refer the term 'streak' as a generalized branch of small-scale structures, which share geometric and kinematic similarity with near-wall streaks and streamwise vortices. Note that LSMs and VLSMs are still streak-like, but are not included in this terminology due to their rather large length scale.

To study the streak spacing in a turbulent boundary layer, 2D velocity fields in multiple wall-parallel planes either measured by 2D planar PIV or sliced from 3D DNS dataset are analyzed. The studied Reynolds number covers a range of $Re_\tau = 440{\sim}2430$. Section 2 gives a brief description of the PIV/DNS dataset. Section 3 provides statistical evidence for the existence of small-scale streaks in flow layer beyond the buffer region. Section 4 deals with a morphological streak identification analysis, a log-normal distribution of the streak spacing with less *Re*-dependency is observed, and an empirical model is developed to account for its wall-normal growth from the buffer layer to the upper bound of the log layer. Finally, a simplified synthetic test is taken in Section 5. It is found that by only considering the distribution of spanwise-spaced streaks, the small-scale part of the spanwise spectra of the streamwise fluctuating velocity can be fairly well restored till the lower bound of the log layer. Concluding remarks are then given in Section 6.

## 2. Description of the PIV/DNS Dataset

### 2.1. Experiment Facilities and PIV Measurement Details

In the present study, both PIV-measured 2D wall-parallel velocity fields and DNS-obtained 3D volumetric velocity fields of a smooth-wall turbulent boundary layer are analyzed. The PIV dataset includes two configurations. One has a small field-of-view (FOV) comparable to most of the previous studies, and the other achieves a rather large FOV (on the order of $\delta$) to clarify the effect of limited FOV on the streak spacing statistics. In the following, $x/y/z$ denotes the streamwise/wall-normal/spanwise direction, and $u/v/w$ the corresponding fluctuating velocity component.

For the first measurement, a flat-plate turbulent boundary layer was developed on the bottom wall of the test section of a low-speed recirculating water channel in Beihang University. The test section of this facility is made of hydraulic-smooth glass and has a size of 3 m in length, 0.7 m in height, and 0.6 m in width. With a typical free-stream velocity $U_\infty$= 0.2 m/s, the free-stream turbulence intensity is about $T_u = 0.5\%$. Boundary layer transition was triggered by a tripping wire with a diameter of 3 mm placed at 0.1 m downstream the test section inlet. The sampling station located at 2.2 m downstream the tripping wire, where the boundary layer develops to full turbulence with satisfying zero-pressure-gradient (ZPG) condition. By changing $U_\infty$, three frictional Reynolds number $Re_\tau = u_\tau\delta/\nu$ = 444, 761 and 1014 were achieved. They are labeled as SE1∼SE3 in Table 2, with 'S' short for small FOV and 'E' for experiment.

The large FOV measurement was performed in a large low-speed recirculating water tunnel in Beihang University. This facility has a main test section with a size of 18 m in length, 1.2 m in height, and 1 m in width, and the typical $T_u$ is about 0.8% when $U_\infty$ = 0.5 m/s. A flat plate with a length of 15 m was vertically positioned in the main test section to develop a thick turbulent boundary layer. This flat plate was assembled from 5 hydraulic-smooth Acrylic plates with lengths of 3 m, widths of 1 m, and thicknesses of 20 mm. Its leading edge had a 4:1 half-elliptical shape to avoid local flow separation. The working surface has a distance of 0.75 m from the tunnel's side wall. For a typical boundary layer thickness $\delta < 0.2$ m or about 25% of the gap, the effect of the side wall is negligible. The water depth was 1.0 m, the wall-parallel PIV sampling region had a vertical span of about 0.268 m and was centered at 0.47 m below the free water surface and 0.53 m above the bottom wall, far enough to neglect the free-surface/bottom-wall effect. A tripping wire with a diameter of 3 mm was glued onto the working surface at 0.4 m downstream the leading edge. The PIV sampling station was 12 m downstream. More details of the setup of this measurement can be found in Wang et al. [79]. Two cases with $Re_\tau$ = 1135 and 2431 were measured, denoted as LE1 and LE2 in Table 2 ('L' for large FOV). Note that due to the long distance of the development, the boundary layer in the measurement station suffered a minor favorable pressure gradient (FPG), the acceleration parameter $K$ ($K = (\nu/U_\infty^2)dU_\infty/dx$) was $0.4 \times 10^{-7}$∼$0.5 \times 10^{-7}$. According to Harun et al. [80], a slight FPG condition will not significantly affect the energetic dynamics of large-scale structures in the outer region but only slightly increase their amplitude modulation degree to near-wall small-scale ones. We thus infer that the present minor FPG condition will not significantly bias the streak spacing statistics from other ZPG cases, this inference will be evidenced later.

Figure 1 shows the wall-normal profiles of both the mean streamwise velocity $\overline{U}^+(y^+)$ and the streamwise velocity fluctuation intensity $u^+_{rms}(y^+)$ obtained by a side-view 2D PIV measurement in $x$-$y$ plane for all the SE and LE cases. Note that the frictional velocity $u_\tau$ are estimated by the Clauser fit of the $\overline{U}^+(y^+)$ profiles with $\kappa$ = 0.41 and $B$ = 5.0 [81,82]. The empirical model of $u^+_{rms}(y^+)$ in Marusic and Kunkel [83] is supplemented in Figure 1b for a comparison. Figure 1b evidences that the minor FPG condition in the LE cases only slightly suppresses $u_{rms}$ in the outer region. Table 2 summarizes the characteristic boundary layer parameters, most of which in the SE and LE cases, i.e., the shape factor $H$ and the inner-scaled edge velocity $U^+_\infty$, are consistent with those in the canonical ZPG turbulent boundary layers well studied in the past [84–88].

Two-dimensional PIV was used to obtain instantaneous 2D velocity fields in multiple wall-parallel $x$-$z$ planes. The flow field was seeded with hollow glass beads whose median diameter was 10 µm and density 1.05 g/mm$^3$, and was illuminated by a double-pulsed laser sheet with thickness of about 1 mm issued from a Nd:YAG laser generator (Beamtech Vlite-500, Beijing, China) at energy output of 200 mJ/pulse. For the small-FOV LE cases, one CCD camera (Imperx ICL-B2520M, Boca Raton, FL, USA) with a resolution of 2456 × 2048 pixels was used for image recording. To guarantee a comparable inner-scaled magnification, a Nikkor 50 mm f/1.8D lens was used for the SE1 case and a Tamron 90 mm f/2.8D lens for the SE2 and SE3 cases. The FOV was 85 × 101 mm$^2$ (streamwise span $\Delta X$ × spanwise span $\Delta Z$) and 36 × 43 mm$^2$, respectively. The corresponding magnification were 0.24, 0.2 and 0.285 wall units/pixel. In the large-FOV LE cases, 8 synchronized CCD cameras (Imperx ICL-B2520M) mounted with Nikkor 50 mm f/1.8D lens were arranged in a 4 × 2 array with 10∼15 mm overlap in the image plane, and jointly provided a total FOV of 636 × 268 mm$^2$. The magnification was 0.39 and 0.96 wall units/pixel in the LE1 and LE2 case, respectively. The inner-scaled FOV are listed in Table 2. To explore the effect of the FOV truncation effect (in Section 4.1), velocity fields in the LE2 case will be sliced to a FOV span $\Delta Z^+$ = 1500, the same to that of the LE1 case when necessary. In both PIV configurations, around 3600 pairs of particle images were recorded at each measurement plane. The sampling repetition rate was 7.5 Hz in the SE cases and 5 Hz in the LE cases. The whole sampling duration $Tu_{\tau}/\delta$, as listed in Table 2, was large enough for the convergence of the second-order statistics of the fluctuating velocity.

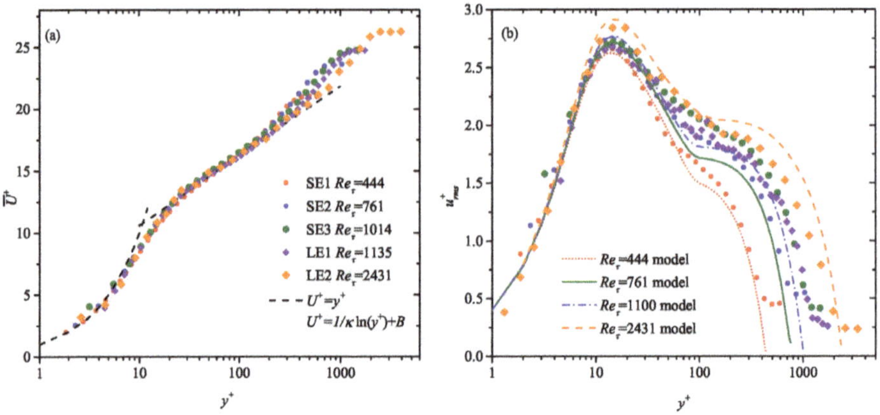


**Figure 1.** Wall-normal variation of (**a**) the mean streamwise velocity $\overline{U}^+(y^+)$ and (**b**) the streamwise velocity fluctuation intensity $u^+_{rms}(y^+)$ in the SE and LE cases. Straight dashed lines in (**a**) indicate the linear law and the log law, respectively. Curves in (**b**) are $u^+_{rms}(y^+)$ predicted by the empirical model of Marusic and Kunkel [83]. The present cases are represented by solid markers listed in Table 2. The same legend will be used in the following unless mentioned specifically.

**Table 2.** Summarization of characteristic boundary layer parameters. SE1~SE3 are small-field-of-view (FOV) particle image velocimetry (PIV) cases; LE1 and LE2 are large-FOV PIV cases; LD0~LD3 are large-FOV direct numerical simulation (DNS) cases from Simens et al. [89] and Sillero et al. [90,91].

| Cases | $U_\infty$ (mm/s) | $Re_\theta$ | $\delta$ (mm) | $H$ | $u_\tau$ (mm/s) | $Re_\tau$ | FOV $\Delta X^+ \times \Delta Z^+$ | Spatial Res. $\Delta x^+ \times \Delta z^+$ | $Tu_\tau/\delta$ | Marker |
|---|---|---|---|---|---|---|---|---|---|---|
| SE1 | 146 | 908 | 75.5 | 1.46 | 6.7 | 444 | 480 × 600 | 6 × 6 | 43 | ● |
| SE2 | 299 | 2044 | 65.8 | 1.39 | 13.1 | 761 | 400 × 500 | 5 × 5 | 97 | ● |
| SE3 | 455 | 3125 | 62.1 | 1.37 | 18.6 | 1014 | 560 × 700 | 7 × 7 | 144 | ● |
| LE1 | 145 | 2983 | 202 | 1.32 | 5.6 | 1135 | 4000 × 1500 | 9 × 9 | 22 | ◆ |
| LE2 | 340 | 5076 | 174 | 1.30 | 13.7 | 2431 | 8900 × 3750 (8900 × 1500) | 23 × 23 | 57 | ◆ |
| LD0 | 999 | 945 | 2.6 | 1.43 | 47.8 | 440 | 2000 × 1500 | 6 × 4 | - | ■ |
| LD1 | 1001 | 3100 | 7.6 | 1.38 | 40.3 | 1100 | 2000 × 1500 | 7 × 4 | - | ■ |
| LD2 | 1002 | 4800 | 11.4 | 1.37 | 38.1 | 1500 | 2000 × 1500 | 7 × 4 | - | ■ |
| LD3 | 1001 | 6500 | 15.4 | 1.36 | 36.8 | 2000 | 2000 × 1500 | 7 × 4 | - | ■ |

An optical flow solver based on the Lucas-Kanade algorithm was used to calculate 2D velocity fields from particle image pairs via GPU acceleration [92,93]. The interrogation window in the final pass was 48 × 48 pix with an overlap of 75%. The spatial resolution was about 6 wall units/vector in the SE cases and increased to 9 and 23 wall units/vector in the LE cases. The straddle time within the image pairs was selected to keep the maximum particle offset around 14~16 pixels in the image plane. The relative error of the velocity measurement was estimated to be less than 1%.

The optical system was mounted on a linear stage, allowing the wall-normal offset of the laser sheet at a resolution of 0.01 mm. A comparison of the $\overline{U}^+(y^+)$ and $u^+_{rms}(y^+)$ profiles obtained by the wall-parallel PIV measurement with those by side-view measurement showed satisfying consistency (not shown here for simplicity). The uncertainty of the laser sheet positioning was estimated to be around $\sigma^+_y$ = 1~3. In the large-FOV LE cases, a 45° inclined reflective mirror with length of 100 mm and width of 10 mm was positioned at 0.8 m downstream the end of the FOV, it reflected the laser sheet towards upstream to provide a large illumination extent without substantially affecting the upstream flow field. Cylindrical lenses with long focus length were used to keep the laser sheet thickness be around 1 mm over a distance of 2 m. The wall-parallel condition was checked by keeping the variation of the height of the laser sheet less than 0.5 mm over a distance of 1.5 m. Table 3 summarizes the wall-parallel planes being measured. According to Klewicki et al. [94] and Marusic et al. [82], the upper bound of the log layer can be estimated as around $y/\delta = 0.15$. The planes above this height are labeled with asterisks $*$. Note that the lowest measurement position was constrained by the laser sheet thickness, the wall reflection and the width of the immersed mirror, and was $y_{min}$ = 3 mm above the wall for the SE cases and $y_{min}$ = 5 mm for the LE cases.

### 2.2. DNS Dataset

Four DNS datasets of a spatially developing turbulent boundary layer over a smooth wall are also analyzed. As shown in Table 2, the LD0 case ('L' for large FOV and 'D' for DNS) with $Re_\tau = 440$ was obtained by Simens et al. [89], and the LD1∼LD3 cases with $Re_\tau = 1100 \sim 2000$ were obtained by Sillero et al. [90,91]. Readers can refer to Simens et al. [89], Sillero et al. [90,91], Borrell et al. [95] for detailed description about these DNS datasets.

Each LD case analyzed here contains 20 snapshots of instantaneous 3D volumetric velocity fields, which are available online (http://torroja.dmt.upm.es/ftp/blayers/). Planar velocity fields in multiple $x$-$z$ planes (as indicated in Table 3) were sliced from these snapshots with streamwise extent of 2000 wall units and spanwise extent covering the whole simulation domain (i.e., 6000 wall units for the LD0 case and about 16,000 wall units for the LD1∼LD3 cases). They were then cut into smaller sections with a size of $\Delta X^+ \times \Delta Z^+ = 2000 \times 1500$, making $\Delta Z^+$ comparable to those in the LE cases. This formed an ensemble of about 80 realizations in the LD0 case and 200 realizations in the LD1∼LD3 cases. As will be shown in Section 4.1 and Appendix B.3, the ensemble size is large enough for the convergence of the probability density function (PDF) of the streak spacing in the log layer and below. One advantage of DNS dataset is that the inner-layer is fully-resolved, which provides an ideal supplement for the PIV experiment which is limited by the lowest measurement plane.

**Table 3.** Summarization of wall-parallel planes being studied. Those planes at $y/\delta > 0.15$ are indicated by asterisks.

| Case | $Re_\tau$ | $\Delta Z^+$ | Wall-Normal Height $y^+$ | | | | | | | | | | | |
|---|---|---|---|---|---|---|---|---|---|---|---|---|---|---|
| SE1 | 444 | 600 | 17 | 24 | 29 | 35 | 47 | 59 | 76 * | 94 * | 118 * | 147 * | 182 * | 223 * |
| SE2 | 751 | 500 | | | | 35 | 46 | 58 | 70 | 93 | 116 * | 150 * | 185 * | 231 * |
| SE3 | 1014 | 700 | | | | | 49 | 65 | 81 | 98 | 131 | 163 * | | 212 * |
| LE1 | 1135 | 1500 | | | | 28 | | 57 | | | 113 | | | 226 * |
| LE2 | 2431 | 1500 | | | | | | | 70 | | | 140 | | 280 |
| LD0∼LD3 | 440∼2000 | 1500 | | | | | | | | 5∼223 | | | | |

## 3. Existence of Small-Scale Streak in Higher Layer

To study the streak spacing beyond the buffer layer, the first issue to be clarified is whether they exist in higher flow layer with statistical significance. Figure 2 illustrates typical instantaneous $u(x,z)$ fields in the LE1 case at $y^+$ = 28 (in the buffer layer) and $y^+$ = 226 (above the upper bound of the log layer). It can be visually identified that small-scale streaks and LSMs are dominant structures at $y^+$ = 28 and $y^+$ = 226, respectively. But structures with length scales far from the local most energetic scale are also observable in both flow layers.

For a quantitative description of such a multi-scale feature, a flow-field scale separation is desired. Fourier-based scale filtering was commonly used for this purpose [48,54], its limitation is the arbitrariness in the selection of the scale cutting-off threshold. Another popular method is the Empirical Mode Decomposition (EMD) and its derivatives [79,96], which empirically separates the length scales without reference to a fixed scale threshold. Nevertheless, EMD-based method usually requires a predetermined mode number, and its physical interpretation is not as clear as

Fourier decomposition. In the present study, Proper Orthogonal Decomposition (POD) is used as an alternative. POD has been used as a scale-filtering tool to isolate large-scale structures from small-scale ones in wall-bounded turbulence [45,97]. In essence, it decomposes a given space-time realization $\mathbf{V}(\mathbf{x},t)$ into a linear combination of a set of orthogonal bases whose spatial and temporal dimension are fully decoupled as:

$$\mathbf{V}(\mathbf{x},t) = \sum_{n=1}^{N} a_n(t)\,\phi_n(\mathbf{x}) = \underbrace{\sum_{n=1}^{s} a_n(t)\,\phi_n(\mathbf{x})}_{\mathbf{V}^L} + \underbrace{\sum_{n=s+1}^{N} a_n(t)\,\phi_n(\mathbf{x})}_{\mathbf{V}^H}. \tag{1}$$

In Equation (1), $a_n(t)$ is the time coefficient of the $n$th mode, $\phi_n(\mathbf{x})$ is the corresponding mode basis function and $N$ is the total number of the POD modes. The decomposition is based on an optimal energy recovery criteria, i.e., the TKE recovery using the POD mode bases is always the best for each level of reconstruction. In this sense, POD decomposes the flow-field ensemble by energy content, in distinct contrast to the scale-based decomposition methods (FFT filtering or EMD). The multi-scale structures in wall-bounded turbulence have different TKE contribution, so that they are projected onto different POD modes.

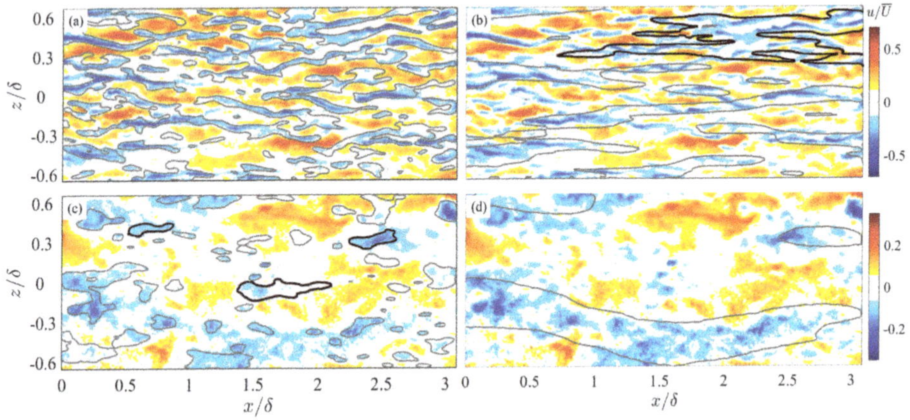


**Figure 2.** A snapshot of streamwise fluctuating velocity field $u/\overline{U}$ (pseudo-color maps) in the LE1 case at (**a**,**b**) $y^+ = 28$ and (**c**,**d**) $y^+ = 226$. Proper orthogonal decomposition (POD)-separated high- and leading-order field, i.e., $u^H$ and $u^L$, are superimposed in (**a**,**c**) and (**b**,**d**) as isolines, respectively. The solid isolines represent low-speed regions with the level of $u^{H|L} = -0.05\overline{U}$. The bold isoline in (**b**) indicates a region of the amalgamation of several small-streak streaks to form a large-scale structure in $u^L$, while bold isolines in (**c**) indicate streaks which are isolated from large-scale motions (LSMs) revealed in $u^L$.

Snapshot POD analysis [98,99] is applied to all the SE and LE cases. As a supplementary illustration, Appendix A illustrates the cumulative TKE contribution of all the POD modes, the characteristic spanwise length scale carried by each mode and the typical mode basis functions in the LE1 case. Since the POD modes are ranked by their relative TKE contribution $E_k = \lambda_k/\Sigma_{n=1}^{N}\lambda_n$ with $\lambda_n$ the eigenvalue of the $n$th mode, a cumulative energy cut-off threshold $P_s = \Sigma_{n=1}^{s}E_n$ can be set to separate all the POD modes into a leading-order group including the first $s$ modes and a high-order group containing the rest $N-s+1$ ones. Similar to Wu and Christensen [45] and Deng et al. [97], velocity field reconstruction using these two mode groups via the right part of Equation (1) is taken. This separates the original full-order $\mathbf{V}$ into a leading-order $\mathbf{V}^L$ and a high-order $\mathbf{V}^H$. In the following, the energy cut-off threshold is set as $P_s = 0.5$, i.e., $\mathbf{V}^L$ and $\mathbf{V}^H$ equally contribute to 50% of the total TKE. Additional tests showed that a moderate change of $P_s$ around 0.5 will not significantly affect the characteristic scales contained in $\mathbf{V}^L$ and $\mathbf{V}^H$. Appendix A further shows that all the modes with spanwise scale larger than $\delta$ are fully sorted into the leading-order group when $P_s = 0.5$.

The scale-separated velocity fields are visualized as the isolines of $u^{H|L} = 0.05\overline{U}$ superimposed onto the full-order $u$ field in Figure 2a–d, respectively. In the near-wall region, the footprint of outer-layer LSMs can be visualized as the amalgamation/coordination of several small-scale streaks to a larger one (see the structure highlighted with bold isolines in Figure 2b for example). While in the log layer and above, $u^H$ captures the core regions of LSMs. Small-scale structures independent from LSMs also appear now and then in $u^H$ (as indicated bold isolines in Figure 2c). They have reduced streamwise extent and expanded spanwise scale if compared to the near-wall streaks.

Figure 3 further shows the pre-multiplied spanwise spectra $k_z\Phi_{uu}(\lambda_z^+)$ of the LE1 case (with $Re_\tau = 1135$, bold solid curves) at various $y^+$. Agreement with those of the LD1 case (with $Re_\tau = 1100$, dashed curves in Figure 3) is observed. Both cases present a quick increase of the most energetic length scale from $\lambda_z^{\Phi+} \sim O(10^2)$ at $y^+ = 28$ (in Figure 3a) to $\lambda_z^{\Phi} \sim \delta$ at $y^+ = 113$ (in Figure 3c). The spectra profiles $k_z\Phi_{u^Lu^L}(\lambda_z^+)$ and $k_z\Phi_{u^Hu^H}(\lambda_z^+)$ of the POD-separated $u^L$ and $u^H$ are also shown in Figure 3 (as thin solid curves): the inner-scaled and outer-scaled spectrum peak can be now distinguished from each other at each flow layer. This evidences that the decoupling of TKE via POD does lead to the separation of length scales. More importantly, a distinct peak always appears in the $k_z\Phi_{u^Hu^H}$ spectra till the upper bound of the log layer. To our regards, this spectra peak is attributed to the streak-liked structures in $u^H$ as visualized in Figure 2, the corresponding peak $\lambda_z$ can thus be interpreted as the characteristic spanwise scale of these structures. This is supported by the observation that $\lambda_z$ of the $k_z\Phi_{u^Hu^H}$ peak always has correspondence with the most probable streak spacing (dashed lines in Figure 3), which will be discussed in Section 4.

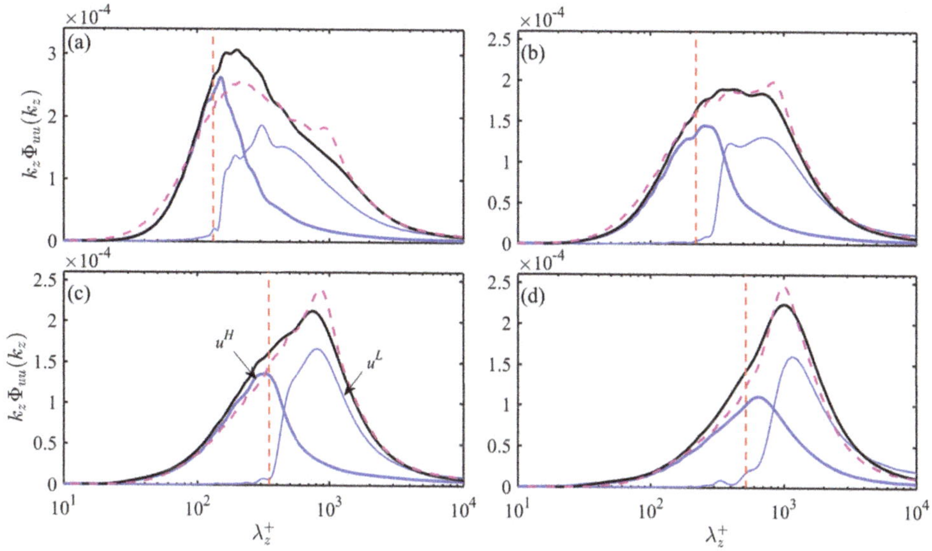


**Figure 3.** Comparison of premultiplied spanwise spectra $k_z\Phi_{uu}$ of the full-order $u$ fields with $k_z\Phi_{u^{L|H}u^{L|H}}$ of the POD-separated leading- or high-order $u^{L|H}$ fields in the LE1 case at **(a)** $y^+ = 28$; **(b)** $y^+ = 57$; **(c)** $y^+ = 113$ and **(d)** $y^+ = 226$. $k_z\Phi_{uu}$ in the LD1 case is also given for a comparison. ——: $k_z\Phi_{uu}$ of LE1; – – : $k_z\Phi_{uu}$ of LD1; ——: $k_z\Phi_{u^{L|H}u^{L|H}}$ of LE1 with $P_s = 0.5$. Vertical dashed lines are the most probable streak spacing $\lambda_{mp}^+$ predicted by the empirical model in Section 4.2.

The prevalence of small-scale streak-liked structures in higher layers can be further evidenced by the map of two-point correlation coefficient $\overline{R}$, which was widely used in previous researches [36,41,43,53,73,91]:

$$\overline{R}_{\chi\chi}(r_x, r_z, y_{ref}) = \frac{\langle \chi(x, y_{ref}, z, t) \cdot \chi(x + r_x, y_{ref}, z + r_z, t) \rangle}{\sigma_\chi^2} \quad (2)$$

In Equation (2), $\chi$ is $u$, $u^L$ or $u^H$, $r_x$ and $r_z$ are the $x/z$ offset from the reference point, $\sigma_\chi$ is the standard deviation of $\chi$, and $\langle\cdot\rangle$ the average over both the spatial and temporal domain. Figure 4 plots $\overline{R}_{u^H u^H}$ and $\overline{R}_{u^L u^L}$ at $y^+ = 28$ and 226 in the LE1 case with $P_s = 0.5$. $\overline{R}_{uu}$ of the full-order $u$ fields are supplemented in Figure 4b,d (as bold isolines). As shown in Figure 4b,d, $\overline{R}_{u^H u^H}$ and $\overline{R}_{u^L u^L}$ in inner and outer region are both characterized as streamwise elongated structures with length scales sufficiently gaped from each other, resembling those instantaneous structures shown in Figure 2. A characteristic spanwise scale $\overline{\lambda_c}$ can be defined as the gap between the two negative valleys as illustrated in Figure 4a. The spanwise scale of $\overline{R}_{u^L u^L}$ and $\overline{R}_{uu}$ are both $\overline{\lambda_c} \sim \delta$, consistent with previous studies that showed the validity of the outer scaling of $\overline{R}_{uu}$ even in the buffer region of high *Re* TBL [41,43,53]. Figure 4a,c shows that $\overline{R}_{u^H u^H}$ present streak-liked pattern in both $y^+ = 28$ and $y^+ = 226$. This provides a statistical evidence for the existence of small-scale streaks beyond the buffer region. Furthermore, $\overline{\lambda_c}$ of $\overline{R}_{u^H u^H}$ is much smaller than those of $\overline{R}_{u^L u^L}$, the the magnitude is rather close to the most probable streak spacing to be shown in Section 4.2.

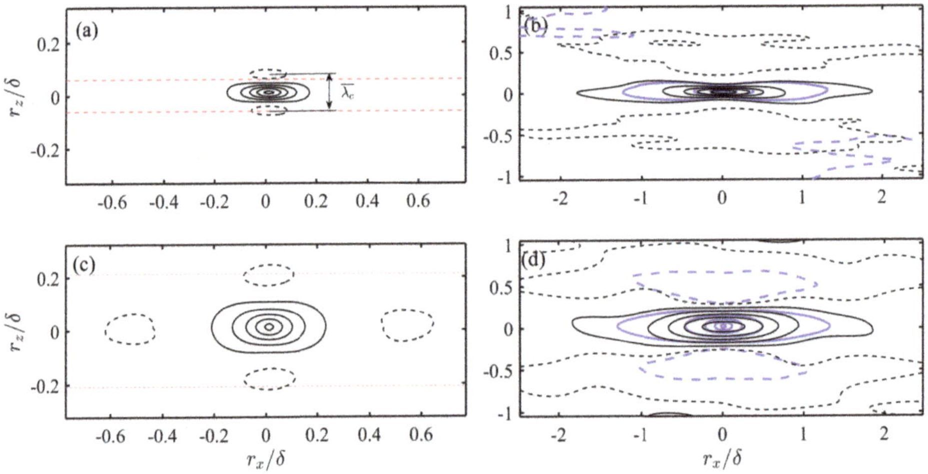


**Figure 4.** Two-point correlation map of POD-separated $u^{L|H}$ fields in the LE1 case with $P_s = 0.5$. (**a**) $\overline{R}_{u^H u^H}$ at $y^+ = 28$; (**b**) $\overline{R}_{u^L u^L}$ at $y^+ = 28$; (**c**) $\overline{R}_{u^H u^H}$ at $y^+ = 226$; (**d**) $\overline{R}_{u^L u^L}$ at $y^+ = 226$. Thin solid/dashed isolines represent positive/negative correlation with contour level uniformly spaced from −0.1 to 0.9 with a gap of 0.2. $\overline{R}_{uu}$ of the full-order $u$ fields are superimposed in (**b**,**d**) as bold solid/dashed isolines. The interval between the horizontal thin dashed lines in (**a**,**c**) indicates the most probable streak spacing $\lambda_{mp}$ predicted by the empirical model in Section 4.2.

## 4. Streak Spacing Based on Morphological Identification

To further study the spanwise spacing of neighboring streaks, the morphological-based streak identification algorithm proposed by [76] is used in this section with slight modification. The essence of this algorithm is to isolate low-speed streak-liked regions by binarizing $u(x, z)$ snapshots with a pre-given velocity deficit threshold and extract their skeletons with the aid of computer vision technique. The streak spacing is then counted as the spanwise gap between two adjacent low-speed streak skeletons only if at least one high-speed streak skeleton is clapped in between. The algorithm details are given in Appendix B.1. It is stressed that this algorithm does not differentiate small-scale structures from large-scale ones, but only finds the nearest gap between two neighboring streak-liked structures. Nevertheless, Figure 2 shows that even the core region of LSMs is clustered with small-scale coherent motions, thus the streak spacing obtained by this algorithm will represent the typical spanwise scale of the smallest energetic structures.

In this algorithm, there are a set of parameters, i.e., the velocity deficit threshold, the non-streak filter, and the skeleton extracting parameters, that should be set manually. As shown in Appendix B.2, their influence on the streak skeleton extraction is rather weak, any moderate change from the selected parameter combination will only lead to a change of the statistics of the streak spacing less than 15%.

This morphological algorithm is applied to all the present studied cases. The ergodic state to account for the streak pattern variation, i.e., streak splitting or merging, is achieved by counting the streak spacing at streamwise stations gaped as $\Delta x^+ \approx 30$ in every snapshot. This forms an ensemble with samples more than $O(10^6)$ for the SE cases, $O(10^7)$ for the LE cases and $O(10^6)$ for the LD cases in the near-wall region. However, due to the reduced streak population (to be discussed in Section 4.2), the ensemble size drops to $O(10^4)$, $O(10^5)$ and $O(10^4)$ in the log layer, respectively. A convergence test is taken in Appendix B.3 to show that for all the studied cases, both the PDF of the streak spacing and the related statistics get acceptable convergence till the upper bound of the log layer.

### 4.1. Streak Spacing Distribution

Figure 5 gives an overview of the wall-normal variation of the PDF of inner-scaled streak spacing $\lambda^+$ in the SE and LD cases. Every PDF profile $P(\lambda^+)$ presents a single peak without a sign of bi-modal pattern even in the upper bound of the log layer. The long tail of $P(\lambda^+)$ extends towards the large value side to form a left-skewed shape. The most probable streak spacing $\lambda^+_{mp}$ increases monotonically with $y^+$, while $P(\lambda^+_{mp})$ decreases, in together with a distinct growth of the long tail. However, a so-called 'truncation effect' is observed in higher layers of the SE cases: due to the limited FOV span ($\Delta Z^+ = 500{\sim}700$), those events with potentially large $\lambda^+$ are not detected, making a remarkable shortening of the long tails of $P(\lambda^+)$ if compared to those in the LD cases. Note that truncation effect has been inferred by Smith and Metzler [30] as 'This result (biased streak spacing) was felt to be a consequence of the narrowness of the data window'. A detailed inspection of all the $P(\lambda^+)$ profiles shows that for the LE and LD cases whose FOV span $\Delta Z^+ = 1500$ is rather large, the truncation effect is minor; while for the LE cases, once $\lambda^+_{mp}$ is far from $\Delta Z^+$, the truncation of the long tail part of $P(\lambda^+)$ will not bias the value of $\lambda^+_{mp}$ but change the overall probability level.

As summarized in Table 1, Lee et al. [34] and Smith and Metzler [30] reported a log-normal distribution of the streak spacing $\lambda^+$ in the near-wall region, while an alternative Rayleigh distribution was claimed by Lin et al. [76]. These two distributions are:

$$P(\lambda^+) = \frac{1}{\lambda^+\sigma\sqrt{2\pi}}\exp\left(-\frac{(\ln\lambda^+ - \mu)^2}{2\sigma^2}\right), \tag{3}$$

$$P(\lambda^+) = \frac{\lambda^+}{s^2}\exp\left(-\frac{\lambda^{+2}}{2s^2}\right), \tag{4}$$

with free parameters ($\mu$, $\sigma$) and $s$, respectively. Both models are evaluated by the raw $P(\lambda^+)$ profiles via a least-square fitting. The fitting determination coefficients $R^2$, as shown in Figure 6a, suggests that the log-normal model outperforms the Rayleigh model everywhere, but the former still presents a performance drop beyond $y^+ \approx 30$, which is more prominent in the SE cases with smaller FOV span. This is attributed to the deteriorated truncation effect that begins to distort the $P(\lambda^+)$ profiles in higher layer.

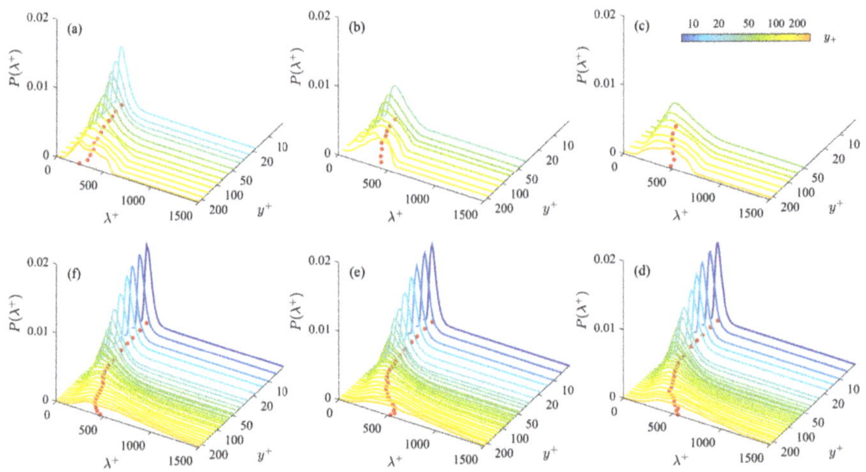


**Figure 5.** Wall-normal variation of the PDF of the streak spacing $P(\lambda^+)$. (**a**) SE1; (**b**) SE2; (**c**) SE3; (**d**) LD1; (**e**) LD2; (**f**) LD3. The local maxima of the probability density function (PDF) are projected on the $\lambda^+$-$y^+$ plane as solid dots.

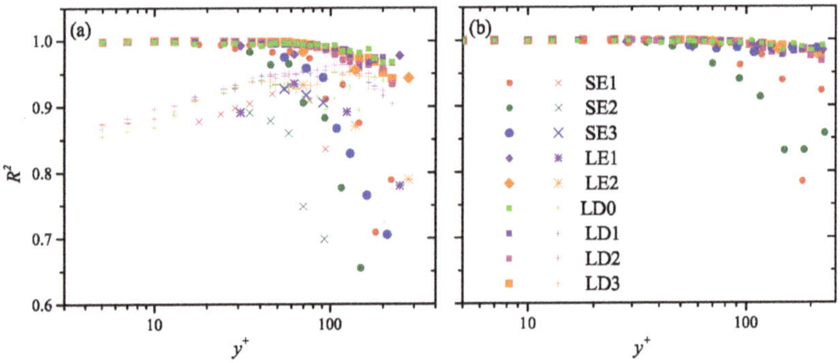


**Figure 6.** Wall-normal variation of the determination coefficient $R^2$ in all the studied cases. (**a**) $R^2$ of both the log-normal fitting via Equation (3) (solid circular, diamond, or rectangle) and the Rayleigh fitting via Equation (4) (diagonal cross, star, and cross); (**b**) $R^2$ of the dimensional constraint log-normal fitting via Equation (5).

A dimensional constraint log-normal fitting is proposed to compensate for the truncation effect. This fitting is based on the observation that in a non-severe truncation case where the most probable $\lambda^+_{mp}$ is far from the FOV span $\Delta Z^+$, only the long tail of $P(\lambda^+)$ close to $\Delta Z^+$ is truncated, in together with the enhancement of the probability level of smaller $\lambda^+$ events. This is clearly seen in Figure 7 which highlights the difference in the $P(\lambda^+)$ profiles between the SE3 case and the LE1/LD1 cases with similar $Re_\tau$. A log-normal fitting can then be applied to the dimensional frequency number distribution $n(\lambda^+)$ instead of the non-dimensional probability $P(\lambda^+)$ via

$$n(\lambda^+) = \frac{\alpha}{\lambda^+ \sigma\sqrt{2\pi}} \exp\left(-\frac{(\ln\lambda^+ - \mu)^2}{2\sigma^2}\right), \; 0 < \lambda^+ < C\Delta Z^+. \quad (5)$$

In Equation (5), only $n(\lambda^+)$ on the left of $\lambda^+ = C\Delta Z^+$ is fitted. The parameter $C$ regulates the fitting range and is manually fixed as 0.8 here, i.e., the right 20% part of the $n(\lambda^+)$ profile is rejected in this fitting. An additional free parameter $\alpha$ appears in Equation (5), it allows the floating of the integral area of the $n(\lambda^+)$ profile.

The validity of this dimensional constraint fitting is illustrated in Figure 7. It shows that even in the log layer ($y^+ \approx 130$), the raw $P(\lambda^+)$ profiles in the LE1/LD1 cases present Gaussian shape with satisfying symmetry in a logarithmically scaled $x$ axis. In contrast, the truncation of the long tails of $P(\lambda^+)$ of the SE3 case (hollow square markers), due to the insufficient FOV span, leads to asymmetrical profiles. The raw $P(\lambda^+)$ profiles (hollow square markers) of the SE3 case at $y^+ \approx 60$ and 130 are then fitted via both canonical log-normal model (Equation (3)) and the dimensional constraint version (Equation (5)). The latter leads to a more reasonable prediction of $P(\lambda^+)$ (diagonal cross markers) if compared to the raw profiles in the LE1/LD1 cases (dashed/solid lines) which are believed to be less affected by the truncation effect.

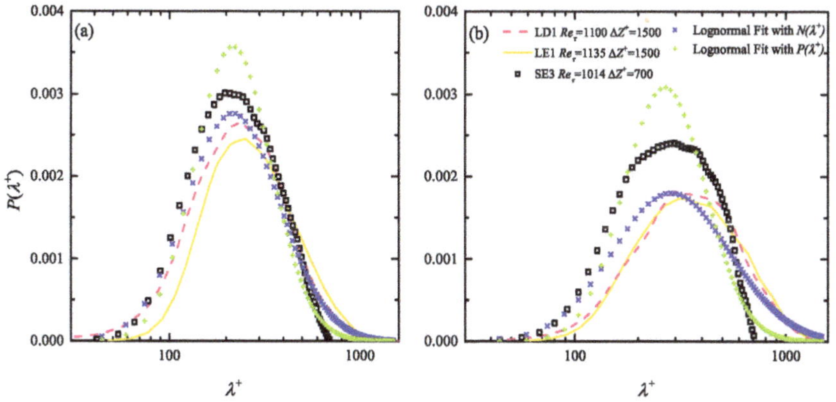


**Figure 7.** Illustration of the feasibility of the dimensional constraint log-normal fitting via Equation (5) in the SE3 case at (**a**) $y^+$ = 65; (**b**) $y^+$ = 131. Dashed and solid curves are $P(\lambda^+)$ at similar $y^+$ in the LD1 and LE1 cases where the field-of-view (FOV) truncation effect are minor and $Re_\tau$ are similar. Rectangle markers are the raw $P(\lambda^+)$ profiles in the SE3 case, diagonal cross markers are the estimations via Equation (5), while cross markers are the estimations via Equation (3).

With this truncation compensation strategy, the performance of the log-normal model, as can be seen in Figure 6b, gets persistently improved. The enhancement of $R^2$ is quite remarkable for the SE cases. For the LE/LD cases, the magnitude of $R^2$ in higher layer ($y^+ > 100$) elevates above 0.98, indicating a good accordance to the log-normal model. Nevertheless, $R^2$ in the SE1/SE2 cases beyond $y^+ = 100$ is still smaller than 0.9, the reason is that the most probable part of these $P(\lambda^+)$ profiles are rather close to $\Delta Z$, making the full compensation of the truncation effect rather difficult.

### 4.2. An Empirical Model for Streak Spacing

Given a log-normal distribution of $P(\lambda^+)$, the mean and the most probable streak spacing, i.e., $\overline{\lambda}^+$ and $\lambda^+_{mp}$, can be determined by the controlling parameters $\mu$ and $\sigma^2$ in Equation (3) or (5):

$$\begin{aligned} \overline{\lambda}^+ &= e^{\mu+\sigma^2/2} \\ \lambda^+_{mp} &= e^{\mu-\sigma^2} \end{aligned} . \tag{6}$$

Figure 8 summarizes $\mu$ and $\sigma^2$ in all the studied cases estimated by the dimensional constraint fitting via Equation (5). Except for the SE1 case with $y^+ > 100$, $\mu$ is independent from $Re$ till $y^+ \approx 220$. In contrast, $\sigma^2$ presents a non-negligible scattering beyond $y^+$=100. Note that the scattering level is $\Delta\sigma^2 \sim O(10^{-1})$, more than one-order smaller than the magnitude of $\mu$; therefore, its contribution to $\overline{\lambda}^+$ and $\lambda^+_{mp}$ in Equation (6) is comparably small. The smaller magnitude of $\mu$ in the SE1 case with

$y^+ > 100$ is attributed to the inability of compensating the truncation effect when $\lambda_{mp}$ is rather close to the FOV span $\Delta Z$. For a test, the LD0 case with a similar $Re_\tau$ is resampled with the same FOV as SE1, i.e., $\Delta Z^+$=600. The magnitudes of $\mu$ (hollow squares with cross markers in Figure 8a) are now comparable to those of the SE1 case, and $\sigma^2$ (in Figure 8b) also get distinctly reduced.

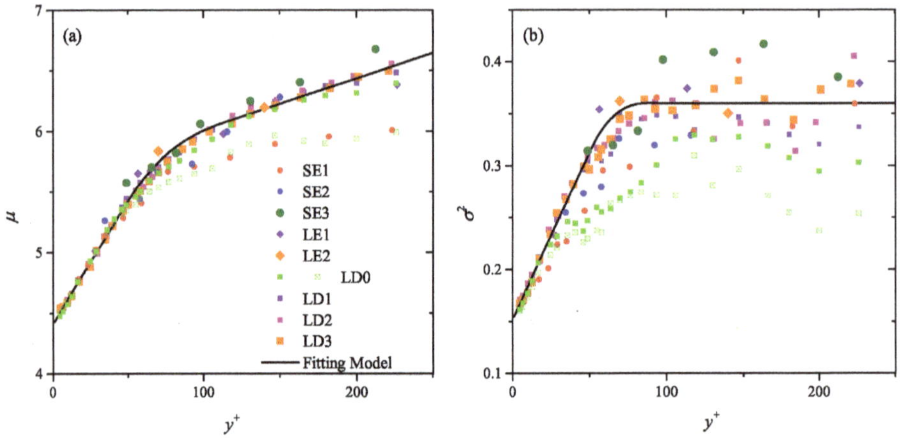


**Figure 8.** Wall-normal variation of the free parameter (**a**) $\mu$ and (**b**) $\sigma^2$ in log-normal distribution. Solid markers are the estimations by dimensional constraint model (Equation (5)) in all the studied cases, and bold solid lines are the two-stage linear model of Equations (7) or (8). Hollow squares with cross markers indicate the truncated LD0 case with $\Delta Z^+ = 600$.

$\mu$ and $\sigma^2$ in the LD cases, which are less affected by the truncation effect, are used to construct an empirical model from the buffer layer to the upper bound of the log layer ($y/\delta \sim 0.15$):

$$\mu = \begin{cases} 0.02y^+ + 4.4, \; 10 < y^+ < 50 \\ 4.2 \times 10^{-3}y^+ + 5.60, \; 100 < y^+ < \min(0.15\delta^+, 220) \end{cases} \tag{7}$$

$$\sigma^2 = \begin{cases} 3.2 \times 10^{-3}y^+ + 0.15, \; 10 < y^+ < 50 \\ 0.36, \; 100 < y^+ < \min(0.15\delta^+, 220) \end{cases} \tag{8}$$

This empirical model includes two linear stages of $\mu(y^+)$ and $\sigma^2(y^+)$, i.e., within $y^+$=10∼50 and beyond $y^+ = 100$, which are bridged by a cubic fitting in the middle. As shown in Figure 8, it fairly predicts $\mu(y^+)$ and $\sigma^2(y^+)$ of the SE and LE cases till the upper bound of the log layer. Using this model, the wall-normal variation of $\overline{\lambda}^+$ and $\lambda^+_{mp}$ can be predicted via Equation (6).

The validity of this empirical model can be evidenced by the following two aspects. Firstly, as shown in Figure 9, $\overline{\lambda}^+(y^+)$ and $\lambda^+_{mp}(y^+)$ of the SE and LE cases (solid dots), which are not used to construct this model, generally collapse onto the model's prediction (bold solid lines) till the upper bound of the log layer. Moreover, $\overline{\lambda}^+$ in the near-wall region reported by most of previous studies [30,72,74,75] (hollow markers in Figure 9a) are also compatible with it. Two exceptions are Nakagawa and Nezu [36] and Lin et al. [76], who reported remarkably smaller $\overline{\lambda}^+$ beyond $y^+ = 30$. To our regards, this might be related with either the condition invoked in the conditional correlation calculation in Nakagawa and Nezu [36] or the insufficient FOV span ($\Delta Z^+ = 320$) in Lin et al. [76]. For the latter, the LD3 case with similar $Re$ is resampled by $\Delta Z^+ = 300$, this leads to a reduced $\overline{\lambda}^+$ (hollow squares with cross markers in Figure 9a) consistent with those in Lin et al. [76].

Secondly, the linear scaling of $\mu$ and $\sigma^2$ in Equations (7) and (8) leads to a two-stage exponential growth of $\overline{\lambda}^+$. Since the mean streak population density $\Pi$, which measures the average number of streaks per unit span, is the inverse of $\overline{\lambda}^+$, a two-stage exponential decay of $\Pi$ is expected. As shown in Figure 10a, $\Pi(y^+)$ obtained by counting the number of the identified streaks in the whole snapshot ensemble, present less scattering among all the studied cases. The general trend follows a two-sectional decay gaped at $y^+ \approx 50$, and fairly agrees with the prediction of the empirical model till the upper bound of the log layer, again evidencing the validity of the latter.

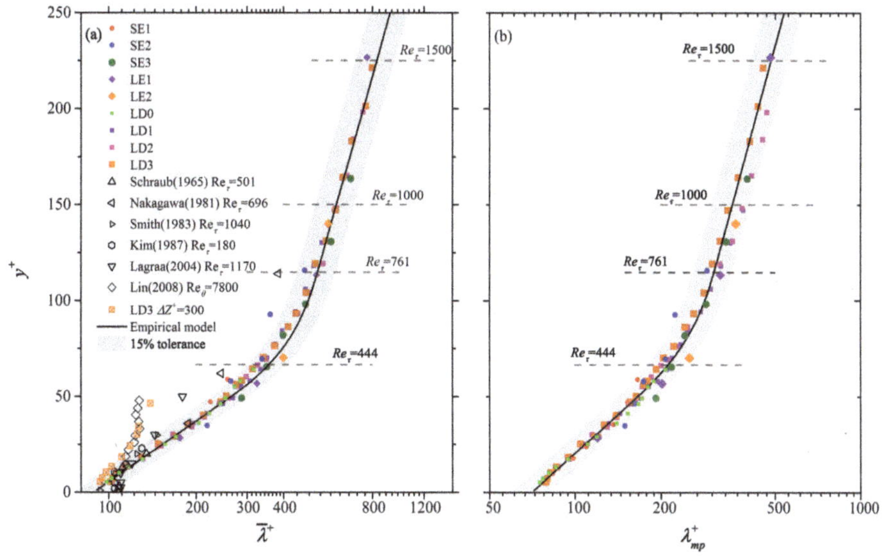


**Figure 9.** Wall-normal variation of (**a**) $\overline{\lambda}^+$ and (**b**) $\lambda^+_{mp}$ till the upper bound of log layer. Solid markers are the estimations by Equation (5) in all the studied cases. Only the data below the upper bound of the log layer, i.e., $y/\delta \sim 0.15$ indicated as dashed horizontal lines for typical $Re_\tau$ in (**a**), is shown. The same in Figures 10 and 11. Hollow markers in (**a**) are $\overline{\lambda}^+$ obtained by previous studies listed in Table 1. Hollow squares with cross markers indicate the truncated LD3 case with $\Delta Z^+ = 300$. Bold solid curves are predictions of the empirical model of Equations (6)∼(8), with shaded regions indicating a ±15% tolerance.

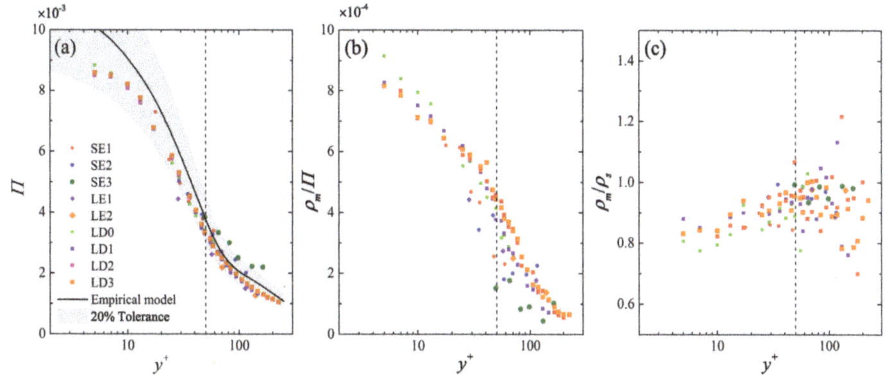


**Figure 10.** Wall-normal variation of (**a**) the mean streak population density $\Pi$; (**b**) the streak merging frequency $\rho_m/\Pi$ and (**c**) the ratio between the streak merging and splitting frequency $\rho_m/\rho_s$ till the upper bound of the log layer ($y/\delta \sim 0.15$). In (**a**), bold solid lines are the prediction of the empirical model of Equations (6)∼(8), with the shaded regions indicating a ±20% tolerance. Vertical dashed lines indicates the flow layer of $y^+ = 50$, the same in Figure 11.

### 4.3. Discussion on the Empirical Model for Streak Spacing

The proposed empirical model (Equations (6)–(8)) describes a *Re*-independent two-stage exponential growth of the mean streak spacing along the wall-normal direction, the second stage of which has a reduced growth rate beyond $y^+ = 50$. For a kinematic explanation of such a growth trend, the wall-normal variation of the streak amplitude $A_s$ is first investigated. Here, $A_s$ measures the peak momentum deficit within one streak. In the present study, it is simplified as the normalized local streamwise fluctuation velocity $u/\overline{U}$ on the identified streak skeleton.

Figure 11a shows the PDF of $A_s$ at typical $y^+$ in the LE1 and LD1 cases with equivalent $Re_\tau$. Both cases present similar $P(A_s)$ profiles with a single peak and a long tail extending towards the left side. With the increase of $y^+$, a shrink of the spread of $A_s$ is seen, in together with the right shift of the most probable value $A_{s,mp}$. Figure 11b shows the wall-normal variation of $A_{s,mp}$ in all the studied cases. It reveals a minor growth of the magnitude of $A_{s,mp}$ below $y^+ = 10$, where $A_{s,mp}$ is mildly correlated with $Re_\tau$. This indicates both the active streak generation events in this region and the *Re*-dependent amplitude modulation effect that is consistent with the previous observation by Bradshaw and Langer [52]. Beyond $y^+ = 10$, $A_{s,mp}$ shows a constant decay till $y^+ = 50$, and then slowly asymptotes to the streak binarization threshold $u/\overline{U} = -0.1$ used in the streak identification algorithm.

The correlation coefficient $R_{A_s,\lambda}$ between $A_s$ and $\lambda^+$, which measures the degree of the relationship between the strength of one streak and its spanwise spacing to the nearest neighborhood, are summarized in Figure 11c. A minor negative correlation, i.e., $R_{A_s,\lambda} \approx -0.1$ is observed above the viscous sublayer, indicating that stronger streaks prefer to be spaced further away from its neighborhood. Such a correlation gradually relaxes towards $R_{A_s,\lambda} = 0$ in higher layers, and the relaxation rate sharply accelerates beyond $y^+ = 50$.

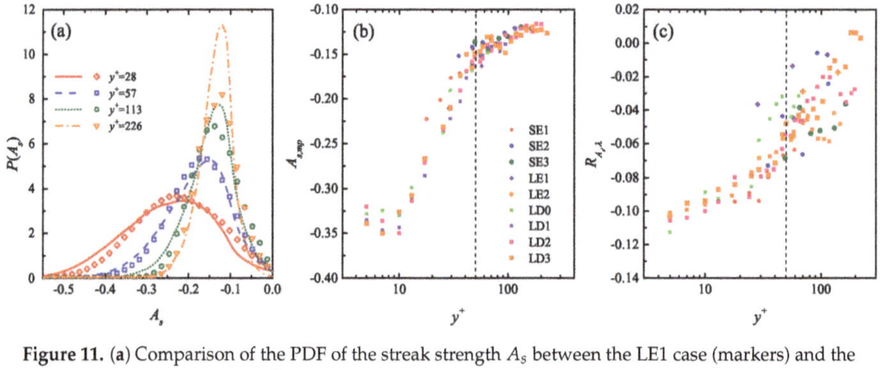


**Figure 11.** (**a**) Comparison of the PDF of the streak strength $A_s$ between the LE1 case (markers) and the LD1 cases (curves) at various $y^+$; (**b**) wall-normal variation of the most probable $A_{s,mp}$; (**c**) wall-normal variation of the correlation coefficient $R_{A_s,\lambda}$ between $A_s$ and $\lambda^+$ in all the studied cases till $y/\delta = 0.15$.

To our regards, the first fast growth stage of $\overline{\lambda^+}(y^+)$ below $y^+ = 50$ can be mainly attributed to the streak merging scenario. Smith and Metzler [30] proposed that the streak merging behavior is most active in the range of $10 < y^+ < 30$. Tomkins and Adrian [41] observed that neighboring streaks merge with each other frequently at $20 < y^+ < 100$, but the merging frequency remarkably drops beyond $y^+ = 100$. Note that for exponential decay of a variable, e.g., the streak population density $\Pi(y^+)$ shown in Figure 10a, the decay rate is proportional to the variable's magnitude. This is the case for the streak merging scenario: The more crowded streak distribution, the more chance for the occurrence of streak merging, thus leads to both the sharp reduction of $\Pi$ and $A_s$ and the quick growth of $\lambda^+$.

Another attractive property of streak merging scenario is that it does not destroy the log-normal distribution of the streak spacing beyond the buffer region, which is clearly shown in Section 4.2. As stated by Smith and Metzler [30], 'a random variable will develop a log-normal distribution when the independent influences cause variations which are proportional to the variable. Thus the log-normal distribution of streak spacing would seem to indicate that the independent physical influences which affect the variations in streak spacing are in some manner dependent up on the relative value of the streak spacing itself.' On considering that the merging rate is strongly dependent on the streak spacing, the streak merging scenario, to our regards, might be a possible candidate for such 'physical influences'.

For a quantitative description, the streak merging events are counted from instantaneous snapshots as where a pattern of two neighboring low-speed streak skeletons converging into one is identified. The related detection algorithm is briefly described in Appendix B.1. Figure 10b summarizes the merging frequency $\rho_m/\Pi$, in which $\rho_m$ is the average number of the streak merging event per unit span. Figure 10c further shows the ratio between the streak merging and splitting frequency $\rho_m/\rho_s$, the latter is counted via a similar scheme. It is clearly shown that $\rho_m/\Pi(y^+)$ of all the studied cases follow a two-sectional decay gapped at about $y^+ = 50$, similar to that of $\Pi(y^+)$. This is consistent with the observation of Smith and Metzler [30] and Tomkins and Adrian [41], and highlights a strong correlation between the streak population and the streak merging frequency: the amalgamation of two neighboring streaks will leave the signature of only one streak in higher layer; as a consequence, the increased streak spacing there will lower the local streak merging frequency.

Interestingly, the streak splitting event, which serves as a counter-acting role of inhibiting the streak spacing growth, has a slightly higher frequency than that of the streak merging event in the near-wall region. However, such an in-equilibrium gradually diminishes with the increase of $y^+$. A detailed examination of instantaneous velocity fields show that new-born streaks through streak splitting always have comparably weaker strength and shorter length; while in a streak merging event, the merged streak tends to pose weaker peak strength but broader width. Therefore, both events contribute to the wall-normal decay of the streak strength, and the latter weighs more to promote the quick growth of the streak spacing in the near-wall region.

For those streaks with stronger strength and gaped further away from others, they have more chance to survive through the active streak instability process in the near-wall region. Recalling that the second stage of the empirical streak spacing model presents a linear growth of $\mu$ with reduced slope but a quasi-constant $\sigma^2$ beyond $y^+ = 100$. Since $\sigma^2$ characterizes the width of the $P(\lambda^+)$ profile, the constant $\sigma^2$ implies a passive streak dynamics in this region: due to the rather large streak spacing, the streak merging/splitting in higher flow layer is inactive; instead, those small-scale streak-liked structures, most of which might be the remnants of near-wall streaks, act as being 'frozen', i.e., they can be either synchronized to larger scales to form the core region of LSMs or gradually dissipated by viscosity.

## 5. Synthetic Simulation of the Spanwise Spectra

In this section, we attempt to restore part of the spanwise spectra of $u$ through synthetic simulation by only considering the spanwise distribution of streaks that is independent of $Re$. One of the practical meanings of this attempt is that it might promote the understandings on how large-scale structures affects the spectra to formulate a $Re$-dependency, and it might provide useful information for the development of the near-wall model in LES.

The idea is to randomly distribute multiple elementary streak units along spanwise direction with spacing determined by the empirical model developed in Section 4.2. For simplicity, only 1D scenario, i.e., the spanwise variation of the $u$ component fluctuation velocity, is considered here. The elementary streak unit follows the model proposed by Hutchins and Marusic [43]:

$$\theta\left(z_i\right) = \pi z_i/\lambda_z, -\frac{3}{2}\lambda_z < z_i < \frac{3}{2}\lambda_z, \tag{9}$$

$$u_s\left(\theta\left(z_i\right)\right) = A_s\left(-\frac{3}{4} - \frac{1}{4}\mathrm{sgn}\left(\cos\left(\theta\right)\right)\right)\cos\left(\theta\right). \tag{10}$$

In this model, $\theta(z_i)$ is the phase angle at $z_i$; $u_s(z_i)$, which is actually a 3/2 periods of cosinusoid modulated by a box function, represents a spanwise profile of one single low-speed streak centering at $z_i = 0$; $\lambda_z$ sets the wavelength of the streak; $A_s$ is the nominal streak amplitude and $\text{sgn}(\cos(\theta))$ returns the sign of $\cos(\theta)$. Figure 12a shows a typical streak unit with $\lambda_z^+ = 100$ and $A_s = 1$. For a given $y^+$, multiple streak units whose $\lambda_z^+$ are randomly generated via the empirical streak spacing model of Equations (3), (7), and (8) are successively added along the spanwise direction till the whole span is full, i.e.,

$$u\left(z^+\right) = \sum_{i=1}^{N} u_{s,i}\left(z^+ - z_i^+\right), z^+ \in [1, 2^{13}]. \tag{11}$$

in which $u_{s,i}$ is the $i$th streak unit centering at $z_i$ with wavelength of $\lambda_{zi}$, and $u(z^+)$ is the full signal with a total length of $2^{13}$ wall units. To avoid severe overlap which causes unexpected wavelength growth, one streak is gaped from its neighborhoods by the following constraint:

$$\left(z_i^+ - z_{i-1}^+\right)^2 \geqslant \lambda_{z,i}^{+2} \text{ and } \left(z_i^+ - z_{i+1}^+\right)^2 \geqslant \lambda_{z,i}^{+2}. \tag{12}$$

Finally, a Gaussian smooth is applied to $u(z^+)$ to eliminate discontinuity. An example of the $u(z^+)$ profile is shown in Figure 12b with $\overline{\lambda}_z^+ = 100$ and $A_s = 1$. Section 4.2 already shows that the streak amplitude $A_s$ is only weakly correlated with the streak spacing $\lambda$ in the near-wall region. Here, we assume $A_s$ to be constant at each flow layer with magnitude equal to the local $u_{rms}$. This actually attributes all the $u$ component TKE to small-scale streaks and ignores the TKE contribution from either large-scale structures or their modulation effect on smaller ones. Although this assumption is far from the real case, it provides an artificial scenario to infer the effect of the unconsidered large-scale motions on the velocity spectra.

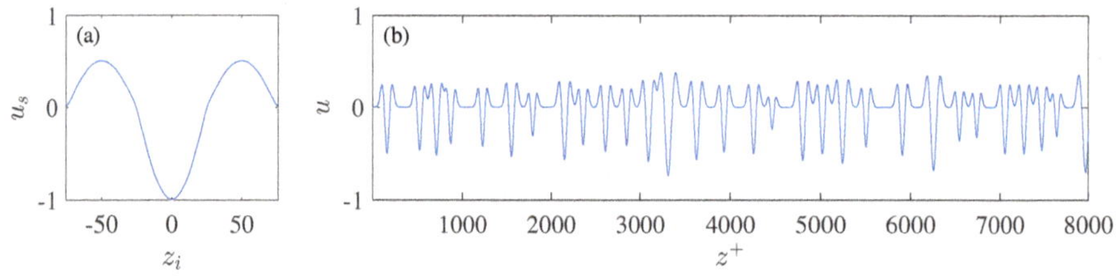


**Figure 12.** (**a**) Profile of the elementary streak unit described by Equations (9) and (10); (**b**) an example of a synthetic signal of $u(z^+)$ with $\overline{\lambda}_z^+ = 100$ and $A_s = 1$.

Figure 13 compares the $k_z\Phi_{uu}$ spectra of the fabricated $u(z^+)$ fields (dashed isolines) to the original ones (pseudo-color maps) in the LD0 and LD3 cases. Combing with other cases that are not shown, it can be concluded that the present simulation, despite its simplicity, is capable of restoring the core region of the inner-layer spectra patch within $y^+ < 50$ and $\lambda_z^+ < 300$. The reason, to our regards, is that the ridge of the inner-layer spectra patch is well predicted by the empirical streak spacing model, which in turn is fully utilized when modeling the $u(z^+)$ fields. More interestingly, Figure 13a show that if the outer-layer spectra patch is absent, the general shape of $k_z\Phi_{uu}$ can be acceptably captured till $y^+ \approx 100$. This describes a scenario of the penetration of small-scale streaks into higher layer, which is further supported by the observation that with the presence of the outer-layer spectra patch, the small-scale part of the $k_z\Phi_{uu}$ spectra on the left side of the mean streak spacing (bold dashed lines in Figure 13b) is roughly predicted till $y^+ \approx 100$.

Since LSMs and their near-wall footprints are not considered in the present synthetic simulation, the yielded $k_z\Phi_{uu}$ significantly differs from the original spectra in the large-scale domain with $\lambda_z^+ > 400$, as is shown in Figure 13. One can get an impression on the *Re*-dependency of these large-scale structures by subtracting the simulated $k_z\Phi_{uu}(\lambda_z^+)$ profiles from the original ones. Recalling that the full $u$ component TKE (measured as $u_{rms}$), part of which is originally carried by large-scale structures, is arbitrarily assigned to small-scale streaks during the fabrication of $u(z^+)$, this leads to an overestimation of the energy content in the small-scale near-wall domain, which becomes more prominent at higher *Re* (comparing the near-wall profiles in Figure 13a,b for an illustration).

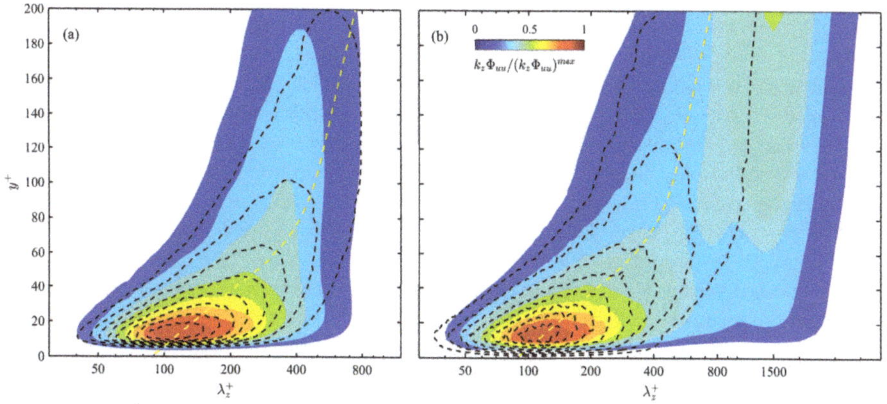


**Figure 13.** Comparison of the premultiplied spanwise spectra $k_z\Phi_{uu}$ simulated by the simplified synthetic method (dashed isolines) with the original one (pseudo-color maps) in (**a**) the LD0 case and (**b**) the LD3 case. $k_z\Phi_{uu}(\lambda_z^+)$ are all normalized by the maximum value in the near-wall region, and the contour labels are evenly spaced from 0.2 to 0.9 with interval of 0.1. Dashed lines are $\overline{\lambda}^+$ predicted by the empirical model in Section 4.2.

Such an overestimation might be improved by assigning not the whole $u_{rms}$ but the streak-contributed portion of $u_{rms}$ to $A_s$. A scale-based decomposition, instead of the energy-based POD filtering used in Section 3, is thus needed to quantify the TKE contribution from streaks. This is an issue to be studied in the future. Nevertheless, on considering that the present synthetic simulation only relies on the knowledge of both the $u_{rms}(y)$ profile that is dependent on *Re* and the streak spacing distribution that seems to be independent of *Re*, the slight difference in the simulated small-scale energy content is acceptable, and will not undermine the practical value of such a test. Of course, more complicated issues, like accounting streaks' streamwise extent and modeling both the dynamical process of the streak instability [4] and their response to outer-layer large-scale structures [54,55], should be taken into consideration. But one of the particular attractions of the present idea is that due to the *Re*-independence of the streak distribution, the modeling of the streak dynamics might be obtained from a low-*Re* DNS database via either the techinque of reduced-order modeling [100,101] or minimum flow unit simulation [102] , and then applied to high-*Re* case through proper scaling.

## 6. Concluding Remarks

In summary, the present work deals with the wall-normal variation of the characteristic lateral length scale of small-scale streak-liked structures in a smooth-wall turbulent boundary layer. The primary aim is to extend the existing knowledge on the streak spacing in the near-wall region to higher flow layers. Morphological analysis on the $u$ component fluctuating velocity is taken in a range of $Re_\tau = 440 \sim 2400$. It is found that the streak spacing $\lambda$ keeps a log-normal distribution till the upper bound of the log layer. The inner-scaled mean and most probable value, i.e., $\overline{\lambda}^+$ and $\lambda_{mp}^+$, follows a two-stage exponential wall-normal growth that is less dependent on *Re* and can be well described by a two-sectional empirical model.

The first fast growth stage of $\overline{\lambda}^+(y^+)$ and $\lambda_{mp}^+(y^+)$ below $y^+ = 50$ can be attributed to the active streak merging event, which results in a quick drop of both the streak strength and the streak population density there. A simplified synthetic simulation, which only models the spanwise distribution of streak elements via the proposed empirical model, fairly restores the core region of the inner-layer $k_z\Phi_{uu}$ spectra patch residing in this region. The second stage beyond $y^+ = 50$ presents a reduced growth rate in $\overline{\lambda}^+(y^+)$ and $\lambda^+_{mp}(y^+)$, consistent with the relaxation of the decay of the streak strength, the streak population density and the streak merging frequency. This suggests that most of the small-scale streaks identified beyond the buffer layer might be the remnants of near-wall structures. Despite of their sparse population, they contribute to the small-scale part of the $k_z\Phi_{uu}$ spectra till $y^+ = 100$, which can be fairly restored by the simplified synthetic simulation.

To our regards, the exponential scaling of the streak spacing proposed here, i.e., $y^+ \propto \ln\overline{\lambda}^+$ and $y^+ \propto \ln\lambda^+_{mp}$ till the upper bound of the log layer, is different from the linear scaling of wall-attached large-scale structures [11,41,67,68]. This suggests that small-scale streaks do not behave in an attached-eddy way. Instead, those structures that survive through the active near-wall streak instability events passively lift to higher layers, either gradually fading out due to viscous dissipation or being synchronized into larger-scale structures. It is believed that more detailed information in this aspect will provide helpful insight into the origin of LSMs, and thus deserves to be studied later.

Finally, since the *Re*-independency of the streak spacing suggests that the amplitude modulation does not alter the geometric characteristics of small-scale structures, this provides a justification for the so-called 'universal' signal that was used by Marusic et al. [55] and Zhang and Chernyshenko [103] to predict the near-wall fluctuating velocity statistics given the knowledge of the log-layer large-scale signal. Nevertheless, the failure of restoring the large-scale part of $k_z\Phi_{uu}$ in the simplified synthetic simulation stresses the accumulated importance of large-scale motions with the increase of *Re*. To fully restore the whole spectra, the geometrical characteristics of these large-scale motions should be modeled properly. Note that the scale separation tool and the morphological identification algorithm used in the present study can be also applied for such a purpose in the future.

**Author Contributions:** All the authors have contributed equally to this research work and the writing of this paper. Conceptualization, J.W. and C.P.; Investigation, W.W.; Writing—original draft, W.W. and C.P.; Writing—review & editing, J.W.

**Funding:** This research was funded by National Natural Science Foundation of China (grant numbers 11490552, 11672020 and 11721202) and the Fundamental Research Funds for the Central Universities of China (grant number YWF-16-JCTD-A-05).

**Acknowledgments:** The authors would like to thank Sillero, J. A. and Jiménez, J. for sharing their turbulent boundary layer DNS dataset.

**Conflicts of Interest:** The authors declare no conflict of interest.

## Appendix A. Scale and Mode Shape of POD Modes

Snapshot POD is used to decompose the fluctuating velocity fields into discrete POD modes. It has been checked that the ensemble size of the SE and LE cases are large enough for a converged decomposition. Since POD is based on an optimal energy recovery criteria, the relationship between the energy content of POD modes and their characteristic length scale should be checked before POD method can be used for scale separation.

The characteristic spanwise scale $\lambda_c^{\phi_n}$ carried by the $n$th rank POD mode can be estimated from $R_{\phi_n^u\phi_n^u}(r_z)$, the two-point correlation of the $u$ component mode basis function $\phi_n^u$. In the present work, $\lambda_c^{\phi_n}$ is defined as the gap between two negative valleys in the $R_{\phi_n^u\phi_n^u}(r_z)$ map, similar to $\overline{\lambda_z^+}$ illustrated in Figure 4. Figure A1 shows $\lambda_c^{\phi_n}$ as a function of the mode rank $n$ at typical $y^+$ in the LE1 case, in together with the cumulative TKE contribution curve $P_n(n)$. In general, $\lambda_c^{\phi_n}$ monotonically decreases with the increase of $n$, indicating a distinct correlation between the energy content and the length scale in the hierarchy of POD mode set.

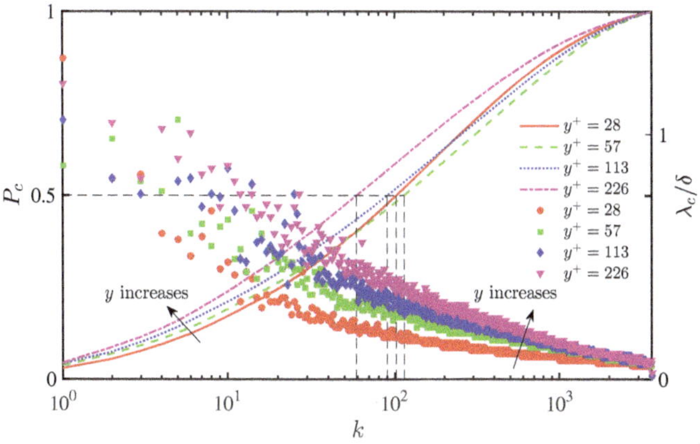


**Figure A1.** The cumulative turbulent kinetic energy (TKE) contribution $P_n$ (lines) and the characteristic spanwise scale $\lambda_c^{\phi_n}$ (solid symbols) carried by each POD mode at various $y^+$ in the LE1 case.

$\phi_n^u$ of the first POD mode ($n = 1$) at $y^+ = 28$ and 226 in the LE1 case are illustrated in Figure A2a,c. Both characterize high- and low-speed strips aligned in spanwise direction with quasi-periodicity. Their spanwise scales are $O(\delta)$, and the streamwise coherence extends beyond $3\delta$. On considering its TKE significance, the first POD mode is regarded as the projection of LSMs onto the mode subspace. The geometrical similarity of $\phi_{n=1}^u$ between the inner layer ($y^+ = 28$) and the outer layer ($y^+ = 226$) implies a scale invariance when LSMs extend their influence into the near-wall region, consistent with the outer-layer spectra patch shown in Figure 13b.

As a comparison, Figure A2b,d show $\phi_s^u$ of a $s$th rank POD mode whose $P_s$ is 0.5 at $y^+ = 28$ and 226, respectively. Recalling that $P_s = 0.5$ is the POD energy cutoff threshold used in Section 3. Figure A1 shows that the mode rank $s$ with $P_s = 0.5$ decreases with the increase of $y^+$, the corresponding $\lambda_c^{\phi_s}$ increases with $y^+$, but the magnitude is always far from $O(\delta)$. Such a small-scale feature can be evidenced by Figure A2b,d, where small-scale streaky pattern of $\phi_s^u$ is observed in both inner and outer layer, in together with a clear tendency of the scale growth. Finally, it can be concluded from Figure A1 that the leading-order $\mathbf{V}^L$ fields constructed in Section 3 includes all the POD modes whose spanwise scales are larger than $\delta$.

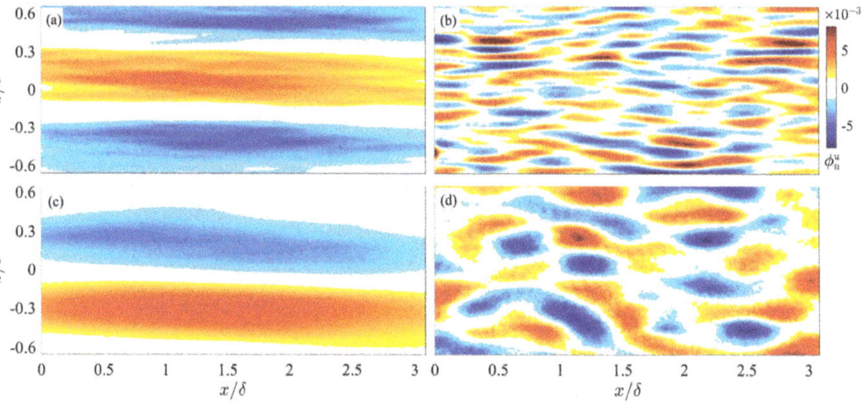


**Figure A2.** Typical mode basis function $\phi_n^u$ in the LE1 case. (**a**) $n = 1$, $y^+ = 28$; (**b**) $n = s = 100$, $y^+ = 28$; (**c**) $n = 1$, $y^+ = 226$; (**d**) $n = s = 61$, $y^+ = 226$. In (**b**,**d**), the mode rank $s$ is chosen to make the POD energy cutoff threshold $P_s = 0.5$.

Other SE and LE cases reveal a similar relation between the TKE content and the length scale in the POD mode set. The only difference is that the streaky pattern in the leading-order POD modes in the SE cases poses smaller length scale due to the FOV limitation.

## Appendix B. Morphological-Based Streak Identification Algorithm

### *Appendix B.1. Algorithm Description*

The morphological method used in Section 4 identifies streak-liked structures from instantaneous $u$ component field. As shown in Figure A3, this algorithm poses 3 steps: binarization, cleaning, and skeletonization.

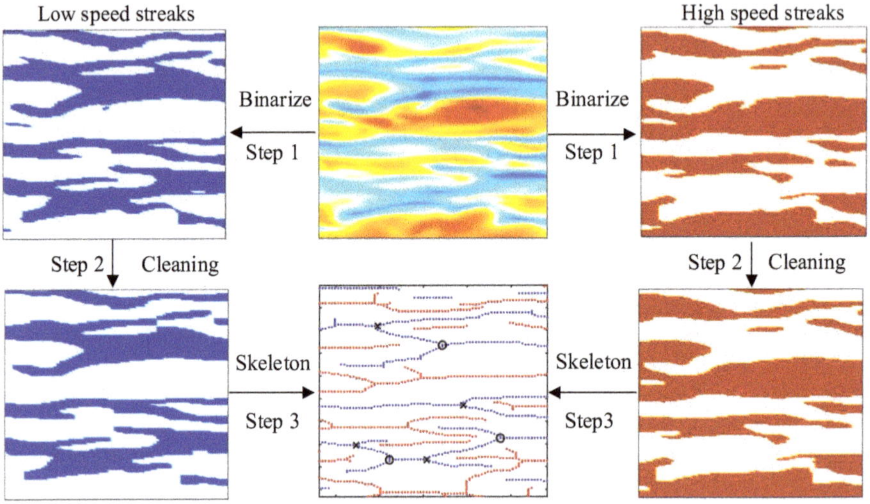


**Figure A3.** Schematic illustration of the morphological-based streak identification algorithm. Circle and cross makers in the final streak skeleton subplot indicate the detected streak merging and splitting event, respectively.

Step 1, binarization. According to the definition of streak described in Section 1, the instantaneous $u$ component fluctuating velocity in wall-parallel $x$-$z$ plane is binarized into low- and high-speed elements $F_i^l$ and $F_i^h$ by:

$$F_i^l = \begin{cases} 1, & F_d < -C_t^l \\ 0, & \text{Otherwise} \end{cases}, \quad F_i^h = \begin{cases} 1, & F_d > C_t^h \\ 0, & \text{Otherwise} \end{cases}. \tag{A1}$$

The detection function $F_d$ in Equation (A1) is the ratio of the fluctuating velocity to the mean velocity of the investigated flow layer: $F_d = u/\overline{U}$. Following previous studies [41,46,48,104], the streak strength threshold $C_t^l$ in Equation (A1) was set to be 0.1. And the ratio between $C_t^h$ and $C_t^l$ was fixed as 0.5 since Smith and Metzler [30] reported that the momentum flux ratio between high- and low-speed streaks was about 0.5.

Step 2, cleaning. A combination of closing and opening operation, based on a rectangle filter template, is applied onto $F^l$ and $F^h$ to fill small holes and remove isolated noise. To simulate the large aspect ratio of streaks, the length/width of this filter template is set to be 50/10 wall units, the same as in Lin et al. [76]. Those connecting regions in $F^l$ whose aspect ratio is smaller than 2 are also discarded, this guarantees that only streamwise elongated connecting regions are considered. After the closing and opening operation, there are still some connecting regions in $F^l$ too small to be taken as streaks. An area cutting-off threshold of 800 wall units$^2$ is then set to reject these small structures.

Step 3, skeletonization. A simple morphological skeleton operation widely used in computer vision [105] is taken to shrink the identified structures into a skeleton. This operation will generate unexpected branches whose length scale is quite small compared to that of the main stem. The spur operation is then followed to trim the branches whose length is smaller than 12 wall units.

The above three procedures are conceptually similar to that of Lin et al. [76]. An additional concern here is that some streaks might not be well separated from others due to the streak merging or splitting event; moreover, the branched pattern can be also caused by the skeleton operation, as mentioned above. Therefore, the location of high-speed streaks (or high-speed regions in higher layers) are used as a supplemental criterion: the gap between two neighboring low-speed skeletons is counted only if there is at least one high-speed skeleton between them. This is consistent with Smith and Metzler [30], who considered streaks to be completely merged if there is no high-speed region between them.

Finally, on considering the possibility that the streaks close to the FOV edge are incompletely captured, the skeletons crossing the streamwise boundary of the FOV are cut-off by a length of 50 wall units. In a similar sense, the skeletons whose distance to the spanwise FOV extent are smaller than 20 wall units are also discarded. The influence of the parameters on the statistics of the streak spacing will be further discussed in Appendix B.2.

After the streak skeleton identification, the streak merging and splitting event can be detected by examining the topology around a node on one streak skeleton where a new branch grows towards either downstream or upstream. Those branches shorter than 50 wall units, which might be caused by local expansion of streaks, are rejected as an additional streak skeleton. The streak skeleton nodes are detected by a connectivity evaluation scheme that is commonly used in computer vision [105]. An example of the detected streak merging/splitting events is shown in Figure A3 for qualitative evidence of the feasibility of this detecting method.

### *Appendix B.2. Effect of Algorithm Parameters on the Streak Spacing Statistics*

In the present morphological algorithm, several parameters need to be manually selected; therefore, their influence on the streak identification should be evaluated carefully. The default baseline parameters are set as: the binary threshold $C_t^l = 0.1$, the size of cleaning structure element $50 \times 10$ wall units, the spur length 12 wall units and the area cutting-off threshold 800 wall units$^2$. This parameter set is used in Section 4 for the streak identification. The effect of each parameter is then inspected by examining the relative change of the mean streak spacing $\overline{\lambda}^+$ from the baseline case. Here, only the SE1 and LD1 cases are illustrated, all the other cases pose similar behavior and are not presented. In the following test, the spanwise FOV span of the LD1 case is truncated to $\Delta Z^+ = 600$ for a direct comparison with the SE1 case.

The streak strength threshold $C_t^l$ defines the boundary of the low-momentum region, and thus directly determines the size and population of the identified streaky structures. Figure A4a gives the relative deviation of $\overline{\lambda}^+$ from the baseline case ($C_t^l = 0.1$) as $C_t^l$ varying from 0.05 to 0.15. For flow layers beneath $y^+ = 10$, $\overline{\lambda}^+$ is only weakly correlated with $C_t^l$; beyond that, the dependency becomes a bit more strong: $\pm25\%$ variation of $C_t^l$ results in about $\pm10\%$ variation of $\overline{\lambda}^+$. Moreover, the correlation between $\overline{\lambda}^+$ and $C_t^l$ quantitatively holds for all the higher layers, this makes the trend of the wall-normal growth of $\overline{\lambda}^+$ decoupled from the selection of $C_t^l$.

Except for $C_t^l$, all the other parameters pose little influence on $\overline{\lambda}^+$. As shown in Figure A4b, the cleaning structure elements with two different sizes, $24 \times 4$ and $72 \times 12$ wall units, are tested. They only result in $\pm6\%$ variation of $\overline{\lambda}^+$ from the baseline case. Two different area cutting-off thresholds of 400 and 1200 wall uints$^2$ (with $\pm50\%$ variation) make the relative change of $\overline{\lambda}^+$ within $\pm3\%$, as shown in Figure A4c. Finally, Figure A4d shows that in the procedure of removing small branches from the main stem of the streak skeleton, ±66% variation of the spur length threshold leads to a variation of $\overline{\lambda}^+$ smaller than ±2%.

In short, only the choice of $C_t^l$ has a relatively large influence on the streak spacing statistics. The default value of $C_t^l = 0.1$ is chosen to be consistent with a majority of previous researches [41,46,48,104].

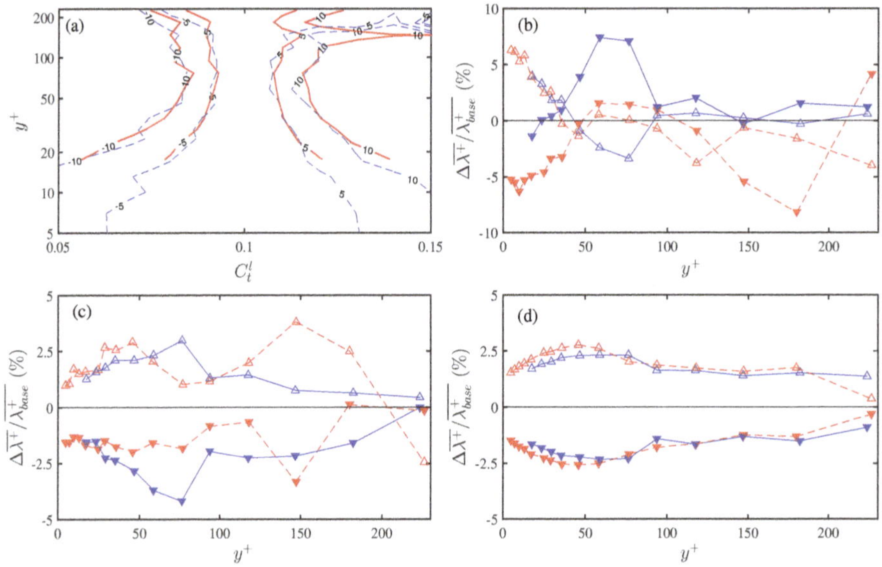

**Figure A4.** Percentage of the change of the mean streak spacing $\overline{\lambda}^+$ from the baseline case due to the change of the following parameters in the morphological algorithm: (**a**) the binary threshold $C_t^l$; (**b**) the size of the cleaning structure element, baseline 48 × 8, △ 72 × 12 and ▾ 24 × 4 wall units$^2$; (**c**) the area cutting-off threshold, baseline 800, △ 400, ▾ 1200 wall units$^2$ and (**d**) the spur length threshold, baseline 12, △ 20, ▾ 4 wall units. In (**a**), solid and dashed isolines indicate the SE1 case and the LD1 case, respectively; in (**b–d**), markers with solid lines indicate the SE1 case and markers with dashed lines the LD1 case.

*Appendix B.3. Effect of Ensemble Size on the Streak Spacing Statistics*

The convergence state of the streak spacing is essential for estimating the related streak statistics. The dependency of the first and second-order statistics of the streak spacing, i.e., $\overline{\lambda}^+$ and $\sigma_{\lambda^+}$, on the number of the analyzed frames $N_f$ is shown in Figure A5 at four $y^+$ in the SE1 case. Note that $\overline{\lambda}^+$ and $\sigma_{\lambda^+}$ is directly calculated from the whole ensemble, instead of being estimated by the log-normal fitting in Section 4.1. It is clearly shown that more snapshots are needed for a stable $\overline{\lambda}^+$ and $\sigma_{\lambda^+}$ at higher flow layer. This is reasonable since the streak population decays with the increase of $y^+$, as is shown in Figure 10. The convergence of $\sigma_{\lambda^+}$ is relatively slower than that of $\overline{\lambda}^+$, but a total ensemble size of around 3000 frames is sufficient. Moreover, it reminds us that careful inspection of the convergence of the PDF profile $P(\lambda^+)$ is critical. Figure A6 compares the insufficiently sampled $P(\lambda^+)$ profiles with the converged ones in both the LE2 and LD3 cases. It is evidenced that half the total ensemble is enough to yield a converged $P(\lambda^+)$ in the log layer ($y^+ \approx 140$). The low-sampled profiles present multi-modal shape; however, increasing the ensemble size will smooth all the non-physical PDF peaks and form a single-peak left-skewed shape. Using this inspection, we have checked that $P(\lambda^+)$ gets acceptable convergence till $y^+ \approx 220$ in all the studied cases.

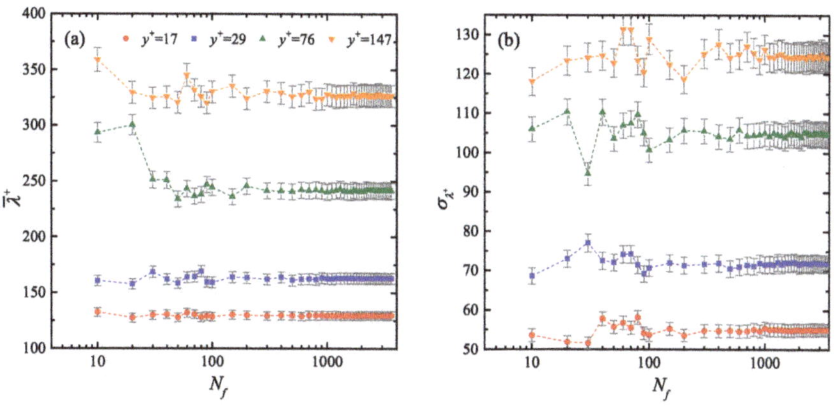


**Figure A5.** Effect of the analyzed frame number $N_f$ on (**a**) the mean streak spacing $\overline{\lambda}^+$ and (**b**) the *r.m.s* of the streak spacing $\sigma_{\lambda^+}$ in the SE1 case at • $y^+ = 17$; ■ $y^+ = 29$; ▲ $y^+ = 76$; ▾ $y^+ = 147$. Error bars indicate ±3% variation.

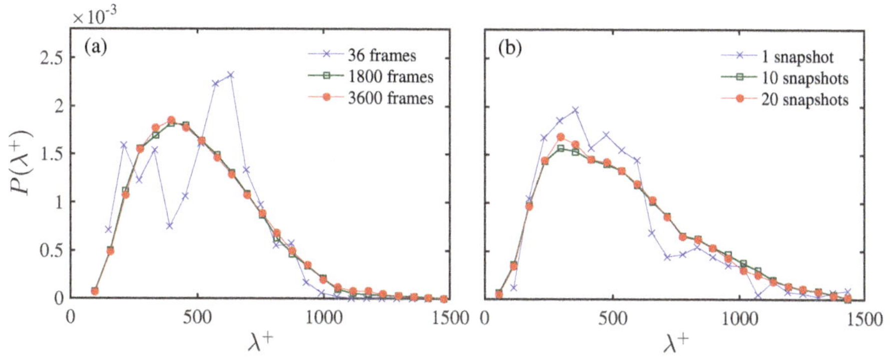


**Figure A6.** Effect of the analyzed frame/snapshot number on the shape of $P(\lambda^+)$ in (**a**) the LE2 case at $y^+ = 140$ and (**b**) the LD3 case at $y^+ = 147$. In (**a**), ×36 frames, ▫ 1800 frames, • 3600 frames; in (**b**), ×1 snapshot, ▫ 10 snapshots, • 20 snapshots. Note that for the LE2 case, 1800 instantaneous frames lead to more than $10^5$ samples of streak spacing. While in the LD3 case, 1 snapshot of the DNS velocity field is sliced into 10 frames with FOV of 2000 × 1500 wall units, the total number of 20 DNS snapshots correspond to 200 frames, with more than $10^4$ samples being recorded.

## References


1. Hama, F.R.; Nutant, J. Detailed flow-field observations in the transition process in a thick boundary layer. In *Proceedings of the Heat Transfer and Fluid Mechanics Institute*; Stanford University Press: Palo Alto, CA, USA, 1954; Volume 16, p. 77.
2. Ferrell, J.K.; Richardson, F.M.; Beatty, K.O., Jr. Dye displacement technique for velocity distribution measurements. *Ind. Eng. Chem.* **1955**, *47*, 29–33. [CrossRef]
3. Jeong, J.; Hussain, F.; Schoppa, W.; Kim, J. Coherent structures near the wall in a turbulent channel flow. *J. Fluid Mech.* **1997**, *332*, 185–214. [CrossRef]
4. Schoppa, W.; Hussain, F. Coherent structure generation in near-wall turbulence. *J. Fluid Mech.* **2002**, *453*, 57–108. [CrossRef]
5. Blackwelder, R.F.; Eckelmann, H. Streamwise vortices associated with the bursting phenomenon. *J. Fluid Mech.* **1979**, *94*, 577–594. [CrossRef]
6. Smith, C.R.; Schwartz, S.P. Observation of streamwise rotation in the near-wall region of a turbulent boundary layer. *Phys. Fluids* **1983**, *26*, 641–652. [CrossRef]

7. Kasagi, N.; Hirata, M.; Nishino, K. Streamwise pseudo-vortical structures and associated vorticity in the near-wall region of a wall-bounded turbulent shear flow. *Exp. Fluids* **1986**, *4*, 309–318. [CrossRef]
8. Hwang, Y. Near-wall turbulent fluctuations in the absence of wide outer motions. *J. Fluid Mech.* **2013**, *723*, 264–288. [CrossRef]
9. Brandt, L. The lift-up effect: The linear mechanism behind transition and turbulence in shear flows. *Eur. J. Mech. B Fluids* **2014**, *47*, 80–96. [CrossRef]
10. Landahl, M.T. On sublayer streaks. *J. Fluid Mech.* **1990**, *212*, 593–614. [CrossRef]
11. Del Álamo, J.C.; Jiménez, J. Linear energy amplification in turbulent channels. *J. Fluid Mech.* **2006**, *559*, 205–213. [CrossRef]
12. Hwang, Y.; Cossu, C. Self-sustained process at large scales in turbulent channel flow. *Phys. Rev. Lett.* **2010**, *105*, 044505. [CrossRef] [PubMed]
13. Butler, K.M.; Farrell, B.F. Three-dimensional optimal perturbations in viscous shear flow. *Phys. Fluids A* **1992**, *4*, 1637–1650 [CrossRef]
14. Trefethen, L.N.; Trefethen, A.E.; Reddy, S.C.; Driscoll, T.A. Hydrodynamic stability without eigenvalues. *Science* **1993**, *261*, 578–584. [CrossRef] [PubMed]
15. Reddy, S.C.; Henningson, D.S. Energy growth in viscous channel flows. *J. Fluid Mech.* **1993**, *252*, 209–238. [CrossRef]
16. Willis, A.P.; Hwang, Y.; Cossu, C. Optimally amplified large-scale streaks and drag reduction in turbulent pipe flow. *Phys. Rev. E* **2010**, *82*, 036321. [CrossRef] [PubMed]
17. Hamilton, J.M.; Kim, J.; Waleffe, F. Regeneration mechanisms of near-wall turbulence structures. *J. Fluid Mech.* **1995**, *287*, 317–348. [CrossRef]
18. Waleffe, F. On a self-sustaining process in shear flows. *Phys. Fluids* **1997**, *9*, 883–900. [CrossRef]
19. Jiménez, J.; Pinelli, A. The autonomous cycle of near-wall turbulence. *J. Fluid Mech.* **1999**, *389*, 335–359. [CrossRef]
20. Kim, J. Physics and control of wall turbulence for drag reduction. *Philos. Trans. Math. Phys. Eng. Sci.* **2011**, *369*, 1396–1411. [CrossRef]
21. Rao, K.N.; Narasimha, R.; Narayanan, M.A.B. The 'bursting' phenomenon in a turbulent boundary layer. *J. Fluid Mech.* **1971**, *48*, 339–352. [CrossRef]
22. Kim, H.; Kline, S.J.; Reynolds, W.C. The production of turbulence near a smooth wall in a turbulent boundary layer. *J. Fluid Mech.* **1971**, *50*, 133–160. [CrossRef]
23. Hwang, Y.; Bengana, Y. Self-sustaining process of minimal attached eddies in turbulent channel flow. *J. Fluid Mech.* **2016**, *795*, 708–738. [CrossRef]
24. Bogard, D.G.; Tiederman, W.G. Characteristics of ejections in turbulent channel flow. *J. Fluid Mech.* **1987**, *179*, 1–19. [CrossRef]
25. Asai, M.; Minagawa, M.; Nishioka, M. The instability and breakdown of a near-wall low-speed streak. *J. Fluid Mech.* **2002**, *455*, 289–314. [CrossRef]
26. Brandt, L.; de Lange, H.C. Streak interactions and breakdown in boundary layer flows. *Phys. Fluids* **2008**, *20*, 024107. [CrossRef]
27. Wang, J.J.; Pan, C.; Zhang, P.F. On the instability and reproduction mechanism of a laminar streak. *J. Turb.* **2009**, *10*, N26. [CrossRef]
28. Hwang, Y.; Willis, A.P.; Cossu, C. Invariant solutions of minimal large-scale structures in turbulent channel flow for $Re_\tau$ up to 1000. *J. Fluid Mech.* **2016**, *802*, R1 [CrossRef]
29. Kline, S.; Reynolds, W.; Schraub, F.; Runstadler, P. The structure of turbulent boundary layers. *J. Fluid Mech.* **1967**, *30*, 741–773. [CrossRef]
30. Smith, C.; Metzler, S. The characteristics of low-speed streaks in the near-wall region of a turbulent boundary layer. *J. Fluid Mech.* **1983**, *129*, 27–54. [CrossRef]
31. Runstadler, P.W.; Kline, S.J.; Reynolds, W.C. *An Experimental Investigation of the Flow Structure of the Turbulent Boundary Layer*; Technical Report MD-8; Department of Mechanical Engineering, Stanford University: Stanford, CA, USA, 1963.
32. Coantic, M. A study of turbulent pipe fow and of the structure of its viscous sublayer. In Proceedings of the 4th Euromech Colloquium, Southampton, UK, March 1967.
33. Gupta, A.; Laufer, J.; Kaplan, R. Spatial structure in the viscous sublayer. *J. Fluid Mech.* **1971**, *50*, 493–512. [CrossRef]

34. Lee, M.K.; Eckelman, L.D.; Hanratty, T.J. Identification of turbulent wall eddies through the phase relation of the components of the fluctuating velocity gradient. *J. Fluid Mech.* **1974**, *66*, 17–33. [CrossRef]
35. Klewicki, J.C.; Metzger, M.M.; Kelner, E.; Thurlow, E.M. Viscous sublayer flow visualizations at $Re_\theta$ = 1,500,000. *Phys. Fluids* **1995**, *7*, 857–863. [CrossRef]
36. Nakagawa, H.; Nezu, I. Structure of space-time correlations of bursting phenomena in an open-channel flow. *J. Fluid Mech.* **1981**, *104*, 1–43. [CrossRef]
37. Jiménez, J.; Moin, P. The minimal flow unit in near-wall turbulence. *J. Fluid Mech.* **1991**, *225*, 213–240. [CrossRef]
38. Jiménez, J.; Del Álamo, J.C.; Flores, O. The large-scale dynamics of near-wall turbulence. *J. Fluid Mech.* **2004**, *505*, 179–199. [CrossRef]
39. Kim, K.C.; Adrian, R.J. Very large-scale motion in the outer layer. *Phys. Fluids* **1999**, *11*, 417–422. [CrossRef]
40. Adrian, R.J.; Meinhart, C.D.; Tomkins, C.D. Vortex organization in the outer region of the turbulent boundary layer. *J. Fluid Mech.* **2000**, *422*, 1–54. [CrossRef]
41. Tomkins, C.D.; Adrian, R.J. Spanwise structure and scale growth in turbulent boundary layers. *J. Fluid Mech.* **2003**, *490*, 37–74. [CrossRef]
42. Ganapathisubramani, B.; Longmire, E.K.; Marusic, I. Characteristics of vortex packets in turbulent boundary layers. *J. Fluid Mech.* **2003**, *478*, 35–46. [CrossRef]
43. Hutchins, N.; Marusic, I. Evidence of very long meandering features in the logarithmic region of turbulent boundary layers. *J. Fluid Mech.* **2007**, *579*, 1–28. [CrossRef]
44. Balakumar, B.J.; Adrian, R.J. Large- and very-large-scale motions in channel and boundary-layer flows. *Philos. Trans. Math. Phys. Eng. Sci.* **2007**, *365*, 665–681. [CrossRef] [PubMed]
45. Wu, Y.; Christensen, K.T. Spatial structure of a turbulent boundary layer with irregular surface roughness. *J. Fluid Mech.* **2010**, *655*, 380–418. [CrossRef]
46. Dennis, D.J.C.; Nickels, T.B. Experimental measurement of large-scale three-dimensional structures in a turbulent boundary layer. Part 2. Long structures. *J. Fluid Mech.* **2011**, *673*, 218–244. [CrossRef]
47. Lee, J.H.; Sung, H.J. Very-large-scale motions in a turbulent boundary layer. *J. Fluid Mech.* **2011**, *673*, 80–120.
48. Lee, J.; Lee, J.H.; Choi, J.I.; Sung, H.J. Spatial organization of large- and very-large-scale motions in a turbulent channel flow. *J. Fluid Mech.* **2014**, *749*, 818–840. [CrossRef]
49. de Giovanetti, M.; Hwang, Y.; Choi, H. Skin-friction generation by attached eddies in turbulent channel flow. *J. Fluid Mech.* **2016**, *808*, 511–538.
50. Marusic, I.; Mckeon, B.J.; Monkewitz, P.A.; Nagib, H.M.; Smits, A.J.; Sreenivasan, K.R. Wall-bounded turbulent flows at high reynolds numbers: Recent advances and key issues. *Phys. Fluids* **2010**, *22*, 065103.
51. Smits, A.J.; Mckeon, B.J.; Marusic, I. High-Reynolds number wall turbulence. *Annu. Rev. Fluid Mech.* **2011**, *43*, 353–375. [CrossRef]
52. Bradshaw, P.; Langer, C.A. Nonuniversality of sublayer streaks in turbulent flow. *Phys. Fluids* **1995**, *7*, 2435–2438. [CrossRef]
53. Hutchins, N.; Marusic, I. Large-scale influences in near-wall turbulence. *Philos. Trans. Math. Phys. Eng. Sci.* **2007**, *365*, 647–664.
54. Mathis, R.; Hutchins, N.; Marusic, I. Large-scale amplitude modulation of the small-scale structures in turbulent boundary layers. *J. Fluid Mech.* **2009**, *628*, 311–337. [CrossRef]
55. Marusic, I.; Mathis, R.; Hutchins, N. Predictive model for wall-bounded turbulent flow. *Science* **2010**, *329*, 193–196. [CrossRef] [PubMed]
56. Mathis, R.; Hutchins, N.; Marusic, I. A predictive inner–outer model for streamwise turbulence statistics in wall-bounded flows. *J. Fluid Mech.* **2011**, *681*, 537–566. [CrossRef]
57. Ganapathisubramani, B.; Hutchins, N.; Monty, J.P.; Chung, D.; Marusic, I. Amplitude and frequency modulation in wall turbulence. *J. Fluid Mech.* **2012**, *712*, 61–91. [CrossRef]
58. Duvvuri, S.; Mckeon, B.J. Triadic scale interactions in a turbulent boundary layer. *J. Fluid Mech.* **2015**, *767*, R4. [CrossRef]
59. Agostini, L.; Leschziner, M. Predicting the response of small-scale near-wall turbulence to large-scale outer motions. *Phys. Fluids* **2016**, *28*, 015107. [CrossRef]
60. Hoyas, S.; Jiménez, J. Scaling of the velocity fluctuations in turbulent channels up to $Re_\tau$ = 2003. *Phys. Fluids* **2006**, *18*, 011702. [CrossRef]

61. Jiménez, J.; Hoyas, S. Turbulent fluctuations above the buffer layer of wall-bounded flows. *J. Fluid Mech.* **2008**, *611*, 215–236. [CrossRef]
62. Townsend, A.A. *The Structure of Turbulent Shear Flow*; Cambridge University Press: Cambridge, UK, 1976; pp. 411–412.
63. Perry, A.E.; Chong, M.S. On the mechanism of wall turbulence. *J. Fluid Mech.* **1982**, *119*, 173–217. [CrossRef]
64. Perry, A.E.; Marusic, I. A wall-wake model for the turbulence structure of boundary layers. Part 1. Extension of the attached eddy hypothesis. *J. Fluid Mech.* **1995**, *298*, 361–388. [CrossRef]
65. Marusic, I.; Perry, A.E. A wall-wake model for the turbulence structure of boundary layers. Part 2. Further experimental support. *J. Fluid Mech.* **1995**, *298*, 389–407. [CrossRef]
66. Marusic, I. On the role of large-scale structures in wall turbulence. *Phys. Fluids* **2001**, *13*, 735–743. [CrossRef]
67. Hwang, Y. Statistical structure of self-sustaining attached eddies in turbulent channel flow. *J. Fluid Mech.* **2015**, *767*, 254–289. [CrossRef]
68. Baars, W.J.; Hutchins, N.; Marusic, I. Self-similarity of wall-attached turbulence in boundary layers. *J. Fluid Mech.* **2017**, *823*. [CrossRef]
69. Walsh, M.J. Riblets as a viscous drag reduction technique. *AIAA J.* **1983**, *21*, 485–486. [CrossRef]
70. Fransson, J.H.M.; Talamelli, A. On the generation of steady streamwise streaks in flat-plate boundary layers. *J. Fluid Mech.* **2012**, *698*, 211–234. [CrossRef]
71. Bai, H.; Zhou, Y.; Zhang, W.; Xu, S.; Wang, Y.; Antonia, R. Active control of a turbulent boundary layer based on local surface perturbation. *J. Fluid Mech.* **2014**, *750*, 316–354. [CrossRef]
72. Schraub, F.A.; Kline, S.J. *A Study of the Structure of the Turbulent Boundary Layer With And Without Longitudinal Pressure Gradients*; Technical Report MD-12; Department of Mechanical Engineering, Stanford University: Stanford, CA, USA, 1965.
73. Bakewell, H.P., Jr.; Lumley, J.L. Viscous sublayer and adjacent wall region in turbulent pipe flow. *Phys. Fluids* **1967**, *10*, 1880–1889. [CrossRef]
74. Kim, J.; Moin, P.; Moser, R. Turbulence statistics in fully developed channel flow at low reynolds number. *J. Fluid Mech.* **1987**, *177*, 133–166. [CrossRef]
75. Lagraa, B.; Labraga, L.; Mazouz, A. Characterization of low-speed streaks in the near-wall region of a turbulent boundary layer. *Eur. J. Mech. B Fluids* **2004**, *23*, 587–599. [CrossRef]
76. Lin, J.; Laval, J.P.; Foucaut, J.M.; Stanislas, M. Quantitative characterization of coherent structures in the buffer layer of near-wall turbulence. Part 1: Streaks. *Exp. Fluids* **2008**, *45*, 999–1013. [CrossRef]
77. Chen, J.; Hussain, F.; Pei, J.; She, Z.S. Velocity–Vorticity Correlation Structure in Turbulent Channel Flow. *J. Fluid Mech.* **2014**, *742*, 291–307. [CrossRef]
78. Toh, S.; Itano, T. Interaction between a large-scale structure and near-wall structures in channel flow. *J. Fluid Mech.* **2005**, *524*, 249–262. [CrossRef]
79. Wang, W.; Pan, C.; Wang, J. Quasi-bivariate variational mode decomposition as a tool of scale analysis in wall-bounded turbulence. *Exp. Fluids* **2018**, *59*, 1. [CrossRef]
80. Harun, Z.; Monty, J.P.; Mathis, R.; Marusic, I. Pressure gradient effects on the large-scale structure of turbulent boundary layers. *J. Fluid Mech.* **2013**, *715*, 477–498. [CrossRef]
81. Pope, S.B.; Pope, S.B. *Turbulent Flows*; Cambridge University Press: Cambridge, UK, 2000; p. 806.
82. Marusic, I.; Monty, J.P.; Hultmark, M.; Smits, A.J. On the logarithmic region in wall turbulence. *J. Fluid Mech.* **2013**, *716*, R3. [CrossRef]
83. Marusic, I.; Kunkel, G.J. Streamwise turbulence intensity formulation for flat-plate boundary layers. *Phys. Fluids* **2003**, *15*, 2461–2464. [CrossRef]
84. Chauhan, K.A.; Monkewitz, P.A.; Nagib, H.M. Criteria for assessing experiments in zero pressure gradient boundary layers. *Fluid Dyn. Res.* **2009**, *41*, 021404. [CrossRef]
85. Jiménez, J.; Hoyas, S.; Simens, M.P.; Mizuno, Y. Turbulent boundary layers and channels at moderate reynolds numbers. *J. Fluid Mech.* **2010**, *657*, 335–360. [CrossRef]
86. Meneveau, C.; Marusic, I. Generalized logarithmic law for high-order moments in turbulent boundary layers. *J. Fluid Mech.* **2013**, *719*, R1. [CrossRef]
87. Wu, Y. A study of energetic large-scale structures in turbulent boundary layer. *Phys. Fluids* **2014**, *26*, 045113. [CrossRef]

88. Meneveau, C.; Marusic, I. Turbulence in the Era of Big Data: Recent Experiences with Sharing Large Datasets. In *Whither Turbulence and Big Data in the 21st Century?* Springer International Publishing: Cham, Switzerland, 2017; pp. 497–507.
89. Simens, M.P.; Jiménez, J.; Hoyas, S.; Mizuno, Y. A High-Resolution Code For Turbulent Boundary Layers. *J. Comput. Phys.* **2009**, *228*, 4218–4231. [CrossRef]
90. Sillero, J.A.; Jiménez, J.; Moser, R.D. One-point statistics for turbulent wall-bounded flows at reynolds numbers up to $\delta^+ \approx 2000$. *Phys. Fluids* **2013**, *25*, 105102. [CrossRef]
91. Sillero, J.A.; Jiménez, J.; Moser, R.D. Two-point statistics for turbulent boundary layers and channels at reynolds numbers up to $\delta^+ \approx 2000$. *Phys. Fluids* **2014**, *26*, 105109. [CrossRef]
92. Champagnat, F.; Plyer, A.; Le Besnerais, G.; Leclaire, B.; Davoust, S.; Le Sant, Y. Fast and accurate PIV computation using highly parallel iterative correlation maximization. *Exp. Fluids* **2011**, *50*, 1169–1182. [CrossRef]
93. Pan, C.; Xue, D.; Xu, Y.; Wang, J.; Wei, R. Evaluating the accuracy performance of Lucas-Kanade algorithm in the circumstance of PIV application. *Sci. China Phys. Mech. Astron.* **2015**, *58*, 104704. [CrossRef]
94. Klewicki, J.; Fife, P.; Wei, T. On the logarithmic mean profile. *J. Fluid Mech.* **2009**, *638*, 73–93. [CrossRef]
95. Borrell, G.; Sillero, J.A.; Jiménez, J. A code for direct numerical simulation of turbulent boundary layers at high reynolds numbers in Bg/P supercomputers. *Comput. Fluids* **2013**, *80*, 37–43. [CrossRef]
96. Agostini, L.; Leschziner, M.; Gaitonde, D. Skewness-induced asymmetric modulation of small-scale turbulence by large-scale structures. *Phys. Fluids* **2016**, *28*, 903–995. [CrossRef]
97. Deng, S.; Pan, C.; Wang, J.; He, G. On the spatial organization of hairpin packets in a turbulent boundary layer at low-to-moderate Reynolds number. *J. Fluid Mech.* **2018**, *844*, 635–668. [CrossRef]
98. Sirovich, L. Turbulence and the dynamics of coherent structures. I—Coherent structures. II—Symmetries and transformations. III—Dynamics and scaling. *Q. Appl. Math.* **1987**, *45*, 561–571. [CrossRef]
99. Berkooz, G.; Holmes, P.; Lumley, J.L. The proper orthogonal decomposition in the analysis of turbulent flows. *Annu. Rev. Fluid Mech.* **1993**, *25*, 539–575. [CrossRef]
100. Bourgeois, J.A.; Noack, B.R.; Martinuzzi, R.J. Generalized phase average with applications to sensor-based flow estimation of the wall-mounted square cylinder wake. *J. Fluid Mech.* **2013**, *736*, 316–350. [CrossRef]
101. Brunton, S.L.; Proctor, J.L.; Kutz, J.N. Discovering governing equations from data by sparse identification of nonlinear dynamical systems. *Proc. Natl. Acad. Sci. USA* **2016**, *113*, 3932–3937 [CrossRef] [PubMed]
102. Yin, G.; Huang, W.X.; Xu, C.X. Prediction of near-wall turbulence using minimal flow unit. *J. Fluid Mech.* **2018**, *841*, 654–673. [CrossRef]
103. Zhang, C.; Chernyshenko, S.I. Quasisteady quasihomogeneous description of the scale interactions in near-wall turbulence. *Phys. Rev. Fluids* **2016**, *1*, 014401. [CrossRef]
104. Baltzer, J.R.; Adrian, R.J.; Wu, X. Structural organization of large and very large scales in turbulent pipe flow simulation. *J. Fluid Mech.* **2013**, *720*, 236–279. [CrossRef]
105. Arcelli, C.; Di Baja, G.S. Skeletons of planar patterns. *Mach. Intell. Pattern Recognit.* **1996**, *19*, 99–143.

*entropy*

MDPI

*Article*

# The Radial Propagation of Heat in Strongly Driven Non-Equilibrium Fusion Plasmas

**Boudewijn van Milligen [1,*], Benjamin Carreras [2], Luis García [2] and Javier Nicolau [2]**

1 Laboratorio Nacional de Fusión, CIEMAT, Av. Complutense 40, 28040 Madrid, Spain
2 Departamento de Física, Universidad Carlos III de Madrid, Av. de la Universidad 30, 28911 Leganés, Madrid, Spain; bacarreras@gmail.com (B.C.); lgarcia@fis.uc3m.es (L.G.); javherna@fis.uc3m.es (J.N.)
* Correspondence: boudewijn.vanMilligen@ciemat.es

Received: 13 December 2018; Accepted: 1 February 2019; Published: 5 February 2019

**Abstract:** Heat transport is studied in strongly heated fusion plasmas, far from thermodynamic equilibrium. The radial propagation of perturbations is studied using a technique based on the transfer entropy. Three different magnetic confinement devices are studied, and similar results are obtained. "Minor transport barriers" are detected that tend to form near rational magnetic surfaces, thought to be associated with zonal flows. Occasionally, heat transport "jumps" over these barriers, and this "jumping" behavior seems to increase in intensity when the heating power is raised, suggesting an explanation for the ubiquitous phenomenon of "power degradation" observed in magnetically confined plasmas. Reinterpreting the analysis results in terms of a continuous time random walk, "fast" and "slow" transport channels can be discerned. The cited results can partially be understood in the framework of a resistive Magneto-HydroDynamic model. The picture that emerges shows that plasma self-organization and competing transport mechanisms are essential ingredients for a fuller understanding of heat transport in fusion plasmas.

**Keywords:** magnetic confinement fusion; turbulence; heat transport

## 1. Introduction

The initial goal of fusion research is to design a system that sustains fusion reactions in a safe manner on Earth, which is a necessary first step towards the development of a fusion reactor, potentially a nearly inexhaustible power source for humankind, free from the pernicious greenhouse effect. Currently, one of the most promising approaches is magnetic confinement, in which the ionized gas or plasma is bound to a strong magnetic field. To avoid end losses, the field lines are bent back on themselves, leading to the typical doughnut-shaped devices called tokamaks and stellarators. The choice of gas is usually a mixture of Deuterium and Tritium, as this combination is easiest to ignite. To achieve sustained fusion reactions, the parameters of the plasma must fulfill the Lawson criterion: $nT\tau > \theta$, where $n$ is the particle density, $T$ the temperature, $\tau$ the confinement time, and $\theta$ a threshold value [1].

To comply with this requirement in the core region of the plasma, the plasma is heated and fueled by various methods. Without entering into details, we note that temperatures achieved in the core of present-day fusion devices range from about 1000 to several times 10,000 eV, corresponding to equivalent temperatures of $10^7$–$10^8$ K. Given such extreme core temperatures, along with the requirement that the walls surrounding the plasma must be kept below the melting temperature of the corresponding materials, it is not unreasonable to state that the temperature gradients created in fusion-grade plasmas are among the highest achieved anywhere on Earth. Hence, the system as a whole is necessarily very far from thermodynamic equilibrium, and standard approaches to study the transport of particles and heat in the plasma must be used with great caution.

Unsurprisingly, the steep gradients, providing an abundance of free energy, trigger the growth of many instabilities, eventually leading to a strongly turbulent state. However, this turbulence is not isotropic, due to the interaction between the dominant confining magnetic field and the ionized plasma, and large-scale coherent structures (known as "zonal flows", analogous to the bands that form in the atmosphere of Jupiter [2]) tend to form spontaneously, which tame the turbulence somewhat. The ensuing complex multi-scale interactions between turbulence and the large-scale structures often leads to a situation best described as a self-organized state. Due to the existence of thresholds for the triggering of instabilities, it has been surmised that fusion-grade plasmas are, in fact, Self-Organized Critical (SOC) systems, and some evidence has been presented that appears to confirm this conjecture [3].

Since the start of fusion development in the 1950s, progress towards raising the achieved values of the parameters of the Lawson criterion has been steady and rather impressive [4]. However, one issue has kept the fusion community from achieving even higher rates of progress: "power degradation". Power degradation is the phenomenon whereby the radial outward transport of heat increases more than linearly with the applied input heating power, thus reducing the efficiency of putative fusion power systems significantly. Of course, considering that the system is non-linear and far from equilibrium, it would be somewhat naive to expect this power scaling to be linear. A full understanding of the mechanisms underlying this phenomenon has so far eluded the community.

In the present work, we will address this issue from the novel viewpoint offered by an analysis technique that was recently introduced in the field of information theory: the transfer entropy. This paper is organized as follows. In Section 2, we describe the diagnostic method and the analysis technique used and show a few highlights from the analysis of data from the TJ-I and W7-X stellarators. In Section 3, we show results from the JET tokamak and proceed to analyze these results in more detail, making estimates of "persistency" and an effective diffusion coefficient and interpreting the results in terms of a Continuous Time Random Walk (CTRW). We then discuss this interpretation in light of the simulations of plasma turbulence, which provide some understanding of the reported observations. In Section 4, we discuss our results in the framework of earlier studies and their significance for the power degradation issue. Finally, in Section 5, we summarize our results, which suggest the existence of minor transport barriers and fast and slow heat transport channels.

## 2. Experiments and Methods

Generally speaking, turbulence in fusion plasmas is not easy to study due to the fact that local measurements in the interior of the plasma are difficult to perform. For example, due to the high temperature of the plasma, inserting physical probes is often unpractical and even undesirable due to the induced perturbations. Other measurement systems yield line-integral rather than local quantities (as is the case with some types of electromagnetic emissions from the plasma), generally not very suited to the analysis of turbulence, or only achieve low sampling rates, insufficient to follow the rapid evolution of turbulence in detail (such as the scattering of laser light known as Thomson scattering). Nevertheless, some local and fast measurements are possible. Here, we will focus on a technique known as Electron Cyclotron Emission (ECE).

ECE is a technique developed in the early days of plasma research and is based on a simple physical principle. In the strongly magnetized and highly ionized plasma, electrons gyrate around the field lines with a frequency $\omega_c = eB/m_e$ and emit radiation at this frequency and higher harmonics. Consequently, the radiation frequency is related to the magnetic field. If the spatial variation of the magnetic field is known, the origin of the emitted radiation can be deduced with good precision, subject to some conditions. The intensity of the detected radiation is directly related to the electron temperature $T_e$, again subject to some conditions [5]. Therefore, the measurement of ECE radiation provides a means to study the evolution of the local electron temperature. By measuring at various emission frequencies simultaneously, one may obtain this information at various locations inside the plasma, which is useful to study both the time-averaged temperature profile and the evolution and propagation of temperature perturbations along the measurement chord. Due to these interesting properties of ECE diagnostics, most present-day magnetic confinement devices are fitted with such systems [6].

To probe the transport properties of a system, it is customary to introduce a small perturbation and observe its propagation. The velocity and spreading of the propagating perturbation can then be related to the convection and diffusion coefficients of the system. However, strongly driven fusion plasmas, far from equilibrium, are typically pervaded by many instabilities and noise. Consequently, it is usually not feasible to track individual perturbations, and a statistical approach is needed.

In recent work, we have found that a technique based on ideas from the field of information theory, the transfer entropy, offers a robust way to address this problem [7]. This nonlinear technique measures the "information transfer" or causal relation between two time series. More specifically, the transfer entropy between discretely sampled signals $y(t_i)$ and $x(t_i)$ quantifies the number of bits by which the prediction of the next sample of signal $x$ can be improved by using the time history of not only the signal $x$ itself, but also that of signal $y$.

In this work, we use a simplified version of the transfer entropy:

$$\mathbb{T}_{Y\to X} = \sum p(x_{n+1}, x_{n-k}, y_{n-k}) \log_2 \frac{p(x_{n+1}|x_{n-k}, y_{n-k})}{p(x_{n+1}|x_{n-k})}. \tag{1}$$

Here, $p(a|b)$ is the probability distribution of $a$ conditional on $b$, $p(a|b) = p(a,b)/p(b)$. The probability distributions $p(a,b,c,\dots)$ are constructed using $m$ bins for each argument, i.e., the object $p(a,b,c,\dots)$ has $m^d$ bins, where $d$ is the dimension (number of arguments) of $p$. The sum in Equation (1) runs over the corresponding discrete bins. The number $k$ can be converted to a "time lag" by multiplying it by the sampling rate. The construction of the probability distributions is done using "course graining", i.e., a low number of bins (here, $m = 3$), to obtain statistically significant results. For more information on the technique, please refer to [8]. The value of the transfer entropy $\mathbb{T}_{Y\to X}$, expressed in bits, can be compared with the total bit range, $\log_2 m$, equal to the maximum possible value of $\mathbb{T}_{Y\to X}$, to help decide whether the transfer entropy is significant or not. The statistical significance of the transfer entropy can be estimated by calculating $\mathbb{T}_{Y\to X}$ for two random (noise) signals [9].

The Transfer Entropy (TE) has proven useful for the study of heat transport in stellarators [10,11]. Due to some remarkable properties, the TE is a powerful technique that provides unprecedented radial detail. First, it is directional, acting as a filter that preferentially selects information components related to (directional) propagation. Second, unlike linear tools such as the cross-correlation or the conditional average, it does not depend on the temporal waveform or even the amplitude of the fluctuations, but merely on the time lag between $x$ and $y$. A comparison between this technique and the cross-correlation was made in previous work [11], and it was concluded that the TE is an exquisitely sensitive tool to study the propagation of perturbations in highly non-linear systems (such as fusion plasmas), in which perturbations tend to be deformed or change shape quickly as they propagate.

The TE is calculated between two signals, in this case between data measured by an ECE channel at a reference position $r_{\rm ref}$ ($Y$ in Equation (1)) and data from an ECE channel at another position, $r$ ($X$ in Equation (1)).

Figure 1 shows an example from the TJ-II stellarator (major radius $R_0 = 1.5$ m) [12], a machine characterized, among other things, by low magnetic shear. The ECE reference channel is taken at $\rho_{\rm ref} \simeq -0.07$, and the other ECE channels are distributed along the minor radius $-1 \le \rho \le 1$. Here, $\rho = 0$ corresponds to the magnetic axis of the torus, while $|\rho| = |r/a| = 1$ corresponds to the minor radius of the torus. By convention, ECE channels with a negative $\rho$ coordinate (the locations of which are indicated in the figure by white circles) are located on the high field side of the magnetic axis.

The two panels in this figure (a and b) correspond to plasmas with a very different level of electron cyclotron heating power, as indicated in the caption. Comparing the low and high power cases shown in the figure, one observes a relatively smooth "plume" of propagating perturbations in the low-power case, propagating outward from $\rho = \rho_{\rm ref}$. The main body of the plume occurs in the range $-0.35 < \rho < -0.07$, although a rather weak continuation of the plume reaches about $\rho \simeq -0.55$, where some stagnation may be visible. This situation would be roughly consistent with "normal" diffusive propagation. However, in the high-power case, the plume clearly stagnates at $\rho \simeq -0.35$, developing a long horizontal "tail"; yet, for $\tau \simeq 0.2$ ms, a second propagation branch appears at $\rho \simeq -0.55$, with an amplitude comparable to or greater than the first branch. Note that this response occurs without any detectable response at $\rho \simeq -0.45$, so that the perturbations seem to have "jumped over" this intermediate position. The perturbations at $\rho \simeq -0.55$ have a stronger causal link to $\rho_{\mathrm{ref}}$ (higher value of TE) than in the low power case. The stronger causal response at $\rho \simeq -0.55$ may be related to power degradation, as perturbations seem to be better able to reach this position and influence turbulence there, possibly implying a more intense radial transport from $\rho_{\mathrm{ref}}$ to $\rho \simeq -0.55$ in the high power case.

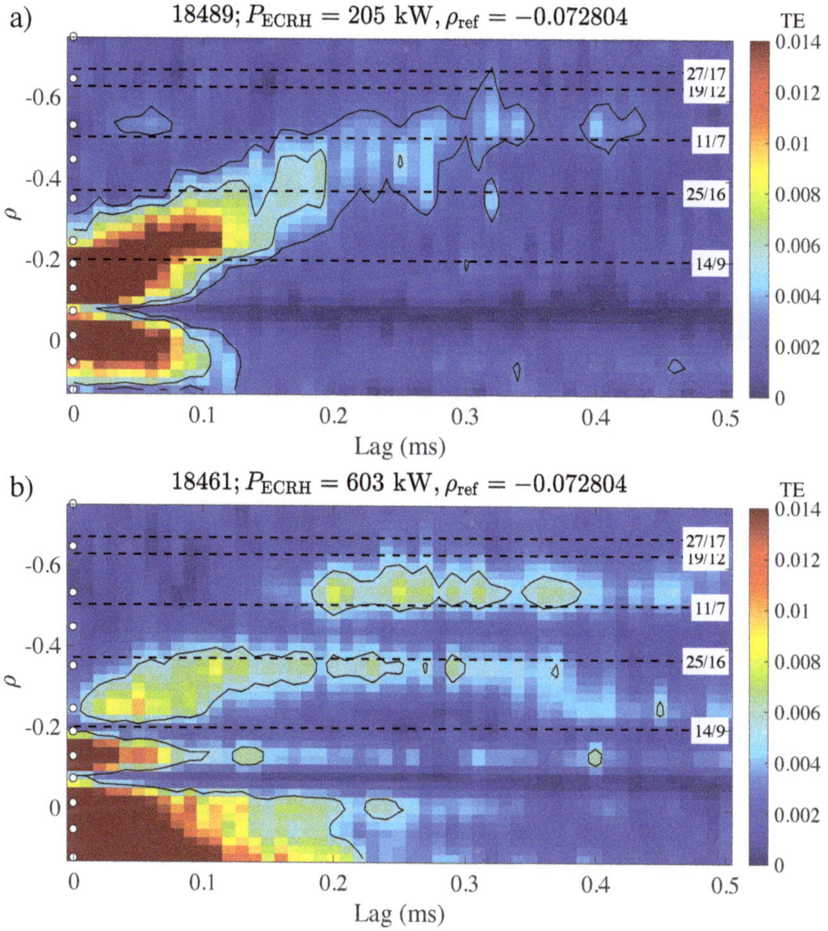


**Figure 1.** Examples of transfer entropy calculated from Electron Cyclotron Emission (ECE) data taken at the TJ-II stellarator, using $\rho_{\mathrm{ref}} \simeq -0.07$, at (**a**) $P_{\mathrm{ECRH}} = 205$ kW and (**b**) $P_{\mathrm{ECRH}} = 603$ kW. The color scale indicates the value of $\mathbb{T}$. ECE channel positions are indicated with white circles. The approximate location of some major rational surfaces is indicated by horizontal dashed lines; the line labels specify the corresponding rotational transform of the magnetic field, $n/m$ (toroidal per poloidal turns). Figure reproduced from [13].

Figure 2 shows similar results from a discharge of the W7-X stellarator (major radius $R_0 = 5.5$ m) [14], also with low magnetic shear, but with a size significantly exceeding that of TJ-II. The number of available ECE channels (again indicated by white dots) is much larger here. Note that the convention regarding the radial coordinate, $|\rho| = |r/a|$, is reversed from TJ-II: here, negative values of $\rho$ correspond to the low field side of the plasma. Due to issues related to data contamination, we only consider data in the range $0 < \rho < 0.85$. Comparing the low and high ECRH power phases, one observes that they have in common that some perturbations propagate outward relatively slowly to the 4/5 rational surface, which acts as a "trapping zone" for these perturbations. In the high power phase, there is an additional branch of radial propagation, faster and more intense (in terms of information transfer), reaching the 9/11 rational surface.

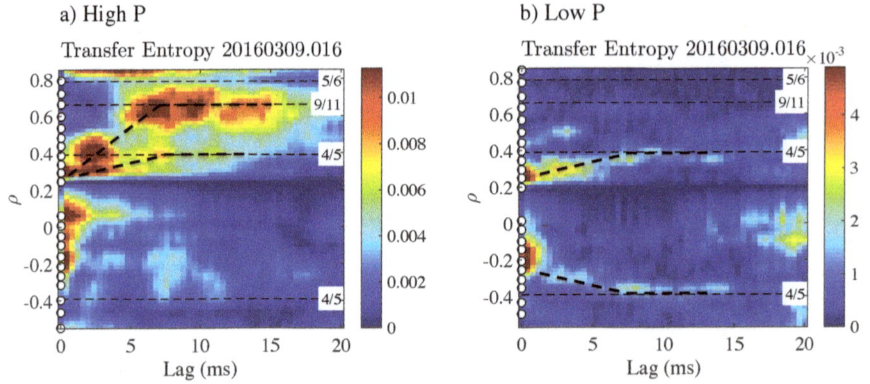


**Figure 2.** Transfer entropy calculated from ECE data taken at the W7-X stellarator, at (**a**) $P_{ECRH} \simeq 2.0$ MW and (**b**) $P_{ECRH} \simeq 0.6$ MW. The color scale indicates the value of $\mathbb{T}$. Radial propagation is indicated with thick dashed lines. Figure reproduced from [11].

We would like to point out the similarity between Figures 1 and 2. Both show the existence of a clear outward propagating "plume" of "information" from the reference position, $\rho_{ref}$. This "plume" has a tendency to stagnate near specific low order rational surfaces, producing horizontally extended structures in the figures. On the other hand, occasionally, especially at high power, information is seen to "arrive" at outward positions without having "passed through" positions further inside, giving the impression of having "jumped over" intermediate positions. In the following, we will further investigate this remarkable phenomenology using a different set of techniques.

## 3. Analysis

In this section, we will analyze high-resolution ECE data from the JET tokamak (major radius $R_0 \simeq 2.96$ m) [15]. JET discharges are usually characterized by sawtooth activity in the core region (reconnection events associated with the $q = 1$ rational surface). These events produce a rapid expulsion of heat from the core, and the resulting heat pulses can be analyzed to obtain information about heat transport [16–18]. In Figure 3, a typical TE graph is shown for $R_{ref} = 3.30$ m, versus time lag and the $R$ value of the other ECE channels. The $R$ range is chosen outside the $q = 1$ surface, in order to allow tracking the propagation of the heat pulses caused by the sawtooth crashes. Different from the results shown in Figures 1 and 2, here, the radius indicated on the ordinate of the graph is the major radius, rather than the normalized minor radius. The reader should be aware that the magnetic axis or plasma center is typically located near the major radius of the torus, $R_0 \simeq 2.96$ m, while the plasma edge is located near $R \simeq 3.85$ m. This example graph shows that overall transport is outward, as indicated by the white dashed line. The velocity of this propagation, given by the slope of this line, is consistent with the typical heat transport coefficients measured in the JET tokamak using other techniques [19].

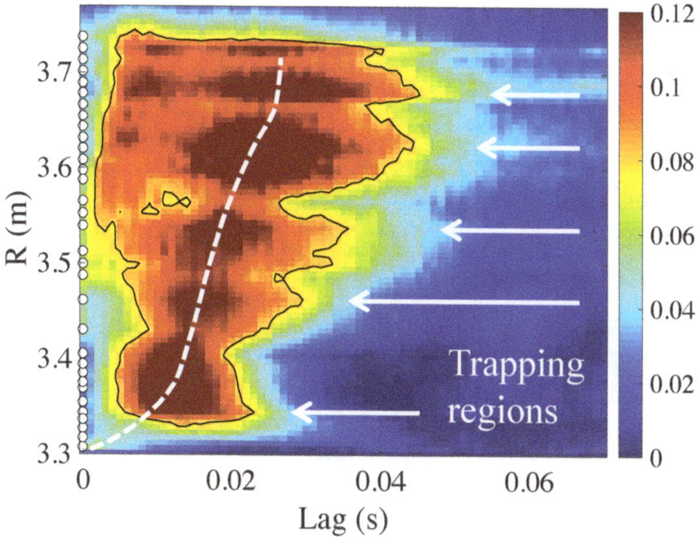


**Figure 3.** Transfer entropy for JET discharge 82,292. $R_{\text{ref}} = 3.30$. The color bar indicates the value of $\mathbb{T}$. White circles indicate the locations of ECE channels. To emphasize the shape of the high TE region, a contour at $\mathbb{T} = 0.08$ is shown (black line). The white dashed line indicates the overall outward propagation. White arrows indicate "trapping regions" (see the text). Figure reproduced from [15].

### 3.1. Radial Modulation of the TE

We draw attention to the fact that the TE shown in Figure 3 is modulated radially. There are well-defined radial zones where the distribution is broader horizontally than elsewhere, as indicated by the white arrows. As before, we interpret these regions as "trapping regions", where outward transport is delayed and heat tends to accumulate. Likewise, there are radial "dips" where the TE is significantly lower. In the framework of sheared flow models, "minor transport barriers" are regions where the zonal flow is high and turbulence is suppressed (fully or partially); these regions would correspond to the observed "dips". The "trapping regions", however, are zones in-between the minor transport barriers, where turbulence is not suppressed, but turbulent vortices exist that tend to trap the propagating heat.

### 3.2. Persistence of Minima

In order to quantify the location of the observed radial minima of the TE, we calculate the average of the TE over the available time lags (or up to a specific maximum time lag), $\langle\mathbb{T}\rangle$. Figure 4 shows an example of this curve for various choices of reference radius. It is observed that the locations of some minima of $\langle\mathbb{T}\rangle$ do not depend on the choice of reference radius, within a reasonable range, but rather are associated with the magnetic configuration (cf. the minimum indicated by the vertical dashed line in Figure 4). The minimum occurring at the reference radius itself has a trivial origin and should be ignored. The location of minima in the graphs of $\langle\mathbb{T}\rangle$ can be subjected to a statistical analysis, based on the set of all available $R_{\text{ref}}$ values for a given discharge. To do so, we count how often each local minimum occurs with respect to the total number of reference radii $R_{\text{ref}}$ studied and express it as a percentage. This number is defined as the "persistence" of any given local minimum.

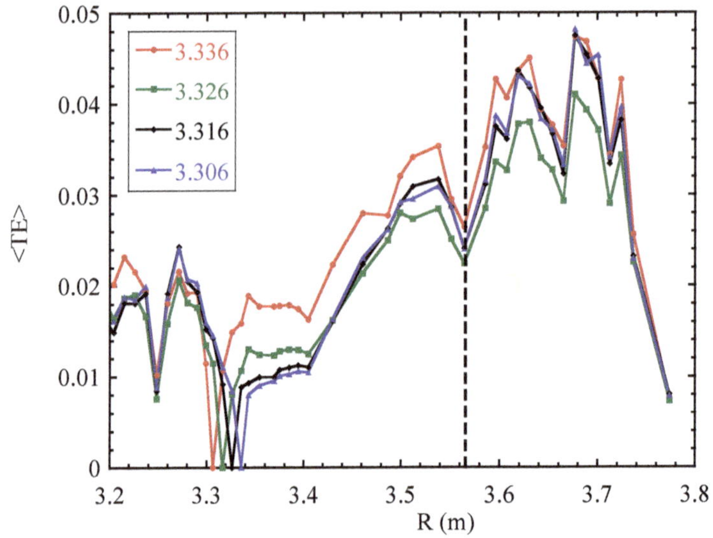


**Figure 4.** Time average of TE over all time lags $0 \le \tau \le 0.2$ s for JET discharge 82,292, for a few reference values $R = R_{\text{ref}}$, as indicated in the legend. Figure reproduced from [15].

### 3.3. *Effective Diffusivity*

It is also possible to estimate an effective diffusion coefficient from the radial propagation of information. Calculating an effective diffusion coefficient is important, as it allows contrasting and comparing the results from this method to traditional estimates of heat transport and is helpful to elucidate the power degradation issue mentioned in the Introduction. Nevertheless, it should be borne in mind that the calculation of an effective diffusion coefficient does not imply that transport is actually diffusive in nature; in fact, as we have argued above, it is unlikely to be so. For each available ECE channel, one can estimate the mean time delay $\langle\tau\rangle$ from:

$$\langle\tau(R)\rangle = \frac{\int \tau \mathbb{T}(R,\tau)d\tau}{\int \mathbb{T}(R,\tau)d\tau} \tag{2}$$

Figure 5 shows an example corresponding to the same case as Figure 4. Using an appropriate reconstruction of the magnetic equilibrium [20], we can convert the ECE measurement location $R$ to a minor radius value $r = a\sqrt{\Psi_N}$, where $\Psi_N$ is the toroidal magnetic flux, normalized such that it equals zero at the magnetic axis and one at the plasma edge (or separatrix).

Then, an effective diffusion coefficient can be defined by:

$$\langle D\rangle = c \cdot \frac{(r - r_0)^2}{\langle\tau(r)\rangle} \tag{3}$$

The coefficient $c$ appearing in this equation is set at $c = \frac{1}{8}$, corresponding to the "time to peak" estimate [21], although slightly different values are sometimes also used in literature [16]. Note that this estimate of the effective diffusion coefficient is not very accurate, for two reasons. First, it is not defined for $r = r_0$ as both the numerator and the denominator of the expression tend to zero, and the radial behavior tends to be dominated by the numerator $(r - r_0)^2$ for small values of $r - r_0$. Therefore, the extracted diffusion coefficient should not be taken too seriously in the region near the reference position. Second, it is defined exclusively on the basis of the time (or phase) delay, whereas a proper recovery of the underlying effective diffusion coefficient would require information about the perturbation amplitude as well. Nevertheless, it may serve as a means to visualize the radial variation of transport, and in this paper, we will use it only for this purpose.

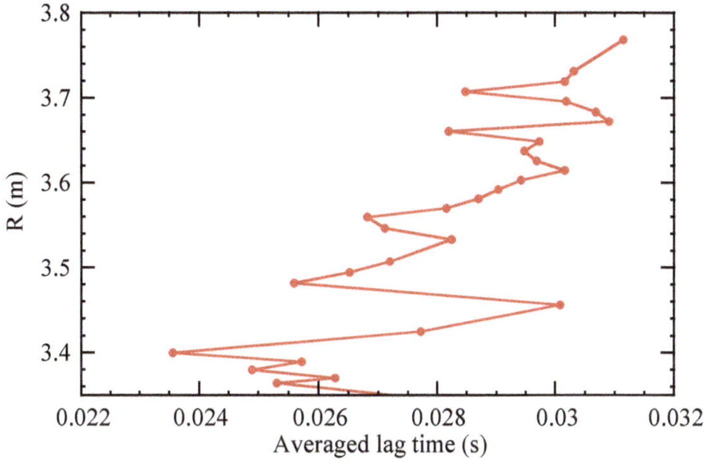


**Figure 5.** Example graph of position $R$ versus the mean lag time $\langle \tau \rangle$, showing radial variation.

The resulting value $\langle D \rangle$ is the mean diffusivity over the interval $[r_0, r]$. To extract the local value, we consider that this mean diffusivity is calculated as follows from the local diffusivity:

$$\langle D \rangle_N = \frac{1}{r_N - r_0} \sum_{i=0}^{N-1} (r_{i+1} - r_i) D(r_i) \tag{4}$$

so that:

$$(r_N - r_0)\langle D \rangle_N - (r_{N-1} - r_0)\langle D \rangle_{N-1} = (r_N - r_{N-1}) D(r_{N-1}), \tag{5}$$

from which the local effective diffusivity $D(r_{N-1})$ follows. Of course, when $\langle D \rangle$ does not depend strongly on $r$, the mean diffusivity and the local diffusivity are nearly the same.

Next, we attempt to correct for the unphysical fact that $D$ tends to zero at $r = r_0$. To do so, we first compute $D_0(r)$, i.e., the local effective diffusion coefficient using $r_0 \simeq 0$. Then, we estimate the corrected local effective diffusion coefficient at different reference radii $r_0$ using:

$$D_{r_0}^{\mathrm{corr}}(r) = D_0(r_0) + D_{r_0}(r) \tag{6}$$

This correction, while still not perfect, should bring the estimated value of the diffusion coefficient closer to the "true" diffusion coefficient, by partially correcting for the unphysical effect mentioned above.

Figure 6b shows an example of the corrected effective diffusion coefficient $D^{\mathrm{corr}}$, along with the location of minima of $\langle \mathbb{T} \rangle$, indicated by bars proportional to the degree of persistence. It may be observed that structures in the $D^{\mathrm{corr}}$ profile are often correlated with persistent minima, suggesting that these minima indeed act as minor transport barriers, affecting radial heat transport. Figure 6a also shows the corresponding profile of the safety factor, $q = m/n$ (from a reconstruction of the magnetic equilibrium by the program EFIT, using magnetics alone; the sawteeth inversion radius, determined from the $T_e$ time traces, is located at $r \simeq 0.47$, close to the $q = 1$ surface). It can be seen, for example, that the barrier at $r \simeq 0.73$ is not far from the point where $q = 3/2$, although uncertainties in the $q$-profile do not allow one to make a definite identification.

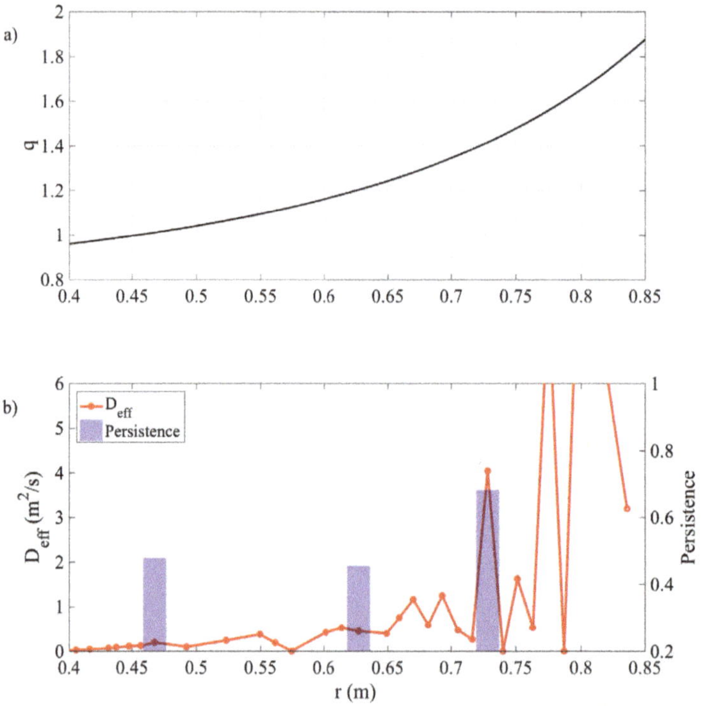


**Figure 6.** (**a**) Profile of the safety factor, $q$, averaged over the time window of interest (9–12 s) and (**b**) corrected effective diffusion and persistence.

### *3.4. Propagation Paths*

Note that Figure 3 shows two branches of propagation. The "slow branch" is indicated by the white dashed line. However, there appears to be a "fast branch", visible for $3.55 < R < 3.74$ m at lag times $\tau < 0.01$ s. In this section, we investigate this issue further.

The transfer entropy $\mathbb{T}(r_{\text{ref}}, r, \tau)$ specifies the improvement of the prediction of the next sample of the signal $x(r, t)$, based on the knowledge of $x(r_{\text{ref}}, t - \tau)$. Hence, it seems reasonable to assume that some kind of "particles" carry this information from $r_{\text{ref}}$ to $r$, taking a time $\tau$ to take this step. In the present context, the "particles" would represent heat, rather than actual particles, of course. The latter description is reminiscent of the continuous time random walk [22].

If one interprets the transfer entropy in this framework, the transfer entropy can be associated with the probability distribution for taking a step $\Delta r = r - r_{\text{ref}}$ in time $\tau$, simply by normalizing $\mathbb{T}_{r_{\text{ref}}}(\Delta r, \tau) = \mathbb{T}(r_{\text{ref}}, r, \tau)$ by a factor $N$, so that the resulting distribution $p_{r_{\text{ref}}}(\Delta r, \tau) = \mathbb{T}_{r_{\text{ref}}}(\Delta r, \tau)/N$ is a probability distribution such that its integral over all relevant $\Delta r$ and $\tau$ equals one. One can then concatenate successive steps of a given particle, drawing the values $(\Delta r, \tau)$ of each step randomly from this probability distribution and study the corresponding compound paths. To reduce the computational load somewhat, we will only consider paths that move strictly outward.

The procedure described above is an iterative procedure, and it allows studying the compound paths statistically. Alternatively, one can use a recursive procedure, by applying a threshold to the step probability distribution. The resulting binary distribution then only states which steps $(\Delta r, \tau)$ are allowed and which are not. Subsequently, all allowed compound outward paths can be followed, using a recursive algorithm, and these can again be subjected to a statistical analysis.

Figure 7 shows the distribution of radial steps. Previous studies involving the analysis of tracer trajectories in simulations of the topological structures in plasma turbulence suggest that the lognormal distribution may play a significant role [23,24], and indeed, the present result seems to be compatible with this idea, as shown by the fitted line.

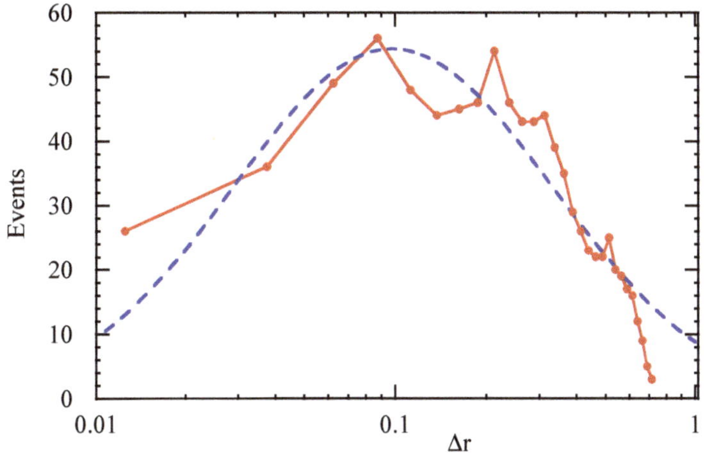


**Figure 7.** Probability distribution of the radial steps taken by particles, according to the TE analysis. The blue dashed curve is a fitted lognormal distribution.

Figure 8 shows the statistical distribution of the times needed to reach the outer edge of the system from an initial position in the core, calculated from a transfer entropy dataset obtained from ECE data (one element of the set, at a single reference radius, being Figure 3), using the recursive method described above. Remarkably, the distributions seem to separate into two distinct classes, namely fast and slow paths, according to the first step taken ($R_2$).

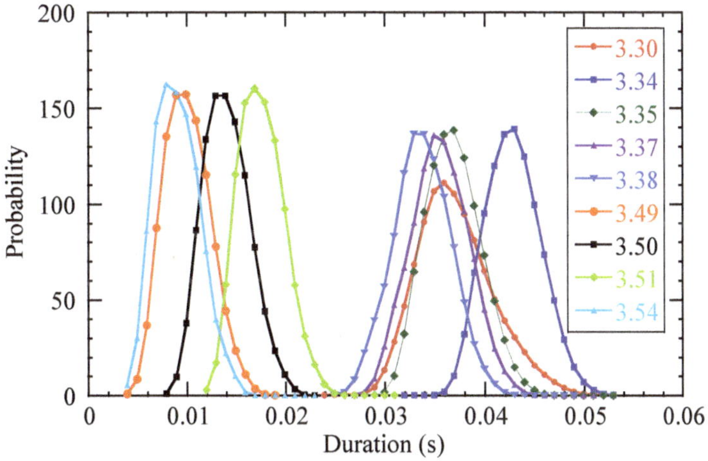


**Figure 8.** Probability distribution of the times needed to reach the outer edge of the system from an initial position in the core. The legend indicates the position of the first step of the compound path ($R_2$).

The figure shows that each individual distribution is roughly Gaussian, as one might expect. Therefore, these distributions are well characterized by their mean and standard deviation. Figure 9 shows the mean and standard deviation of the durations of the compound paths to reach the edge of the system as a function of the first step taken. The graph separates into two clear classes ($R_2 < 3.45$ and $R_2 > 3.53$), while there is a narrow transition region in-between.

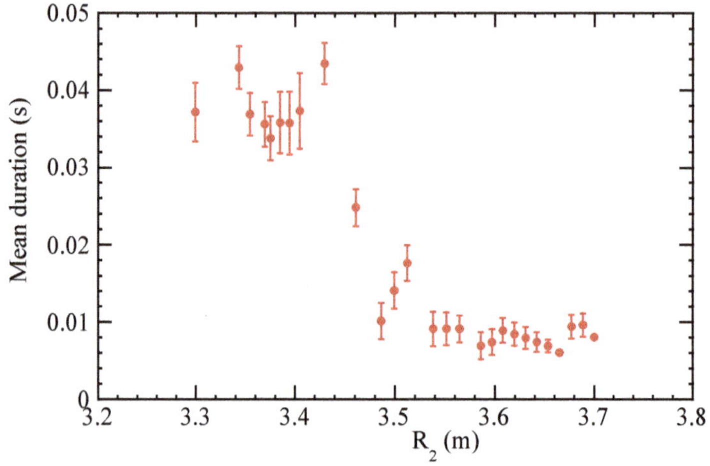


**Figure 9.** Mean and standard deviation of the durations of the compound paths to reach the edge of the system as a function of the first step taken ($R_2$).

Figure 10 shows some examples of the fast and slow paths. The slow paths are reminiscent of a directed random walk, while the fast paths include some very long jumps, which suggests they could be Lévy flights [25]. Future work may be able to clarify this point. In any case, the result of this analysis is that radial heat transport in these plasmas appears to be characterized by different transport channels, with different propagation velocities. Presumably, the plasma is able to vary the relative importance of these channels in order to achieve the mentioned self-organization of radial transport.

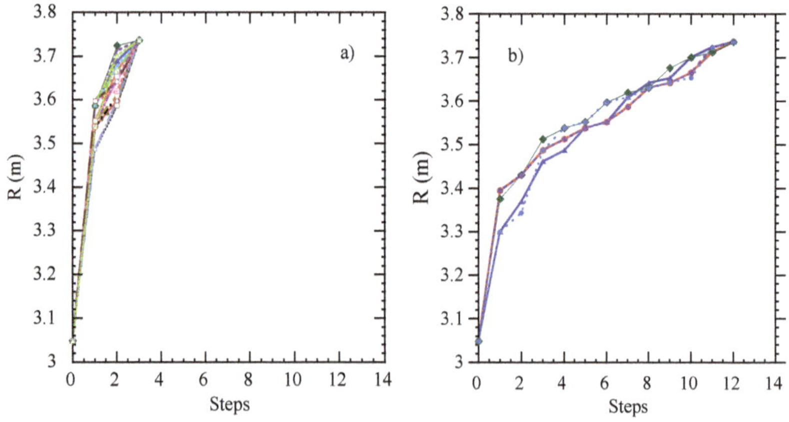


**Figure 10.** Mean compound paths to reach the edge of the system, for (**a**) paths in the "fast" group and (**b**) paths in the "slow" group of Figure 9.

### 3.5. Modeling

As noted in the Introduction, the plasmas considered here are confined by a magnetic field. Inside the plasma, the magnetic field lines lie on surfaces of constant flux, which have a toroidal topology. The mean field line twist on each surface is such that $\Delta\phi = q\Delta\theta$, on average, where $\Delta\phi$ is the angle in the toroidal direction (long way around the torus) and $\Delta\theta$ is the angle in the poloidal direction (short way around the torus). On each flux surface, $q$ is constant. When $q$ takes a rational value, the magnetic field lines close on themselves after a finite number of turns. This is where turbulent vortices, which are elongated along the direction of the field line and therefore have a filamentary structure, are preferentially located.

The turbulent flow velocity of the plasma can be expressed as $V = b \times \nabla\Phi$, where $\Phi$ is a stream function (proportional to the electrostatic potential) and $b$ is a unit vector in the toroidal (field) direction. Theoretically, transport barriers may arise as a consequence of zonal flows generated by turbulence. The mechanics of the interaction between turbulent fluctuations and zonal flows is well understood: fluctuations may generate flows through Reynolds stress [26], and the shear in these flows then suppresses the fluctuations [27]. The complexity of these interactions has been clarified using simplified models [28], and it has been found that sheared flow regions are preferentially formed near rational surfaces.

Figure 11 shows the radial structure of an electrostatic fluctuation potential near a rational surface, arbitrarily placed at $r/a = 0.5$, and the associated sheared flow in a very simple slab model.

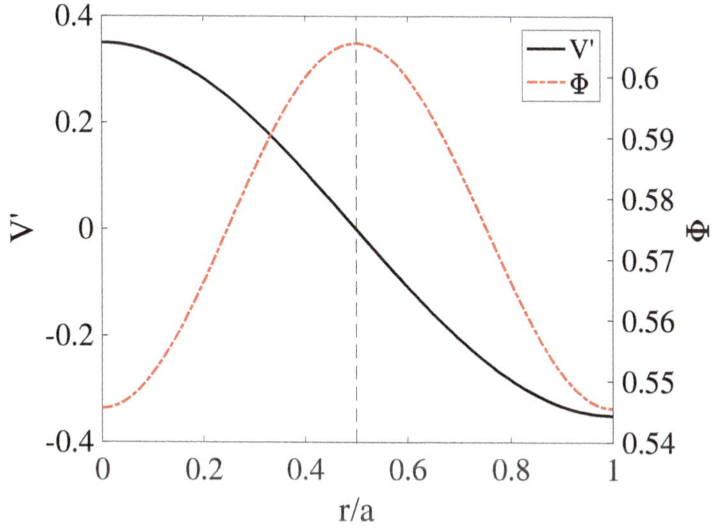


**Figure 11.** Potential fluctuation ($\Phi$) and shear of the generated flow ($V' = dV/dr$) for a simple nonlinear slab model. The vertical dashed line shows the position of the singular surface. Figure reproduced from [15].

This figure is no more than a cartoon, shown to illustrate the idea of the association between fluctuations, rational surfaces, and sheared flow. If the instability eigenfunction $\Phi$ is symmetric with respect to the rational surface, the flow shear $|V'| = |dV/dr|$ will peak off the rational surface, at a distance of the order of the width of the turbulent vortices. Likewise, an antisymmetric eigenfunction will place the flow shear peak at the rational surface. Each type of instability will generate its own structure, possibly modulated by the presence of other structures nearby, and the actual situation can be rather convoluted. Nevertheless, the central idea is that the sheared flow regions are usually located near singular surfaces.

The plasma is pervaded by many types of instability. However, the fact that we detect minor transport barriers associated with rational surfaces provides a hint with regard to the underlying mechanism. Therefore, we have turned to a resistive MHD turbulence model to interpret experimental results [29]. Thus, we have been able to show that the spontaneously arising turbulence in this model generates sheared flow regions that act as minor transport barriers [30]. Injecting tracers to better understand the effect of the turbulence and the sheared flow regions on transport, we have observed that some of the tracers are trapped in the turbulent vortices, while others, with higher kinetic energies, perform rapid radial excursions, "jumping over" the barriers. As the system is driven more strongly (by increasing heating power levels), on average, tracers are endowed with higher energies, so that more tracers will be able to "jump" the minor barriers. In fact, this is the mechanism we proposed to explain the degradation of confinement in the TJ-II stellarator [13]. Likewise, in the framework of the present study, we observe the existence of minor transport barriers and two classes of "particles": slow and fast, or "diffusive" and "jumping", which seems to fit nicely with these ideas.

Figure 12 shows a snapshot of a typical modeling result obtained with the mentioned resistive MHD model in stellarator-like (low shear) cylindrical geometry. The area of the graph corresponds to a region of the poloidal-radial $(\theta, r)$ plane at constant toroidal angle ($\phi$ = constant). The graph shows vortices (trapping regions), such as the poloidally periodic structures seen near $r/a = 0.7$, related to a corresponding rational surface. Also visible are zonal flow regions (horizontally elongated structures with predominantly horizontal flow velocities in both directions), on both sides of the vortex sequence.

In previous work, we have successfully applied the transfer entropy to turbulence simulations of this type. This effort yielded a qualitatively similar picture as the reported experimental results, with "trapping zones" and radial "jumps" [10,11]. We also verified the calculation of the effective diffusivity from the TE and compared it to traditional estimates for such simulations [31].

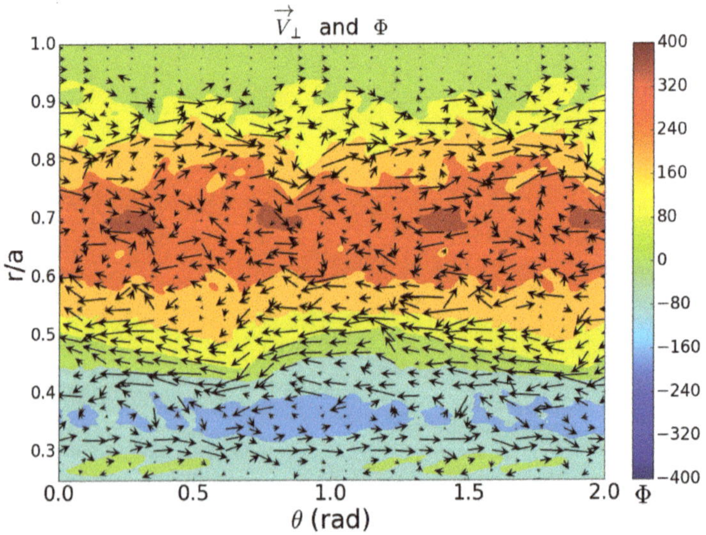


**Figure 12.** Modeling results, showing zonal flow regions and vortices.

## 4. Discussion

It has long been known that magnetically confined plasmas occasionally develop spontaneous transport barriers. Early work carried out at the RTP tokamak clearly demonstrated the existence of a multiplicity of such transport barriers throughout the plasma, whose location was found to be close to low order rational surfaces [32]. Subsequently, a simplified so-called "$q$-comb" transport model was developed to interpret the observations, based on radially localized reductions of the heat diffusion coefficient, coinciding with low order rational surfaces [33]. However, this and similar work has not led to a general incorporation of mechanisms associated with rational surfaces in heat transport models for fusion plasmas, probably due to the fact that further experimental evidence for these minor transport barriers, associated with rational surfaces, has been difficult to obtain.

Under specific conditions, plasmas can also develop so-called Internal Transport Barriers (ITBs) [34], which arise only transiently, but are much stronger than the "minor transport barriers" that are the focus of this paper. In tokamaks, strong ITBs can be established by creating a core reversed magnetic shear region, while the location of the ITB appears correlated with integral values of the safety factor, $q$ [35]. The impact of ITBs on heat transport has been studied in some detail at, e.g., Alcator C-Mod [36] and JET [19,37], showing that the heat diffusivity drops strongly in the ITB region. ITBs have also been obtained and studied in stellarators [38], and here, too, a relationship with the magnetic configuration is suggested. The existence of ITBs is widely acknowledged and supported by experimental evidence on many machines.

A localized transport barrier (i.e., a local reduction of heat flux) implies a local change of slope of the temperature profile. Given the general turbulent state of the plasma and the prevailing measurement resolution and errors, such rather localized changes of slope are usually not easy to detect. Even with strong ITBs, it is often difficult to delimit the precise location of the ITB, based on the temperature profile alone. Hence, it is not very surprising that minor transport barriers usually go undetected. As a result, many transport models completely ignore their possible existence and do not contemplate any effects that explicitly depend on the rational values of the rotational transform.

In our recent series of papers, using a novel method to detect minor transport barriers based on the transfer entropy, we have tried to show that such barriers occur quite frequently, even in plasmas with no easily discernible "steps" in the temperature profile, and they tend to be associated with low order rational surfaces [10,11,13]. By studying the barriers at different heating power levels, we have been able to observe a change in the characteristics of transport (an increased importance of heat "jumping" over the minor barriers) that suggests that these minor barriers could in fact play a prime role in the understanding of the important and ubiquitous phenomenon of power degradation.

To recall, power degradation is the phenomenon that the energy confined in the plasma ($W$) increases less than linearly with the heating power. In all magnetic confinement devices where the scaling of the energy confinement time ($\tau_E = W/P$, subject to some caveats and corrections) with heating power ($P$) has been studied, it is found that it scales like $\tau_E \propto P^{\alpha_P}$, where $\alpha_P = -0.6 \pm 0.1$ [39–43]. The fact that this scaling holds across the board for the main types of magnetic fusion devices (tokamaks and stellarators) indicates that it must be due to a very basic mechanism, common to these devices.

Our analysis suggests that transport does not involve a single mechanism, but various competing mechanisms, whose relative importance depends on the drive. Hence, describing transport via a single diffusion coefficient (or a similar simplified description) may not be adequate to capture the physics underlying power degradation.

In previous work, we have made use of a resistive MHD model [29] to understand both the detected minor transport barriers and the "jumping" behavior [10,11,13]. While this model does not capture all details of turbulence in fusion-grade plasmas, it does allow a precise analysis of the effect of MHD-type turbulence, which typically is associated with low order rational surfaces. In view of the fact that our analyses seem to indicate that low order rational surfaces play an important role, it makes sense to use this type of model to gain further insight. The modeling results seem to indicate that sheared flow layers tend to form near low order rational surfaces as a consequence of plasma self-organization. These sheared flow layers tend to suppress turbulence locally, leading to minor transport barriers [2]. Near these barriers, turbulent vortices form where radially propagating "particles" can get trapped. The observed "jumping" behavior is also reproduced by the modeling results and could be associated with the coupling between MHD turbulence associated with different rational surfaces or, more generally, "avalanches". The observations indicate that the "jumping" behavior increases in intensity when the heating power is increased, suggesting an explanation for the phenomenon of power degradation mentioned in the Introduction.

We note that the suggested association with low order rational surfaces may apply only under specific circumstances (namely, those where the resistive MHD model we used are relevant; typically, stellarators). Recent theoretical [44] and experimental [45] work on tokamaks suggests the existence of a so-called $E \times B$ "staircase" in hot plasmas, largely analogous to the ideas we propose here, but only loosely connected to rational surfaces, if at all. We conclude from this work that magnetically confined fusion plasmas have a general tendency to self-organize by forming sheared flow layers and minor transport barriers, with characteristics that may depend somewhat on the underlying turbulence mechanisms.

In previous work, we have studied transport from the particle perspective by injecting tracer particles in the turbulent flow computed using the mentioned resistive MHD model [23,24,30,46]. Depending on the energy of the tracer particles, some are trapped by the turbulent vortices, while others, typically with more energy, escape the vortices and end up in the zonal flow regions near the vortices, which constitute a barrier for radial transport. Only particles with the highest energies are able to jump over the barriers [30]. These tracer particle dynamics are consistent with the dynamical picture offered by the transfer entropy analysis presented here.

## 5. Conclusions

This work highlights the non-linear and complex nature of heat transport in strongly driven fusion plasmas. Using a relatively novel analysis method, the transfer entropy, we have shown that heat transport in magnetic fusion devices exhibits qualitatively similar properties in two stellarators and one tokamak. Analysis based on the use of the transfer entropy demonstrates the existence of radially localized zones that can be described as "minor barriers" and associated "trapping regions". A measure was introduced to quantify the "persistence" of local radial TE minima, associated with the minor barriers. We also devised a simple technique to obtain a crude estimate of the effective local heat diffusivity from the TE. The resulting effective heat diffusivity was found to be compatible with traditional estimates, while showing radial variations that appear to be associated with the previously identified minor barriers.

In previous work on two stellarators, we found that the "minor barriers" appear to be associated with low order rational surfaces. In the tokamak case, the relation with low order rational surfaces was less clear [15]. Heat transport was found to be able to "jump over" these minor barriers to some degree, and as heating power was raised, the "jumping behavior" was shown to increase in intensity [11,13,15], providing a possible explanation for the ubiquitous phenomenon of "power degradation" observed in magnetically confined fusion plasmas.

In the present work, we have extended the analysis by reinterpreting the transfer entropy in terms of a continuous time random walk. This approach revealed the existence of clearly separated "fast" and "slow" transport channels (which also appears to be in accordance with a recent more traditional analysis reported in [47]). We interpret the "slow" channel in terms of the usual diffusive transport, whereas the "fast channel" would be associated with the "jumping" behavior mentioned above. In terms of CTRW terminology, the former would be associated with the standard random walk, whereas the latter would correspond to Lévy walks.

The methodology used here does not allow making quantitative statements about the relative importance of the "fast" and "slow" transport channels. This important issue is left to future work, as is the question of particle transport (as compared to heat transport). Furthermore, so far, we have focused on fusion plasmas with relatively low heating power (L-mode plasmas), the reason being that it is often easier to obtain a steady state in L-mode, while the absence of violent instabilities associated with the H-mode edge transport barrier (so-called edge localized modes) further facilitates the analysis. It is clear, however, that it would be important to extend this work also to H-mode plasmas.

**Author Contributions:** Investigation, B.v.M., B.C., L.G. and J.N.; Writing—original draft, B.v.M.

**Funding:** Research sponsored in part by the Ministerio de Economía y Competitividad of Spain under Project No. ENE2015-68206-P and ENE2015-68265-P. This work has been carried out within the framework of the EUROfusion Consortium and has received funding from the Euratom research and training program 2014-2018 and 2019-2020 under Grant Agreement No. 633053. The views and opinions expressed herein do not necessarily reflect those of the European Commission.

**Acknowledgments:** The authors would like to acknowledge pleasant and fruitful collaborations with the experimental teams of TJ-II, W7-X and JET.

**Conflicts of Interest:** The authors declare no conflict of interest.

## References

1. Freidberg, J. *Plasma Physics and Fusion Energy*; Cambridge University Press: Cambridge, UK, 2007.
2. Diamond, P.; Itoh, S.I.; Itoh, K.; Hahm, T. Zonal flows in plasma—A review. *Plasma Phys. Control. Fusion* **2005**, *47*, R35. [CrossRef]
3. Sánchez, R.; Newman, D.; Carreras, B. Mixed SOC diffusive dynamics as a paradigm for transport in fusion devices. *Nucl. Fusion* **2001**, *41*, 247. [CrossRef]
4. International Fusion Research Council. Status report on fusion research. *Nucl. Fusion* **2005**, *45*, A1. [CrossRef]
5. Hutchinson, I. *Principles of Plasma Diagnostics*; Cambridge University Press: New York, NY, USA, 2002.
6. Hartfuß, H.J.; Geist, T. *Fusion Plasma Diagnostics With Mm-Waves: An Introduction*; Wiley: Hoboken, NJ, USA, 2013. [CrossRef]
7. Schreiber, T. Measuring information transfer. *Phys. Rev. Lett.* **2000**, *85*, 461. [CrossRef] [PubMed]
8. van Milligen, B.; Birkenmeier, G.; Ramisch, M.; Estrada, T.; Hidalgo, C.; Alonso, A. Causality detection and turbulence in fusion plasmas. *Nucl. Fusion* **2014**, *54*, 023011. [CrossRef]
9. van Milligen, B.; Carreras, B.; García, L.; Martin de Aguilera, A.; Hidalgo, C.; Nicolau, J.; The TJ-II Team. The causal relation between turbulent particle flux and density gradient. *Phys. Plasmas* **2016**, *23*, 072307. [CrossRef]
10. van Milligen, B.; Nicolau, J.; García, L.; Carreras, B.; Hidalgo, C.; The TJ-II Team. The impact of rational surfaces on radial heat transport in TJ-II. *Nucl. Fusion* **2017**, *57*, 056028. [CrossRef]
11. van Milligen, B.; Hoefel, U.; Nicolau, J.; Hirsch, M.; García, L.; Carreras, B.; Hidalgo, C.; The W7-X Team. Study of radial heat transport in W7-X using the Transfer Entropy. *Nucl. Fusion* **2018**, *58*, 076002. [CrossRef]
12. Alejaldre, C.; Alonso, J.; Almoguera, L.; Ascasíbar, E.; Baciero, A.; Balbín, R.; Blaumoser, M.; Botija, J.; Brañas, B.; de la Cal, E.; et al. First plasmas in the TJ-II flexible Heliac. *Plasma Phys. Control. Fusion* **1999**, *41*, A539. [CrossRef]
13. van Milligen, B.; Carreras, B.; Hidalgo, C.; Cappa, Á.; The TJ-II Team. A possible mechanism for confinement power degradation in the TJ-II stellarator. *Phys. Plasmas* **2018**, *25*, 062503. [CrossRef]
14. Wolf, R.; Ali, A.; Alonso, A.; Baldzuhn, J.; Beidler, C.; Beurskens, M.; Biedermann, C.; Bosch, H.S.; Bozhenkov, S.; Brakel, R.; et al. Major results from the first plasma campaign of the Wendelstein 7-X stellarator. *Nucl. Fusion* **2017**, *57*, 102020. [CrossRef]
15. van Milligen, B.; Carreras, B.; de la Luna, E.; Solano, E.R.; The JET Team. Radial variation of heat transport in L-mode JET discharges. *Nucl. Fusion* **2018**, in press. [CrossRef]
16. Fredrickson, E.; Callen, J.; McGuire, K.; Bell, J.; Colchin, R.; Efthimion, P.; Hill, K.; Izzo, R.; Mikkelsen, D.; Monticello, D.; et al. Heat pulse propagation studies in TFTR. *Nucl. Fusion* **1986**, *26*, 849. [CrossRef]
17. Lopes Cardozo, N. Perturbative transport studies in fusion plasmas. *Plasma Phys. Control. Fusion* **1995**, *37*, 799. [CrossRef]
18. Mantica, P.; Ryter, F. Perturbative studies of turbulent transport in fusion plasmas. *Comptes Rendus Physique* **2006**, *7*, 634. [CrossRef]
19. Mantica, P.; Gorini, G.; Imbeaux, F.; Kinsey, J.; Sarazin, Y.; Budny, R.; Coffey, I.; Dux, R.; Garbet, X.; Garzotti, L.; et al. Perturbative transport experiments in JET low or reverse magnetic shear plasmas. *Plasma Phys. Control. Fusion* **2002**, *44*, 2185. [CrossRef]
20. Brix, M.; Hawkes, N.C.; Boboc, A.; Drozdov, V.; Sharapov, S.E.; JET-EFDA Contributors. Accuracy of EFIT equilibrium reconstruction with internal diagnostic information at JET. *Rev. Sci. Instrum.* **2008**, *79*, 10F325. [CrossRef] [PubMed]

21. Soler, M.; Callen, J. On measuring the electron heat diffusion coefficient in a tokamak from sawtooth oscillation observations. *Nucl. Fusion* **1979**, *19*, 703. [CrossRef]
22. Klafter, J.; Blumen, A.; Shlesinger, M. Stochastic pathway to anomalous diffusion. *Phys. Rev. A* **1987**, *35*, 3081. [CrossRef]
23. Carreras, B.; García, L.; Llerena, I. Tracer particle trapping times in pressure-gradient-driven turbulence in toroidal geometry and their connection to the dynamics of large-scale cycles. *Plasma Phys. Control. Fusion* **2010**, *52*, 105005. [CrossRef]
24. García, L.; Llerena Rodríguez, I.; Carreras, B. Width and rugosity of the topological plasma flow structures and their relation to the radial flights of particle tracers. *Nucl. Fusion* **2015**, *55*, 113023. [CrossRef]
25. Balescu, R. *Aspects of Anomalous Transport in Plasmas*; Institute of Physics: Bristol, UK, 2005.
26. Carreras, B.; Lynch, V.; García, L. Electron diamagnetic effects on the resistive pressure-gradient-driven turbulence and poloidal flow generation. *Phys. Plasmas* **1991**, *3*, 1438. [CrossRef]
27. Biglari, H.; Diamond, P.; Terry, P. Influence of sheared poloidal rotation on edge turbulence. *Phys. Plasmas* **1990**, *2*, 1. [CrossRef]
28. Diamond, P.; Liang, Y.M.; Carreras, B.; Terry, P. Self-regulating shear flow turbulence: a paradigm for the L to H transition. *Phys. Rev. Lett.* **1994**, *72*, 2565. [CrossRef] [PubMed]
29. García, L.; Carreras, B.; Lynch, V.; Pedrosa, M.; Hidalgo, C. Sheared flow amplification by vacuum magnetic islands in stellarator plasmas. *Phys. Plasmas* **2001**, *8*, 4111. [CrossRef]
30. García, L.; Carreras, B.; Llerena, L. Relation of plasma flow structures to passive particle tracer orbits. *Nucl. Fusion* **2017**, *57*, 116013. [CrossRef]
31. Nicolau, J.H.; García, L.; Carreras, B.A.; van Milligen, B. Applicability of transfer entropy for the calculation of effective diffusivity in heat transport. *Phys. Plasmas* **2018**, *25*, 102304. [CrossRef]
32. Lopes Cardozo, N.; Hogeweij, G.; de Baar, M.; Barth, C.; Beurskens, M.; De Luca, F.; Donné, A.; Galli, P.; van Gelder, J.; Gorini, G.; et al. Electron thermal transport in RTP: filaments, barriers and bifurcations. *Plasma Phys. Control. Fusion* **1997**, *39*, B303. [CrossRef]
33. Schilham, A.; Hogeweij, G.; Lopes Cardozo, N. Electron thermal transport barriers in RTP: experiment and modelling. *Plasma Phys. Control. Fusion* **2001**, *43*, 1699. [CrossRef]
34. Wolf, R. Internal transport barriers in tokamak plasmas. *Plasma Phys. Control. Fusion* **2003**, *45*, R1. [CrossRef]
35. Joffrin, E.; Challis, C.; Conway, G.; Garbet, X.; Gude, A.; Günter, S.; Hawkes, N.; Hender, T.; Howell, D.; Huysmans, G.; et al. Internal transport barrier triggering by rational magnetic flux surfaces in tokamaks. *Nucl. Fusion* **2003**, *43*, 1167. [CrossRef]
36. Wukitch, S.; Boivin, R.; Bonoli, P.; Fiore, C.; Granetz, R.; Greenwald, M.; Hubbard, A.; Hutchinson, I.; In, Y.; Irby, J.; et al. Double transport barrier experiments on Alcator C-Mod. *Phys. Plasmas* **2002**, *9*, 2149. [CrossRef]
37. Marinoni, A.; Mantica, P.; Eester, D.V.; Imbeaux, F.; Mantsinen, M.; Hawkes, N.; Joffrin, E.; Kiptily, V.; Pinches, S.D.; Salmi, A.; et al. Analysis and modelling of power modulation experiments in JET plasmas with internal transport barriers. *Plasma Phys. Control. Fusion* **2006**, *48*, 1469. [CrossRef]
38. Fujisawa, A. Transport barriers and bifurcation characteristics in stellarators. *Plasma Phys. Control. Fusion* **2002**, *44*, A1. [CrossRef]
39. Stroth, U.; Murakami, M.; Dory, R.; Yamada, H.; Okamura, S.; Sano, F.; Obiki, T. Energy confinement scaling from the international stellarator database. *Nucl. Fusion* **1996**, *36*, 1063. [CrossRef]
40. Carreras, B. Progress in anomalous transport research in toroidal magnetic confinement systems. *IEEE Trans. Plasma Sci.* **1997**, *25*, 1281. [CrossRef]
41. Doyle, E.; Houlberg, W.; Kamada, Y.; Mukhovatov, V.; Osborne, T.; Polevoi, A.; Bateman, G.; Connor, J.; Cordey, J.; Fujita, T.; et al. Chapter 2: Plasma confinement and transport. *Nucl. Fusion* **2007**, *47*, S18. [CrossRef]
42. Dinklage, A.; Maaßberg, H.; Preuss, R.; Turkin, Y.A.; Yamada, H.; Ascasibar, E.; Beidler, C.; Funaba, H.; Harris, J.H.; Kus, A.; et al. Physical model assessment of the energy confinement time scaling in stellarators. *Nucl. Fusion* **2007**, *47*, 1265. [CrossRef]
43. Hirsch, M.; Baldzuhn, J.; Beidler, C.; Brakel, R.; Burhenn, R.; Dinklage, A.; Ehmler, H.; Endler, M.; Erckmann, V.; Feng, Y.; et al. Major results from the stellarator Wendelstein 7-AS. *Plasma Phys. Control. Fusion* **2008**, *50*, 053001. [CrossRef]
44. Dif-Pradalier, G.; Hornung, G.; Garbet, X.; Gendrih, P.; Grandgirard, V.; Latu, G.; Sarazin, Y. The $E \times B$ staircase of magnetised plasmas. *Nucl. Fusion* **2017**, *57*, 066026. [CrossRef]

45. Hornung, G.; Dif-Pradalier, G.; Clairet, F.; Sarazin, Y.; Sabot, R.; Hennequin, P.; Verdoolaege, G. $E \times B$ staircases and barrier permeability in magnetised plasmas. *Nucl. Fusion* **2017**, *57*, 014006. [CrossRef]
46. Carreras, B.; Lynch, V.; Zaslavsky, G. Anomalous diffusion and exit time distribution of particle tracers in plasma turbulence model. *Phys. Plasmas* **2001**, *8*, 5096. [CrossRef]
47. van Berkel, M.; Vandersteen, G.; Zwart, H.; Hogeweij, G.; Citrin, J.; Westerhof, E.; Peumans, D.; de Baar, M. Separation of transport in slow and fast time-scales using modulated heat pulse experiments (hysteresis in flux explained). *Nucl. Fusion* **2018**, *58*, 106042. [CrossRef]

*entropy*

MDPI

*Article*

# Coherent Structure of Flow Based on Denoised Signals in T-junction Ducts with Vertical Blades

**Jing He, Xiaoyu Wang and Mei Lin ***

School of Energy and Power Engineering, Xi'an Jiaotong University, Xi'an 710049, China; 15172383587@163.com (J.H.); wdy94820@163.com (X.W.)
* Correspondence: janeylinm@mail.xjtu.edu.cn; Tel./Fax: +86-29-82667936

Received: 24 December 2018; Accepted: 15 February 2019; Published: 21 February 2019

**Abstract:** The skin friction consumes some of the energy when a train is running, and the coherent structure plays an important role in the skin friction. In this paper, we focus on the coherent structure generated near the vent of a train. The intention is to investigate the effect of the vent on the generation of coherent structures. The ventilation system of a high-speed train is reasonably simplified as a T-junction duct with vertical blades. The velocity signal of the cross duct was measured in three different sections (upstream, mid-center and downstream), and then the coherent structure of the denoised signals was analyzed by continuous wavelet transform (CWT). The analysis indicates that the coherent structure frequencies become abundant and the energy peak decreases with the increase of the velocity ratio. As a result, we conclude that a higher velocity ratio is preferable to reduce the skin friction of the train. Besides, with the increase of velocity ratio, the dimensionless frequency $St$ of the high-energy coherent structure does not change obviously and $St = 3.09 \times 10^{-4}$–$4.51 \times 10^{-4}$.

**Keywords:** T-junction; denoise; coherent structure; continuous wavelet transform

## 1. Introduction

It is well known that the energy consumed by the frictional resistance increases sharply as a train's speed increases. Due to the high speed of the train, the air passing over its surface is turbulent. The frictional resistance generated by turbulence is much higher than that of laminar flow, which is closely related to the coherent structure [1]. The coherent structure is a structure that is a recognizable orderly large-scale movement. The position and time when it is triggered are not certain, but once triggered, it will develop in a quasi-periodic way and it plays an important role in the transport of mass, momentum, heat and energy [2–5]. Therefore, it is necessary to study the coherent structures generated by the air passing over the surface of train to lower the frictional resistance.

The windows of a high-speed train cannot be opened because of the high running velocity, so ventilation openings are an indispensable feature of the train. Their main function is to exchange the inside air with the outside environment, and to import some cold external air for cooling the power equipment. Studying the coherent structure of the flow near the ventilation port could give some guidance for the drag reduction of the train, the selection of materials and the placement of the ventilation ports. In this paper, the flow area near the vent was simplified as the internal flow of the T-junction with a small scale ratio of 1:4.8. The velocity ratio, which is the ratio of the suction velocity of ventilation to the velocity of the train, was proposed to ensure similar flow patterns. More details about the physical model of T-junction have been reported by Su et al. [6].

The T-junction is widely used in the industry field, and many studies have focused on divergent flow and/or combined flow, laminar flow and/or turbulent flow. Beneš et al. [7] dealt with numerical solutions of the laminar and turbulent flows of Newtonian and non-Newtonian fluids in branched channels with two outlets. The results showed that the EARSM turbulence model is capable of capturing secondary flows in rectangular cross-section channel. Neofytou et al. [8] performed numerical investigations on the shear-thinning and shear-thickening effects of flow in a T-junction of rectangular ducts. The results demonstrated the extent of the effect of the *Re* number on the velocity profiles at different positions in the domain for both Newtonian and non-Newtonian cases. Wu et al. [9] investigated the breakup dynamics of ferrofluid droplets under magnetic fields in a microfluidic T-junction. Chen et al. [10] computed a global linear sensitivity analysis of a complex flow through a pipe T-junction. They found that when $Re \geq 320$, the T-junction contains four instances of a bubble-type vortex-breakdown-like flow feature, where the dynamics are highly sensitive to spatially localized feedback, especially near the boundaries of the recirculation regions. Lu et al. [11] discussed how a T-junction, as a separator in an emerging thermodynamic cycle, affects the cycle efficiency.

Wavelet analysis is a powerful tool for studying turbulence characteristics. The continuous wavelet transform (CWT) has been used by many scholars to analyze the coherent structures in flows. Baars et al. [12] studied the modulation mechanism of small scale flows by large-scale motions, and the results revealed that the time shift in frequency modulation is smaller than that in amplitude modulation. Besides, the wavelet Morlet was used in their study to analyze turbulent signals. Bulusu et al. [13] detected the coherent structure of a curved artery model by means of CWT and decomposition (or Shannon) entropy. They concluded that the optimal wavelet-scale search was driven by a decomposition entropy-based algorithmic approach and proposed a threshold-free coherent structure detection method. The method was successfully utilized in the detection of secondary flow structures in three clinically-relevant blood flow scenarios. CWT was used to detect and establish the length and time scales of the largest horizontal coherent structures existing in a shallow open channel flow by Kanani et al. [14]. They found CWT was particularly well suited to determine the average time and length scales of the structures. Individual horizontal coherent structures whose characteristic times approximately twice larger than the value of average time scale could be identified with the method. Both the Reynolds and wavelet methods were utilized to analyze a solitary wave, a gravity wave, a density current and a low-level jet in the stable atmospheric boundary layer by Ferreres et al. [15]. The results indicated that wavelet analysis had the capacity to distinguish the different scales involved in all these events. Wang et al. [16] investigated the multi-resolution characteristics of velocity components using Morelet and db3 wavelets. It was found that the energy cascade of the drag reducing flow was greatly suppressed with fewer energy branching with the existence of polymer additives. Sarma et al. [17] discussed the self-similarity properties of turbulence in magnetized DC glow discharge plasmas by evaluating the Hurst exponent from wavelet variance plots. The exact frequency responsible for the chaotic behavior could be further determined. In addition, the Morlet wavelet is widely used to analyze turbulent or velocity signals [14,18,19], which indicates that the Morlet wavelet could be one of the options for the CWT in our study.

Several papers about simplified T-junction models of the vents of high-speed trains have been published by the members of our group. They mainly focused on the distribution and high-order statistics of the velocity in cross ducts, as well as the distribution of pressure in branch ducts [20–23]. Atzori et al. [24] studied the coherent structure of a square duct, and the results showed that the large-scale structure attached to a horizontal wall has the same size as that of a vertical wall. As the Reynolds number increases, the size of the coherent structure is bigger. Su et al. [6] studied multi-resolution coherent structures in T-junctions without blades, but they did not eliminate the edge-effect of the CWT. In fact, there are typically blades in the vents of the train to prevent waste from entering the train cabin. In this study, four blades were set at the entrance of the branch duct, and denoising was performed. Besides, the edge-effect that may lead to a wrong result in CWT was eliminated. Thus the results were closer to a real situation. This is beneficial for controlling the flow resistance generated by high-speed train vents and ultimately improving the train efficiency.

## 2. Experimental Setup

To investigate the influencing factors of the flow near the vents of high-speed trains, an experimental system in the form of T-junction was built to mimic a real situation. This experiment was carried out in a low-speed wind tunnel with a maximum speed of 57 m/s. A schematic diagram of experimental system is illustrated in Figure 1. The cross duct has a length of 36.5 D and its cross-sectional area is $z \times y$ = 161.7 × 143.3 mm$^2$. The branch duct is 14.2 D long and its cross-sectional area is a square with a side length of D = 110 mm. To simulate the grating of the train vent, four vertical blades were mounted equidistantly on the inlet section of the branch duct. Each of them is 2.0 × 21.0 × 110.0 mm$^3$ in size, as shown in Figure 2. In order to ensure that the turbulence is fully developed as soon as possible, a 2.0 mm rod was placed in the entrance of cross duct. The coordinate system is used herein, the origin of the coordinate is at the inlet center of the branch duct (as shown in Figure 3). The $x$, $y$ and $z$ axes are aligned with the streamwise, wall-normal and spanwise directions of the cross duct, respectively.

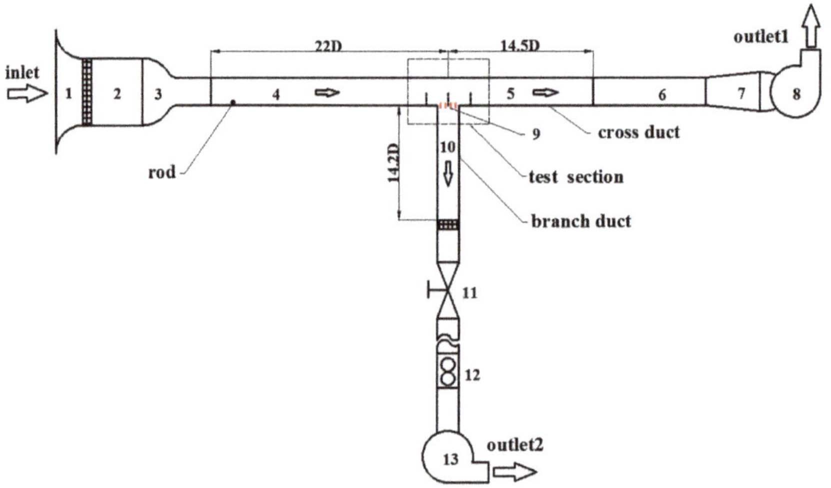


**Figure 1.** Schematic diagram of experimental system: 1-Entrance, 2-Settling chamber, 3-Contraction section, 4-Front section of cross duct, 5-Back section of cross duct, 6-Connect section, 7-Expansion section, 8-Fan, 9-Blades, 10-Branch duct, 11-Valve, 12- Glass rotameter, 13-Fan.

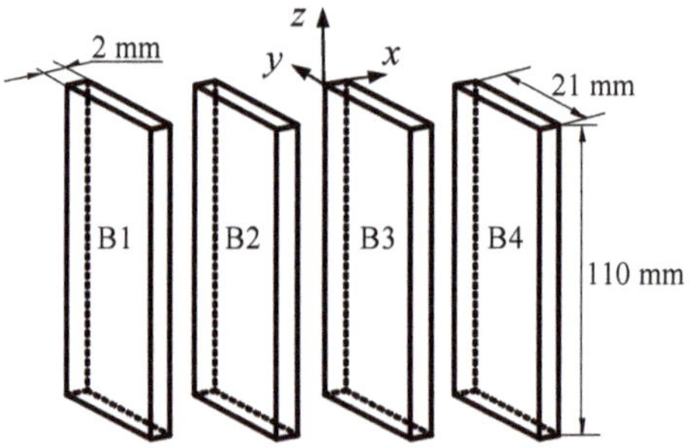


**Figure 2.** Schematic diagram of the blades.

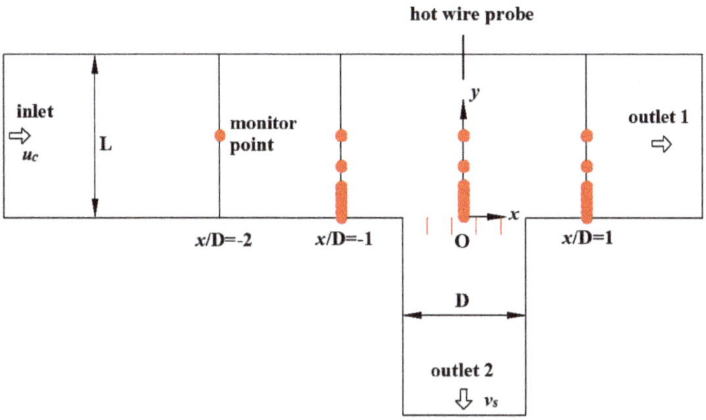


**Figure 3.** Sketch of the measurement points.

In this experiment, the flows at different positions and different cross sections were analyzed by varying the velocity of the cross duct and the velocity ratio $R$ (the ratio of the branch velocity to the cross velocity). The cross velocity was measured by an IFA300 hot wire anemometer and regulated by changing the cross flow using the high-power inverter fan. The sampling frequency was 50 kHz. The sampling time was 40.6 s, and the uncertainty of the cross velocity was $\pm1.0\%$ after calibration. When the cross velocity was constant, the velocity ratio was varying with the velocity of the branch duct which was controlled by a low-power inverter fan. The flow rate of the branch duct was measured by a glass rotameter, and the uncertainty of the bulk branch velocity was $\pm4.3\%$. More details on the experimental setup have been published in Yin et al. [20] and Su et al. [6].

The centre velocities of the cross duct in the experiment were 30, 40, and 50 m/s, respectively, and the velocity ratios were 0.08, 0.13, and 0.18. The experimental conditions are listed in Table 1. The measurement sections were $x/\mathrm{D} = -1$, $x/\mathrm{D} = 0$, $x/\mathrm{D} = 1$ and eight measurement points were selected varying from $y/\mathrm{L} = 0.0070$ to 0.5000 at each section as presented in Figure 3. Note, L is the height of the cross duct. All processing and analysis of data were carried out with the MATLAB software.

**Table 1.** Experimental conditions.

| Case# | $u_c$ (m/s) | $v_s$ (m/s) | $R$ |
|---|---|---|---|
| 1 | 30 | 3.9 | 0.13 |
| 2 | 40 | 3.2 | 0.08 |
| 3 | 40 | 5.2 | 0.13 |
| 4 | 40 | 7.2 | 0.18 |
| 5 | 50 | 6.5 | 0.13 |

In order to ensure the reliability of the experimental system, the velocity distribution curve obtained at $x/\mathrm{D} = -3$ was compared with the results obtained by Gessner et al. [25].

The results are shown in Figure 4a and consistency between our experiment and Gessner's was found. The wall velocity distribution curve at $x/\mathrm{D} = -5$ was compared with the results of Schultz and Flack [26], and the result is shown in Figure 4b. There is also a good conformity. Detailed information can be found in Su et al. [6].

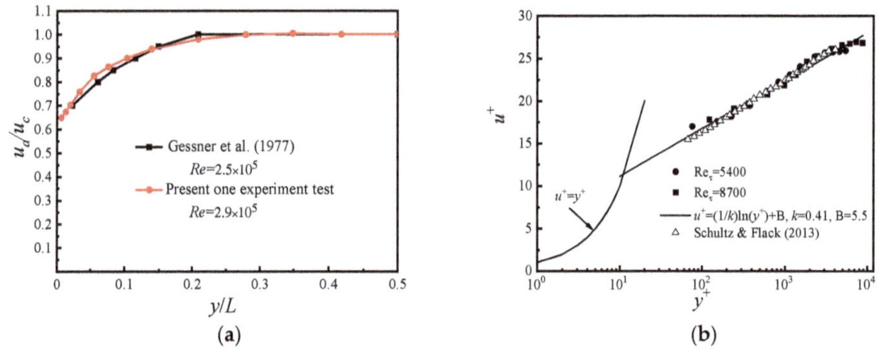


**Figure 4.** (**a**) Comparison of velocity distribution between our experiment and Gessner's et al.; (**b**) Comparison of wall velocity distribution between our experimental data and results of Schultz and Flack.

The accuracy of the hot-wire anemometry suffers from several problems, such as the heat conduction to the wall, calibration at low velocities, spatial resolution (due to the wire length *l*) and determination of the wall position (due to heat conduction) and the risk of probe damage. The error due to the spatial resolution is maximized at only 13% at $y/\mathrm{L} = 0.0070$, then sharply reduces to 6% at $y/\mathrm{L} = 0.014$. Therefore, it could be thought that the error is systematic due to the same hot-wire probe, and this does not adversely affect the results near the T-junction [6]. The inaccuracy because of the blockage effects can be negligible for $y^+$ values larger than 20 wall units, and the minimum $y^+$ in our study is 76, so the effect of blockage can be negligible [27].

## 3. Processing of Data

In this paper, the coherent structures were mainly analyzed by performing CWT on the obtained signals. CWT means that a convolution was executed on the obtained signals and the selected wavelet basis function where the scale dilation *a* is continuous. Its expression is shown in Equation (1):

$$W_f(a,b) = \int_{-\infty}^{+\infty} f(t)\psi_{a,b}(t)dt = \langle f(t), \psi_{a,b}(t)\rangle \tag{1}$$

where:

$$\left\{\psi_{a,b}(t) = \frac{1}{\sqrt{a}}\psi\left(\frac{t-b}{a}\right) | a > 0, b \in R\right\} \tag{2}$$

The variables *a* and *b* in these equations are commonly called the scale dilation and translation parameters, respectively [28]. $f(t)$ is the obtained signal. The translation parameter, *b*, corresponds to the position of the wavelet basis function in the analyzed signal in time, and the scale dilation parameter, $a$, reflects the degree of dilation of the wavelet. The *a* has the following correspondence with frequency:

$$f = f_c \times \frac{f_s}{a} \tag{3}$$

where $f$ is the corresponding scale frequency, which reflects the frequency (scale) of the signal at the certain moment and scale dilation a. $f_c$ is the center frequency of wavelet basis function, and $f_s$ is the sampling frequency.

To analyze coherent structure, the CWT power spectrum of a signal is necessary. It can be defined as follows [14,18,29]:

$$E_c(a,b) = \int_{-\infty}^{\infty} \left|W_f(a,b)\right|^2 \tag{4}$$

*3.1. Denoising*

During the acquisition of turbulent signals, the obtained signals are actually the mixture of real signals and noise signals due to the influence of measuring devices and ambient noises. The obtained signals can be expressed as the following form: $o(t) = s(t) + n(t)$, where the $o(t)$ is the obtained signal, the term $s(t)$ is the ideal real signal and the $n(t)$ is the noise signal caused by the measurement process. When CWT is performed on the obtained signal to analyze coherent structures, the presence of noise may affect the results, so denoising is a necessity. The aim is to remove the noise signal $n(t)$ and obtain the ideal real signal to the utmost extent.

Many denoise methods can be applied, such as Gaussian filters, wavelet filters and deep learning. The Gaussian filter method is more suitable for stationary signals, and deep learning is suitable for images. The wavelet method can decompose signals into the frequency domain and the time domain, and will not smooth out the instantaneous component of signals during the decomposition and reconstruction process. Turbulent signals have the characteristics of being instantaneous and pulsating. It is important to maintain the spikes and abrupt changes during the denoising process, so the wavelet filter is a powerful tool for denoising turbulent signals. Common wavelet denoising methods are based on modulus maxima, translation invariant, and threshold. Among them, the method based on threshold can fully consider the propagation characteristics of signals and noises at different scales. Meanwhile it is simple to calculate and distinguish the signals from the noises. Therefore, the wavelet threshold method was used in this paper to denoise the signals.

In the actual processing, it is widely believed that the noises are distributed in the high frequency band. Thus, only the high frequency coefficients need to be processed. The parameters that need to be considered for wavelet threshold denoising include the threshold function, threshold, decomposition layer number and wavelet basis function. The process is described as follows:

Step (1):Wavelet decomposition of signals. Select the basis function and the number of decomposition layers to perform wavelet decomposition on signals;

Step (2):Perform threshold denoising on the high frequency coefficients. The high frequency coefficients of each decomposition layer are processed using the selected threshold and threshold function;

Step (3):Signal reconstruction. Wavelet reconstruction is performed with the processed high frequency coefficient and the low frequency coefficient of the largest decomposition layer.

The noises are often regarded as white noises, that is, obeying a Gaussian distribution. Corresponding to the process of denoising using the method of wavelet threshold, that is, the energy of real signals can be centralized on a few wavelet coefficients, while the energy of noises is distributed on most wavelet coefficients. By choosing an appropriate threshold, most of the noises can be removed and the best approximation of the ideal real signals can be obtained. Commonly used threshold functions are the hard threshold and soft threshold function. In a hard threshold function, the coefficients with absolute values lower than the threshold are set to zero; in soft one, beyond that, the coefficients with absolute values higher than the threshold are shrunk. The hard threshold function will be discontinuous at the threshold, which will bring about the Gibbs phenomenon, and furthermore distorts the reconstruction. Thus there may be false coherent structures because of the distortion. The soft threshold function has a constant error due to shrinking and the reconstruction result is not accurate. Thus the energy value of coherent structure may be decreased and the boundary of the coherent structure is not clear. Therefore an improved threshold function is necessary. It should provide a smooth transition at the threshold position and should be extremely approximate away from the threshold. Many improved functions have been proposed [29–32]. The sigmoid function proposed by Yi et al. [33] was chosen in this paper, and its expression is described as follows:

$$\overline{d_{i,j}} = \begin{cases} (|d_{i,j}| - t) - \left[\frac{2}{1+e^{\beta\left(\frac{|d_{i,j}|-t}{t}\right)}}\right] & , \quad |d_{i,j}| \geq t \\ 0 & , \quad |d_{i,j}| < t \end{cases} \tag{5}$$

where: β = 10, the $d_{i,j}$ is the coefficient obtained by CWT, and the $t$ is the value of threshold. This function can overcome the shortcomings of the hard threshold and soft threshold functions to some extent, and offers better denoising performance.

The threshold is also one of the important parameters in the denoising process. An excessive threshold will eliminate useful components in the signals, and a smaller threshold will lead to insufficient denoising. The commonly used threshold $t = \sigma\sqrt{2\ln N}$ was proposed by Donoho [34]. Here $\sigma$ is the standard deviation of the noise and $N$ is the length of the signal. The aforesaid threshold is a fixed threshold, and the same value is used in different decomposition layers, which may erase some useful signals. In this paper, the threshold proposed by Lu [30] was chosen:$t = \sigma\sqrt{2\ln N}/\log 2(j+1)$, where $j$ is the layers in the decomposition. The thresholds vary with the decomposition layers, and are consistent with the propagation characteristics of noise.

An appropriate basis function is the key point of obtaining the desired features of signals. An ideal basis function should satisfy the following characteristics: biorthogonality, compact support, regularity and symmetry. Combining the above characteristics and the results of Zhang [35], the sym5 wavelet was used as basis function in denoising for this article. In the actual processing, the noises are often regarded as white noises, that is, obeying the Gaussian distribution, so the value of the decomposition layer is often chosen based on the theory of verification of white noise That is, if the wavelet coefficients at a certain decomposition layer are verified to satisfy the characteristics of white noise, the current layer is the goal decomposition layer-number. The "white noise test" proposed by Zhang et al. [36] was used for reference in denoising for this study, but errors may occur when the sample size is small, so an improved method based on K-S test was used and finally two layers were chosen for denoising.

### 3.2. Eliminating of Edge Effect

When executing CWT to analyze signals, the ideal length of the signal is infinite, but the measured signal has a finite length, so errors occur at the start and the end of the transformed signal. These errors are called the edge effect or the cone of influence (COI) [37–39]. The edge effect causes false peaks at the edges, affecting the acquisition of correct results, therefore, it is necessary to eliminate the edge effect. The radius of COI depends on the selected wavelet basis function and scale. As the scale increases, the area affected by COI also increases.

For CWT analysis of turbulence, the most common wavelet basis function is the Morlet wavelet [7,14,18]. It is a complex-valued wavelet and can provide not only phase information, but also amplitude information. Furthermore, it has a good local balance in both the time and frequency domains [6], so the Morlet wavelet was used in the processing of CWT in this study. Its definition is:

$$\varphi(t) = \pi^{-1/4} e^{i\omega_0 t} e^{-t^2/2} \tag{6}$$

where $\omega_0$ is the dimensionless frequency and here taken to be $\omega_0$ = 6 following the recommendations of Torrence et al. [37].

According to the definition of Torrence [37], the radius of COI of Morlet wavelet is the e-folding time ($e^{-2}$) of the wavelet power autocorrelation at each scale. That means the wavelet power for a discontinuity at the edge drops by a factor of $e^{-2}$ and this ensures that the edge effects are negligible beyond this point. In addition, referring to the definitions of Boltezar et al. [38] and Mayer et al. [39], the radius of COI was finally selected as $2a$ ($a$ is the scale of CWT).

The steps for eliminating the edge effects are described as follows:

Step (1):Extension. Extend the signal to be analyzed by the abovementioned length $2a$;
Step (2):CWT transform. Perform CWT on extended signals obtained after Step (1);
Step (3):Truncation. Truncate the transformed signal after Step (2) to the same length as the original signal.

## 4. Results and Discussion

The wavelet coefficients obtained by CWT represent the information of coherent structures with different scales in turbulence [18], and the power spectrum represents the energy of coherent structures with different scales [14]. In this section, CWT was used to obtain the power spectrum of the denoised fluctuating velocity signals to analyze the coherent structure. Besides, the edge effect generated on CWT was eliminated. The cross velocity and velocity ratio were varied to observe the power spectrum differences in different sections and positions. A total of 16384 data points were selected for analysis, and previous studies have shown that this amount of data is sufficient [7,40]. Figures 5–9 are the wavelet power spectra of the fluctuating velocity under different parameters, which mirror the time-frequency characteristics of the coherent structures. In all following figures, the abscissa refers to time, and the ordinate refers to frequency transformed from scale dilation $a$. The colorbar on the right expresses the wavelet power defined in Equation (4). The wider the spectrogram band along the abscissa is, the larger the scale of the coherent structure is.

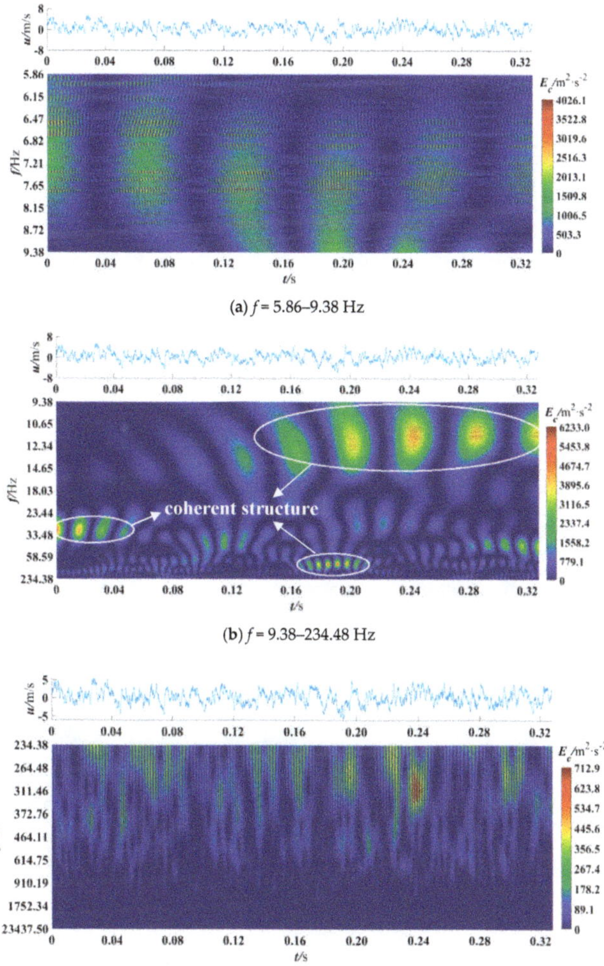


(a) $f$ = 5.86–9.38 Hz

(b) $f$ = 9.38–234.48 Hz

(c) $f$ = 234.38–23437.5 Hz

**Figure 5.** Wavelet power spectrum at $x/\mathrm{D} = 0$, $R = 0.13$, $u_c = 40$ m/s, $y/\mathrm{L} = 0.0070$.

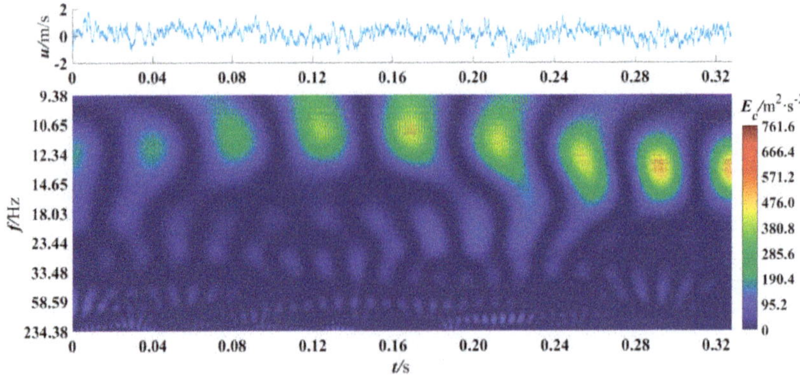


(**a**) $y$/L = 0.5000

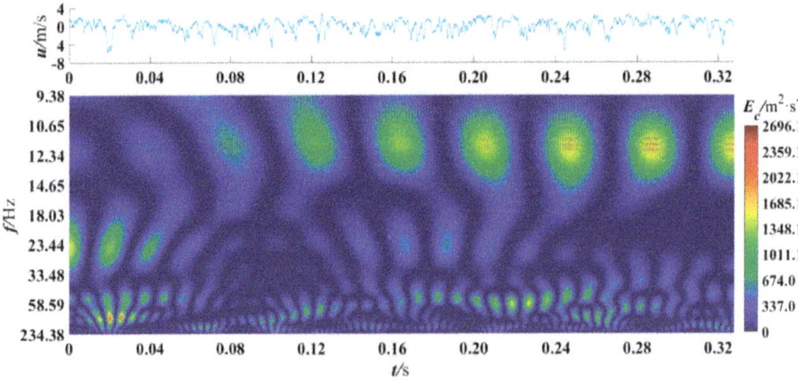


(**b**) $y$/L = 0.1396

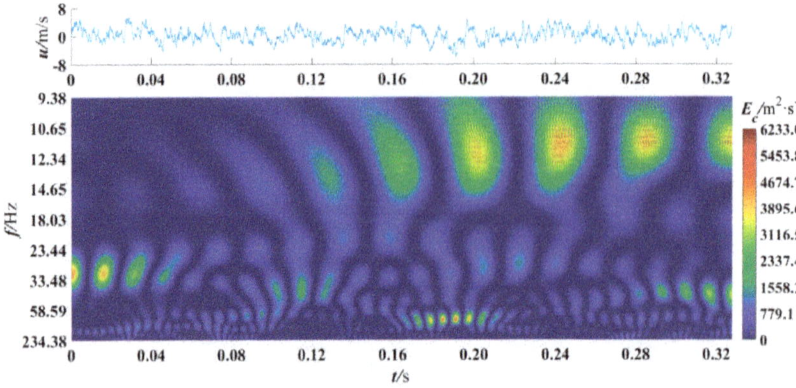


(**c**) $y$/L = 0.0070

**Figure 6.** Wavelet power spectrum at $x/D = 0$, $R = 0.13$, $u_c = 40$ m/s.

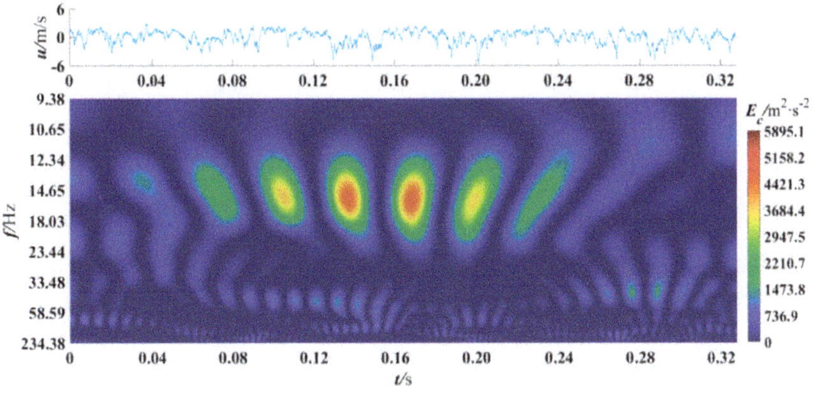


(**a**) $x$/D = −1

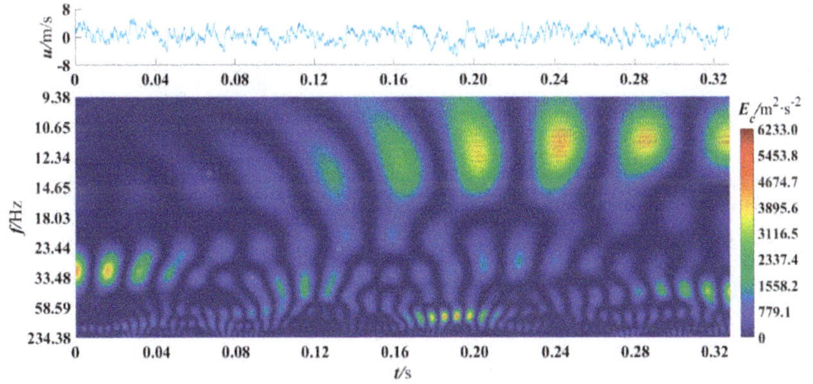


(**b**) $x$/D = 0

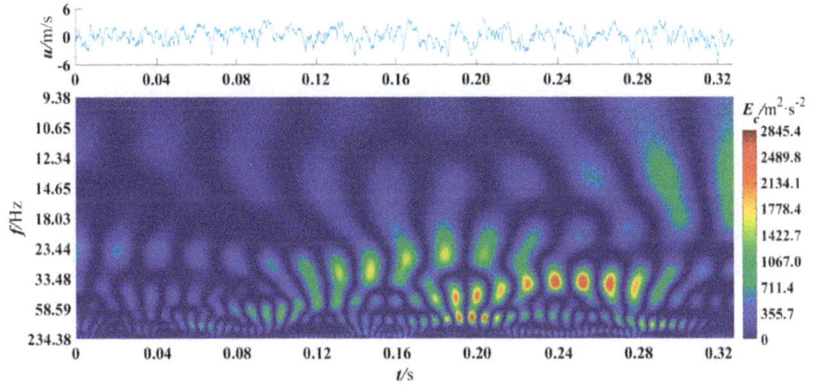


(**c**) $x$/D = 1

**Figure 7.** Wavelet power spectrum at $u_c$ = 40 m/s, $R$ = 0.13, $y$/L = 0.0070.

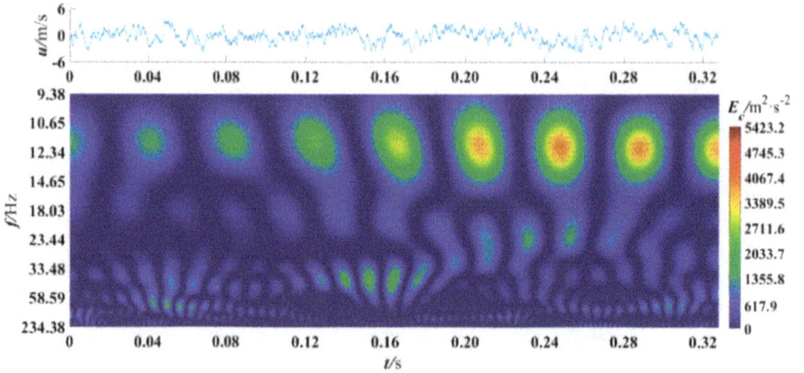


(**a**) $u_c$ = 30 m/s

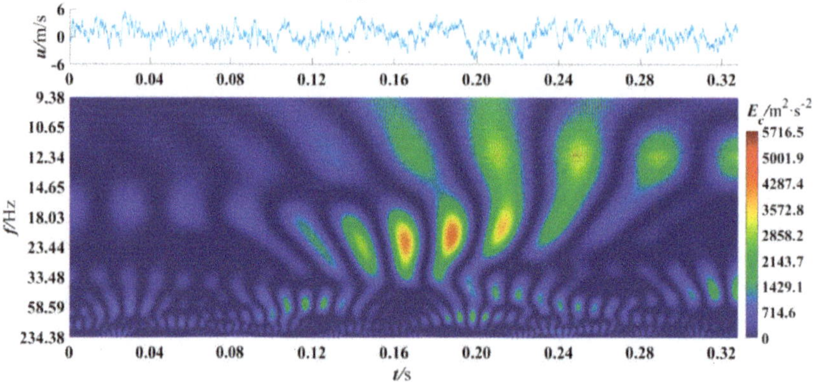


(**b**) $u_c$ = 40 m/s

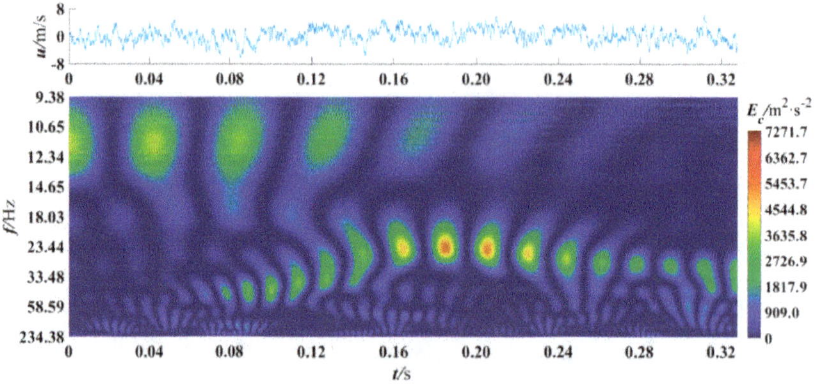


(**c**) $u_c$ = 50 m/s

**Figure 8.** Wavelet power spectrum at $x/D = 0$, R = 0.13, $y/L = 0.0070$.

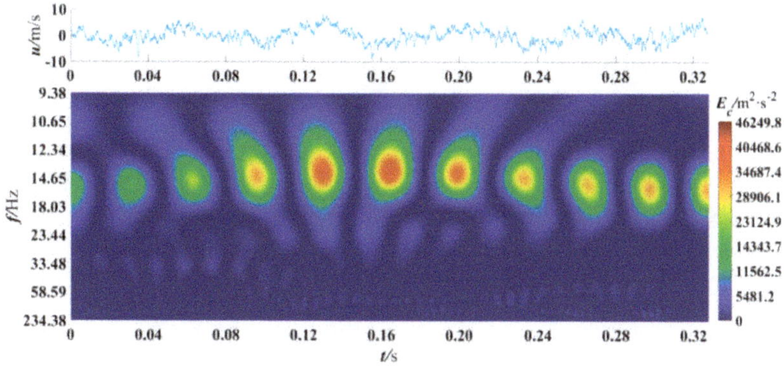


(a) $R = 0.08$

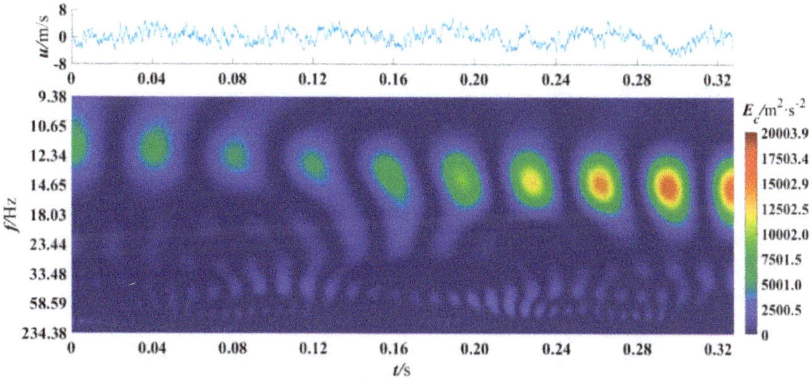


(b) $R = 0.13$

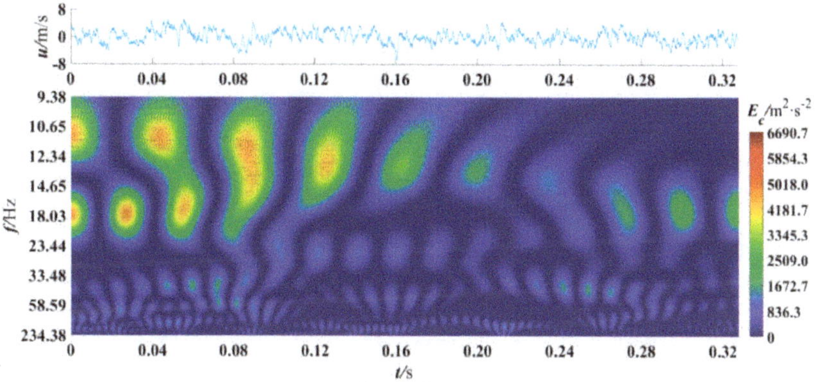


(c) $R = 0.18$

**Figure 9.** Wavelet power spectrum at $x/\mathrm{D} = 0$, $u_c = 40\mathrm{m/s}$, $y/\mathrm{L} = 0.0070$.

The Strouhal number is often used to give a non-dimensional description of the flow characteristics of a periodic flow [41,42]. Coherent structures have quasi-periodic characteristics, so the Strouhal number ($St$) is employed to describe the frequency characteristics of coherent structures. The $St$ is defined as follows:

$$St = \frac{yf}{u_c} \tag{7}$$

where $y$ is the distance from the wall, and the $f$ is the main frequency component of the coherent structure. Figure 5 shows the power spectrum under different frequency ranges at the first measured point, $y/\mathrm{L} = 0.0070$, where the minimum value is 5.86 Hz and the maximum is 2,3437.50 Hz and other parameters are $u_c = 40$ m/s, $R = 0.13$, $x/\mathrm{D} = 0$.

In Figure 5a, large-scale high-energy coherent structures can be found over the whole time period with a scale frequency range of 5.86–9.38 Hz, but they are not very clear. The high-energy coherent structure was defined as those whose value of energy accounts for more than 33% of the maximum value, as shown in Figure 5b. There, except for the five large-scale coherent structures with the dominant scale frequency of 12.34 Hz, there are three coherent structures at 0–0.04 s with a dominant scale frequency of 30 Hz. Coherent structures with a scale frequency of 60 Hz can be found at 0.10–0.12 s and 0.29–0.32 s. There are also coherent structures at 0.17–0.20 s whose scale frequency is more than 100 Hz. The presence of large-scale coherent structures is hardly observed in Figure 5c. What's more, the peak value of energy is $E_c = 712.9$ m$^2$/s$^2$, or 7.7 times lower than that in Figure 5b. The above description shows that coherent structures with large energy exist with a frequency of 9.38–234.38 Hz, and they play a vital role in the transport of mass and energy. Therefore, the frequency of the coherent structures studied in this paper ranges from 9.38–234.38 Hz (corresponding to scale dilation $a$ of 200–5000), which is consistent with the results of Su et al. [6]. This may indicate that the presence of blades does not remarkably change the frequency range of coherent structures.

Figure 6 shows the wavelet power spectrum at different locations away from the wall ($y/\mathrm{L} = 0.0070$–$0.5000$). They were all measured with $x/\mathrm{D} = 0$, $R = 0.13$, $u_c = 40$ m/s. In Figure 6a, at $y/\mathrm{L} = 0.5000$, the coherent structures with the maximum energy value of $E_c = 761.6$ m$^2$/s$^2$ appear at 0.08–0.32 s, whose dominant scale frequency is 9.38–14.65 Hz. In Figure 6b, at $y/\mathrm{L} = 0.1396$, besides the coherent structures of 12.34 Hz at 0.12–0.32s, the coherent structure of 23.44 Hz appears in 0–0.04 s. The coherent structures with the largest energy value appear in the high-frequency band of about 100 Hz. The peak value of energy is 2696.1 m$^2$/s$^2$. In Figure 6c, at $y/\mathrm{L} = 0.0070$, the scale frequency of the coherent structures is wider, including low-frequency of 9.38–14.65 Hz, 24–40 Hz and some high-frequency coherent structures of 100 Hz and greater than 100 Hz. The peak value of energy is 6233.0 m$^2$/s$^2$, which is 7.2 times higher than that in Figure 6a and 1.3 times higher than that in Figure 6b. The similarity among the three figures is the periodic distribution of coherent structures: they are all periodic in term of time but in an uncertain period. That is consistent with the characteristic of quasi-periodic of the coherent structures. And the tendency of large-scale coherent structures separating into small-scale coherent structures can also be seen. That means the event of energy cascade occurring. However, in areas far from the wall, the coherent structure contains less energy and the frequency range is not as big as the near wall area. That indicates that there are more coherent structures triggered near the wall ($y/\mathrm{L} = 0.0070$). Therefore, in the following figures, the distribution of coherent structures in the near-wall region was mainly concerned.

Figure 7 presents wavelet power spectra at different cross sections. The other parameters are $u_c = 40$ m/s, $R = 0.13$, $y/\mathrm{L} = 0.0070$. In Figure 7a, at $x/\mathrm{D} = -1$, there are six large-scale high-energy coherent structures with a frequency of 12.34–18.03 Hz at 0.08–0.24 s, and the peak value of the energy is 5895.1 m$^2$/s$^2$. In Figure 7b, at $x/\mathrm{D} = 0$, the coherent structures span a wider scale frequency range. Within 0.16–0.32 s, there are six high-energy coherent structures with the scale frequency about 12.34 Hz. There are four coherent structures with a frequency of 30 Hz at 0–0.04 s, and six small-scale high-frequency (more than 100 Hz) coherent structures at 0.17–0.20 s. Coherent structures with a scale frequency of about 60 Hz also exist at 0.10–0.12 s and 0.29–0.32 s, respectively. The peak value of energy is 6233.0.4 m$^2$/s$^2$.

In Figure 7c, at $x/\mathrm{D} = 1$, the coherent structures are mainly concentrated in the middle-frequency and high-frequency above 23.44 Hz at 0.12–0.28 s, and coherent structures with a wide scale frequency range (33.48–234.38 Hz) are periodically passed. At 0.30 s and 0.32 s, two coherent structures with the abundant scale frequency components of 12.34–18.03 Hz can be seen if looking carefully. In summary, upstream, the coherent structures are mainly low-frequency large-scale structures, and *St*

= $3.09 \times 10^{-4}$–$4.51 \times 10^{-4}$, which indicates that without the influence of blades. At the mid-center and downstream, the frequency components of the coherent structures are richer, and $St = 3.09 \times 10^{-4}$, $7.50 \times 10^{-4}$, $15.00 \times 10^{-4}$ and $25.00 \times 10^{-4}$. That may because the suction of the branch duct accelerates the energy cascade. Therefore, with the existence of the branch duct suction, the $St$ in a square section is mainly $3.09 \times 10^{-4}$ –$25.00 \times 10^{-4}$. To reduce the flow resistance, the formation of coherent structures with the corresponding scale frequency at the upstream and mid-center locations should be suppressed.

Figure 8 illustrates the wavelet power spectrum at the near wall point under different cross velocity conditions and a given velocity ratio ($R = 0.13$). It can be seen from Figure 8a that, at $u_c$ = 30 m/s ($Re_c = u_c \cdot L/\nu = 2.85 \times 10^5$, $\nu$ is the kinematic viscosity), there are six clear large-scale coherent structures periodically passed through at 0.12–0.32 s, and their scale frequency range is 10.65–14.65 Hz. There are four small-scale coherent structures at 0.14–0.16 s and their scale frequency range is 33.48–58.59 Hz. The coherent structure with the highest energy appears at 0.24 s with an energy peak value of 5423.2 $m^2/s^2$. In Figure 8b, at $u_c$ = 40 m/s ($Re_c = 3.8 \times 10^5$) five large-scale coherent structures with high-energy are present at 0.18–0.32 s. The phenomenon can be seen at 0.20–0.24 s that the low-frequency coherent structures tends to break into higher-frequency ones. At 0.12–0.24 s, coherent structures with a scale frequency band of 18.03–33.48 Hz are periodically passed through. The coherent structures with highest energy are present at 0.16–0.18 s and their energy value is 5716.5 $m^2/s^2$.

In Figure 8c, at $u_c$ = 50 m/s ($Re_c = 4.75 \times 10^5$) there are four large-scale coherent structures at 0–0.12 s with a dominant scale frequency of 11 Hz, and the highest energy coherent structures appear at 0.16–0.20 s with the scale frequency band of 23.44–33.48 Hz. A number of small-scale coherent structures with frequency of 50 Hz and 33 Hz are seen at 0.08–0.32 s. As the cross velocity increases, the peak value of energy also increases.

With the increase of $Re_c$, the frequency band of the coherent structures is wider, which are $St$ = $3.55 \times 10^{-4}$–$4.88 \times 10^{-4}$ for $Re_c = 2.85 \times 10^5$, $St = 2.66 \times 10^{-4}$–$8.37 \times 10^{-4}$ for $Re_c = 3.8 \times 10^5$ and $St$ = $2.20 \times 10^{-4}$–$10.00 \times 10^{-4}$ for $Re_c = 4.75 \times 10^5$. Therefore, it could be concluded that if the speed of train is increased without changing the vent speed, the energy consumed by the friction near the ventilation would also be increased. Comparing to the results in Ref. [6], the periodicity of coherent structure is more obvious and the value of energy is lower. This may indicate that the existence of blades could rectify the flow and lower the intensity of the turbulence. Thus we recommend that an appropriate number of blades could be helpful to reduce the skin friction of trains. The appropriate number needs to be studied further.

Figure 9 presents the wavelet power spectrum at a given cross velocity $u_c$ = 40 m/s ($Re_c = 3.8 \times 10^5$) under different velocity ratios. From Figure 9a, at low velocity ratio $R = 0.08$, it is demonstrated that the coherent structures periodically pass through over the whole time period, and their dominant scale frequency is 14.65 Hz. The coherent structures with highest value of energy appear from 0.13 to 0.16 s, and the maximum value is 46249.8 $m^2/s^2$. In Figure 9b, at $R = 0.13$, coherent structures with lower energy are spread from 0 to 0.16 s, and their dominant scale frequency is gradually increased. In addition, high-energy coherent structures periodically exist at 0.16–0.32 s, and their dominant scale frequency is 14.65 Hz. In Figure 9c, at high velocity ratio, $R = 0.18$, there are coherent structures periodically passing through at the time period of 0–0.16 s, and the dominant scale frequencies of them are 11 Hz for large-scale and 18.03 Hz for mid-scale respectively. Coherent structures spanned the frequency band of 9.38–23.44 Hz can be clearly seen at 0.05–0.08 s. This indicates that there is a tendency for large-scale coherent structures to break into higher frequency and smaller scale ones, which is called an energy cascade. Viewing the above three figures, it can be concluded that as the velocity ratio increases, the energy peak of the coherent structures decreases, and the energy cascade phenomenon becomes obvious. This may be because a smaller velocity ratio means a higher velocity in the cross duct when the cross velocity is the same, and the higher velocity could thin the laminar sub-layer near the entrance of the branch duct and motivate more violent turbulence, which in turn generates more coherent structures. Moreover, when the velocity ratio is smaller, the variance of the pulsation velocity is larger. At $R$ = 0.08, the variance of the pulsation velocity is $7.31 m^2/s^2$, $3.86 m^2/s^2$ at $R$ = 0.13, and $3.18 m^2/s^2$ at $R$ = 0.18. That means bursting events could increase under a lower velocity ratio. Therefore, increasing the velocity ratio may be advantageous to accelerate the energy cascade and suppress the generation of high-energy coherent structures. The acceleration and suppression suggest that the drag could be reduced by increasing the velocity ratio of the train. Besides, the local $St$ doesn't change with the increase of the velocity ratio. The $St$ of the coherent structure that contains high energy and with a large scale is $3.09 \times 10^{-4}$–$4.51 \times 10^{-4}$.

From Figures 5–9, we can summarize that, at the position of $y/L$ = 0.0070 and $x/D$ = 0, the main frequency of coherent structure is $St = 2.20 \times 10^{-4}$–$10.00 \times 10^{-4}$. There more large-scale coherent structures triggered near the wall ($y/L$ = 0.0070), which means an intense turbulent event, and the quasi-periodic characteristics of the coherent structures have no relation with the intensity of the turbulent event. Because of the suction of the branch duct, the scales of coherent structures become more abundant, and the maximum frequency reaches $St = 25.00 \times 10^{-4}$. This means the existence of the suction accelerates the mass and energy transport process. The higher cross velocity is helpful to the generation of coherent structures and the blades at the entrance of branch duct could weaken the process of coherent structure triggering. The higher velocity ratio would result in lower energy coherent structure peaks.

There is a secondary flow because of the square shape of the duct which is related to the secondary shear Reynolds stress and centrifugal force. The interaction of the motion of sweep and injection generated from the side wall influences the wall-shear stress and heat transfer performance. The velocity fluctuation in the outer region is stronger than that in the circular pipe [43], so the coherent structures in rectangular ducts may be different from those in circular pipes. Accurate coherent structure results should be studied further considering the influence of secondary flow.

## 5. Conclusions

In this paper, the vent of a high-speed train is simplified as a T-junction duct with vertical blades. The velocities at three different locations, i.e., upstream, mid-center and downstream, were measured by a hot wire anemometer. The velocity signals were denoised with the wavelet threshold denoising method, wherein the threshold function is improved. The wavelet power spectrum was obtained by CWT, and the coherent structures in the T-junction under different conditions were analyzed, while the COI was eliminated. The following three main conclusions are drawn:

1. The coherent structures in the upstream region of a T-junction are mainly low-frequency. There are more abundant frequency components of coherent structures at the mid-center and downstream. The $St$ of the coherent structure in this study is ranging from $3.09 \times 10^{-4}$ to $25.00 \times 10^{-4}$. The energy of coherent structures is the highest at the mid-center, and suppressing the formation of low-frequency coherent structures at the upstream and mid-center may be beneficial to reduce the drag force.
2. With the increase of $Re_c$, the energy peak of the coherent structures also increases and the frequency range of coherent structures is more abundant. The dimensionless frequency $St$ changes from $3.55 \times 10^{-4}$–$4.88 \times 10^{-4}$ to $2.20 \times 10^{-4}$–$10.00 \times 10^{-4}$. Therefore, the energy consumed by friction may be increased with the improvement of speed of train. The existence of blades is helpful to reduce the skin friction of the train.
3. When the velocity ratio increases, the energy peak of coherent structures decreases, and the energy cascade phenomenon becomes obvious. Therefore, the drag force and skin friction of a high-speed train could be reduced by increasing the velocity ratio. The dimensionless frequency $St$ of the high-energy coherent structure does not change obviously and $St = 3.09 \times 10^{-4}$–$4.51 \times 10^{-4}$.

**Author Contributions:** All authors contributed equally to this work.

**Funding:** This work was supported by the National Natural Science Foundation of China (Grant No. 51376145 and 51876146).

**Conflicts of Interest:** The authors declare no conflict of interest.

## References

1. Kravchenko, A.G.; Choi, H.; Moin, P. On the relation of near-wall streamwise vortices to wall skin friction in turbulent boundary layers. *Phys. Fluids* **1993**, *5*, 3307–3309. [CrossRef]
2. Kline, S.J.; Reynolds, W.C.; Schraub, F.A.; Runstadler, P.W. The structure of turbulent boundary layers. *J. Fluid Mech.* **1967**, *30*, 741–773. [CrossRef]
3. Kim, H.T.; Kline, S.J.; Reynolds, W.C. The production of turbulence near a smooth wall in a turbulent boundary layer. *J. Fluid Mech.* **1971**, *50*, 133–160. [CrossRef]
4. Hussain, A.K.M.F. Coherent structures and Turbulence. *J. Fluid Mech.* **1986**, *173*, 303–356. [CrossRef]
5. Sirovich, L. Turbulence and the dynamics of coherent structures. I. Coherent structures. *Q. Appl. Math.* **1987**, *45*, 561–571. [CrossRef]
6. Su, B.; Yin, Y.T.; Li, S.C.; Lin, M.; Wang, Q.W.; Guo, Z.X. Wavelet analysis on the turbulent flow structure of a T-junction. *Int. J. Heat Fluid Flow.* **2018**, *73*, 124–142. [CrossRef]
7. Beneš, L.; Louda, P.; Kozel, K.; Keslerova, R.; Štigler, J. Numerical simulations of flow through channels with T-junction. *Appl. Math. Comput.* **2013**, *219*, 7225–7235. [CrossRef]
8. Neofytou, P.; Housiadas, C.; Tsangaris, S.G.; Stubos, A.K.; Fotiadis, D.I. Newtonian and power-law fluid flow in a T-junction of rectangular ducts. *Theor. Comp. Fluid Dyn.* **2014**, *28*, 233–256. [CrossRef]
9. Wu, Y.N.; Fu, T.T.; Ma, Y.G.; Li, H.Z. Active control of ferrofluid droplet breakup dynamics in a microfluidic T-junction. *Microfluid. Nanofluid.* **2015**, *18*, 19–27. [CrossRef]
10. Chen, K.K.; Rowley, C.W.; Stone, H.A. Vortex dynamics in a pipe T-junction: Recirculation and sensitivity. *Phys. Fluids* **2015**, *27*, 034107. [CrossRef]
11. Lu, P.; Deng, S.; Li, Z.; Shao, Y.W.; Zhao, D.P.; Xu, W.C.; Zhang, Y.; Wang, Z. Analysis of pressure drop in T-junction and its effect on thermodynamic cycle efficiency. *Appl. Energy* **2018**, *231*, 468–480. [CrossRef]
12. Baars, W.J.; Talluru, K.M.; Hutchins, N.; Marusic, I. Wavelet analysis of wall turbulence to study large-scale modulation of small scales. *Exp. Fluids* **2015**, *56*, 188. [CrossRef]
13. Bulusu, K.V.; Plesniak, M.W. Shannon entropy-based wavelet transform method for autonomous coherent structure identification in fluid flow field data. *Entropy* **2015**, *17*, 6617–6642. [CrossRef]
14. Kanani, A.; Silva, A.M.F.D. Application of continuous wavelet transform to the study of large-scale coherent structures. *Environ. Fluid Mech.* **2015**, *15*, 1293–1319. [CrossRef]
15. Ferreres, E.; Soler, M.R. Analysis of turbulent exchange and coherent structures in the stable atmospheric boundary layer based on tower observations. *Dynam. Atmos. Oceans* **2013**, *64*, 62–78. [CrossRef]
16. Wang, Y.; Yu, B.; Wu, X.; Wang, P. POD and wavelet analyses on the flow structures of a polymer drag-reducing flow based on DNS data. *Int. J. Heat Mass Transfer.* **2012**, *55*, 4849–4861. [CrossRef]
17. Sarma, B.; Chauhan, S.S.; Wharton, A.M.; Iyengar, A.N.S. Continuous wavelet transform analysis for self-similarity properties of turbulence in magnetized DC glow discharge plasma. *J. Plasma Phys.* **2013**, *79*, 885–891. [CrossRef]
18. Farge, M. Wavelet transforms and their applications to turbulence. *Annu. Rev. Fluid Mech.* **1992**, *24*, 395–457. [CrossRef]
19. Toge, T.D.; Pradeep, A.M. Experimental investigation of stall inception of a low speed contrarotating axial flow fan under circumferential distorted flow condition. *Aerosp. Sci. Technol.* **2017**, *70*, 534–548. [CrossRef]
20. Yin, Y.T.; Lin, M.; Xu, X.F.; Qiao, X.Y.; Wang, L.B.; Zeng, M. Mean velocity distributions under flow in the mainstream of a T pipe junction. *Mech. Chem. Eng. Trans.* **2015**, *45*, 1099–1104.
21. Yin, Y.T.; Chen, K.; Qiao, X.Y.; Lin, M.; Lin, Z.M.; Wang, Q.W. Mean pressure distributions on the vanes and flow loss in the branch in a T pipe junction with different angles. *Energy Procedia* **2017**, *105*, 3239–3244. [CrossRef]
22. Wu, B.; Yin, Y.T.; Lin, M.; Wang, L.B.; Zeng, M.; Wang, Q.W. Mean pressure distributions around a circular cylinder in the branch of a T-junction with/without vanes. *Appl. Therm. Eng.* **2015**, *88*, 82–93. [CrossRef]

23. Lin, M.; Qiao, X.Y.; Yin, Y.T.; Lin, Z.M.; Wang, Q.W. Probability density function of streamwise velocity fluctuation in turbulent T-junction flows. *Energy Procedia* **2017**, *105*, 5005–5010. [CrossRef]
24. Atzori, M.; Vinuesa, R.; Adrián, L.; Schlatter, P. Characterization of turbulent coherent structures in square duct flow. *J. Physics: Conf. Series* **2018**, *1001*, 012008. [CrossRef]
25. Gessner, F.B.; Po, J.K.; Emery, A.F. Measurements of Developing Turbulent Flow in a Square Duct. In *Turbulent Shear Flows, I*; Launder, B.E., Schmidt, F.E., Eds.; Springer: Berlin/Heidelberg, Germany, 1979; pp. 119–136.
26. Schultz, M.P.; Flack, K.A. Reynolds-number scaling of turbulent channel flow. *Phys. Fluids.* **2013**, *25*, 025104. [CrossRef]
27. Vinuesa, R.; Nagib, H.M. Enhancing the accuracy of measurement techniques in high Reynolds number turbulent boundary layers for more representative comparison to their canonical representations. *Eur. J. Mech. B-Fluids* **2016**, *55*, 300–312. [CrossRef]
28. Najmi, A.H.; Sadowsky, J. The continuous wavelet transform and variable resolution time-frequency analysis. *Johns Hopkins APL Tech. Dig.* **1997**, *18*, 134–140.
29. Jiang, X.M.; Mahadevan, S. Wavelet spectrum analysis approach to model validation of dynamic systems. *Mech. Syst. Signal Process.* **2011**, *25*, 575–590. [CrossRef]
30. Lu, J.Y.; Lin, H.; Ye, D.; Zhang, Y.S. A new wavelet threshold function and denoising application. *Math. Probl. Eng.* **2016**, *2016*.
31. Liu, B.S.; Sun, J.; Du, R.J.; Zhao, L.Y. Wavelet De-noising Algorithm and Application Based on Improved Threshold Function. In *International Conference on Control Engineering and Mechanical Design (CEMD 2017)*; Li, C., Ed.; ASME: New York, NY, USA, 2017; p. 10.
32. Cui, H.M.; Zhao, R.M.; Hou, Y.L. Improved threshold denoising method based on wavelet transform. *Phys. Procedia* **2012**, *33*, 1354–1359.
33. Yi, T.H.; Li, H.N.; Zhao, X.Y. Noise smoothing for structural vibration test signals using an improved wavelet thresholding technique. *Sensors* **2012**, *12*, 11205–11220. [CrossRef] [PubMed]
34. Baussard, A.; Nicolier, F.; Truchetet, F. Rational multiresolution analysis and fast wavelet transform: Application to wavelet shrinkage denoising. *Signal Process.* **2004**, *84*, 1735–1747. [CrossRef]
35. Zhang, B.; Wang, T.; Gu, C.G.; Dai, Z.Y. Comparison and Selection of Wavelet Functions in Turbulent Signal Processing. *J. Eng. Thermophys. (Beijing, China)* **2011**, *32*, 585–588.
36. Zhang, J.X.; Zhong, Q.H.; Dai, Y.P. The Determination of the Threshold and the Decomposition Order in Thereshold De-Nosing Method Based on Wavelet Transform. In Proceedings of the Chinese Society for Electrical Engineering, Beijing, China, February 2004; pp. 118–122.
37. Torrence, C.; Compo, G.P. A practical guide to wavelet analysis. Bull. *Am. Meteorol. Soc.* **1998**, *79*, 61–78. [CrossRef]
38. Boltezar, M.; Slavic, J. Enhancements to the continuous wavelet transform for damping identifications on short signals. *Mech. Syst. Signal Process.* **2004**, *18*, 1065–1076. [CrossRef]
39. Meyers, S.D.; Kelly, B.G.; O'Brien, J.J. An introduction to wavelet analysis in oceanography and meteorology: With application to the dispersion of Yanai waves. *Mon. Weather Rev.* **1993**, *121*, 2858–2866. [CrossRef]
40. Thacker, A.; Loyer, S.; Aubrun, S. Comparison of turbulence length scales assessed with three measurement systems in increasingly complex turbulent flows. *Exp. Therm. Fluid Sci.* **2010**, *34*, 638–645. [CrossRef]
41. Okajima, A. Strouhal numbers of rectangular cylinders. *J. Fluid Mech.* **1982**, *123*, 379–398. [CrossRef]
42. Fey, U.; König, M.; Eckelmann, H. A new Strouhal–Reynolds-number relationship for the circular cylinder in the range $47 < Re < 2 \times 10^5$. *Phys. Fluids* **1998**, *10*, 1547–1549.
43. Vidal, A.; Vinuesa, R.; Schlatter, P.; Nagibet, H.M. Reprint of: Influence of corner geometry on the secondary flow in turbulent square ducts. *Int. J. Heat Fluid Flow* **2017**, *67*, 94–103. [CrossRef]

*entropy*

MDPI

*Article*

# Multifractal and Chaotic Properties of Solar Wind at MHD and Kinetic Domains: An Empirical Mode Decomposition Approach

**Tommaso Alberti [1,*], Giuseppe Consolini [1], Vincenzo Carbone [2], Emiliya Yordanova [3], Maria Federica Marcucci [1] and Paola De Michelis [4]**

1. INAF-Istituto di Astrofisica e Planetologia Spaziali, via del Fosso del Cavaliere 100, 00133 Rome, Italy; giuseppe.consolini@inaf.it (G.C.); federica.marcucci@iaps.inaf.it (M.F.M.)
2. Dipartimento di Fisica, Università della Calabria, Ponte P. Bucci, 87036 Rende, Italy; vincenzo.carbone@fis.unical.it
3. Swedish Institute of Space Physics, 75121 Uppsala, Sweden; eya@irfu.se
4. Istituto Nazionale di Geofisica e Vulcanologia, via di Vigna Murata 605, 00143 Rome, Italy; paola.demichelis@ingv.it

* Correspondence: tommaso.alberti@inaf.it

Received: 21 February 2019; Accepted: 14 March 2019; Published: 25 March 2019

**Abstract:** Turbulence, intermittency, and self-organized structures in space plasmas can be investigated by using a multifractal formalism mostly based on the canonical structure function analysis with fixed constraints about stationarity, linearity, and scales. Here, the Empirical Mode Decomposition (EMD) method is firstly used to investigate timescale fluctuations of the solar wind magnetic field components; then, by exploiting the local properties of fluctuations, the structure function analysis is used to gain insights into the scaling properties of both inertial and kinetic/dissipative ranges. Results show that while the inertial range dynamics can be described in a multifractal framework, characterizing an unstable fixed point of the system, the kinetic/dissipative range dynamics is well described by using a monofractal approach, because it is a stable fixed point of the system, unless it has a higher degree of complexity and chaos.

**Keywords:** solar wind; scaling properties; fractals; chaos

## 1. Introduction

The interplanetary space is permeated by a supersonic and super-Alfvénic plasma known as solar wind which develops a strong turbulent character during its expansion phase [1]. Due to the presence of a mean magnetic field, solar wind low-frequency fluctuations are usually described within the magnetohydrodynamic (MHD) framework [2–4]. These fluctuations show turbulent properties that are characterized by a quasi-Kolmogorov energy scaling [5–8]. Indeed, magnetic energy density seems to follow a spectral decay as $E(k) \sim k^{-5/3}$, although the theoretical scaling derived from MHD equations suggests a slightly different spectral exponent, e.g., $E(k) \sim k^{-3/2}$ [9,10] for Alfvènic turbulence as it should be in the case of solar wind. Turbulence is a phenomenon showing the presence of small scale fluctuations in the velocity and pressure fields (for fluids), as well as in the magnetic field (for plasmas), and an increased rate of mixing of mass and momentum [1,11]. Turbulent flows exhibit characteristic phenomena like coherent structures in the flow and intermittency. Coherent structures are usually defined as regions of concentrated vorticity where phase correlation exists with a typical lifetime larger than that of the stochastic fluctuations surrounding them, while intermittency is the manifestation of sudden field changes, modifying the shape of the probability distribution functions of field gradients (e.g., velocity and temperature in fluids, magnetic in plasmas) [12,13], and resulting in an anomalous scaling of the field increments [14]. As the analytic and numerical solution of such flows is expensive, investigators rely on models to simulate and simplify their dynamics. Such turbulence models include two-equation models (like the $k$-$\epsilon$ model and the $k$-$\omega$ one [15]), Reynolds stress models (like the Speziale-Sarkar-Gatski model [16] and the Mishra-Girimaji model [17]), along with models in Large Eddy Simulations [18]. Similar models have been also developed for describing turbulent features in plasmas like two-dimensional hybrid-Vlasov simulations [19], compressible Hall MHD direct numerical simulations [20], and shell models [21].

In the framework of turbulence, several phenomena inside the MHD/inertial domain are described, by using the nonlinear energy cascade via the Yaglom law, an exact relation for the scaling of the third-order moment of fluctuations [22,23], or by analyzing the role of the intermittency in changing scaling properties of magnetic fluctuations within a multifractal approach [14,24–26]. Both previous findings are derived from the structure function analysis, through which scaling laws can be investigated (e.g., [27]), exploiting Kolmogorov's universality assumptions (e.g., [28]). More specifically, a turbulent flow is sustained by a persistent source of energy which is rapidly dissipated via the so-called nonlinear energy cascade [29], converting the kinetic energy into internal energy through viscous processes. Indeed, turbulence causes the formation of eddies at different scales and energy is transferred from large- to small-scale structures through an inertial and inviscid mechanism [5,28], i.e., the nonlinear energy cascade. According to Kolmogorov's theory, if all possible symmetries of the Navier-Stokes equation are restored in a statistical sense, then the turbulent flow is self-similar at small scales and has a finite mean energy dissipation rate $\epsilon$ such that, at very high (but not infinite) Reynolds numbers, all small-scale statistical properties are uniquely and universally determined by the length scale $\ell$, the mean energy dissipation rate $\epsilon$, and the kinematic viscosity $\nu$ (e.g., [28]). By simply exploiting dimensional arguments, these assumptions imply that the energy spectrum at large wavenumbers assumes a universal form as

$$E(k) = F(\nu)\epsilon^{2/3}k^{-5/3}. \tag{1}$$

In the limit of infinite Reynolds numbers, Equation (1) becomes independent by the viscosity $\nu$ such that

$$E(k) = C\epsilon^{2/3}k^{-5/3}. \tag{2}$$

being $C$ a universal dimensionless constant [28]. The above assumptions are only valid for all those scales which are smaller than the integral scale $L$, where long-range correlations between particles are found, and are greater than the dissipative scale $\ell_D$, where viscosity dominates and the turbulent kinetic energy is dissipated, i.e., $\ell_D \ll \ell \ll L$ (e.g., [5,28]). Particularly, the dissipation of kinetic energy mostly takes place at the so-called Kolmogorov microscale defined as

$$\eta = \left(\frac{\nu^3}{\epsilon}\right)^{1/4} \tag{3}$$

which is well separated from the integral scale $L$, corresponding to the size of the eddies when they are formed. These two scales mark the extrema of the energy cascade: since eddies with size $L$ are much larger than the dissipative eddies with size $\eta$, kinetic energy is not dissipated at large scales but it is essentially transferred to smaller scales where viscous effects become dominant. Within this range, where nonlinear interactions between eddies take place, inertial effects are larger than viscous ones such that it is usually named "inertial range". Due to the large separation between $L$ and $\eta$, the dissipation rate is primarily determined by the large scales since viscous effects at Kolmogorov scales rapidly dissipate energy. Thus, the overall rate of dissipation is only controlled by the nonlinear scale-to-scale transfer such that the dissipation rate is approximately given as

$$\epsilon = \frac{u^3}{L} \tag{4}$$

being $u$ the bulk velocity of the flow, and consequently

$$\frac{\eta}{L} = Re^{-3/4} \tag{5}$$

where $Re$ is the Reynolds number. Equation (5) can be used as a measure of the number of scales within the inertial range (i.e., the extension of this range of scales) which only depends by the Reynolds number of the flow.

Similar assumptions and scalings can be also derived by using MHD equations for describing plasma dynamics and, particularly, within the same Kolmogorov's assumptions of isotropy, homogeneity, and stationarity, and exploiting dimensional arguments, the energy spectrum at large wavenumbers behaves as

$$E(k) = C\,(\epsilon v_0)^{1/2} k^{-3/2}. \tag{6}$$

where $C$ a universal dimensionless constant and $v_0$ is the rms component of the total turbulent velocity [9,10].

Kolmogorov assumptions break down beyond a scale $\ell_b \sim \ell_i$, being $\ell_i$ the ion inertial length, where the MHD/inertial description cannot be used [30,31], and for which a steeper slope of the energy density spectrum is found [32]. The second power-law domain at small scales can be explained by invoking several possible dispersive/kinetic phenomena as wave-wave coupling, Landau damping, Kinetic Alfvén Waves (KAWs), and so on [33–35].

Solar wind magnetic field fluctuations, including spectral features like the Kolmogorov energy scaling, are usually investigated by following a statistical approach of magnetic increments

$$\delta\mathbf{b}(\mathbf{x}, \ell) \doteq \mathbf{b}(\mathbf{x}, \ell) - \mathbf{b}(\mathbf{x}) \tag{7}$$

requiring that the statistics of these increments is invariant under arbitrary translations

$$\langle \delta\mathbf{b}(\mathbf{x}+\mathbf{r}, \ell) \rangle = \langle \delta\mathbf{b}(\mathbf{x}, \ell) \rangle. \tag{8}$$

In this way, it is possible to derive an exact law for the scaling of the mean-square magnetic field increments between two points separated by a distance $\ell$ according to which $\langle (\delta\mathbf{b}(\ell))^2 \rangle \sim \ell^{1/2}$, from which the Iroshnikov-Kraichnan $k^{-3/2}$ scaling law can be simply derived by noting that $\langle (\delta\mathbf{b}(\ell))^2 \rangle$ is related to the magnetic energy [9,10]. Moreover, from dimensional analysis it is easy to recover the second-order structure function which is equal to $S_2(l) = \langle (\delta\mathbf{b}(\ell))^2 \rangle = C_{IK}(\epsilon\, c_A)^{1/2} \ell^{1/2}$, where $c_A$ is the Alfvén speed and $C_{IK}$ is a universal constant. More generally, generalized structure functions can be calculated from any arbitrary and finite order $q$ from increments such that

$$S_q(\ell) \doteq \langle |\ \delta\mathbf{b}(\mathbf{x}, \ell)\ |^q \rangle. \tag{9}$$

They are extensively used to investigate scaling properties of both velocity and magnetic field fluctuations (e.g., [27]), mostly devoted to the characterization of intermittency and self-similarity properties of solar wind turbulence [14,36–38]. In this context, different multifractal models (e.g., $\beta$-model, $p$-model, and their variations (e.g., [24,26])) have been proposed to explain the evolution of intermittency across the heliosphere, with the solar wind becoming more multifractal in nature when leaving the Sun [1]. In particular, the analysis of scaling exponents $\zeta(q)$ of the $q$-th order structure function shows that different turbulent scenarios can develop, being mainly characterized by a nonlinear dependence of $\zeta(q)$ with $q$ (e.g., [39–42]).

This paper approaches the study of the self-similarity properties of solar wind magnetic field fluctuations at different timescales by using a novel method to evaluate structure functions at different orders $q$. It is based on the Empirical Mode Decomposition (EMD) which allows to correctly derive the timescales embedded into the analyzed time series, as well as to better evaluate structure functions by using local (non-constant) timescales [43,44]. The results evidence a different behavior of magnetic field fluctuations at MHD/inertial and kinetic/dissipative scales. While the former are characterized by a multifractal character, the latter show a monofractal scaling. In a dynamical system framework, both behaviors can be seen as corresponding to two different fixed points: an unstable saddle for the MHD/inertial domain, and a stable node for the dissipative one. The solar wind magnetic field fluctuations undergo a saddle-node bifurcation when moving from the MHD/inertial down to kinetic/dissipative scales.

## 2. Data

We consider solar wind magnetic field measurements from ESA-Cluster mission on 10 January 2004 from 05:30 UT to 06:30 UT. This period is characterized by a high speed (i.e., it is a fast stream, $v \sim 540$ km/s), a mean magnetic field intensity ($B \sim 11$ nT), and a high density ($n \sim 14$ cm$^{-3}$). We used combined magnetic field data from the fluxgate magnetometer (FGM) and the experiment called "spatiotemporal analysis of field fluctuations" (STAFF) onboard Cluster 3 spacecraft to obtain a resolution of data equals to 450 Hz. For computational purposes, the time resolution has been reduced of a factor 4, moving it to $\Delta t = 8.9$ ms.

Figure 1 shows the magnetic field intensity (B) and the three magnetic field components ($B_x$, $B_y$, $B_z$) for the selected time interval in the GSE reference system.

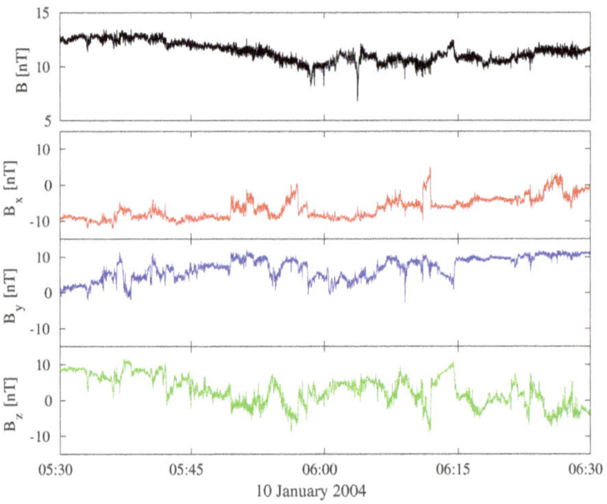


**Figure 1.** Solar wind magnetic field measurements during the time interval 05:30–06:30 UT on 10 January 2004. Data are obtained from Cluster 3 at the time resolution of 8.9 ms.

## 3. Methods

### *3.1. The Empirical Mode Decomposition (EMD): A Brief History*

During the past decades, several decomposition procedures have been suggested to investigate scale variability of time series. Most common methods rely on Fourier-based techniques, Wavelet transforms and/or eigenfunction analysis (e.g., [45]). These methods, by choosing a decomposition basis in a mathematical space with requirements of completeness and orthogonality, allow to obtain oscillating components, with fixed scales and amplitudes, embedded inside time series (e.g., [45]). However, neither stationarity nor linearity is assured when natural phenomena are investigated, unless scaling law theory is mostly derived by exploiting these two requirements (e.g., [27,28]). Recently, the Empirical Mode Decomposition (EMD) has been developed to provide a suitable decomposition method for time series by exploiting their local properties, allowing us to reduce mathematical assumptions by using a completely adaptive and a posteriori decomposition procedure where the number of the extracted empirical modes depends on the signal complexity [43]. The EMD carries out a finite set of embedded modes, usually named Intrinsic Mode Functions (IMFs), from a given time series $x(t)$ by using an iterative process known as sifting process. The main steps of this process can be summarized as follows:

1. evaluate the mean of a signal $x(t)$ and subtract from it to produce a zero-mean signal $x_m(t) = x(t) - \langle x(t) \rangle$;
2. find local maxima and minima of $x_m(t)$;
3. use a cubic spline to evaluate the upper ($e_{max}(t)$) and lower ($e_{min}(t)$) envelopes from local maxima and minima, respectively;
4. evaluate the mean envelope $e_m(t)$ and subtract from $x_m(t)$ to have $h(t) = x_m(t) - e_m(t)$;
5. check if $h(t)$, often called detail or "candidate" IMF, is an IMF that is, check if the number of zero crossings and local extrema differs at most by one and if the local mean is zero;
6. if $h(t)$ is an IMF, then store it ($c_k(t) = h(t)$), else repeat steps from 1 to 5 on the signal $x_h(t) = x_m(t) - h(t)$ until an IMF is obtained.

Once the decomposition is complete, i.e., when no more IMFs can be extracted from $x(t)$, the time series $x(t)$ can be written as

$$x(t) = \sum_{k=1}^{N} c_k(t) + r(t) \tag{10}$$

where $r(t)$ is the residue of the decomposition, a non-oscillating function of time [43]. Mathematically, the sifting process stops only when the number of iterations $n \to \infty$; numerically, it can be stopped after $n'$ iterations by defining a stopping criterion [46] like the Cauchy convergence criterion [43], according to which the sifting algorithm stops when $\sigma_{n'} < \sigma_0$, being $\sigma_{n'} = \sum_{j=1}^{T} \frac{|h_{n'}(t_j) - h_{n'-1}(t_j)|^2}{h_{n'-1}^2(t_j)}$, where $h_{n'}$ is the $n'$ detail and $T$ the length of the time series $x(t)$, and $\sigma_0$ is a threshold value which usually varies between 0.2 and 0.3 [43], or by the threshold method proposed in Reference [47] in which two thresholds, $\theta_1$ and $\theta_2$, are chosen to guarantee globally small fluctuations and, in the meanwhile, to take into account locally large excursions. In this way, by defining $\sigma(t) = \left| \frac{2h_{n'}(t)}{e_{max}(t) - e_{min}(t)} \right|$, the sifting process is iterated until $\sigma(t) < \theta_1$ for a prescribed fraction $1 - \alpha$ of the total duration, and $\sigma(t) < \theta_2$ for the remaining fraction, being typically $\theta_1 = 0.05$ and $\theta_2 = 10\,\theta_1$ [47,48]. More details about the sifting process and its features can be found in several previous works (e.g., [43,47–49]).

The EMD is a fundamental step for providing non-stationary oscillating components which can be used as inputs for the Hilbert Spectral Analysis (HSA), which permits us to investigate amplitude-frequency modulation embedded in time series (e.g., [43,50]). Through the Hilbert Transform (HT), which is a linear mathematical operator that takes each IMF $c_k(t)$ and produces a function $H[c_k](t)$ by convolution with the function $\frac{1}{\pi t}$, each empirical mode can be written as modulated both in amplitude and in frequency

$$c_k(t) = a_k(t) \Re \left\{ exp \left[ i2\pi \int_0^t f_k(t') dt' \right] \right\} \tag{11}$$

where $a_k(t)$ and $f_k(t)$ are the instantaneous amplitude and frequency of the $k$-th empirical mode, respectively, and $\Re$ is the real part of the exponential. The HT allows to investigate non-stationary features of the time series, being $f_k(t)$ a function of time, and also its nonlinear behavior, due to the time-dependence of $a_k(t)$ (e.g., [43,51]). Derived from both $a_k(t)$ and $f_k(t)$, the instantaneous local energy content $E(t, f)$ is studied by contouring the squared-amplitude in a time-frequency plane, i.e., by defining the so-called Hilbert-Huang spectrum [43]. Then, an intermittency measure, similar to that defined by using wavelet analysis, can be introduced as

$$DS(f) = \frac{1}{n\Delta t} \int_t \left[ 1 - \frac{H(t', f)}{h(f)} \right]^2 dt' \tag{12}$$

where $h(f) = \langle H(t', f) \rangle_t$ and $n\Delta t$ is the length of the time series (e.g., [43]). It is often called Degree of Stationarity (DS) (e.g., [43]) and a time series is statistically stationary if DS = 1.

Figure 2 reports the degree of stationarity for the three magnetic field components. A clear increase in the stationary character of time series is found when approaching the frequency $f_b = 0.4$ Hz $\sim f_i$, being $f_i$ the Doppler-shifted ion cyclotron frequency. This suggests that a high non-stationary behavior characterizes the inertial regime, where MHD processes govern the dynamics of the system, while dissipative processes are characterized by a nearly-stationary dynamics as also previously observed (e.g., [52]). The non-stationary character observed in the MHD/inertial domain could be a counterpart of the intermittent nature of fluctuations in the inertial range.

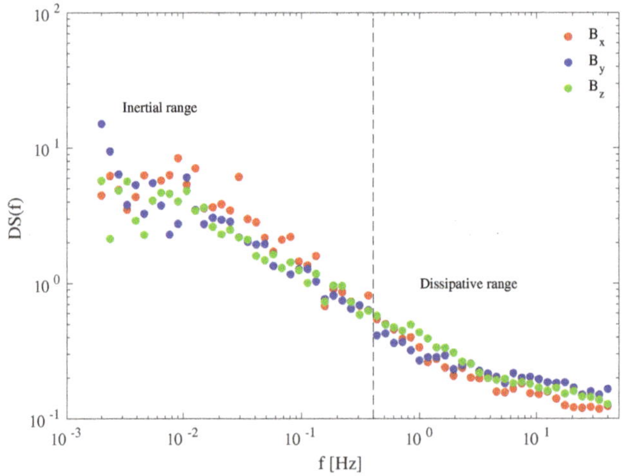


**Figure 2.** Degree of stationarity (DS) of the three different magnetic field components during the selected time interval. The dashed line refers to the Doppler-shifted ion cyclotron frequency ($f_i \sim 0.4$ Hz).

### 3.2. The EMD-Based Multifractal Analysis

Recently, a method capable of detecting the fractal dimension of a time series by partitioning the time and scale domain of a signal into fractal dimension regions has been proposed. This method, which is similar to the Wavelet Transform Modulus Maxima (WTMM), is an EMD-based multifractal analysis. It is named EMD-based dominant amplitude multifractal formalism (DAMF) [44] and it has been proposed to investigate singularities and (multi)fractal behavior of time series. The EMD-DAMF method can be summarized in the following steps:

1. derive instantaneous amplitude $a_k(t)$ and mean timescale $\tau_k = \langle f_k(t) \rangle_t^{-1}$ of each empirical mode;
2. determine the dominant amplitude coefficients $u_{j,k}$ over a time support $I_{j,k}$ around the $j$-th local maximum

$$u_{j,k} \doteq \sup_{k' \leq k} \left\{ \max \left\{ |a_{k'}(t \in I_{j,k})| \right\} \right\} \tag{13}$$

   with $j = 1, \ldots, N_k$, being $N_k$ the number of local maxima of $a_k(t)$, and $k = 1, \ldots, N$;
3. evaluate the $q$-th-order structure function $S_q(\tau_k)$

$$S_q(\tau_k) = \frac{1}{N_k} \sum_{j=1}^{N_k} \left\{ u_{j,k} \right\}^q ; \tag{14}$$

4. estimate the scaling exponent $\zeta(q)$ as the linear slope, in a log-log space, of $S_q(\tau_k)$ vs. $\tau_k$, such that

$$S_q(\tau_k) \sim \tau_k^{\zeta(q)}; \quad (15)$$

5. derive the singularity strengths $\alpha$ and spectrum $f(\alpha)$ by using the Legendre transform of the scaling exponents $\zeta(q)$ as usual

$$\alpha = \frac{d\zeta(q)}{dq} \quad \& \quad f(\alpha) = \alpha q - \zeta(q). \quad (16)$$

The main novelty introduced by this method is that structure functions $S_q(\tau_k)$ are derived by exploiting the local features of empirical modes such that local extrema can be used to correctly calculate differences/increments between two points, instead of considering a fixed timescale as for canonical structure function analysis. Moreover, timescales are not fixed a priori but they are derived from the EMD analysis of time series considering a finite set of multiresolution coefficients $u_{j,k}$. This allows us to have a limited (and small) number of points in the scaling range such that the scaling exponents can be better evaluated and visually inspected.

## 4. Results from the EMD-Based Multifractal Analysis

Figures 3 and 4 report the EMD-DAMF results at MHD/inertial and kinetic/dissipative scales for each magnetic field component, respectively. In each figure, the second-order structure function $S_2(\tau)$ is shown in the upper panel (multiplied by $\tau^{1/2}$ and $\tau^{3/2}$ to have a compensated structure function), the scaling exponents $\zeta(q)$ are reported in the middle panel, and the singularity spectrum $f(\alpha)$ is displayed in the lower panel.

The EMD-DAMF analysis at MHD/inertial scales (Figure 3), i.e., corresponding to the inertial range, can be carried out by considering empirical modes with mean timescales in the range 2–500 s (or $f \in (10^{-3}, 0.4)$ Hz). Indeed, as shown by the second-order structure function $S_2(\tau)$ a scale-break is found when $f = f_b = 0.4$ Hz $\sim f_i$. As expected from structure function theory of the MHD/inertial domain (e.g., [9,10]), the second-order structure function behaves as $\tau^{1/2}$, suggesting that the Fourier energy spectral density decays as $f^{-3/2}$ (or $k^{-3/2}$ assuming Taylor's hypothesis) (e.g., [8–10]).

This result supports the common view according to which energy is injected at large scales (i.e., larger than a typical injection scale $L$) and transferred to small scales (i.e., smaller than a dissipative scale $\ell_D$) through nonlinear interactions and phenomena taking place at scales $\ell$, being $\ell_D \ll \ell \ll L$ (e.g., [3,4,24,26]). Here, $\ell_D$ stands for the dissipation scale (equivalent to Kolmogorov's scale in fluid turbulence). This result has been obtained by considering the "true" timescales which are embedded in the raw time series and extracted via an adaptive procedure, with no assumptions on the stationarity of oscillating components.

By considering structure functions $S_q(\tau)$ with 2 s $< \tau <$ 500 s, the scaling exponents are derived and shown in the middle panel of Figure 3 for the three magnetic field components. From a theoretical point of view, assuming homogeneity, isotropy and scale-invariance of the time series we should obtain $\zeta(q) = q/4$ in the case of Alfvènic MHD turbulence [10]. Our results show a different behavior with scaling exponents characterized by a nonlinear convex trend with the moment order $q$ like $\zeta(q) \sim q/4 + \varphi(q/4)$ [26]. This deviation is the fingerprint of the occurrence of anomalous scaling features, i.e., of an intermittency phenomenon, in the nonlinear energy cascade of the magnetic field, suggesting nonlinear two-point correlations in the real space [39–41]. Interestingly, different scaling exponents are obtained for the different magnetic field components indicating the existence of an anisotropy of the scaling features in the different directions which may reflect the anisotropic nature of the fluctuation field [7]. Moreover, the observed nonlinear scaling of $\zeta(q)$ suggests that the probability distribution functions (PDFs) of increments at MHD/inertial scales are characterized by multifractal scaling features. This aspect can be clearly seen by looking at the singularity spectrum $f(\alpha)$

(see Figure 3, lower panel) which shows a wide range of singularities ($0.1 < \alpha < 0.8$) for all components. Wider singularities are found for the $B_y$ component, while a narrower spectrum is found for $B_z$.

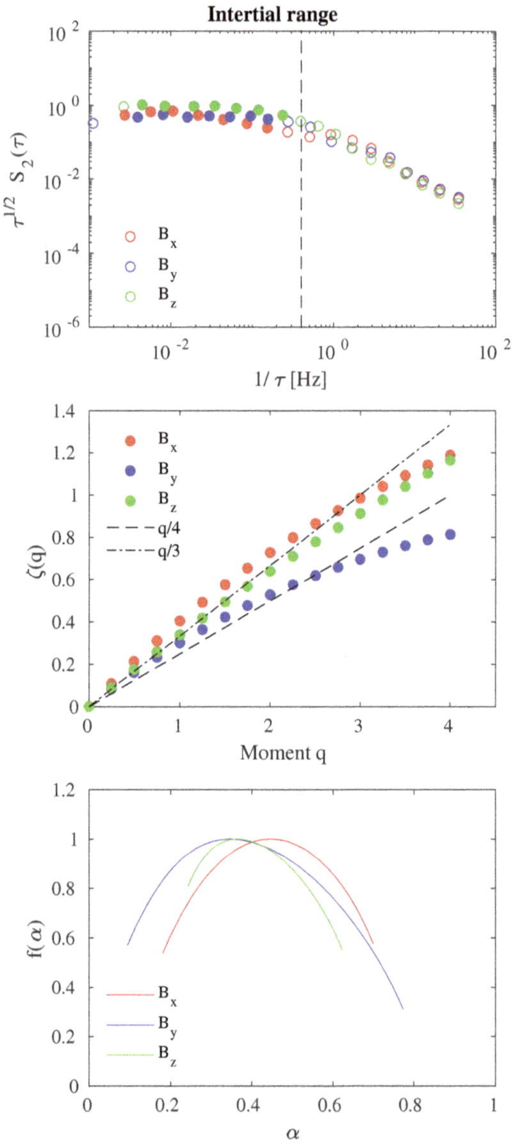


**Figure 3.** Empirical Mode Decomposition-Dominant Amplitude Multifractal Formalism (EMD-DAMF) results for the inertial range: compensated second-order structure function $S_2(\tau)$ (**upper panel**), scaling exponents $\zeta(q)$ (**middle panel**), and singularity spectrum $f(\alpha)$ (**lower panel**). Red, blue and green symbols refer to the $B_x$, $B_y$, and $B_z$ solar wind magnetic field components, respectively. Filled symbols in the upper panel refer to the magnetohydrodynamic (MHD)/inertial scales where a clear Iroshnikov-Kraichnan (IK) spectrum is found. The dashed and dashed-dotted lines in the middle panel refer to $\zeta(q) = q/4$ and $\zeta(q) = q/3$, respectively.

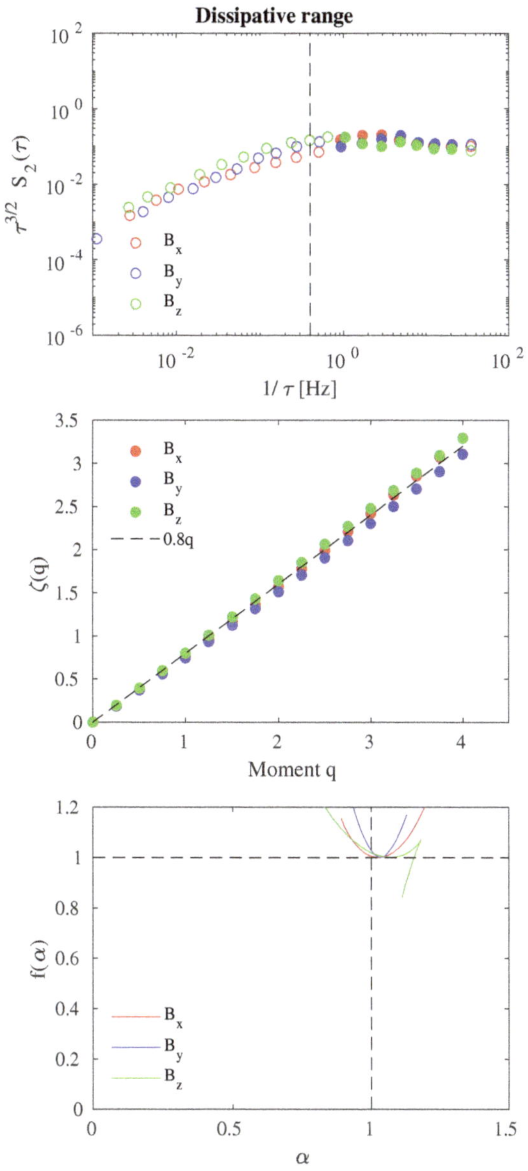


**Figure 4.** EMD-DAMF results for the dissipative range: compensated second-order structure function $S_2(\tau)$ (**upper panel**), scaling exponents $\zeta(q)$ (**middle panel**), and singularity spectrum $f(\alpha)$ (**lower panel**). Red, blue and green symbols refer to the $B_x, B_y$, and $B_z$ solar wind magnetic field components, respectively. Filled symbols in the upper panel refer to the kinetic/dissipative scales. The dashed line in the middle panel refers to $\zeta(q) = 0.8\,q$.

A clear different behavior is found when approaching the dissipative range (see Figure 4), i.e., moving towards higher frequencies ($f > f_b = 0.4$ Hz). A different scaling law is recovered, moving to a greater scaling exponent ($\tau^{3/2}$) and, consequently, a steeper slope for the energy spectral density, decaying as $\sim f^{-5/2}$, is recovered. This suggests that different physical processes operate inside this dynamical regime occurring on small scales. From a fractal point of view, magnetic field fluctuations at kinetic/dissipative scales seem to behave as a monofractal system with a Hurst exponent (i.e., $\zeta(1)$) $\sim 0.8$ [52,53], quite similar for all magnetic field components (Figure 4, middle panel). This behavior is confirmed by the singularity spectrum $f(\alpha)$ (see Figure 4, lower panel) which collapses near the point $(\alpha, f(\alpha)) = (1,1)$. Our findings suggest the absence of intermittency at dissipative scales, which is well in agreement with previous works where generalized Hilbert spectra were used [52–54].

The difference in the scaling properties between the two ranges of scales can be linked to the different physical processes operating in both the inertial and dissipative domains. On one hand, the inertial range is characterized by the nonlinear interactions between eddies of different size, causing their fragmentation to smaller and smaller ones until viscous effects become dominant (it is worthwhile to remark that eddies must not be thought of as real vortices, but as a description of the triadic interaction between modes). Conversely, when approaching the Kolmogorov scale $\eta$ wave-particle mechanisms and small-scale structures (like current sheets) become the most prominent features which characterize the dissipative processes [31]. Indeed, while the inertial range physics is mostly dominated by large-scale phenomena like plasma instabilities and it is characterized by an inhomogeneous nonlinear transfer of energy, resulting in the generation of localized small-scale structures with scale-dependent features [8,14,36], the dissipative range physics is mainly characterized by several dispersive phenomena generated by velocity-space effects and electron dynamics, driven by wave-wave coupling, scattering processes, and damping mechanisms [1,31,34,35].

Moreover, our results seem to confirm the robustness of the EMD-based method in investigating scaling features of solar wind fluctuations. In addition, by using the EMD-DAMF approach we are able to carry out structure function analysis on both positive and negative $q$, allowing us to derive the whole singularity spectrum $f(\alpha)$ such that accurate intermittency measures can be found. Conversely, generalized Hilbert spectra, unless based on the EMD and HSA procedures, cannot be evaluated for $q < 0$, thus permitting only a partial detection of singularities (only the increasing branch of $f(\alpha)$ can be obtained) [55]. Although the difference in the intermittent properties between MHD/inertial and kinetic/dissipative domains remains an open question (e.g., [35]), our results can help to accurately measure scaling exponents and singularities with fewer a priori mathematical assumptions with respect to previous analysis, thus providing useful constraints for modeling purposes.

## 5. Chaotic Measures and Phase-Space Analysis

A dynamical system, like the solar wind, can be also investigated following a chaotic approach, mostly based on looking at the dimensionality of its phase-space. A system is defined to be chaotic if its dimension is a non-integer value [56]. Different measures have been introduced to quantify the presence and degree of chaos [57]. Particularly, the correlation dimension $D_2$, useful for determining the fractional dimensions of fractal objects, is estimated by embedding a time series $x(t)$ in a time-delayed $m$-component state vector as

$$\mathbf{X}_k = \{x_1(t_k), x_2(t_k), \ldots, x_m(t_k)\} \tag{17}$$

where $x_l(t_k) = x(t_k + (l+1)\Delta)$, $m$ is usually named embedding dimension, and $\Delta$ is a time delay. Then, the correlation integral can defined as

$$C(\rho, m) = \lim_{N_s \to \infty} \frac{1}{N_s^2} \sum_{i=1}^{N_s} \sum_{j=1}^{N_s} \Theta(\rho - |\mathbf{X}_i - \mathbf{X}_j|) \tag{18}$$

where $N_s$ is the number of considered phase-space states, $\Theta$ is the Heavyside step function, and $\rho$ is the phase-space threshold distance between two points. If $\rho \to 0$, a power-law behavior is found for the correlation integral as $C(\rho, m) \sim \rho^{D_2}$, where $D_2$ is defined as

$$D_2 = \lim_{\rho \to 0} \frac{\log C(\rho, m)}{\log \rho}. \tag{19}$$

As the embedding dimension $m$ increases, the correlation dimension will converge to its true value. Specifically, if $D_2 = m$ then the system will explore the whole phase-space; conversely, if $D_2 < m$ a strange attractor will characterize the phase-space dynamics. Of course, both $m$ and $\Delta$ need to be properly chosen. Their choice is crucial for a correct estimation of the correlation dimension in the case of chaotic systems [56–58]. Generally, the choice of the time delay $\Delta$ corresponds to the first minimum of the autocorrelation function of the time series, while the choice of the embedding dimension $m$ falls on the lowest value at which $D_2$ approaches from a constant value [57].

Figure 5 shows the behavior of the correlation dimension $D_2(\tau)$ as a function of the mean frequency of each empirical mode, derived as the inverse of the mean timescale $\tau$. This allows us to investigate how the dynamical behavior changes when moving from MHD/inertial to kinetic/dissipative scales.

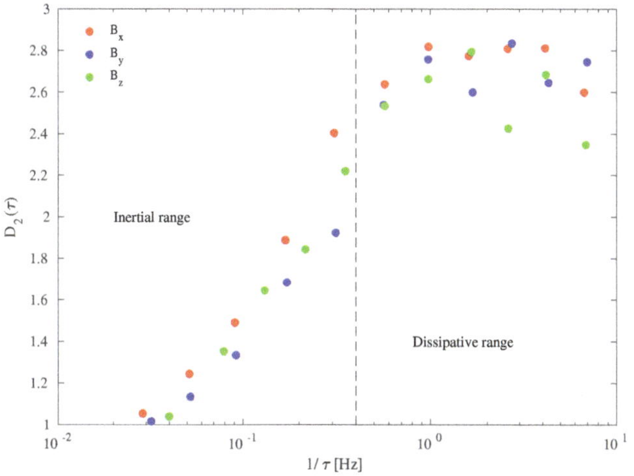


**Figure 5.** Correlation dimension $D_2$ of the different empirical modes as function of the mean frequency $(1/\tau)$. The vertical dashed line separates the inertial range from the kinetic/dissipative one.

The dimensionality of the system clearly exhibits a scale-dependent behavior characterized by an increase in the values of $D_2(\tau)$ with the mean frequency, approaching from a constant value $D_2(\tau) \sim 2.7$ for $f > f_b = 0.4$ Hz. This suggests that magnetic field fluctuations are characterized by a superposition of processes working on different timescales and with different dimensionality. While MHD processes can be described by using a low-dimensional dynamical system, since $D_2 < 2$, the kinetic/dissipative domain dynamics cannot be represented as a linear system since at least three system variables ($D_2 > 2$) are needed to describe processes (perhaps dissipation) occurring at these scales.

An interesting result is the continuous change of the correlation dimension moving from MHD/inertial to kinetic/dissipative domains, which suggests that a single correlation dimension is not capable of describing the complexity features of solar wind magnetic field fluctuations at the MHD/inertial scales, while a single correlation dimension seems to describe kinetic features. This can be interpreted as the signature of the intermittent nature of fluctuations in the MHD/inertial domain, where a hierarchy of dimensions is necessary to describe the complex nature of the nonlinear energy cascade. Conversely, at the kinetic/dissipative scales where dissipation may occur, the correlation dimension seems to converge to a single value of $D_2 \sim 2.7$. This is the temporal counterpart of the multifractal nature of turbulence in the MHD/inertial domain and of the monofractal nature of the dissipative regime, as also shown in Section 4.

We can characterize a dynamical system by looking at its phase-space dynamics in order to recover the existence of fixed points and their nature, as well as to investigate the presence of (strange) attractors [57]. Since by using the EMD we are able to decompose our time series into oscillating functions [48,59–61], we choose to reconstruct empirical modes according to the different dynamical regimes investigated. We can investigate the dynamics at the MHD/inertial and kinetic/dissipative scales in a separate way by defining

$$R_I(t) = \sum_{f_k \in f_I} c_k(t) \tag{20}$$

$$R_D(t) = \sum_{f_k \in f_D} c_k(t) \tag{21}$$

as the reconstructions of empirical modes with characteristic mean frequencies inside the intertial ($f_I$) and kinetic/dissipative ($f_D$) domains. In detail,

$$f_I \doteq \left\{ f_k \mid 10^{-3}\text{Hz} < f_k < f_b \right\} \tag{22}$$

$$f_D \doteq \{ f_k \mid f_k > f_b \} \tag{23}$$

being $f_b = 0.4$ Hz.

Figure 6 reports the phase-space portraits for the two different dynamical regimes, i.e., inertial scales (left panels) and kinetic/dissipative ones (right panels). The different symbols identify different phase-space trajectories starting at different phase-space positions, identified by a black symbol, and ending with a magenta one. The results look quite interesting and can be interpreted in dynamical system framework.

The dynamics seems to be characterized by an unstable orbit at inertial scales, so that the associated fixed point can be classified as a saddle. Indeed, starting from different phase-space positions each trajectory moves along an unstable manifold such that the system will approach the (unstable) fixed point being repelled on different (and opposite) phase-space points. Thus, the set is a repeller. This hyperbolic equilibrium point does not have any center manifolds, and, near it, the orbits of the system resemble hyperbolas. Conversely, the dynamics at kinetic/dissipative scales is characterized by a set which is an attractor since all phase-space trajectories tend to move towards the stable fixed point, which can be identified as a node. This fixed point, due to its fractal dimension and structure (see Section 4), is a chaotic strange attractor, extremely sensitive to initial conditions. By considering two arbitrarily close initial phase-space positions near the attractor, after several time steps they will move on phase-space positions far apart, and after other several time iterations will lead to phase-space positions which are arbitrarily close together. Thus, the dynamics never depart from the attractor [56].

The obtained results seems to suggest a new view of the dynamics of the solar wind at different scales from the MHD/inertial domain down to the kinetic/dissipative one. The system undergoes a saddle-node bifurcation, a local bifurcation in which two fixed points collide and annihilate each other, with an unstable fixed point (saddle) and a stable one (node). This means that both the inertial and kinetic/dissipative ranges can be seen as fixed points of the governing system equations, one unstable and the other stable. In this way, the phenomenological model of the Richardson cascade [24,26,28,29] can be interpreted in the different context of the dynamical system theory. Energy is injected at a scale $L$, which represents a stable fixed point of the system; then, when nonlinear interactions develop, corresponding to changes in one or more dynamical bifurcation parameters, the dynamics of the system changes, moving towards an unstable fixed point (i.e., the inertial regime) which, due to its repeller nature, forces the system to explore the available phase-space until a stable fixed point (i.e., the kinetic/dissipative domain) is reached (Figure 7). In a simple conceptual model, a bifurcation parameter could be the timescale of the different processes operating inside the MHD/inertial and kinetic/dissipative domains such that the dynamics of the system, represented in our case by the magnetic field components $B_i$, can be seen as solely dependent on $\tau$

$$\dot{B}_i = g(B_i, B_j, \tau). \tag{24}$$

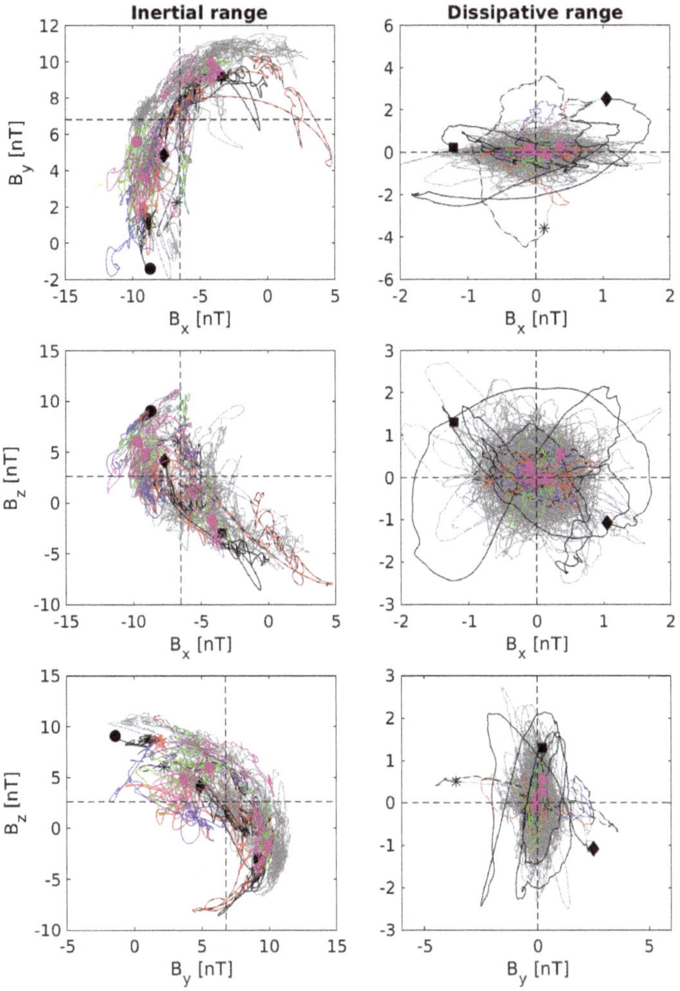


**Figure 6.** Phase-space portraits for the MHD/inertial range dynamics (**left panels**) and for the kinetic/dissipative range one (**right panels**). Symbols mark different phase-space trajectories with colors corresponding to different time instants (each trajectory starts with a black symbol and ends with a magenta one).

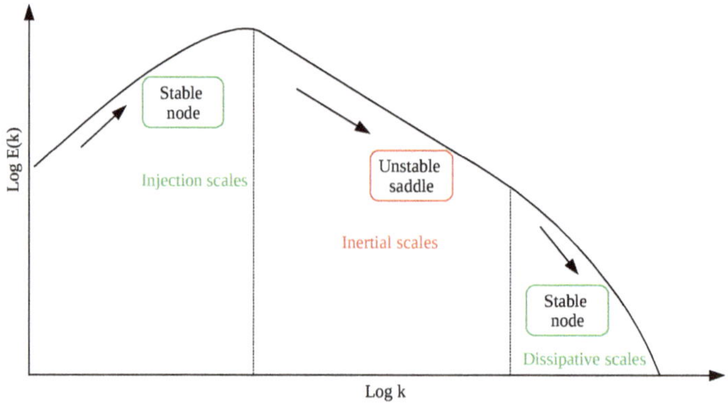


**Figure 7.** A sketch of the different dynamical regimes.

The system is surely characterized by a chaotic dynamics due its dimension (our system is described by three variables) with nonlinear interactions between the different variables (i.e., $g(B_i, B_j, \tau)$ is a nonlinear function of $B_i, B_j$) as required when describing turbulent features (e.g., [1,24,26]).

## 6. Conclusions

Solar wind magnetic field fluctuations at different scales have been investigated by employing both a multifractal and a chaotic approach. The multifractal analysis has been performed by using a novel formalism, the EMD-based dominant amplitude multifractal formalism, through which increments are derived by using local properties of fluctuations at different scales obtained by using the Empirical Mode Decomposition method. The results suggest that MHD fluctuations show an intermittent character, well described in the framework of classical multifractal models (like the $p$-model (e.g., [26])); conversely, magnetic field fluctuations at kinetic scales (i.e., beyond the ion inertial length) show a monofractal behavior, in agreement with previous findings (e.g., [52–54]).

The phase-space dynamics of the two ranges of scales, i.e., inside the MHD/inertial and kinetic/dissipative domains, is characterized by a different degree of chaos, because the system is more chaotic when moving from the MHD down to the kinetic scales. An unstable manifold is recovered for the MHD scales, characterizing an unstable saddle for the magnetic field dynamics. Conversely, a stable manifold, corresponding to a stable node, is found at kinetic scales, suggesting the occurrence of a saddle-node bifurcation passing from MHD down to kinetic scales. These results can open the way to new perspectives in approaching scale-to-scale dynamics of solar wind magnetic field fluctuations as well as in deriving conceptual models to explain the observed dynamical regimes.

**Author Contributions:** Conceptualization, T.A., G.C. and V.C.; Data curation, E.Y.; Investigation, G.C. and V.C.; Methodology, T.A. and G.C.; Writing—original draft, T.A.; Writing—review & editing, T.A., G.C., V.C., E.Y., M.F.M. and P.D.M.

**Funding:** This research received no external funding.

**Acknowledgments:** We acknowledge the Cluster FGM and STAFF P.I.s and teams and the ESA-Cluster Active Archive team for making available the data used in this work.

**Conflicts of Interest:** The authors declare no conflict of interest.

## Abbreviations

The following abbreviations are used in this manuscript:

| | |
|---|---|
| DS | Degree of Stationarity |
| EMD | Empirical Mode Decomposition |
| EMD-DAMF | Empirical Mode Decomposition-Dominant Amplitude Multifractal Formalism |
| ESA | European Space Agency |
| FGM | Fluxgate Magnetometer |
| GSE | Geocentric Solar Ecliptic |
| HSA | Hilbert Spectral Analysis |
| HT | Hilbert Transform |
| IMF | Intrinsic Mode Function |
| KAW | Kinetic Alfvén Wave |
| MHD | Magnetohydrodynamics |
| STAFF | Spatio Temporal Analysis of Field Fluctuations |
| WTMM | Wavelet Transform Modulus Maxima |

## References

1. Bruno, R.; Carbone, V. Turbulence in the solar wind. In *Lecture Notes in Physics*; Springer: Heidelberg, Germany, 2016; pp. 267, ISBN 978-3-319-43439-1.
2. Matthaeus, W.H.; Goldstein, M.L. Measurement of the rugged invariants of magnetohydrodynamic turbulence in the solar wind. *J. Geophys. Res.* **1982**, *87*, 6011–6028. [CrossRef]
3. Marsch, E. Turbulence in the solar wind. In *Reviews in Modern Astronomy*; Klare, G., Ed.; Springer: Berlin, Germany, 1990; pp. 145–156, ISBN 978-3-642-76750-0.
4. Petrosyan, A.; Balogh, A.; Goldstein, M.L.; Léorat, J.; Marsch, E.; Petrovay, K.; Roberts, B.; von Steiger, R.; Vial, J.C. Turbulence in the solar atmosphere and solar wind. *Space Sci. Rev.* **2010**, *156*, 135–238. [CrossRef]
5. Kolmogorov, A.N. The local structure of turbulence in incompressible viscous fluid for very large Reynolds numbers. *Dokl. Akad. Nauk SSSR* **1941**, *30*, 301–305. [CrossRef]
6. Obukhov, A.M. On the distribution of energy in the spectrum of turbulent flow. *Dokl. Akad. Nauk SSSR* **1941**, *32*, 22–24.
7. Dobrowlny, M.; Mangeney, A.; Veltri, P. Fully developed anisotropic hydromagnetic turbulence in interplanetary plasma. *Phys. Rev. Lett.* **1980**, *45*, 144–147. [CrossRef]
8. Tu, C.-Y.; Marsch, E. Evidence for a "background" spectrum of solar wind turbulence in the inner heliosphere. *J. Geophys. Res.* **1990**, *95*, 4337–4341. [CrossRef]
9. Iroshnikov, P.S. Turbulence of a conducting fluid in a strong magnetic field. *Sov. Astron.* **1964**, *7*, 556–571.
10. Kraichnan, R.H. Intertial range spectrum of hydromagnetic turbulence. *Phys. Fluids* **1965**, *8*, 1385–1387. [CrossRef]
11. Pope, S.B. *Turbulent Flows*; Cambridge University Press: Cambridge, UK, 2000; p. 771, ISBN 9780511840531.
12. Kolmogorov, A.N. A refinement of previous hypotheses concerning the local structure of turbulence in a viscous incompressible fluid at high Reynolds number. *J. Fluid Mech.* **1962**, *13*, 82–85. [CrossRef]
13. Mandelbrot, B.B. Intermittent turbulence in self-similar cascades: Divergence of high moments and dimension of the carrier. *J. Fluid Mech.* **1974**, *62*, 331–358. [CrossRef]
14. Marsch, E.; Tu, C.-Y. Intermittency, non-Gaussian statistics and fractal scaling of MHD fluctuations in the solar wind. *Nonlin. Process. Geophys.* **1997**, *4*, 101–124. [CrossRef]
15. Menter, F.R. Two-equation eddy-viscosity turbulence models for engineering applications. *AIAA J.* **1994**, *32*, 1598–1605. [CrossRef]
16. Speziale, C.G.; Sarkar, S.; Gatski, T. B. Modelling the pressure-strain correlation of turbulence: An invariant dynamical systems approach. *J. Fluid Mech.* **1991**, *227*, 245–272. [CrossRef]
17. Mishra, A.A.; Girimaji, S.S. Toward approximating non-local dynamics in single-point pressure-strain correlation closures. *J. Fluid Mech.* **2017**, *811*, 168–188. [CrossRef]
18. Sagaut, P. *Large Eddy Simulation for Incompressible Flows: An Introduction*; Springer Science Business Media: Berlin, Germany, 2006; pp. 493, ISBN 978-3-540-26403-3.

19. Valentini, F.; Califano, F.; Veltri, P. Two-dimensional kinetic turbulence in the solar wind. *Phys. Rev. Lett.* **2010**, *104*, 205002. [CrossRef]
20. Servidio, S.; Carbone, V.; Primavera, L.; Veltri, P.; Stasiewicz, K. Compressible turbulence in hall magnetohydrodynamics. *Planet. Space Sci.* **2007**, *55*, 2239–2243. [CrossRef]
21. Carbone, V.; Veltri, P. A shell model for anisotropic magnetohydrodynamic turbulence. *Geophys. Astrophys. Fluid Dyn.* **1990**, *52*, 153–181. [CrossRef]
22. Yaglom, A.M. On the local structure of the temperature field in a turbulent flow. *Dokl. Akad. Nauk SSSR* **1949**, *69*, 743–746.
23. Gogoberidze, G.; Perri, S.; Carbone, V. The Yaglom law in the expanding solar wind. *Astrophys. J.* **2013**, *769*, 111. [CrossRef]
24. Meneveau, C.; Sreenivasan, K.R.V. Simple multifractal cascade model for fully developed turbulence. *Phys. Rev. Lett.* **1987**, *59*, 1424–1427. [CrossRef]
25. Burlaga, L.F. Multifractal structure of the interplanetary magnetic field: Voyager 2 observations near 25 AU, 1987–1988. *Geophys. Res. Lett.* **1991**, *18*, 69–72. [CrossRef]
26. Carbone, V. Cascade model for intermittency in fully developed magnetohydrodynamic turbulence. *Phys. Rev. Lett.* **1993**, *71*, 1546–1548. [CrossRef] [PubMed]
27. Marsch, E.; Liu, S. Structure functions and intermittency of velocity fluctuations in the inner solar wind. *Ann. Geophys.* **1993**, *11*, 227–238.
28. Frisch, U. *Turbulence. The Legacy of A. N. Kolmogorov*; Cambridge University Press: Cambridge, UK, 1995; p. 296, ISBN 0-521-45713-0.
29. Richardson, L.F. *Weather Prediction by Numerical Process*; Cambridge University Press: Cambridge, UK, 2007; pp. 250, ISBN 978-3798510746.
30. Saharoui, F.; Goldstein, M.L.; Robert, P.; Khotyaintsev, Y.V. Evidence of a cascade and dissipation of solar-wind turbulence at the electron gyroscale. *Phys. Rev. Lett.* **2009**, *102*, 231102. [CrossRef]
31. Alexandrova, O.; Lacombe, C.; Mangeney, A.; Grappin, R.; Maksimovic, M. Solar wind turbulent spectrum at plasma kinetic scales. *Astrophys. J.* **2012**, *760*, 121. [CrossRef]
32. Saharoui, F.; Huang, S.Y.; Belmont, G.; Goldstein, M.L.; Retinò, A.; Robert, P.; De Patoul, J. Scaling of the electron dissipation range of solar wind turbulence. *Astrophys. J.* **2013**, *777*, 15. [CrossRef]
33. Marsch, E. Kinetic physics of the solar corona and solar wind. *Living Rev. Sol. Phys.* **2006**, *3*, 1. [CrossRef]
34. Schekochihin, A.A.; Cowley, S.C.; Dorland, W.; Hammet, G.W.; Howes, G.G.; Quataert, E.; Tatsuno, T. Astrophysical gyrokinetics: Kinetic and fluid turbulent cascades in magnetized weakly collisional plasmas. *Astrophys. J. Suppl.* **2009**, *182*, 310–377. [CrossRef]
35. Narita, Y. Space-time structure and wavevector anisotropy in space plasma turbulence. *Living Rev. Sol. Phys.* **2018**, *15*, 2. [CrossRef] [PubMed]
36. Sorriso-Valvo, L.; Carbone, V.; Veltri, P.; Consolini, G.; Bruno, R. Intermittency in the solar wind turbulence through probability distribution functions of fluctuations. *Geophys. Res. Lett.* **1999**, *26*, 1801–1804. [CrossRef]
37. Bruno, R.; Carbone, V.; Veltri, P.; Pietropaolo, E.; Bavassano, B. Identifying intermittency events in the solar wind. *Planet. Space Sci.* **2001**, *49*, 1201–1210. [CrossRef]
38. Matthaeus, W.H.; Wan, M.; Servidio, S.; Greco, A.; Osman, K.T.; Oughton, S.; Dmitruk, P. Intermittency, nonlinear dynamics and dissipation in the solar wind and astrophysical plasmas. *Phil. Trans. Ser. A* **2015**, *373*, 20140154. [CrossRef]
39. Carbone, V. Scaling exponents of the velocity structure functions in the interplanetary medium. *Ann. Geophys.* **1994**, *12*, 585. [CrossRef]
40. Carbone, V.; Veltri, P.; Bruno, R. Experimental evidence for differences in the extended self-similarity scaling laws between fluid and magnetohydrodynamic turbulent flows. *Phys. Rev. Lett.* **1995**, *75*, 3110–3113. [CrossRef]
41. Politano, H.; Pouquet, A.; Carbone, V. Determination of anomalous exponents of structure functions in two-dimensional magnetohydrodynamic turbulence. *Europhys. Lett.* **1998**, *43*, 516. [CrossRef]
42. Carbone, V.; Marino, R.; Sorriso-Valvo, L.; Noullez, A.; Bruno, R. Scaling laws of turbulence and heating of fast solar wind: The role of density fluctuations. *Phys. Rev. Lett.* **2009**, *103*, 061102. [CrossRef]
43. Huang, N.E.; Shen, Z.; Long, S.R.; Wu, M.C.; Shih, H.H.; Zheng, Q.; Yen, N.; Tung, C.C.; Liu, H.H. The empirical mode decomposition and the Hilbert spectrum for nonlinear and non-stationary time series analysis. *Proc. R. Soc. Lon. Ser. A* **1998**, *454*, 903–995. [CrossRef]

44. Welter, G.S.; Esquef, P.A.A. Multifractal analysis based on amplitude extrema of intrinsic mode functions. *Phys. Rev. E* **2013**, *87*, 032916. [CrossRef]
45. Chatfield, C. *The Analysis of Time Series: An Introduction*; Chapman and Hall/CRC: London, UK, 2016; p. 352, ISBN 9781584883173.
46. Alberti, T. Multivariate empirical mode decomposition analysis of swarm data. *Il Nuovo Cimento* **2018**, *41*, 113.
47. Rilling, G.; Flandring, P.; Goncalves, P. On empirical mode decomposition and its algorithms. In Proceedings of the IEEE-EURASIP Workshop on Nonlinear Signal and Image Processing NSIP-03, Grado, Italy, 8–11 June 2003.
48. Flandring, P.; Rilling, G.; Goncalves, P. Empirical mode decomposition as a filter bank. *IEEE Signal Process. Lett.* **2004**, *11*, 2.
49. Alberti, T.; Consolini, G.; De Michelis, P.; Laurenza, M.; Marcucci, M.F. On fast and slow Earth's magnetospheric dynamics during geomagnetic storms: A stochastic Langevin approach. *J. Space Weather Space Clim.* **2018**, *8*, A56. [CrossRef]
50. Alberti, T.; Consolini, G.; Lepreti, F.; Laurenza, M.; Vecchio, A.; Carbone, V. Timescale separation in the solar wind-magnetosphere coupling during St. Patrick's Day storms in 2013 and 2015. *J. Geophys. Res.* **2017**, *122*, 4266–4283. [CrossRef]
51. Vecchio, A.; Lepreti, F.; Laurenza, M.; Alberti, T.; Carbone, V. Connection between solar activity cycles and grand minima generation. *Astron. Astrophys.* **2017**, *599*, A058. [CrossRef]
52. Consolini, G.; Alberti, T.; Yordanova, E.; Marcucci, M.F.; Echim, M. A Hilbert-Huang transform approach to space plasma turbulence at kinetic scales. *J. Phys. Conf. Ser.* **2017**, *900*, 012003. [CrossRef]
53. Carbone, F.; Sorriso-Valvo, L.; Alberti, T.; Lepreti, F.; Chen, C.H.K.; Nemecek, Z.; Safránková, J. Arbitrary-order Hilbert Spectral Analysis and intermittency in solar wind density fluctuations. *Astrophys. J.* **2018**, *859*, 27. [CrossRef]
54. Kiyani, K.H.; Osman, K.T.; Chapman, S.C. Dissipation and heating in solar wind turbulence: From the macro to the micro and back again. *Phil. Trans. R. Soc. A* **2015**, *373*, 20140155. [CrossRef]
55. Huang, Y.X.; Schmitt, F.G.; Lu, Z.M.; Liu, Y.L. An amplitude-frequency study of turbulent scaling intermittency using Empirical Mode Decomposition and Hilbert Spectral Analysis. *Europhys. Lett.* **2008**, *84*, 40010. [CrossRef]
56. Takens, F. Detecting strange attractors in turbulence. In *Lecture Notes in Mathematics*; Rand, D.A., Young, L.-S., Eds.; Springer: Berlin, Germany, 1981; pp. 336–381.
57. Grassberger, P.; Procaccia, I. Characterization of strange attractors. *Phys. Rev. Lett.* **1983**, *50*, 346–349. [CrossRef]
58. Consolini, G.; Alberti, T.; De Michelis, P. On the forecast horizon of magnetospheric dynamics: A scale-to-scale approach. *J. Geophys. Res.* **2018**, *123*, 9065–9077. [CrossRef]
59. Alberti, T.; Lepreti, F.; Vecchio, A.; Bevacqua, E.; Capparelli, V.; Carbone, V. Natural periodicities and northern hemisphere-southern hemisphere connection of fast temperature changes during the last glacial period: EPICA and NGRIP revisited. *Clim. Past* **2014**, *10*, 1751–1762. [CrossRef]
60. Alberti, T.; Piersanti, M.; Vecchio, A.; De Michelis, P.; Lepreti, F.; Carbone, V.; Primavera, L. Identification of the different magnetic field contributions during a geomagnetic storm in magnetospheric and ground observations. *Annal. Geophys.* **2016**, *34*, 1069–1084. [CrossRef]
61. Piersanti, M.; Alberti, T.; Bemporad, A.; Berrilli, F.; Bruno, R.; Capparelli, V.; Carbone, V.; Cesaroni, C.; Consolini, G.; Cristaldi, A.; et al. Comprehensive analysis of the geoeffective solar event of 21 June 2015: Effects on the magnetosphere, plasmasphere, and ionosphere systems. *Sol. Phys.* **2017**, *292*, 169. [CrossRef]

*entropy*

MDPI

*Article*

# About Universality and Thermodynamics of Turbulence

**Damien Geneste [1], Hugues Faller [1,*], Florian Nguyen [2], Vishwanath Shukla [1], Jean-Philippe Laval [2], Francois Daviaud [1], Ewe-Wei Saw [1,3] and Bérengère Dubrulle [1]**

1. SPEC, CEA, CNRS, Université Paris-Saclay, CEA Saclay, 91191 Gif-sur-Yvette, France; genestedam@gmail.com (D.G.); research.vishwanath@gmail.com (V.S.); Francois.daviaud@cea.fr (F.D.); ewsaw3@gmail.com (E.-W.S.); berengere.dubrulle@cea.fr (B.D.)
2. CNRS, ONERA, Arts et Metiers ParisTech, University of Lille, Centrale Lille, FRE 2017-LMFL-Laboratoire de Mécanique des Fluides de Lille—Kampé de Fériet, F-59000 Lille, France; Florian.Nguyen@univ-lille1.fr (F.N.); jean-philippe.laval@univ-lille1.fr (J.-P.L.)
3. School of Atmospheric Sciences, Sun Yat-sen University, Guangzhou 510275, China

* Correspondence: hugues.faller@normalesup.org; Tel.: +33-169-083-015

Received: 26 February 2019; Accepted: 20 March 2019; Published: 26 March 2019

**Abstract:** This paper investigates the universality of the Eulerian velocity structure functions using velocity fields obtained from the stereoscopic particle image velocimetry (SPIV) technique in experiments and direct numerical simulations (DNS) of the Navier-Stokes equations. It shows that the numerical and experimental velocity structure functions up to order 9 follow a log-universality (Castaing et al. *Phys. D Nonlinear Phenom.* 1993); this leads to a collapse on a universal curve, when units including a logarithmic dependence on the Reynolds number are used. This paper then investigates the meaning and consequences of such log-universality, and shows that it is connected with the properties of a "multifractal free energy", based on an analogy between multifractal and thermodynamics. It shows that in such a framework, the existence of a fluctuating dissipation scale is associated with a phase transition describing the relaminarisation of rough velocity fields with different Hölder exponents. Such a phase transition has been already observed using the Lagrangian velocity structure functions, but was so far believed to be out of reach for the Eulerian data.

**Keywords:** turbulence; intermittency; multifractal; thermodynamics

## 1. Introduction

A well-known feature of any turbulent flow is the Kolmogorov-Richardson cascade by which energy is transferred from large to small length scales until the Kolmogorov length scale below which it is removed by viscous dissipation. This energy cascade is a non-linear and an out-of-equilibrium universal process. Moreover, the corresponding non-dimensional energy spectrum $E(k)/\epsilon^{2/3}\eta^{5/3}$ is an universal function of $k\eta$, where $\eta = (\nu^3/\epsilon)^{1/4}$ is the Kolmogorov length scale, $\epsilon$ the mean energy dissipation rate per unit mass, and $\nu$ the kinematic viscosity. Every used quantity is identified with its definition in a nomenclature available in Table 1. However, there seems to be little dependences on the Reynolds number, boundary, isotropy or homogeneity conditions [1]. In facts, the energy spectrum is based upon a quantity, the velocity correlation that is quadratic in velocity. Nevertheless, it is now well admitted that the universality does not carry over for statistical quantities that involve higher order moments. For example, the velocity structure functions of order $p$, given by $S_p(\ell) = \langle \|\mathbf{u}(\mathbf{x}+\mathbf{r}) - \mathbf{u}(\mathbf{x})\|^p \rangle_{\mathbf{x},\|\mathbf{r}\|=\ell}$ are not universal, at least when expressed in units of the Komogorov scale $\eta$ and velocity $u_K = (\nu\epsilon)^{1/4}$ (see below, Section 3.2 for an illustration).

**Table 1.** Nomenclature.

| Symbol | Mathematical Definition | Interpretation |
|---|---|---|
| $\mathbf{u}(\mathbf{x},t)$ | $\in \mathbb{R}^3 \times \mathbb{R} \to \mathbb{R}^3$ | Velocity field |
| $k$ | $\in \mathbb{R}_+$ | Wavenumber |
| $E(k)$ | $\mathrm{FT}\Big(\langle u_i(\mathbf{x}+\mathbf{r},t)u_i(\mathbf{x},t)\rangle_{\mathbf{x},\Vert\mathbf{r}\Vert=\ell,t}\Big)$ | Energy spectrum |
| $k_f$ | $\in \mathbb{R}_+^*$ | Forcing wavenumber |
| $N_x$ | $\in \mathbb{N}$ | Grid size in direction $x$ |
| $\nu$ | $\in \mathbb{R}_+^*$ | Kinematic viscosity |
| $\epsilon$ | $\in \mathbb{R}_+^*$ | Mean dissipation power per unit mass |
| $\eta$ | $\left(\frac{\nu^3}{\epsilon}\right)^{\frac{1}{4}}$ | Kolmogorov scale |
| $u_K$ | $(\nu\epsilon)^{\frac{1}{4}}$ | Kolmogorov velocity |
| $u_0$ | $\in \mathbb{R}_+^*$ | Characteristic velocity |
| $L_0$ | $\in \mathbb{R}_+^*$ | Characteristic length |
| Re | $\frac{u_0 L_0}{\nu}$ | Reynolds number |
| $\lambda$ | $\sqrt{\frac{\langle \mathbf{u}^2\rangle_{\mathbf{x},t}}{\langle \nabla\mathbf{u}^2\rangle_{\mathbf{x},t}}}$ | Taylor length |
| $u^{\mathrm{rms}}$ | $\sqrt{\langle \mathbf{u}^2\rangle_{\mathbf{x},t} - \langle \mathbf{u}\rangle^2_{\mathbf{x},t}}$ | Root mean squared velocity |
| $R_\lambda$ | $\frac{\lambda u^{\mathrm{rms}}}{\nu}$ | Taylor Reynolds number |
| $\Delta x$ | $\in \mathbb{R}_+^*$ | SPIV spatial resolution |
| $p$ | $\in [1,9]$ | Power |
| $\ell$ | $\in \mathbb{R}_+^*$ | Scale |
| $L$ | $\in \mathbb{R}_+^*$ | Inertial large scale |
| $\delta_\ell u(\mathbf{x},t)$ | $\langle \Vert\mathbf{u}(\mathbf{x}+\mathbf{r},t)-\mathbf{u}(\mathbf{x},t)\Vert\rangle_{\Vert\mathbf{r}\Vert=\ell}$ | Velocity increment at scale $\ell$ |
| $\Phi(\mathbf{x})$ | $\exp(-\Vert\mathbf{x}\Vert^2/2)/(2\pi)^{\frac{3}{2}}$ | Wavelet filter |
| $\Phi_\ell(\mathbf{x})$ | $\ell^{-3}\Phi(\mathbf{x}/\ell)$ | Wavelet filter at scale $\ell$ |
| $G_{ij}(\mathbf{x},\ell,t)$ | $\int \nabla_j\Phi_\ell(\mathbf{r})\, u_i(\mathbf{x}+\mathbf{r},t)\mathrm{d}\mathbf{r}$ | Wavelet transform of $\nabla\mathbf{u}$ |
| $\delta W(\mathbf{x},\ell,t)$ | $\ell \max_{ij} \lvert G_{ij}(\mathbf{x},\ell,t)\rvert .$ | Wavelet velocity increment |
| $S_p(\ell)$ | $\begin{cases} \langle(\delta_\ell u)^p\rangle_{\mathbf{x},t} & \text{In theory} \\ \langle(\delta W(\mathbf{x},\ell,t))^p\rangle_{\mathbf{x},t} & \text{For data analysis}\end{cases}$ | Velocity structure function |
| $\tilde{S}_p(\ell)$ | $\frac{S_p}{S_3^{p/3}}$ | Relative structure function |
| $h(\mathbf{x},t)$ | $\in \mathbb{R}^3\times\mathbb{R}\to[-1,1]$ | Local Hölder exponent |
| $C(h)$ | $\mathbb{P}\left(\log(\lvert\delta_\ell u\rvert/u_0) = h\log(\ell/L_0)\right) \sim (\ell/L_0)^{C(h)}$ | Multifractal Spectrum |
| $\eta_h$ | $L_0\mathrm{Re}^{-\frac{1}{1+h}}$ | Multifractal regularization scale |
| $\kappa$ | $\in \mathbb{R}_+^*$ | Intermittency parameter |
| $\tau(p)$ | $\kappa p(3-p)$ | Lognormal Intermittency correction |
| $\zeta(p)$ | $\frac{p}{3}+\tau(p)$ | Scaling exponent |
| $\theta(\ell)$ | $\frac{\log(L/\ell)}{\log(\mathrm{Re})}$ | Rescaled length |
| $\tau(p,\theta)$ | $\begin{cases} \tau(p) & \text{if} \quad \theta \le \frac{1}{1+h_{\max}} \\ p(\theta-\frac{1}{3})+C(-1+\frac{1}{\theta}) & \text{if} \quad \frac{1}{1+h_{\max}} \le \theta \le \frac{1}{1+h_{\min}}\end{cases}$ | General intermittency correction |
| $\tau(p,\ell)$ | $\tau(p,\theta(\ell))$ | General intermittency correction |
| $\gamma(\mathrm{Re})$,$\beta(\mathrm{Re})$ | $\mathbb{R}_+\to\mathbb{R}$ | Fitting functions |
| $G$ | $\mathbb{R}^2\to\mathbb{R}$ | General function from Castaing [2] |
| $A_p$, $K_0$ | $\gamma(\mathrm{Re})\log\left(\frac{S_p}{A_p u_K^p}\right) = G\left(p,\gamma(\mathrm{Re})\log(\ell K_0/\eta)\right)$ | Universal parameters |
| $H$ | $\mathbb{R}^2\to\mathbb{R}$ | New general function |
| $S_{0p}$ | $\beta(\mathrm{Re})\left(\frac{\log(\tilde{S}_p/S_{0p})}{\log(L_0/\eta)}\right) = H\left(p,\beta(\mathrm{Re})\frac{\log(\ell/\eta)}{\log(L_0/\eta)}\right)$ | Universal parameter |
| $a$, $b$ | $C(h) = \frac{(h-a)^2}{2b}$ | Parabolic fit |
| $\beta_0$ | $1/\beta(R_\lambda) \sim \beta_0/\log(R_\lambda)$ | Parameter |
| $\tau_{p,\mathrm{univ}}$ | $\frac{\tau(p,\ell)}{\log(\ell/L)}$ for $\ell$ in Inertial range | Intermittency correction from general rescaling |
| $\mu_\ell(\mathbf{x})$ | $\frac{\delta W(\mathbf{x},\ell)^3}{<\delta W(\mathbf{y},\ell)^3>_{\mathbf{y}}}$ | Spatial scale dependent measure |
| $S(E)$ | $\mathbb{P}\left[\log(\mu_\ell) = E\log(\ell/\eta)\right] \sim e^{\log(\ell/\eta)S(E)}$ | Large deviation function of $\log(\mu_\ell)$ |
| $k_B$ | $\in \mathbb{R}_+^*$ | Boltzmann constant |
| $T$ | $1/k_B p$ | Temperature |
| $E$ | $\log(\mu_\ell)$ | Energy |
| $N$ | $\log(\mathrm{Re})$ | Number of degrees of freedom |
| $V$ | $\log(\ell/\eta)$ | Volume |
| $P$ | $\tau(p,\ell)$ | Pressure |
| $F$ | $\log(\tilde{S}_{3p})$ | Free energy |

The mechanism behind this universality breakage is identified in [3], where a generalization of the Kolmogorov theory is introduced, based on the hypothesis that a turbulent flow is multifractal. In this model, the velocity field is locally characterized by a Hölder exponent $h$, such that $\delta_\ell u(\mathbf{x}) \equiv \langle \|\mathbf{u}(\mathbf{x}+\mathbf{r}) - \mathbf{u}(\mathbf{x})\| \rangle_{\|\mathbf{r}\|=\ell} \sim \ell^{h(\mathbf{x})}$; here $h$ is a stochastic function that follows a large deviation property [4] $\mathbb{P}\left(\log(|\delta_\ell u|/u_0) = h \log(\ell/L_0)\right) \sim (\ell/L_0)^{C(h)}$, where $u_0$ (resp. $L_0$) is the characteristic integral velocity (resp. length), and $C(h)$ is the multifractal spectrum. Velocity fields with $h < 1$ are rough in the limit $\ell \to 0$. Indeed they are at least not differentiable. In real flows, any rough field with $h > -1$ can be regularized at sufficiently small scale (the "viscous scale") by viscosity. The first computation of such dissipative scale was performed by Paladin and Vulpiani [5], who showed that it scales with viscosity like $\eta_h \propto \nu^{1/(1+h)}$, thereby generalizing the Kolmogorov scale, which corresponds to $h = 1/3$. Such a dissipative scale fluctuates in space and time (along with $h$), resulting in non-universality for high order moments, at least when expressed in units of $\eta$ and $u_K$.

A few years later, Frisch and Vergassola [6] claimed that the universality of the energy spectrum can be recovered, if the fluctuations of the dissipative length scale are taken into account by introducing a new non-dimensionalisation procedure. The new prediction was that $\log\left(E(k)\epsilon^{-\frac{2}{3}}\eta^{-\frac{5}{3}}\right) / \log(\mathrm{Re})$ should be a universal function of $\log(k\eta)/\log(\mathrm{Re})$, where Re is the Reynolds number. This claim was examined by Gagne et al., later using data from the Modane wind tunnel experiments [7]. They further suggested that the prediction can be extended to the velocity structure functions $S_p$, so that $\log(S_p(\ell)/u_K^p)/\log(\mathrm{Re})$ should be a universal function of $\log(\ell/\eta)/\log(\mathrm{Re})$, at any given $p$. They found good agreement for $p$ up to 6. The velocity measurements, in the above experiments, were performed using hot wire anemometry, which provide access to only one component of velocity. To our knowledge, no further attempts have been made to check the claim with more detailed measurements.

The purpose of the present paper is to reexamine this claim. However, now using the velocity fields obtained from the Stereoscopic Particle Image Velocimetry (SPIV) in experiments and the direct numerical simulations (DNS) of the Navier-Stokes equations (NSE). We show that the numerical and experimental velocity structure functions up to order 9 follow a log-universality [7]; they indeed collapse on a universal curve, if we use units that include $\log(\mathrm{Re})$ dependence. We then investigate the meaning and consequences of such a log-universality, and show that it is connected with the properties of a "multifractal free energy", based on an analogy between multifractal and thermodynamics (see [8] for summary). This framework uses co-existing velocity fields with different Hölder exponents which are regularized at variable scales. We show that in such a framework, this fluctuating dissipation length scale is associated with a phase transition describing the relaminarisation of velocity fields.

## 2. Experimental and Numerical Setup

### 2.1. Experimental Facilities and Parameters

We use experimental velocity field described in [9]. The radial, axial and azimuthal velocity are measured in a Von Kármán flow, using Stereoscopic Particle Image Velocimetry technique at different resolutions $\Delta x$. The Von Kármán flow is generated in a cylindrical tank of radius $R = 10$ cm through counter-rotation of two independent impellers with curved blades. The flow was maintained in a turbulent state at high Reynolds number by two independent impellers, rotating at various frequencies. Figure 1 shows the sketch of the experimental setup. The five experiments are performed in conditions so that the non-dimensional mean energy dissipation per unit mass is constant. The viscosity is monitored using mixture of water and glycerol, so as to vary the Kolmogorov length $\eta$. Table 2 summarizes the different parameters; $R_\lambda = \lambda u^{\mathrm{rms}}/\nu$ is the Reynolds number based on the Taylor length scale $\lambda = \sqrt{\frac{\langle \mathbf{u}^2 \rangle}{\langle \nabla \mathbf{u}^2 \rangle}}$, the root mean squared velocity $u^{\mathrm{rms}}$ and the kinematic viscosity $\nu$.

All velocity measurements are performed in a vertical plane that contains the rotation axis. The case (A) corresponds to measurements over the whole plane contained in between the two impellers, and extending from one side to the other side of the cylinder. Its resolution is 5 to 10 times coarser than similar measurements performed by zooming on a region centered around the symmetry point of the experiment (on the rotation axis, half way in between the two impellers), over a square window of size 4 cm × 3 cm. Since the flow is not homogeneous, statistics in this central region may differ from statistics computed over the whole tank. This explains the strong difference of $R_\lambda$ between (A) and (B,C). The little differences between (B) and (C) are explained by the different experimental resolutions used.

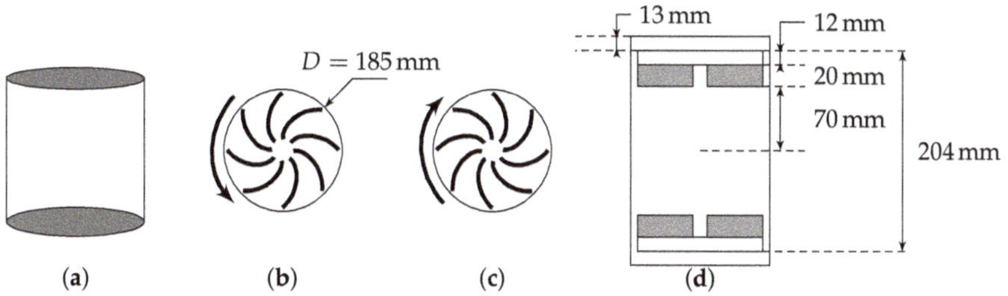


**Figure 1.** Von Kármán swirling flow generator. (**a**) normal view, bottom (**b**) and top (**c**) impellers rotating -both seen from the center of the cylinder, and (**d**) sketch with the relevant measures. A device not shown here maintains the temperature constant during the experiment. Both impellers are counter-rotating.

**Table 2.** Parameters for the 5 experiments realized (A, B, C, D andE). F is the rotation frequency of the discs, Re refers to the Reynolds number based on the diameter of the tank, $R_\lambda$ is the Reynolds based on the Taylor micro-scale. $\eta$ gives the estimated Kolmogorov length according to the experiment and $\Delta x$ refers to the spatial resolution of SPIV measurements. The second last column gives the number of frames over which are calculated the statistics. Except for (E), the Reynolds are much larger than those available with DNS. Table adapted from [10].

| Case | Frequency (Hz) | Glycerol Part | Re | $R_\lambda$ | $\eta$ (mm) | $\Delta x$ | Frames | Symbol |
|---|---|---|---|---|---|---|---|---|
| A | 5 | 0% | $3 \times 10^5$ | $1,9 \times 10^3$ | 0.02 | 2.4 | $3 \times 10^4$ | ○ |
| B | 5 | 0% | $3 \times 10^5$ | $2,7 \times 10^3$ | 0.02 | 0.48 | $3 \times 10^4$ | □ |
| C | 5 | 0% | $3 \times 10^5$ | $2,5 \times 10^3$ | 0.02 | 0.24 | $2 \times 10^4$ | ◇ |
| D | 1 | 0% | $4 \times 10^4$ | $9,2 \times 10^2$ | 0.08 | 0.48 | $1 \times 10^4$ | △ |
| E | 1.2 | 59% | $6 \times 10^3$ | $2,1 \times 10^2$ | 0.37 | 0.24 | $3 \times 10^4$ | ⋆ |

### *2.2. Direct Numerical Simulation*

The direct numerical simulations (DNS), based on pseudo-spectral methods, are performed in order to compare with our experimental data. The DNS runs with $R_\lambda = 25$, $R_\lambda = 80$, $R_\lambda = 90$ and $R_\lambda = 138$ are performed using the NSE solver VIKSHOBHA [10], whereas the run with $R_\lambda = 53$ is carried out using another independent pseudo-spectral NSE solver. The velocity field $\mathbf{u}$ is computed on a $2\pi$ triply-periodic box.

Turbulent flow in a statistically steady state is obtained by using the Taylor-Green type external forcing in the NSE at wavenumber $k_f = 1$ and amplitude $f_0 = 0.12$, the value of viscosity is varied in order to obtain different values of $R_\lambda$ (see Ref. [10] for more details).

## 3. Theoretical Background

### *3.1. Velocity Increments vs. Wavelet Transform (WT) of Velocity Gradients*

The classical theories of Kolmogorov [11,12] are based on the scaling properties of the velocity increment, defined as $\delta_\ell u(\mathbf{x}, t) = \langle ||\mathbf{u}(\mathbf{x}+\mathbf{r}, t) - \mathbf{u}(\mathbf{x}, t)|| \rangle_{||\mathbf{r}||=\ell}$ where $\ell = ||\mathbf{r}||$ is the distance over which the increment is taken. As pointed out by [8], a more natural tool to characterize the local scaling properties of the velocity field is the wavelet transform of the tensor $\partial_j u_i$, defined as:

$$G_{ij}(\mathbf{x},\ell,t) = \int \nabla_j \Phi_\ell(\mathbf{r})\, u_i(\mathbf{x}+\mathbf{r},t)\mathrm{d}\mathbf{r} \tag{1}$$

where $\Phi_\ell(\mathbf{x}) = \ell^{-3}\Phi(\mathbf{x}/\ell)$ is a smooth function, non-negative with unit integral. In what follows, we choose a Gaussian function $\Phi(\mathbf{x}) = \exp(-\|\mathbf{x}\|^2/2)/(2\pi)^{\frac{3}{2}}$ such that $\int \Phi(\mathbf{r})\mathrm{d}\mathbf{r} = 1$. We then compute the wavelet velocity increments as

$$\delta W(\mathbf{x},\ell,t) = \ell \max_{ij} |G_{ij}(\mathbf{x},\ell,t)| \tag{2}$$

This formulation is especially well suited for the analysis of the experimental velocity field, as it naturally allows to average out the noise. It has been verified that the wavelet-based approach yields the same values for the scaling exponents as those computed from the velocity increments [10].

### 3.2. K41 and K62 Universality

In the first theory of Kolmogorov [11], the turbulence properties depend only on two parameters: the mean energy dissipation per unit mass $\epsilon$ and the viscosity $\nu$ . The only velocity and length unit that one can build using these quantities are the Kolmogorov length $\eta = (\nu^3/\epsilon)^{1/4}$ and velocity $u_K = (\epsilon\nu)^{1/4}$. The structure functions are then self-similar in the inertial range $\eta \ll \ell \ll L_0$, where $L_0$ is the integral scale, and follow the universal scaling:

$$S_p(\ell) \equiv \langle(\delta_\ell u)^p\rangle \sim u_K^p \left(\frac{\ell}{\eta}\right)^{p/3} \tag{3}$$

which can also be recast into:

$$\tilde{S}_p(\ell) \equiv \frac{S_p(\ell)}{(S_3(\ell))^{p/3}} = C_p \tag{4}$$

where $C_p$ is a (non universal) constant.

This scaling is typical of a global scale symmetry solution, and was criticized by Landau, who considered it incompatible with observed large fluctuations of the local energy dissipation. Kolmogorov then built a second theory (K62), in which fluctuations of energy dissipation were assumed to follow a log-normal statistics, and taken into account via an intermittency exponent $\kappa$ and a new length scale $L$, thereby breaking the global scale invariance. The resulting velocity structure functions then follow the new scaling:

$$S_p(\ell) \sim (\epsilon\ell)^{p/3} \left(\frac{\ell}{L}\right)^{\kappa p(3-p)} \tag{5}$$

which implies a new kind of universality involving the relative structure functions $\tilde{S}_p$ as:

$$\tilde{S}_p(\ell) \equiv \frac{S_p(\ell)}{(S_3(\ell))^{p/3}} \sim A_p \left(\frac{\ell}{L}\right)^{\tau(p)} \tag{6}$$

where $\tau(p) = \kappa p(3-p)$ and $A_p$ is a constant. Such a formulation already predicts an interesting universality, if $L = L_0$, as we should have:

$$\left(\frac{L_0}{\eta}\right)^{\tau(p)} \tilde{S}_p(\ell) \sim A_p \left(\frac{\ell}{\eta}\right)^{\tau(p)} \tag{7}$$

Therefore, we should be able to collapse all structure functions, at different Reynolds number by plotting $(\frac{L_0}{\eta})^{\tau(p)}\tilde{S}_p$ as a function of $\frac{\ell}{\eta}$, given that $L_0/\eta \sim \mathrm{Re}^{3/4}$. There is however no clear prediction about the value of $L$ and we show in the data analysis (Section 4) that $L$ differs from $L_0$.

The relation (7) shows that $\log\left(\left(\frac{L_0}{\eta}\right)^{\tau(p)}\tilde{S}_p\right)$ is a linear function of $\log(\frac{\ell}{\eta})$. In principle, such universal scaling is not valid outside the inertial range, i.e., for example when $\ell < \eta$. To be more general than previously thought, it can however be shown using the multifractal formalism as first shown by [6].

### 3.3. *Multifractal and Fluctuating Dissipation Length*

For the multifractal (MFR) model, it is assumed that the turbulence is locally self-similar, so that there exists a scalar field $h(\mathbf{x}, \ell, t)$, such that

$$h(\mathbf{x}, t, \ell) = \frac{\log(\delta_\ell u(\mathbf{x}, t)/u_0)}{\log(\ell/L)} \tag{8}$$

for a range of scales in a suitable "inertial range" $\eta_h \ll \ell \ll L$, where $L$ is a large inertial scale, $\eta_h$ a cut-off length scale, and $u_0$ a characteristic large-scale velocity. This scale $\eta_h$ is a generalization of the Kolmogorov scale, and is defined as the scale where the local Reynolds number $\ell|\delta_\ell u|/\nu$ is equal to 1. Writing $\delta_\ell u = u_0(\ell/L)^h$ leads to the expression of $\eta_h$ as a function of the global Reynolds number $\mathrm{Re} = u_0 L/\nu$ as $\eta_h \sim L\mathrm{Re}^{-1/(1+h)}$. This scale thus appears as a fluctuating cut-off which depends on the scaling exponent and therefore on $x$. This is the generalization of the Kolmogorov scale $\eta \sim L\mathrm{Re}^{-3/4} \equiv \eta_{\frac{1}{3}}$, and was first proposed in [5]. Below $\eta_h$, the velocity field becomes laminar, and $\delta_\ell u \propto \ell$. When the velocity field is turbulent, $h \equiv \log(\delta_\ell u/u_0)/\log(\ell/L)$ varies stochastically as a function of space and time. Also, if the turbulence is statistically homogeneous, stationary and isotropic, $h$ only depends on $\ell$, the scale magnitude. Therefore, formally, $h$ can be regarded as a continuous stochastic process labeled by $\log(\ell/L)$. By Kramer's theorem [13], one sees that as in the limit $\ell \to 0$, $\log(L/\ell) \to \infty$, we have:

$$\mathbb{P}\left[\log(\delta_\ell u/u_0) = h\log(\ell/L)\right] \sim e^{\log(\ell/L)C(h)} = \left(\frac{\ell}{L}\right)^{C(h)} \tag{9}$$

where $C(h)$ is the rate function of $h$, also called multifractal spectrum. Formally, $C(h)$ can be interpreted as the co-dimension of the set where the local Hölder exponent at scale $\ell$ is equal to $h$. Using Gärtner-Elis theorem [13], one can connect C and the velocity structure functions as:

$$S_p(\ell) = \langle(\delta_\ell u)^p\rangle = \int_{h_{\min}}^{h_{\max}} u_0^p \left(\frac{\ell}{L}\right)^{ph+C(h)} dh \tag{10}$$

To proceed further and make connection with previous section, we set $\epsilon = u_0^3/L$ so that $S_p(\ell)$ can now be written:

$$S_p(\ell) = (\epsilon\ell)^{p/3}\int_{h_{\min}}^{h_{\max}} \left(\frac{\ell}{L}\right)^{p(h-1/3)+C(h)} dh \sim (\epsilon\ell)^{p/3}\left(\frac{\ell}{L}\right)^{\tau(p)} \tag{11}$$

This shows that $\tau(p)$ is the Legendre transform of the rate function $C(h+1/3)$, i.e., $\tau(p) = \min_h(p(h-1/3)+C(h))$, and equivalently, that $C(h)$ is the Legendre transform of $\tau(p)$. Because of this, it is necessarily convex. The set of points where $C(h) \leq 3$, represents the set of admissible or observable $h$, is therefore necessarily an interval, bounded by $-1 \leq h_{\min}$ and $h_{\max} \leq 1$.

As noted by [6], the scaling exponent $\zeta(p) = p/3 + \tau(p)$ defined via Equation (11) is only constant in a range of scale where $\ell > \eta_h$ for any $h \in [h_{\min}, h_{\max}]$. For small enough $\ell$, this condition is not met anymore, since as soon as $\ell < \eta_h$, all velocity fields corresponding to $h$ are "regularized", and do not contribute anymore to intermittency since they scale like $\ell$. This results in a slow dependence of $\zeta(p)$ with respect to the scale, which is obtained via the corrected formula:

$$S_p(\ell) = (\epsilon\ell)^{p/3} \int\limits_{\eta_h \leq \ell} \left(\frac{\ell}{L}\right)^{p(h-1/3)+C(h)} \mathrm{d}h \sim (\epsilon\ell)^{p/3} \left(\frac{\ell}{L}\right)^{\tau(p,\ell)} \tag{12}$$

To understand the nature of the correction, we can compute the value of $h$ such that $\ell = \eta_h$. This gives: $h(\ell) = -1 + \log(\mathrm{Re})/\log(L/\eta_h)$. With $\theta = \log(L/\ell)/\log(\mathrm{Re})$, Equation (12) can be rewritten as:

$$\tilde{S}_p(\ell) \equiv \frac{S_p(\ell)}{(S_3(\ell))^{p/3}} = \int\limits_{-1+1/\theta}^{h_{\max}} \left(\frac{\ell}{L}\right)^{p(h-1/3)+C(h)} dh \sim \exp\left(-\theta\tau(p,\theta)\log(\mathrm{Re})\right) \tag{13}$$

where $\tau(p,\theta) = \tau(p)$ when $\theta \leq 1/(1+h_{\max})$ and $\tau(p,\theta) = p(\theta - 1/3) + C(-1+1/\theta)$ when $1/(1+h_{\max}) \leq \theta \leq 1/(1+h_{\min})$. As discussed by [6], this implies a new form of universality that extends beyond the inertial range, into the so-called extended dissipative range, as;

$$\frac{\log(\tilde{S}_p)}{\log(\mathrm{Re})} = -\tau(p,\theta)\theta, \quad \theta = \log(L/\ell)/\log(\mathrm{Re}) \tag{14}$$

If the scale $L$ is constant and equal to $L_0$, the integral scale, then we have $\mathrm{Re} = (L_0/\eta)^{4/3}$ and the multifractal universality implies that $\log(\tilde{S}_p)/\log(L_0/\eta)$ is a function of $\log(\ell/\eta)/\log(L_0/\eta)$. When the function is linear, we thus recover the K62 universality. The multifractal universality is thus a *generalization* of the K62 universality.

This form of universality is however not easy to test, as the scale $L$ is not known a priori, and may still depend on Re. In what follows, we demonstrate a new form of universality that allows more freedom upon $L$ and encompass both K62 and multifractal universality.

### 3.4. General Universality

Using the hypothesis that turbulence maximizes some energy transfer in the scale space, Castaing [2] suggested a new form of universality for the structure functions, that reads:

$$\gamma(\mathrm{Re})\log\left(\frac{S_p(\ell)}{A_p u_K^p}\right) = G\left(p, \gamma(\mathrm{Re})\log(\ell K_0/\eta)\right) \tag{15}$$

where $A_p$ and $K_0$ are universal constants and $\beta$ and $G$ are general functions, $G$ being linear in the inertial range, $G(p,x) \sim \tau(p)x$. The validity of this universal scaling was checked by Gagne and Castaing [7] on data obtained from the velocity fields measured in a jet using hot wire anemometry. They found good collapse of the structure functions at different Taylor Reynolds $R_\lambda$, provided $\gamma(\mathrm{Re})$ is constant at low Reynolds numbers and follows a law of the type: $\gamma(\mathrm{Re}) \sim \gamma_0/\log(R_\lambda/R_*)$, where $R_*$ is a constant, whenever $R_\lambda > 400$. Since we have $R_\lambda \sim \mathrm{Re}^{1/2}$ and $(L_0/\eta) \sim \mathrm{Re}^{3/4}$, we can rewrite Equation (15) as:

$$\beta(\mathrm{Re})\left(\frac{\log(\tilde{S}_p(\ell)/S_{0p})}{\log(L_0/\eta)}\right) = H\left(p, \beta(\mathrm{Re})\frac{\log(\ell/\eta)}{\log(L_0/\eta)}\right) \tag{16}$$

where $S_{0p}$ are some constants and $\beta$ and $H$ are general functions. Compared to the K62 or MFR universality Formulas (7) or (14), we see that Formula (16) is a generalization of these two universality with $L = L_0$. It allows however more flexibility than K62 or MFR universality through the function $\beta(\mathrm{Re})$, which is a new fitting function. We test these predictions in Section 4 and provide a physical interpretation of (16) in Section 5.

## 4. Check of Universality Using Data Analysis

The various universality are tested using the velocity structure functions based on the wavelet velocity increments Equation (2), in order to minimize the noise in the experimental data. We define:

$$S_p(\ell) = \langle |\delta W(\mathbf{x}, \ell, t)|^p \rangle_{\mathbf{x},t} \tag{17}$$

We then apply this formula to both experimental data (Table 2) and numerical data (Table 3), to get wavelet velocity structure functions at various scales and Reynolds numbers.

**Table 3.** Parameters for the DNS. $R_\lambda$ is the Reynolds based on the Taylor micro-scale. $\eta$ is the Kolmogorov length. The third column gives resolution of the simulation through $k_{\max}\eta$, where $k_{\max} = N_x/3$ is the maximum wavenumber. The fourth column gives the grid size; notice that the dimensionless length of the box is $2\pi$. Here, $\ell_{\min}$ is the smallest scale available for the calculations of the wavelets. $k_f$ is the forcing wavenumber. The Sample columns gives the number of points (frames $\times$ grid size) over which the statistics are computed.

| $R_\lambda$ | $\eta$ | $k_{\max}\eta$ | $N_x \times N_y \times N_z$ | $\ell_{\min}/\eta$ | Samples | Symbol |
|---|---|---|---|---|---|---|
| 25 | 0.079 | 3.35 | $128^3$ | 0.635 | 5000 | ⋆ |
| 53 | 0.034 | 8.5 | $768^3$ | 0.31 | 105,000 | △ |
| 80 | 0.020 | 1.68 | $256^3$ | 1.22 | 270,000 | □ |
| 90 | 0.017 | 5.7 | $1024^3$ | 0.36 | 10,000 | ◇ |
| 138 | 0.009 | 1.55 | $512^3$ | 1.37 | 12,000 | ○ |

### 4.1. Check of K41 Universality

The K41 universality (3) can be checked by plotting:

$$\log\left(\frac{S_p}{u_K^p}\right) = \mathcal{F}\left(\log\left(\frac{\ell}{\eta}\right)\right) \tag{18}$$

This is shown in Figure 2 for both experimental and numerical data. Obviously, the data do not collapse on a universal curve, meaning that K41 universality does not hold. This is well known, and is connected to intermittency effects [14].

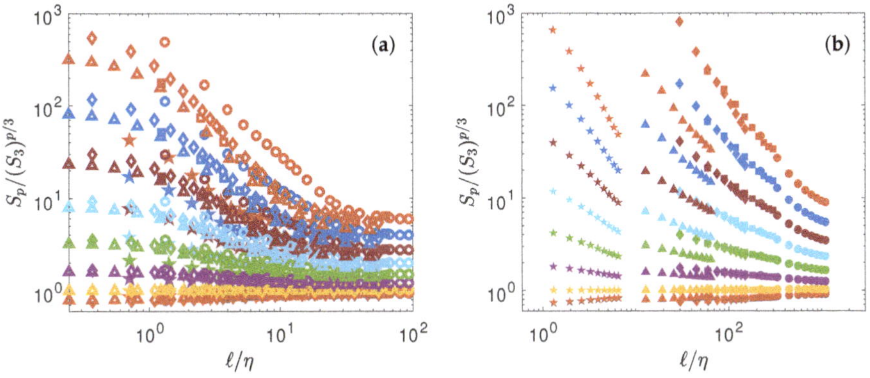


**Figure 2.** Test of K41 universality Equation (4). (**a**) Numerical data (**b**) Experimental data. The structure functions have been shifted by arbitrary factors for clarity and are coded by color: $p = 1$: blue symbols; $p = 2$: orange symbols; $p = 3$: yellow symbols; $p = 4$: magenta symbols; $p = 5$: green symbols; $p = 6$: light blue symbols; $p = 7$: red symbols; $p = 8$: blue symbols; $p = 9$: orange symbols. For K41 universality to hold, all the function should be constant, for a given $p$.

### 4.2. Check of K62 Universality

The K62 universality (7) can be checked by plotting:

$$\log\left[\left(\frac{L_0}{\eta}\right)^{\tau(p)}\tilde{S}_p\right] = \mathcal{F}\left(\log\left(\frac{\ell}{\eta}\right)\right) \quad (19)$$

The collapse depends directly on $\tau(p)$, the intermittency exponents. Obtaining the best collapse of all curves is in fact a way to fit the best scaling exponents $\tau(p)$. We thus implement a minimization algorithm that provides the values of $\tau(p)$ that minimized the distance between the curve and the line of slope $\tau(p)$. The values of $\tau(p)$ are reported in Table 4. The best collapse is shown on Figure 3a for the DNS, and Figure 3b for the experiment. The collapse is better for experiments than for the DNS. However, in both cases, there are significant differences in between points at different $R_\lambda$, at larger scales, showing that universality is not yet reached.

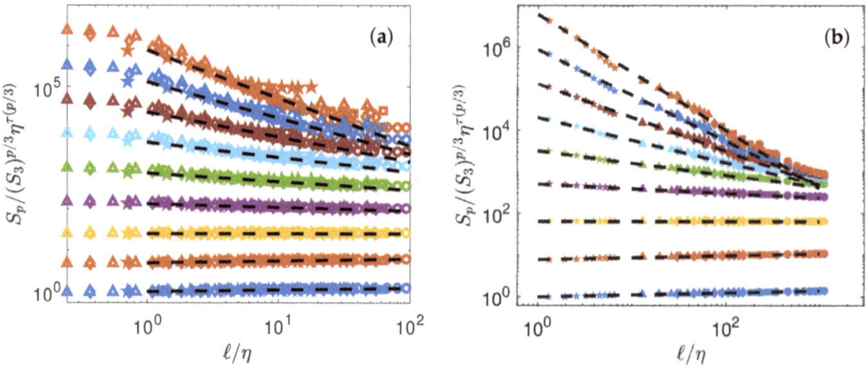


**Figure 3.** Test of K62 universality Equation (7). (**a**) Numerical data (**b**) Experimental data. The structure functions are shifted by arbitrary factors for clarity and are coded by color: $p = 1$: blue symbols; $p = 2$: orange symbols; $p = 3$: yellow symbols; $p = 4$: magenta symbols; $p = 5$: green symbols; $p = 6$: light blue symbols; $p = 7$: red symbols; $p = 8$: blue symbols; $p = 9$: orange symbols. The dashed lines are power laws with exponents $\tau(p) = \zeta(p) - \zeta(3)p/3$, with $\zeta(p)$ shown in Figure 4a.

**Table 4.** Scaling exponents $\tau(p)$ and $\zeta(p)$ found by the collapse method based on K62 universality for experimental data (subscript EXP) or numerical data (subscript DNS). The subscript SAW refers to the values obtained by [9]. The exponents $\tau_{\mathrm{EXP}}(p)$(red square) and $\tau_{\mathrm{DNS}}$ (blue circle) have been computed through a least square algorithm.

| Exponent\Order | $p=1$ | $p=2$ | $p=3$ | $p=4$ | $p=5$ | $p=6$ | $p=7$ | $p=8$ | $p=9$ |
|---|---|---|---|---|---|---|---|---|---|
| $\zeta_{\mathrm{SAW}}/\zeta_{\mathrm{SAW}}(3)$ | 0.36 | 0.69 | 1 | 1.29 | 1.55 | 1.78 | 1.98 | 2.17 | 2.33 |
| $\zeta_{\mathrm{DNS}}$ | 0.31 | 0.58 | 0.80 | 0.98 | 1.12 | 1.23 | 1.26 | 1.25 | 1.23 |
| $\zeta_{\mathrm{EXP}}$ | 0.32 | 0.58 | 0.80 | 0.98 | 1.12 | 1.23 | 1.32 | 1.39 | 1.44 |
| $\tau_{\mathrm{DNS}}$ | 0.04 | 0.05 | 0 | −0.09 | −0.21 | −0.37 | −0.61 | −0.88 | −1.17 |
| $\tau_{\mathrm{EXP}}$ | 0.05 | 0.05 | 0 | −0.09 | −0.21 | −0.36 | −0.54 | −0.74 | −0.96 |

### 4.3. Check of General Universality

We can now check the most general universality, by plotting:

$$\beta(\mathrm{Re})\left(\frac{\log(\tilde{S}_p/S_{0p})}{\log(L_0/\eta)}\right) = H\left(p, \beta(\mathrm{Re})\frac{\log(\ell/\eta)}{\log(L_0/\eta)}\right) \quad (20)$$

In this case, best collapse is obtained by fitting two families of parameters: $S_{0p}$, $\beta(\mathrm{Re})$ that are obtained through a procedure of minimization. We take the DNS at $R_\lambda = 138$ as the reference case, and find for both DNS and experiments, the values of $\beta(\mathrm{Re})$ and $S_{0p}$ that best collapse the curves. The corresponding collapses are provided in Figure 5. The collapses are good for any value of Re, except for the DNS at the lowest Reynolds number, which does not collapse in the far dissipative range.

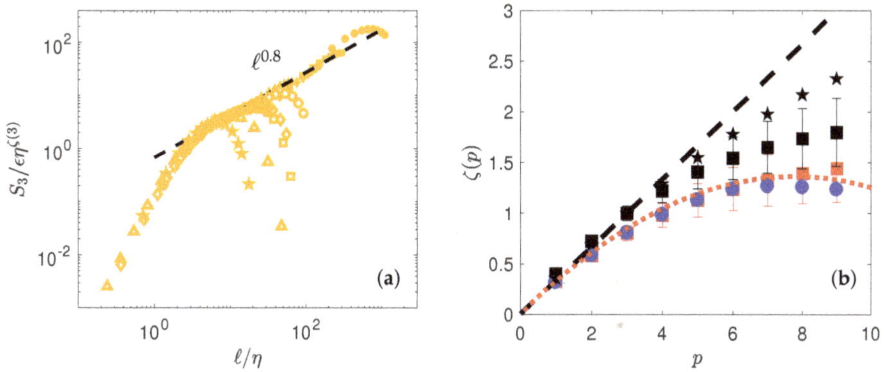


**Figure 4.** (**a**) Determination of $\zeta(3)$ by best collapse using both DNS (open symbols) and experiments (filled symbols). The black dashed line is $\ell^{0.8}$. (**b**) Scaling exponents $\zeta(p)$ of the wavelet structure functions of $\delta W$ as a function of the order, from Table 4, for DNS (blue circle) and experiments (red square) . The red dotted line is the function $\min_h(hp + C(h))$ with $C(h)$ given by $C(h) = (h-a)^2/2b$, with $a = 0.35$ and $b = 0.045$. The black stars correspond to $\zeta_{\mathrm{SAW}}(p)/\zeta_{\mathrm{SAW}}(3)$ (see Table 4), while the black squares correspond to $\zeta_{\mathrm{EXP}}(p)/\zeta_{\mathrm{EXP}}(3)$.

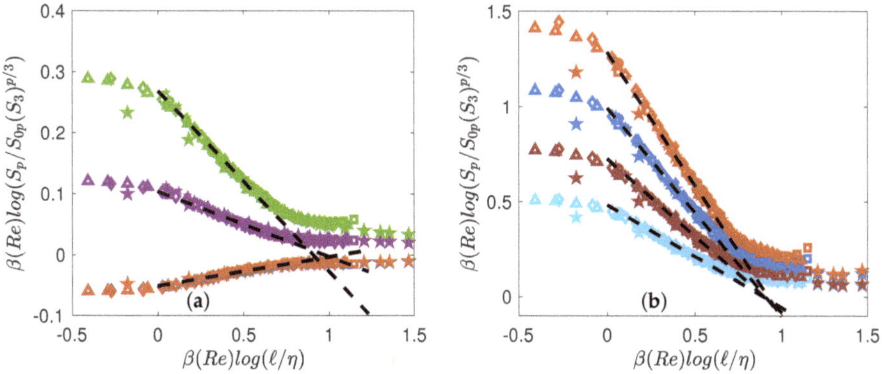


**Figure 5.** Test of general universality Equation (20) using both DNS (open symbols) and experiments (filled symbols). The functions are coded by color. (**a**) $p = 1$: blue symbols; $p = 2$: orange symbols; $p = 4$: magenta symbols; $p = 5$: green symbols; (**b**) $p = 6$: light blue symbols; $p = 7$: red symbols; $p = 8$: blue symbols; $p = 9$: orange symbols. The functions have been shifted by arbitrary factors for clarity. The dashed lines are power laws with exponents $\tau(p) = \zeta(p) - \zeta(3)p/3$, with $\zeta(p)$ shown in Figure 4a.

### 4.4. Function $\beta(Re)$

Motivated by earlier findings by [7], we plot in Figure 6 the value $1/\beta$ as a function of $R_\lambda$.

Our results are compatible with $1/\beta \sim \beta_0/\log(R_\lambda)$, with $\beta_0 \sim 4/3$ over the whole range of Reynolds number. For comparison, we provide also on Figure 6 the values found by Gagne and Castaing [7] in jet of liquid Helium, shifted by an arbitrary factor to make our values coincide with them at large Reynolds number. This shift is motivated by the fact that $\beta(\mathrm{Re})$ is determined up to a constant, depending upon the amplitude of the structure functions used as reference. At large Reynolds, our values are compatible with theirs. At low Reynolds, however, we do not observe the saturation of $1/\beta$ that is observed in the jet experiment of [7]. An interpretation of the meaning of $\beta(\mathrm{Re})$ is provided in Section 5.

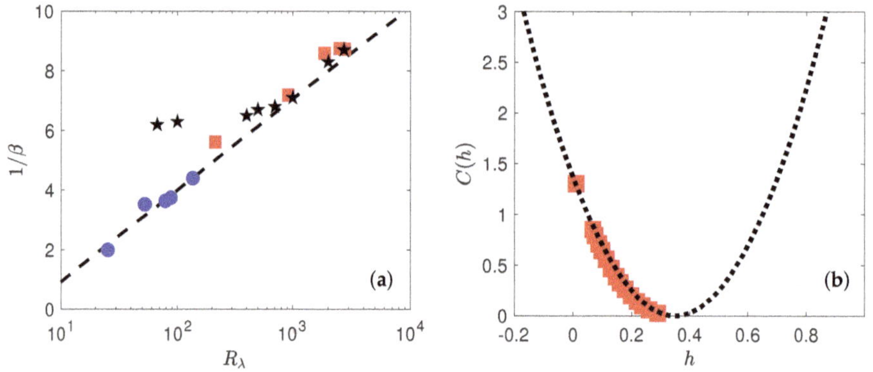


**Figure 6.** (**a**) Variation of $1/\beta(\mathrm{Re})$ versus $\log(R_\lambda)$ in experiments (red square) and DNS (blue circle) when using the DNS at $R_\lambda = 138$ as the reference case. Black stars correspond to the values found by Gagne and Castaing in [7] shifted by an arbitrary factor to coincide the values at large Reynolds. The black dashed line is $(4/3)\log(R_\lambda/5)$. (**b**) Multifractal spectrum $C(h)$ for the experiments. The spectrum is obtained by taking inverse Legendre transform of the scaling exponents $\zeta(p)$ shown in Figure 4. The dotted line is a parabolic fit $C(h) = (h-a)^2/2b$ with $a = 0.35$ and $b = 0.045$.

### 4.5. Scaling Exponents

Our Collapse method enables us to obtain the scaling exponents of the structure functions $\zeta(p)$ by the following two methods:

(i) Using the K62 universality, we get $\tau(p)$, and then $\zeta(p) = \zeta(3)p/3 + \tau(p)$. These estimates still depend on the value of $\zeta(3)$, which is not provided by the K62 universality plot. To obtain them, we use a minimization procedure on both experimental $\log(S_3/u_K^3)$ from the one hand, and the numerical $\log(S_3/u_K^3)$ on the other hand (see Figure 4a), to compute $\zeta(3)$ as the value that minimizes the distance between the curve and a straight line of slope $\zeta(3)$. The values so obtained are reported in Table 4, and are used to compute $\zeta(p)$ from $\tau(p)$. In Table 4, two different methods are used to process the experimental data. The subscript SAW refers to the values obtained by [9] on the same set of experimental data, using velocity increments and Extended Self-Similarity technique [15]. The quantities with subscript EXP are computed through a least square algorithm upon $\tau(p)$, minimizing the scatter of the rescaled structure functions $\log\left[\left(\frac{L_0}{\eta}\right)^{\tau(p)} \tilde{S}_p\right]$ with respect to the line $(\ell/\eta)^{\tau(p)}$. DNS data have been processed the same way as EXP.

(ii) Using the general universality, we may also get $\tau_{p,\mathrm{univ}}$ by a linear regression on the collapse curve. Please note that since the data are collapsed, this provides a very good estimates of this quantity, with the lowest possible noise. In practice, we observe no significant differences with the two estimates; therefore, we only report the values obtained by following the first method.

The corresponding values are plotted in Figure 4 and summarized in Table 4. Please note that for both DNS and experiments, the value of $\zeta(3)$ is different from 1, which is apparently incompatible with the famous Kolmogorov 4/5th law that predicts $\zeta(3) = 1$. This is because we use *absolute* values of wavelet increments, while the Kolmogorov 4/5th law uses signed values. We have checked that using unsigned values, we obtain a scaling that is closer to 1, but with larger noise. Note also that when we consider the relative value $\zeta(p)/\zeta(3)$, we obtain values that are close to the values obtained [9] on the same set of experimental data, using velocity increments and Extended Self-Similarity technique [15].

### 4.6. Multifractal Spectrum

From the values of $\zeta(p)$, one can get the multifractal spectrum $C(h)$ by performing the inverse Legendre transform:

$$C(h) = \min_p [ph + \zeta(p)] \tag{21}$$

Practically, this allows to use the following formula:

$$C\left(\frac{\mathrm{d}\,\zeta(p)}{\mathrm{d}\,p}\Big|_{p^*}\right) = \zeta(p^*) - p^* \frac{\mathrm{d}\,\zeta(p)}{\mathrm{d}\,p}\Big|_{p^*} \tag{22}$$

To estimate $C$, we thus first perform a polynomial interpolation of order 4 on $\zeta(p)$, then derivate the polynomial to estimate $\frac{\mathrm{d}\,\zeta(p)}{\mathrm{d}\,p}$, thus get $C$ through Equation (22). The result is provided in Figure 6b for both the DNS and the experiment.

The curve looks like a portion of parabola, corresponding to a log-normal statistics for the wavelet velocity increments. Specifically, fitting by the shape:

$$C(h) = \frac{(h-a)^2}{2b} \tag{23}$$

we get $a = 0.35$ and $b = 0.045$. This parabola also provides a good fit of the scaling exponents, as shown in Figure 4 by performing Legendre transform of $C(h)$ given by Equation (23).

## 5. Thermodynamics and Turbulence

### 5.1. Thermodynamical Analogy

Multifractal obeys a well-known thermodynamical analogy [8,16,17] that will be useful to interpret and extend the general universality unraveled in the previous section. Indeed, considering the quantity:

$$\mu_\ell = \frac{|\delta W_\ell|^3}{\langle |\delta W_\ell|^3 \rangle} \tag{24}$$

By definition $\mu_\ell$ is positive definite and $\langle \mu_\ell \rangle = 1$ for any $\ell$. It therefore can be interprated as a scale dependent measure. It then also follows a large-deviation property as:

$$\mathbb{P}\left[\log(\mu_\ell) = E \log(\ell/\eta)\right] \sim e^{\log(\ell/\eta) S(E)} \tag{25}$$

where $S(E)$ is the large deviation function of $\log(\mu_\ell)$ and has the meaning of an energy while $\log(\ell/\eta)$ has the meaning of a volume, and $\log(\mu_\ell)/\log(\ell/\eta)$ is an energy density. With the definition of $\mu_\ell$, it is easy to see that $S$ is connected to $C$, the large deviation function of $|\delta W_\ell|$. In fact, since in the inertial range where $\langle |\delta W_\ell|^3 \rangle \sim \ell^{\zeta(3)}$, we have $S(E) = C(3h - \zeta(3))$. By definition, we also have:

$$\tilde{S}_{3p} = \frac{S_{3p}}{S_3^p} = \langle e^{p \log(\mu_\ell)} \rangle \tag{26}$$

so that $\tilde{S}_{3p}$ is the partition function associated with the variable $\log(\mu_\ell)$, at the pseudo-inverse temperature $p = 1/k_B T$. Taking the logarithm of the partition function $\tilde{S}_{3p}$, we then get the free energy $F$ as:

$$F = \log(\tilde{S}_{3p}) \tag{27}$$

By the Gärtner-Elis theorem, $F$ is the Legendre transform of $S$: $F = \min_E(pE - S(E))$. The free energy a priori depends on the temperature $T = 1/k_B p$, on the volume $V = \log(\ell/\eta)$ and on the number of degrees of freedom of the system $N$. If we identify $N = (1/\beta(\text{Re}))\log(L_0/\eta)$, we see that the general universality means:

$$F(T,V,N) = NF(T,\frac{V}{N},1) \tag{28}$$

i.e., can be interpreted as *extensivity* of the free energy.

The thermodynamic analogy is thus meaningful and is summarized in Table 5. It can be used to derive interesting prospects.

**Table 5.** Summary of the analogy between the multifractal formalism of turbulence and thermodynamics.

| | **Thermodynamics** | **Turbulence** |
|---|---|---|
| Temperature | $k_B T$ | $1/p$ |
| Energy | $E$ | $\log(\mu_\ell)$ |
| Number of d.f. | $N$ | $\log(\text{Re}) = \log(L_0/\eta)/\beta_0$ |
| Volume | $V$ | $\log(\ell/\eta)$ |
| Pressure | $P$ | $\tau(p,\ell)$ |
| Free energy | $F$ | $\log(\tilde{S}_{3p})$ |

*5.2. Multifractal Pressure and Phase Transition*

Given our free energy, $F = \log(\tilde{S}_{3p})$, we can also compute the quantity conjugate to the volume, i.e., the multifractal pressure as: $P = \partial F/\partial V$. In the inertial range, where $\tilde{S}_p \sim \ell^{\tau(p)}$, we thus get $P = \tau(p)$, which only depends on the temperature. Outside the inertial range, $P$ has the meaning of a local scaling exponent that also depends upon the scale, i.e., on the volume $V$ and on $N$ (Reynolds number). Using our universal functions derived in Figure 5, we can then compute empirically the multifractal pressure $P$ and see how it varies as a function of $T$, $V$ and $N$. It is provided in Figure 7 for $R_\lambda = 25$ and $R_\lambda = 53$, and in Figure 8 for $R_\lambda = 90$ and $R_\lambda = 138$. We see that at low Reynolds number, the pressure decreases monotonically from the dissipative range, reaches a lowest points and then increases towards the largest scale. There is no clear flat plateau that would correspond to an "inertial" range. In contrast, at higher Reynolds number, a plateau appears for $p = 1$ to $p = 4$ when going towards the largest scale, the value of the plateau corresponding to $\tau_{\text{DNS}}$. The plateau transforms into an inflection point for $p \geq 5$ making the derivative $\partial P/\partial V$ change sign. This is reminiscent of a phase transition occurring in the inertial range, with coexistence of two phases: one "laminar" and one "turbulent". We interpret such a phase transition as the result of the coexistence of region of flows with different Hölder exponents, with areas where the flow has been regularized due to the action of viscosity, because of the random character of the dissipative scale (see below).

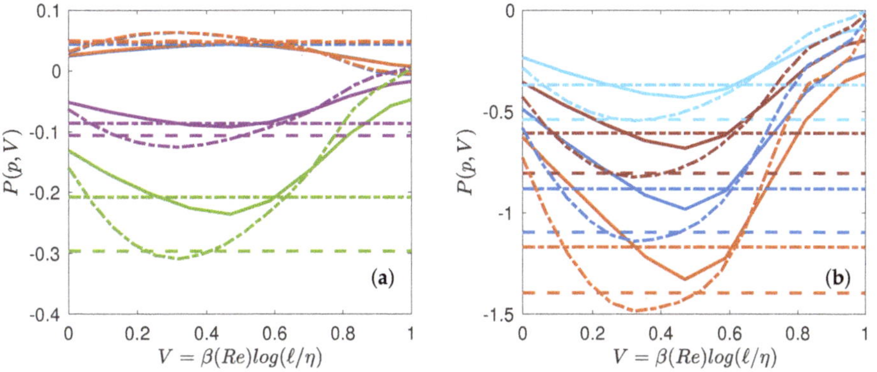


**Figure 7.** Multifractal equation of state of turbulence. Multifractal pressure as a function of the volume for $R_\lambda = 25$ (line) , $R_\lambda = 53$ (dashed-dotted line). The functions are coded by color. (**a**) $p = 1$: blue symbols; $p = 2$: orange symbols; $p = 4$: magenta symbols; $p = 5$: green symbols; (**b**) $p = 6$: light blue symbols; $p = 7$: red symbols; $p = 8$: blue symbols; $p = 9$: orange symbols. The colored dotted line (resp. dashed dotted line) are values corresponding to $P(p, V) = \tau_{\mathrm{EXP}}(p)$ (resp. $\tau_{\mathrm{DNS}}(p)$), that are reported in Table 4.

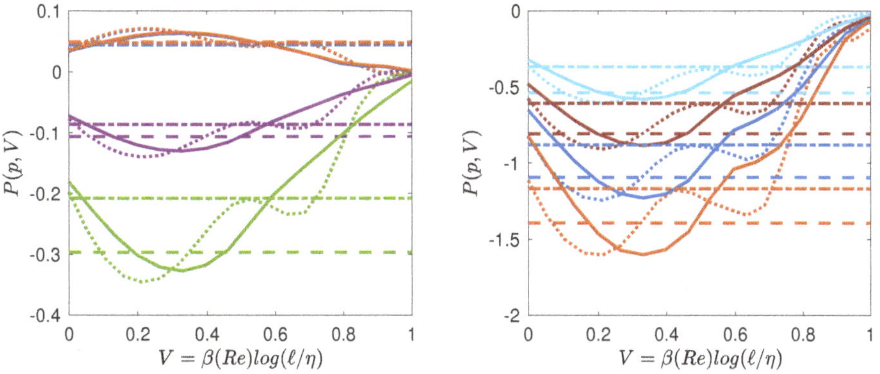


**Figure 8.** Same as Figure 7 for $R_\lambda = 90$ (line), $R_\lambda = 138$ (dotted line). Note the inflexion point appearing in the curves.

## 6. Conclusions

We show that a deep analogy exists between multifractal and classical thermodynamics. In this framework, one can derive from the usual velocity structure function an effective free energy that respects the classical extensivity properties, provided one uses several degrees of freedom (given by $N = 1/\beta(\mathrm{Re})$) that scales like $\log(R_\lambda)$. This number is much smaller than the classical $N \sim \mathrm{Re}^{9/4}$ that is associated with the number of nodes needed to discretize the Navier-Stokes equation down to the Kolmogorov scale. It would be interesting to see whether this number is also associated with the dimension of a suitable "attractor of turbulence". Using the analogy, we also find the "multifractal" equation of state of turbulence, by computing the multifractal "free energy" $F$ and "pressure" $P = \partial F/\partial V$. We find that for large enough $R_\lambda$ and $p$ (the temperature), the system obeys a phase transition, with coexistence of phase like in the vapor-liquid transition. We interpret this phase transition as the result of the coexistence of region of flows with different Hölder exponents, with areas where the flow is relaminarized due to the action of viscosity, because of the random character of the dissipative scale. We note that this kind of phenomenon has already been observed in the context of Lagrangian velocity increments, using the local scaling exponent $\zeta(p, \Delta t) = \mathrm{d}(\log(S_p(\Delta t)))/\mathrm{d}(\log(\Delta t))$ [18]. The phase transition is then associated with the existence of a fluctuating dissipative time scale. It is further shown that in a multifractal without fluctuating dissipative time scale, the local exponent decreases monotonically from dissipative scale to large scale, implying a disappearance of the phase transition [19].

**Author Contributions:** Conceptualization, B.D.; Data curation, D.G., H.F., F.N., V.S., J.-P.L., F.D., E.-W.S. and B.D.; Formal analysis, D.G., H.F., V.S. and B.D.; Funding acquisition, F.D. and B.D.; Project administration, B.D.; Supervision, B.D.; Writing—original draft, D.G., H.F. and B.D.; Writing—review & editing, H.F., F.N., V.S., J.-P.L., F.D., E.-W.S. and B.D.

**Funding:** H.F. is supported by a CFR. F.N. is supported by a fellowship from the ENS. This work has been supported by the Labex PALM (project Interdist) and by the ANR EXPLOIT, grant agreement no. ANR-16-CE06-0006-01. Part of this work was granted access to the resources of IDRIS under the allocation 2A310096 made by GENCI (Grand Equipement National de Calcul Intensif).

**Conflicts of Interest:** The authors declare no conflict of interest.

## References

1. Dubrulle, B. Beyond Kolmogorov cascades. *J. Fluid Mech. Perspect.* **2019**. [CrossRef]
2. Castaing, B. Conséquences d'un principe d'extremum en turbulence. *J. Phys. Fr.* **1989**, *50*, 147–156. [CrossRef]
3. Frisch, U.; Parisi, G. On the singularity structure of fully developed turbulence. In *Turbulence and Predictability in Geophysical Fluid Dynamics and ClimateDynamics*; Gil, M., Benzi, R., Parisi, G., Eds.; Elsevier: Amsterdam, The Netherlands, 1985; pp. 84–88.
4. Eyink, G.L. Turbulence Theory. Course Notes, The Johns Hopkins University. 2007–2008. Available online: http://www.ams.jhu.edu/~eyink/Turbulence/notes/ (accessed on 09/24/2018).
5. Paladin, G.; Vulpiani, A. Anomalous scaling laws in multifractal objects. *Phys. Rev.* **1987**, *156*, 147–225. [CrossRef]
6. Frisch, U.; Vergassola, M. A Prediction of the Multifractal Model: The Intermediate Dissipation Range. *Europhys. Lett. (EPL)* **1991**, *14*, 439–444. [CrossRef]
7. Castaing, B.; Gagne, Y.; Marchand, M. Log-similarity for turbulent flows? *Phys. D Nonlinear Phenom.* **1993**, *68*, 387–400. [CrossRef]
8. Muzy, J.F.; Bacry, E.; Arneodo, A. Wavelets and multifractal formalism for singular signals: Application to turbulence data. *Phys. Rev. Lett.* **1991**, *67*, 3515. [CrossRef] [PubMed]
9. Saw, E.W.; Debue, P.; Kuzzay, D.; Daviaud, F.; Dubrulle, B. On the universality of anomalous scaling exponents of structure functions in turbulent flows. *J. Fluid Mech.* **2018**, *837*, 657–669. [CrossRef]
10. Debue, P.; Shukla, V.; Kuzzay, D.; Faranda, D.; Saw, E.W.; Daviaud, F.; Dubrulle, B. Dissipation, intermittency, and singularities in incompressible turbulent flows. *Phys. Rev. E* **2018**, *97*, 053101. [CrossRef] [PubMed]
11. Kolmogorov, A.N. The local structure of turbulence in incompressible viscous fluids for very large Reynolds number. *Dokl. Akad. Nauk SSSR [Sov. Phys.-Dokl.]* **1941**, *30*, 913. [CrossRef]
12. Kolmogorov, A.N. A refinement of previous hypotheses concerning the local structure of turbulence in a viscous incompressible fluid at high Reynolds number. *J. Fluid Mech.* **1962**, *13*, 82. [CrossRef]
13. Touchette, H. The large deviation approach to statistical mechanics. *Phys. Rep.* **2009**, *478*, 1–69. [CrossRef]
14. Frisch, U. Turbulence: The Legacy of A.N. Kolmogorov Cambridge University Press: Cambridge, UK, 1995.
15. Benzi, R.; Ciliberto, S.; Tripiccione, R.; Baudet, C.; Massaioli, F.; Succi, S. Extended self-similarity in turbulent flows. *Phys. Rev. E* **1993**, *48*, R29–R32. [CrossRef]
16. Bohr, T.; Rand, D. The entropy function for characteristic exponents. *Phys. D Nonlinear Phenom.* **1987**, *25*, 387–398. [CrossRef]
17. Rinaldo, A.; Maritan, A.; Colaiori, F.; Flammini, A.; Rigon, R.; Rodriguez-Iturbe, I.; Banavar, J.R. Thermodynamics of fractal networks. *Phys. Rev. Lett.* **1996**, *76*, 3364. [CrossRef] [PubMed]

18. Arneodo, A.; Benzi, R.; Berg, J.; Biferale, L.; Bodenschatz, E.; Busse, A.; Calzavarini, E.; Castaing, B.; Cencini, M.; Chevillard, L.; et al. Universal Intermittent Properties of Particle Trajectories in Highly Turbulent Flows. *Phys. Rev. Lett.* **2008**, *100*, 254504. [CrossRef] [PubMed]
19. Biferale, L.; Boffetta, G.; Celani, A.; Devenish, B.J.; Lanotte, A.; Toschi, F. Multifractal Statistics of Lagrangian Velocity and Acceleration in Turbulence. *Phys. Rev. Lett.* **2004**, *93*, 064502. [CrossRef] [PubMed]

*entropy*

MDPI

*Article*

# The Influence of Internal Intermittency, Large Scale Inhomogeneity, and Impeller Type on Drop Size Distribution in Turbulent Liquid-Liquid Dispersions

**Wioletta Podgórska**

Faculty of Chemical and Process Engineering, Warsaw University of Technology, 00-645 Warsaw, Poland; wioletta.podgorska@pw.edu.pl

Received: 1 March 2019; Accepted: 25 March 2019; Published: 28 March 2019

**Abstract:** The influence of the impeller type on drop size distribution (DSD) in turbulent liquid-liquid dispersion is considered in this paper. The effects of the application of two impellers, high power number, high shear impeller (six blade Rushton turbine, RT) and three blade low power number, and a high efficiency impeller (HE3) are compared. Large-scale and fine-scale inhomogeneity are taken into account. The flow field and the properties of the turbulence (energy dissipation rate and integral scale of turbulence) in the agitated vessel are determined using the k-ε model. The intermittency of turbulence is taken into account in droplet breakage and coalescence models by using multifractal formalism. The solution of the population balance equation for lean dispersions (when the only breakage takes place) with a dispersed phase of low viscosity (pure system or system containing surfactant), as well as high viscosity, show that at the same power input per unit mass HE3 impeller produces much smaller droplets. In the case of fast coalescence (low dispersed phase viscosity, no surfactant), the model predicts similar droplets generated by both impellers. In the case of a dispersed phase of high viscosity, when the mobility of the drop surface is reduced, HE3 produces slightly smaller droplets.

**Keywords:** drop breakage; drop coalescence; local intermittency; turbulent flow; population balance equation; high efficiency impeller; Rushton turbine

---

## 1. Introduction

Liquid-liquid dispersions in a turbulent flow are common in many applications in chemical, petroleum, pharmaceutical, and food industries. Processes involving liquid-liquid dispersions include suspension polymerization, extraction, and heterogeneous reactions. The rate of a heterogeneous chemical reaction is often controlled by mass transfer. Mass transfer is also the base of the extraction process. The efficiency of mass transfer strongly depends on the interfacial area determined by drop size distribution, which in turn is controlled by drop breakage and coalescence processes. Drop size distribution also determines the quality of the product obtained in suspension polymerization. Droplet breakage, which is a short-duration process, i.e., the process characterized by time scales smaller than time constants of related turbulent events, can be strongly influenced by internal intermittency (also called local or fine-scale intermittency) [1,2]. Internal intermittency also affects the coalescence process. Internal intermittency results from vortex stretching, which leads to the formation of regions of space characterized by high vorticity surrounded by nearly irrotational fluid. Small scale intermittency can be deduced from probability distribution functions of velocity gradients and differences [3,4]. From the distribution of the velocity derivatives, it is evident that the energy associated with large wave numbers (small length scales) is very unevenly distributed. Dissipation associated with increasing wavenumbers becomes increasingly concentrated in small regions [5,6]. It means that there are regions and periods of activity and quiescence. This spotty distribution in time and space manifests in an anomalous scaling of fluctuating quantities. Two scaling laws of special interest are those for a velocity increment over a distance, *r*:

$$\langle (\delta\, u(r))^p \rangle \sim r^{\zeta_p}, \tag{1}$$

and for energy dissipation, ε, averaged over a ball of a size, *r*:

$$\langle \varepsilon_r^p \rangle \sim r^{\tau_p}. \tag{2}$$

The exponent of the structure function, $\zeta_p$, differs from $p/3$ predicted by Kolmogorov theory and the discrepancy between $\zeta_p$ and $p/3$ increases with increasing $p$. For positive $p$, the exponents in Equations (1) and (2) are related by $\zeta_p = p/3 + \tau_{p/3}$ [7,8]. The intermittent character of turbulence can be modeled using multifractal formalism [4,8]. There are theoretical arguments for this formalism related to the nonlinear character of Navier-Stokes equations. There exists a strange attractor for Navier-Stokes equation (N-S) and solutions attracted to the strange attractor correspond to the turbulence. Instantaneous realization of the flow or any instantaneous solution of an N-S equation can be treated as an object consisting of various objects related to fractal sets embedded in physical space. The N-S equations in the zero viscosity limit are invariant under the following group of rescaling transformations: $x_i' = \lambda x_i$, $u_i' = \lambda^h u_i$, and $t' = \lambda^{1-h} t$, provided that $< \eta > < r, r' < L$ and $L >> < \eta >$, where $r = \sqrt{x_i^2}$. $L$ is the integral scale of turbulence, $< \eta >$ is the Kolmogorov microscale, and $h$ is a scaling exponent. When the viscous term is neglected at high Reynolds numbers, there are infinitely many scaling groups, labeled by their scaling exponent, $h$, which can be any real number [8]. When one considers energy dissipation, then $\varepsilon_r/\varepsilon_L \propto (r/L)^{\alpha - d_s}$, where $\varepsilon_L$ is the average of ε over a box of a size, $L$; $\alpha$ is a scaling exponent (also called a multifractal exponent or singularity strength); and $d_s$ is the space dimension. Scaling exponents for velocity, $h$, and for dissipation, $\alpha$, are related by $h = \alpha/3$ [4,8]. The transformation for dynamic pressure, $p$, can be neglected because the pressure can be eliminated from the Navier-Stokes equation [8]. However, in turbulent flow, the breakage of droplets with a size from the inertial subrange results from dynamic pressure fluctuations, thus the scaling law for pressure is of interest. The pressure transforms as $u_i^2$, or scales as $p' = \lambda^{2\alpha/3} p$ and the local normal pressure stresses in the inertial subrange acting on droplets of a size, $d$, are [2]:

$$p(d,\alpha) = C_p \rho_C [\langle\varepsilon\rangle d]^{2/3} \left(\frac{d}{L}\right)^{\frac{2}{3}(\alpha-1)}. \tag{3}$$

The velocity increment over a distance, *r*, is:

$$u_r = [\langle\varepsilon\rangle r]^{1/3} \left(\frac{d}{L}\right)^{\frac{\alpha-1}{3}}. \tag{4}$$

At pure breakage (i.e., when coalescence is negligible), the maximum stable drop size, $d_{\max}$ (for dispersed phase of low viscosity), results from the balance of pressure stresses given by Equation (3) and shape restoring stresses given by $\sigma/d$, where $\sigma$ is an interfacial tension [2]:

$$d_{\max} = C_x^{\frac{5}{3+2\alpha}} L \left(\frac{\sigma}{\rho_C \langle\varepsilon\rangle^{2/3} L^{5/3}}\right)^{\frac{3}{3+2\alpha}}. \tag{5}$$

For viscous drops, the additional stabilizing stress (viscous stress) should be taken into account, thus $d_{\max}$ is given by [2]:

$$d_{\max} = C_x \frac{\sigma^{0.6}}{\rho_C^{0.6} \langle\varepsilon\rangle^{0.4}} \left(\frac{d_{\max}}{L}\right)^{0.4(1-\alpha)} \left[1 + \beta_\mu \mu_D \left(\frac{d_{\max}}{L}\right)^{\frac{\alpha-1}{3}} \frac{\langle\varepsilon\rangle^{1/3} d_{\max}^{1/3}}{\sigma}\right]^{0.6}. \tag{6}$$

where $\rho_C$ is a continuous phase density, $\mu_D$ is a dispersed phase viscosity, and $C_x$ and $\beta_\mu$ are constants. Equations (5) and (6) do not give any reference to time. Drop size evolution in time can be predicted by solving the population balance equation with suitable breakage and coalescence models.

Models of breakage and coalescence are usually based on a classical Kolmogorov theory of turbulence that neglects intermittency. One of the first and most popular breakage models was proposed by Coulaloglou and Tavlarides [9]. The authors assumed that the droplet would be broken if the kinetic energy transmitted from eddies to the drop is larger than the drop surface energy. The fraction of eddies interacting with the droplet that have a kinetic energy larger than the surface energy is equal to the fraction of eddies that have velocities larger than the corresponding fluctuating velocity. It was assumed that only energies associated with velocity fluctuations of a scale smaller than the drop diameter tend to disperse the drop. A Gaussian distribution of turbulent velocity was assumed. Chatzi and Lee [10] and Chatzi et al. [11] assumed that the probability density of the kinetic energy of eddies is described by three-dimensional Maxwell distribution. Narsimhan et al. [12] treated droplets as one-dimensional simple harmonic oscillators. According to their model the oscillations of a drop are induced by the arrival of eddies of different scales and frequencies, and the number of arriving eddies is assumed to be a Poisson process. Konno et al. [13] assumed that breakage is caused by nonisotropic turbulence inside the impeller-disc edge and isotropic turbulence outside the impeller-disc edge. The breakage frequency in the region of isotropic turbulence was derived by using assumptions similar to those proposed by Coulaloglou and Tavlarides [9], but the probability density function of relative velocity was represented by Maxwell distribution. In the nonisotropic turbulent region, regularity in the direction of droplet elongation was observed. Therefore, breakage frequency was derived under the assumption that large energy-containing eddies are responsible for drop deformation and disruption. Martinez-Bazan et al. [14] based their model on a purely kinematic idea. They postulated that the acceleration of the fluid particle interface during deformation is proportional to the difference between the deformation and restoring stresses. All the models were derived for droplets of a size corresponding to the inertial subrange of scales. There is a group of breakage models based on a concept of collisions between droplets and eddies [15–17]. In recent years, these models, which were also formulated for the inertial subrange, were extended by using a wide energy spectrum [18–21]. The important question that appears when a breakage model is formulated is whether the droplet is broken by eddies smaller than the droplet, eddies of a size comparable to the drop diameter, or eddies larger than the droplet. According to Hinze [22], the droplet is disrupted by eddies of the same scale. Larger eddies only convey the drop, while smaller eddies are too weak to disperse the drop. In Coulaloglou and Tavlarides' model [9], eddies smaller than the drop are responsible for breakage, while Andersson and Andersson [23,24] argue that eddies of a size approximately equal to and up to three times larger than the drop are responsible for dispersion. There are also breakage models taking into account the increased viscosity of the dispersed phase [25,26]. These models were further modified by Maaβ and Kraume [27], who assumed that the two mechanisms of breakage operate simultaneously (breakage induced by pressure fluctuations and breakage induced by two-dimensional elongational flow). All these breakage models neglect intermittency. However, as was discussed earlier, the local intermittency can have a profound effect on breakage and a noticeable effect on coalescence. Multifractal breakage models taking into account internal intermittency [2,28] allow the scale effect on the drop size to be explained; they explain the drift of the exponent on the Weber number from −0.6 to −0.93 in dimensionless relation for a maximum stable drop:

$$\frac{d_{\max}}{D} \propto We^{-0.6\left(\frac{1}{1-0.4(1-\alpha)}\right)}. \tag{7}$$

when the multifractal exponent changes from 1 to the infimum value of 0.12. They also explain the slow drift of transient drop size distributions at long agitation times. In these models, the concept of drop-eddy collision is not used. Models for droplets smaller than the Kolmogorov scale were also formulated using multifractal formalism [2]. Multifractal breakage and coalescence models allow proper predictions to be made of the changes of the drop size distribution both for short and long agitation times [2,28–33].

The coalescence process can be considered as an interaction triangle consisting of the continuous phase flow and two fluid particles. A continuous phase flow can be split into the external flow responsible for droplet collisions and the internal flow responsible for film drainage between colliding droplets [34]. The frequency of collisions of droplets of a size from the inertial subrange is based on the relative drop velocity, which is calculated as the characteristic velocity variation in the basic flow over a distance, $d$ [35]. Another possible assumption is that colliding drops take the velocity of an eddy of the same size [9,15]. The efficiency of collisions depends on the drop surface mobility, drop size, surface deformation, etc. In pure liquid-liquid systems, partially mobile interfaces can be assumed [29,30,34]. In many models, immobilized interfaces are assumed [9,36,37]. Immobilization may be caused by the surfactant presence or high dispersed phase viscosity. A mobility parameter dependent on the viscosity ratio, $\mu_D/\mu_C$, can also be introduced to model the coalescence of droplets of a relatively high viscosity [15,31,32,38].

Immiscible liquids are often contacted in stirred vessels. Therefore, the geometry of the tank and the type of the used impeller are of great importance for producing a desired drop size distribution. Impellers can be classified as producing shear or flow. Radial disc turbines, like a Rushton turbine, commonly used for liquid-liquid systems, produce strong radial flow as well as intense turbulence. They can produce a high interfacial area. Hydrofoil impellers, such as Lightnin A310 or Chemineer HE3, produce axial or mixed flow and are especially good for systems differing in the density of the continuous and disperse phase. They have blades mounted at a shallow angle to reduce drag at the leading edges, and provide intensive axial flow with small power requirements. They are able to achieve a suspended state at a lower rotational speed than disc turbines. Therefore, they are particularly suitable for solid-liquid systems [39,40]. However, it was shown that the low power number high flow agitators, like HE3, can be used for liquid-liquid dispersions and produce smaller droplets than high power numbers, high shear agitators at the same power input per unit mass (i.e., the same average energy dissipation rate in the tank, $\overline{\langle\varepsilon\rangle}$) [41,42]. Therefore, in this paper, the influence of the impeller type on drop size distribution is presented. Two types of impellers are considered: Six-blade Rushton turbine (RT) and three-blade high efficiency impeller (HE3). The distribution of the locally averaged properties of the turbulence (including energy dissipation rate, $\varepsilon$, and integral scale of turbulence, $L$) are determined using the computational fluid dynamics CFD method. The distribution of these properties in the stirred tank affects the drop breakage and coalescence rates.

Both processes (breakage and coalescence) are taken into account in this paper. Breakage takes place in practice only in the zone of the highest energy dissipation rate (impeller zone). The zone in the agitated tank where coalescence is privileged depends on the drop deformation in the contact area and on the mobility of the drop interfaces. The rates of both the breakage and coalescence depend on the mean power input per unit mass, and on a strong local and instantaneous variability of the energy dissipation rate related to the internal intermittency. Multifractal formalism was applied to model fine-scale intermittency.

## 2. Breakage and Coalescence Models

The time evolution of drop size distribution in a stirred tank is predicted by solving the population balance equation. A population of droplets of a volume, $v$, and diameter, $d$, ($v = \pi d^3/6$) from the inertial subrange of turbulence is considered. The macroscopic population balance equation (averaged in the external phase space) formulated in the volume domain (for one internal coordinate corresponding to the drop volume) for chemically equilibrated liquid-liquid dispersion (with no mass transport) and batch operation is given by:

$$\begin{aligned}\frac{\partial n(v,t)}{\partial t} &= \frac{1}{2}\int_0^v h(v-v',v')\lambda(v-v',v')n(v-v')n(v')\mathrm{d}v' - n(v,t)\int_0^\infty h(v,v')\lambda(v,v')n(v',t)\mathrm{d}v' \\ &+\int_v^\infty \beta(v,v')\nu(v')g(v')n(v',t)\mathrm{d}v' - g(v)n(v,t)\end{aligned} \quad (8)$$

where $n(v,t)$ is the number density of drops of a volume, $v$, at time, $t$ ($\mathrm{m}^{-6}$). The drop breakage rate $g(d) = g(v)$ ($\mathrm{s}^{-1}$) in intermittent turbulent flow was developed by summing up the contributions to the break-up frequency from all vigorous eddies [2]:

$$g(d) = \int_{\alpha_{\min}}^{\alpha_x} g(\alpha,d)P(\alpha)\mathrm{d}\alpha = C_g\sqrt{\ln\left(\frac{L}{d}\right)}\frac{\langle\varepsilon\rangle^{1/3}}{d^{2/3}}\int_{\alpha_{\min}}^{\alpha_x}\left(\frac{d}{L}\right)^{\frac{(\alpha+2-3f(\alpha))}{3}}\mathrm{d}\alpha. \quad (9)$$

$P(\alpha)$ is a probability density for $\alpha$ in a box of a length, $r$; $g(\alpha,d)$ is the characteristic frequency of eddies of a size, $d$, labeled by a scaling exponent, $\alpha$. Vigorous eddies that can disperse the drop are characterized by a multifractal exponent, $\alpha$, from the range ($\alpha_{\min}$, $\alpha_x$). The most vigorous eddies are characterized by $\alpha_{\min}$. This value is difficult to measure and entails the extrapolation procedure. It was approximated for tails of the probability density of dissipation in boxes of a size $r$, $E_r$, normalized by the overall dissipation, $E_t$. The tails of distribution of ($E_r/E_t$) were found to be of the square-root exponential type and $\alpha_{min} = 0.12$ [4]. The upper bound of the integral in Equation (9), $\alpha_x$, results from the balance of stresses acting on the droplet and characterizes the weakest eddies that can disperse the drop [2,28]. The multifractal spectrum, $f(\alpha)$, is for practical reasons approximated by a polynomial [2] fitted to the experimental spectrum [4]. Thus, $f(\alpha)$ is given by:

$$f(\alpha) = a + b\alpha + c\alpha^2 + d\alpha^3 + e\alpha^4 + f\alpha^5 + g\alpha^6 + h\alpha^7 + i\alpha^8, \quad (10)$$

where $a = -3.4948$, $b = 18.721473$, $c = -55.918539$, $d = 120.90274$, $e = -162.54397$, $f = 131.51049$, $g = -62.572242$, $h = 16.1$, and $i = -1.7264619$. The constant, $C_g$, in Equation (9) is equal to $C_g = 0.0035$. Depending on the liquid-liquid system, different stresses act on droplets. When the dispersed phase viscosity is low, the only stabilizing stress that opposes the disruptive turbulent stress given by Equation (3) is the shape restoring stress associated with interfacial tension, $\sigma$, $\tau_\sigma \propto \sigma/d$. Thus, the multifractal exponent, $\alpha_x$, resulting from the stress balance is given by:

$$\alpha_x = \frac{2.5\ln[(L\langle\varepsilon\rangle^{0.4}\rho_C^{0.6})/(C_x\sigma^{0.6})]}{\ln(L/d)} - 1.5, \quad (11)$$

where the constant is $C_x = 0.23$. High dispersed phase viscosity, $\mu_D$, increases the stabilizing effect. The viscous stress inside the drop is generated when a drop deforms. Thus, there are viscous and interfacial tension stresses that oppose the turbulent disruptive stress [2]. The droplet must be elongated to the elongation at burst during a time period smaller than the Lagrangian time macroscale. The weakest eddies that can disperse the viscous drop are thus labeled by the following multifractal exponent:

$$\alpha_x = 3\frac{\ln\left\{2\left[\frac{\beta_\mu C_x^{5/3}\mu_D}{\rho_C\langle\varepsilon\rangle^{1/3}L^{1/3}d} + \sqrt{\left(\frac{\beta_\mu C_x^{5/3}\mu_D}{\rho_C\langle\varepsilon\rangle^{1/3}L^{1/3}d}\right)^2 + \frac{4C_x^{5/3}\sigma}{\langle\varepsilon\rangle^{2/3}L^{2/3}\rho_C d}}\right]^{-1}\right\}}{\ln\left(\frac{L}{d}\right)}. \quad (12)$$

In this case, $\alpha_x$ depends on the interfacial tension and dispersed phase viscosity. Furthermore, the new constant, $\beta_\mu$ ($\beta_\mu = 1.91$), appears. When surfactant is present in the system, an additional disruptive stress that adds to the turbulent stress given by Equation (3) may be generated. This extra stress is due to the difference between the dynamic interfacial tension of the fresh surface (exposed during drop deformation under the action of pressure fluctuations), $\sigma_{t\to 0}$, and static interfacial tension, $\sigma$ [28]. This extra stress is observed when surfactant can be easily removed from the surface [28,33], but is not observed when surface active additive is strongly grafted to the surface [43]. The multifractal exponent characterizing the weakest eddies that can disperse the drop covered with surfactant, which can be removed from the surface during its deformation, is given by [28]:

$$\alpha_x = \frac{2.5\ln[(L\langle\varepsilon\rangle^{0.4}\rho_C^{0.6})/(C_x(2\sigma-\sigma_{t\to 0})^{0.6})]}{\ln\left(\frac{L}{d}\right)} - 1.5, \tag{13}$$

In all cases, binary breakage (number of daughter drops, $\nu(v') = 2$) was assumed. It was also assumed that breakage into drops differing much in volume is more probable than breakage into equal drops. The daughter distribution function, $\beta(v, v')$, based on the surface energy increase was used [15].

For comparison, a breakage model that neglects intermittency will be used. For this purpose, Coulaloglou and Tavlarides' model [9] was chosen:

$$g(d) = C_1\frac{\langle\varepsilon\rangle^{1/3}}{d^{2/3}}\exp\left(-C_2\frac{\sigma}{\rho_C\langle\varepsilon\rangle^{2/3}d^{5/3}}\right), \tag{14}$$

The constants that are most often used are $C_1 = 0.00481$ and $C_2 = 0.08$ [44].

The coalescence rate depends on the drop collision frequency and coalescence efficiency. The average collision rate in a turbulent field is calculated using the method of steepest descent [30]. The function, $h(v, v') = h(d, d')$ ($m^3s^{-1}$), appearing in the population balance equation is expressed as:

$$h(d,d') = \sqrt{\frac{8\pi}{3}}\langle\varepsilon\rangle^{1/3}\left(\frac{d+d'}{2}\right)^{7/3}\left(\frac{d+d'}{2L}\right)^{0.026}. \tag{15}$$

The coalescence efficiency, $\lambda(v, v') = \lambda(d, d')$, is determined by the ratio of the average film drainage time, $t_c(d, d')$, and average interaction time $t_i(d, d')$:

$$\lambda(d,d') = \exp\left(-C\frac{t_c(d,d')}{t_i(d,d')}\right), \tag{16}$$

where C is a non-dimensional coefficient. The film drainage time depends on the mobility of drop interfaces. For pure liquid-liquid systems and a low dispersed phase viscosity, drop interfaces remain partially mobile and film drainage is controlled by the flow inside the drop. The average drainage time in intermittent turbulent flow for deformed droplets with partially mobile interfaces can be expressed as follows [30]:

$$t_c = \frac{\mu_D\tilde{a}R_{eq}^{2/3}}{4\sigma R_L^{1/2}}\left(\frac{1}{h_c}\left(\frac{d_{jk}}{L}\right)^{0.016} - \frac{1}{\tilde{h}_0}\left(\frac{d_{jk}}{L}\right)^{-0.01}\right). \tag{17}$$

The film radius, $\tilde{a}$, is derived under the assumption that the whole kinetic energy is transformed into excess surface energy, and the initial film thickness, $\tilde{h}_0$, results from a comparison of the turbulent velocity and drainage rate [29,31]. The critical (rupture) film thickness, $h_c$, is calculated from a comparison of the van der Waals radial force per unit volume and the pressure gradient responsible for the film thinning rate [34]. $R_L$ is a radius of a larger drop. The equivalent radius for unequal droplets is defined as $R_{eq} = dd'/(d + d')$ and $d_{jk} = (d + d')/2$.

The interaction time, $t_i$, is usually smaller than, or of the order of the time scale for two droplets to pass one another, $t_{ext}$. For intermittent turbulent flow, the average time scale, $t_{ext}$, is then given by:

$$t_{ext} = \frac{d_{jk}^{2/3}}{\langle \varepsilon \rangle^{1/3}} \left( \frac{L}{d_{jk}} \right)^{0.052}. \tag{18}$$

However, for droplets of low viscosity, the interaction time can be estimated as the time resulting from a droplet bouncing [29]:

$$t_i = \frac{1}{2} \left( \frac{8}{3} \frac{R_S^3 (\rho_D/\rho_C + 0.75)\rho_C}{\sigma(1+\xi^3)} \right)^{1/2}, \xi = \frac{R_S}{R_L}. \tag{19}$$

When the dispersed phase viscosity is high the drop interfaces are immobilized. Different cases can be considered: Undeformed droplets, deformed droplets with a film radius resulting from the balance between the pressure caused by external force and Laplace pressure, and deformed droplets with a film radius proportional to the radius of the smaller droplet [31,32]. It was shown that drops of a high viscosity differing in size behave in a completely different way. In this paper, large deformed droplets and parallel-sided film are considered. In this case, the interaction time, $t_i = t_{ext}$, and is given by Equation (18). The film drainage controlled by Laplace pressure is assumed and the drainage time is calculated as follows [32]:

$$t_c = \frac{3\mu_C \rho_C R_{eq}^4 \langle \varepsilon \rangle^{2/3} d_{jk}^{2/3}}{16\sigma^2 h_c^2} \left( \frac{d_{jk}}{L} \right)^{0.026} \tag{20}$$

## 3. Geometry and CFD Model

The properties of turbulence for tanks equipped with one of the impellers, six-blade Rushton turbine (RT) or three blade high efficiency impeller (HE3), are determined using CFD. The impellers are shown in Figure 1. The image of HE3 is taken from [45]. Simulations are performed for a tank of a diameter, $T$ = 0.15 m, and height, $H = T$, completely filled and closed. The stirred tank is flat bottomed and fully baffled (four equally spaced baffles of a width equal to $T/10$). It was assumed that the impeller diameter to tank diameter ratio is $D/T$ = 0.4 and the impeller clearance is $C/T = 1/4$ for HE3. The high efficiency impeller has a uniform blade width equal to 0.01 m. The blade angle is 30 degrees at the hub. The tip chord angle is 15 degrees. The blade is bent at 50% of its length. The thickness of the blade is equal to 0.002 m. The Rushton turbine has a diameter of $D = 0.5T$. A disc diameter is equal to 0.75 $D$, a blade thickness and disk thickness are 0.01 $D$. The impeller clearance is $C/T = 1/2$.

The unstructured tetrahedral meshes with approximately 400,000 cells for a tank equipped with a Rushton turbine and 600,000 cells for a high efficiency impeller were generated using Mixsim software. Steady state 3D simulations were performed using the finite volume package, Ansys Fluent. The multiple reference frame approach and standard $k$-$\varepsilon$ model with standard wall functions were used. The SIMPLE algorithm was used for pressure-velocity coupling. The PRESTO scheme was used for pressure interpolation, and the second order upwind scheme was used for the momentum, kinetic energy, and energy dissipation rate equations.

The CFD simulations were performed to obtain power numbers, $P_o$, pumping capacity (flow number, $Fl$), as well as the normalized mean energy dissipation rate and normalized integral scale of turbulence in the impeller and bulk zones ($\phi_{imp}$, $\phi_{bulk}$, $L_{imp}/D$, $L_{bulk}/D$). These values were then used in the circulation flow model (the dispersion circulates through the impeller and bulk zones). A multifractal model allows the probability of stresses characterized by different multifractal exponents, $\alpha$, for a given average energy dissipation rate calculated for a given zone to be predicted. Such a model gives excellent results in predicting drop size evolution as was shown in previous papers [30,33,43]. The details of the flow pattern were not used.

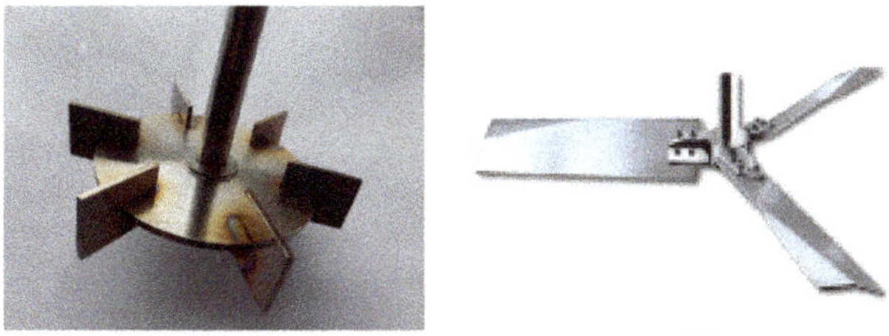

(**a**) (**b**)

**Figure 1.** Impellers: (**a**) Rushton turbine and (**b**) high efficiency impeller.

## 4. Results and Discussion

CFD simulations were performed for a high efficiency impeller (HE3) for an impeller speed of $N$ = 800 rpm. The operating fluid was water. The presence of an organic phase was not taken into account in these simulations. It was justified by low values of the dispersed phase volume fraction ($\varphi$ = 0.001 for the pure breakage case and $\varphi$ = 0.05 for the coalescence case). For the tank equipped with a Rushton turbine, simulations were performed for the impeller speed that was expected to give the same power input per unit mass ($N$ = 213 rpm). These simulations allowed the power and flow numbers for both impellers to be determined: $P_o$ = 0.34, $Fl$ = 0.39 for the high efficiency impeller, and $P_o$ = 4.98, $Fl$ = 0.74 for the Rushton turbine. These values agree well with the measured ones ($P_o$ = 0.305 and $Fl$ = 0.41 for HE3 of D/T = 0.46 [46], $P_o$ = 0.3 for HE3 of D/T = 0.39 and an impeller clearance equal to T/4 [40]). Power number values for RT reported by different authors are in the range of 4.6 to 6.3. According to Bujalski et al., the correlation based on experimental measurements of the power number, $P_o$, depends on the tank size and the impeller disc thickness [47]. For $T$ = 0.15 m and a disc thickness equal to 0.01 $D$, the power number should be equal to 5.5. However, this correlation was obtained for vessels of a diameter from 0.22 m to 1.83 m. The flow generated by HE3 at the clearance of $T/4$ has a strong axial component directed to the base. Between the impeller hub and the tank base, there is a weak reverse flow. Local values of the integral scale, $L$, were determined using calculated local values of the energy dissipation rate, $\varepsilon$, and turbulent kinetic energy, $k$, $L = (2k/3)^{3/2}/\varepsilon$. The contours of $L$ for both types of impellers are presented in Figure 2. In both cases, there is a distinct difference in the integral length scale in the impeller zone and the bulk, though the average values of $L$ in these zones do not differ as much as the values of the energy dissipation rate do.

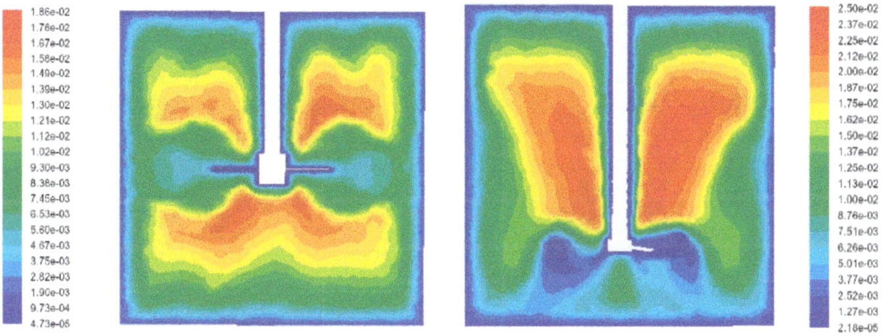


(**a**) (**b**)

**Figure 2.** Contours of the integral scale for: (**a**) Rushton turbine and (**b**) HE3 impeller (plane θ = 45° between baffles).

Earlier studies of the author [48,49] have shown that the multiple zone model of the tank (10-zones) predicts similar drop size distributions as the 2-zone model, provided that the impeller zone is properly defined. For example, for the Rushton turbine of $D/T = 1/3$, only part of the impeller stream should be included into the impeller zone. However, in the case of RT of D/T = 1/2 that is considered in the present paper, the whole discharge region is included to the impeller zone. This impeller stream region together with the impeller swept volume (extended here 3 mm above and 3 mm below impeller blades) occupies a fraction, $x_{imp} = 0.095$, of the tank volume. To determine the mean values of the energy dissipation rate and the scale of large eddies, auxiliary surfaces were created and the surface integrals function was used. Surfaces were separated by 1 mm in the impeller zone. Such densely created surfaces enabled us to better define the limits of the impeller zone. For the bulk zone, the surfaces were separated by 2 mm. The relative properties of the turbulence for the Rushton turbine of $D/T = 1/2$ are as follows: $\phi_{imp} = \langle\varepsilon\rangle_{imp}/\overline{\langle\varepsilon\rangle} = 6.1$, $\phi_{bulk} = 0.465$, $L_{imp}/D = 0.0806$, $L_{bulk}/D = 0.13$, where $\overline{\langle\varepsilon\rangle} = \rho P_o N^3 D^5/(\rho V)$. In the case of HE3, the impeller zone is defined as a cylinder of a radius of $r = 0.034$ m (slightly larger than the impeller radius, $R = 0.03$ m) and positioned between $z = 0.0325$ m and $z = 0.0425$ m. The volume fraction of this zone is equal to $x_{imp} = 0.0137$, the normalized mean energy dissipation rate in this zone is $\phi_{imp} = 45$, while in the bulk it is $\phi_{bulk} = 0.389$. The normalized mean integral scales of the turbulence are $L_{imp}/D = 0.0573$ in the impeller zone, and $L_{bulk}/D = 0.187$ in the bulk. The spatial distribution of the energy dissipation rate in the impeller zone has been reported for the RT by many researchers. However, most works were devoted to impellers of a diameter of $D = T/3$. The percentage of the total energy dissipated in the swept impeller and the impeller stream regions reported by different authors vary from 42% to 70% [50]. For an impeller diameter of $D = T/2$, the normalized energy dissipation rate in the impeller zone is $\phi_{imp} = 5.93$ and the impeller volume fraction is $x_{imp} = 0.105$ according to the Okamoto correlation [51]. Thus, the percentage of the total energy dissipation in the impeller zone is 62.3%. Zhou and Kresta [52] measured a 43.5% dissipation in the control volume containing impeller swept and impeller stream zones and occupying 10% of the tank volume. The percentage of the total energy dissipation in the impeller region predicted in the present work (57.95%) lies between these literature values. The used $k$-$\varepsilon$ model gives reasonable results. The Reynolds stress model (RSM) used in previous work [48] was able to predict the characteristic properties of the energy dissipation rate profiles. For example, it predicts that the rate of the energy dissipation in the impeller stream for a radial position of $r/R = 1.325$ is much higher than at smaller and larger radial distances. It agrees with the PIV measurements of Baldi and Yianneskis [53], who observed a similar jump at a radial position of $r/R = 1.32$. The $k$-$\varepsilon$ model does not predict any jump. It predicts the decrease of $\varepsilon$ with the increase of the distance from the impeller blades. However, the normalized mean energy dissipation rate in the impeller stream predicted by both turbulence models differs only by 2%. The difference between the pumping capacity predicted by both models was smaller than 1.4%.

The population balance equation was solved for three liquid-liquid systems. In the first case, the dispersed phase of low viscosity is considered ($\mu_D = 0.001$ Pa·s). The density of the continuous phase is assumed to be $\rho_C = 1000\ \mathrm{kg/m^3}$, and the interfacial tension is $\sigma = 0.035$ N/m. The second liquid-liquid system is characterized by $\mu_D = 0.5$ Pa·s, $\rho_C = 1000\ \mathrm{kg/m^3}$, and $\sigma = 0.035$ N/m. For both these liquid-liquid systems, pure breakage (dispersed phase volume fraction, $\phi = 0.001$) as well as breakage together with coalescence ($\phi = 0.05$) were simulated. In the calculations, the constant, $C$, in Equation (16) defining the coalescence efficiency is assumed to be $C = 0.5$ for the first liquid-liquid system characterized by partially mobile interfaces and drainage and interaction times are calculated from Equations (17) and (19), respectively. For the system with a drop surface mobility decreased due to a high dispersed phase viscosity, Equations (20) and (18) are used to estimate the drainage and interaction times and the constant is equal to $C = 0.1$. The Hamaker constant, which influences the critical film thickness, is assumed to be $A = 10^{-20}$ J (characteristic for pure liquid-liquid systems). The third liquid-liquid system contains surfactant, which is easily removed from the surface. Because of the surfactant presence, the coalescence is not observed when starting from big droplets (very slow coalescence could be observed after an impeller speed reduction—see [28,33]), and additional disruptive stress appears due to the interfacial tension difference between the freshly exposed interface and the interface covered by surfactant [28]. In this case, a multifractal exponent characterizing the weakest eddies is calculated from Equation (13). The interfacial tension values are $\sigma = 0.0233$ N/m and $\sigma_{t\to 0} = 0.0255$ N/m (as measured for toluene/1 mM sodium dodecyl sulfate SDS aqueous solution [33]). An initial drop size of $d = 3$ mm was assumed in the calculations.

Figure 3 shows the transient drop size distributions predicted for both types of impeller at the same power input per unit mass for conditions when only the breakage of droplets of a low viscosity (liquid-liquid system 1) takes place. The drop size distribution at short agitation times is much wider for the HE3. However, the mean Sauter diameter, $d_{32}$, is only slightly larger for HE3 than for RT. This is because the largest volume fraction of drops is formulated by smaller droplets in the tank equipped with the HE3 than in that equipped with the RT. The higher the power input per unit mass (and thus the higher mean energy dissipation in the tank), the smaller the observed $d_{32}$ for the HE3 in comparison with $d_{32}$ for the RT (even after a few minutes of agitation). However, the largest droplets are still bigger for the HE3 than for the RT at short agitation times. After long agitation times, droplets produced by the HE3 are much smaller than droplets produced by the RT.

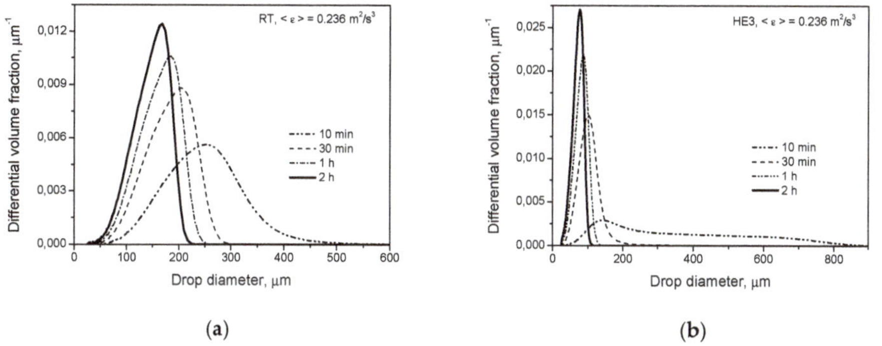


**Figure 3.** Influence of the impeller type on transient drop size distributions for a dilute system ($\varphi$ = 0.001) with a dispersed phase of low viscosity ($\mu_D = 0.001$ Pa·s): (**a**) Rushton turbine; (**b**) high efficiency impeller, HE3.

A comparison of the drop size distributions and mean sizes of drops produced by different impellers after 2 h of agitation is shown in Figure 4. The described behavior of droplets can be explained by the smaller impeller zone volume and larger $\phi_{imp}$ in the tank equipped with a high efficiency impeller. Very large droplets that are easily broken in both systems have a greater chance of appearing in the zone of the high energy dissipation rate and, therefore, high turbulent disruptive stresses in the tank with the RT ($x_{imp,RT} > x_{imp,HE3}$). However, the final drop size is determined by the magnitude of $\langle\varepsilon\rangle_{imp}$ and this is much higher for the HE3.

Breakage and coalescence models were previously verified experimentally for Rushton turbines of different $D/T$ ratios. A comparison between the measured (literature as well as our own experiments) and the predicted distributions (with the energy dissipation rate distribution based on the experimental Okamoto correlation as well as being predicted using the CFD) one can find elsewhere [2,30–33,54]. Some experimental results from the literature (for long agitation times) for the RT, $D = T/2$, for low as well as high dispersed phase viscosity [13,54–57] are presented in Figure 4b and Figure 6b. In the case of Arai et al.'s [55] experiments, it was assumed that the Sauter diameter is $d_{32} = 0.6d_{\max}$ ($d_{\max}$ is the diameter of the maximum stable drop size) for a dispersed phase of low viscosity and $d_{32} = 0.5d_{\max}$ for a dispersed phase of high viscosity. When the power number was not measured, the value of $P_o = 4.98$ was used to estimate the power input per unit mass. Additional information is shown in Figure 4b and Figure 7b. Good agreement between the model predictions and experimental data was obtained. The slopes of the lines in Figure 4b, obtained by solving the PBE with the multifractal breakage model derived for the dispersed phase of low viscosity (Equations (9) and (11)), are −0.535

for the Rushton turbine and −0.547 for the high efficiency impeller, respectively. These values are closer to the exponent of −0.617 resulting from Equation (5) for the minimum multifractal exponent, $\alpha = \alpha_{min} = 0.12$ (characterizing the most vigorous turbulent events and highest stresses), than to the −0.4 predicted for the most probable events characterized by $\alpha \approx 1$ (see Equation (5) for $\alpha = 1$). A slope of −0.4 is also predicted when intermittency is neglected and no multifractal exponent is introduced. The limiting value of the maximum stable drop size, $d_{max} \propto \varepsilon^{-0.617}$, corresponds to $d_{max} \propto We^{-0.93}$. Such a low exponent on the Weber number was first observed by Konno and Saito [58]. The exact value of this exponent is predicted by the multifractal breakage model. As was shown in previous papers, the multifractal breakage model not only predicts this exponent, but it also predicts transient drop size distributions very well [2,28,43], both for short agitation times, when the most probable stresses determine the drop size, and for long agitation times, when droplets of a diameter, $d$, are not disrupted by stresses, $p(d, \alpha = 1)$, but can be broken by stresses characterized by $\alpha < 1$. The really stable droplet size is given by $d_{32} \propto \varepsilon^{-0.617}$ (because the Sauter diameter, $d_{32}$, is proportional to $d_{max}$). The dashed line in Figure 4b shows the Sauter diameter predicted for the Rushton turbine by Coulaloglou and Tavlarides' model [9], which does not take into account local intermittency. In this case, the slope of the line is close to −0.4. Figure 5a shows a comparison of the transient drop size distributions predicted by Coulaloglou and Tavlarides' model for the Rushton turbine and high efficiency impeller for short ($t$ = 10 min) and long ($t$ = 2 h) agitation times. This model also predicts that the HE3 produces broader DSD at short agitation times and smaller drops at long agitation times. Figure 5b shows the distributions predicted for the Rushton turbine by different models. One can see that Coulaloglou and Tavlarides' model predicts smaller droplets. It can also be observed in Figure 4b.

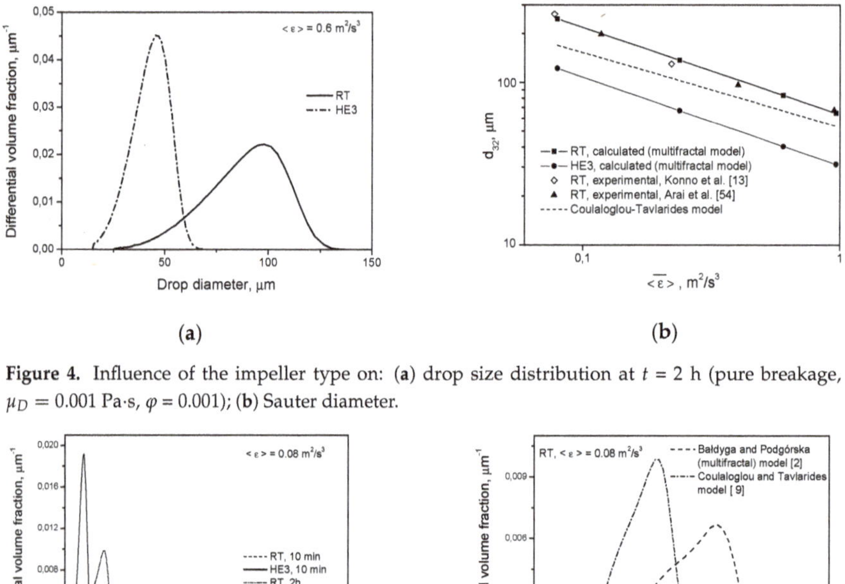


**Figure 4.** Influence of the impeller type on: (**a**) drop size distribution at $t$ = 2 h (pure breakage, $\mu_D = 0.001$ Pa·s, $\varphi = 0.001$); (**b**) Sauter diameter.

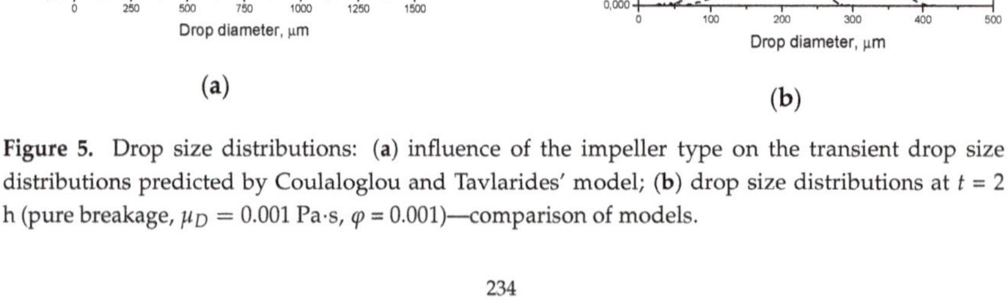

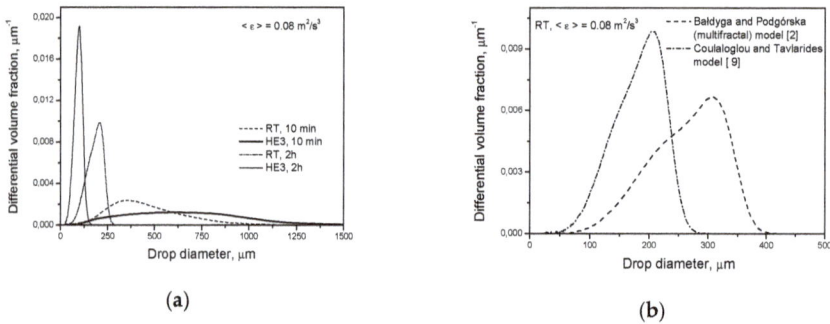

**Figure 5.** Drop size distributions: (**a**) influence of the impeller type on the transient drop size distributions predicted by Coulaloglou and Tavlarides' model; (**b**) drop size distributions at $t$ = 2 h (pure breakage, $\mu_D = 0.001$ Pa·s, $\varphi = 0.001$)—comparison of models.

Similar trends as those shown in Figures 3 and 4 are observed for the breakage of droplets of high viscosity, Figures 6 and 7, i.e., smaller droplets produced by the HE3 impeller. Calculations were performed using Equations (9) and (12). The slopes in Figure 7b are $-0.365$ for the RT and $-0.354$ for the HE3. As can be seen from Equation (6), the dependence of $d_{\max}$ on the energy dissipation rate is more complicated than for low viscosity, but again the limiting size is determined by $\alpha_{\min} = 0.12$. The slope for the case when intermittency is neglected can be predicted by setting $\alpha = 1$. In this case, $(d_{\max}/L)^{0.4(1-\alpha)}$ and $(d_{\max}/L)^{(1-\alpha)/3}$ disappear. Additionally, when 1 is small in comparison with the second term in a square bracket (i.e., when the effect of the shape restoring stress due to interfacial tension is small in comparison with the viscous stabilizing stress), the resulting slope can be estimated as $-0.25$. Thus, again, $-0.365$ (or $-0.354$) is smaller than $-0.25$, which shows that intermittency is important.

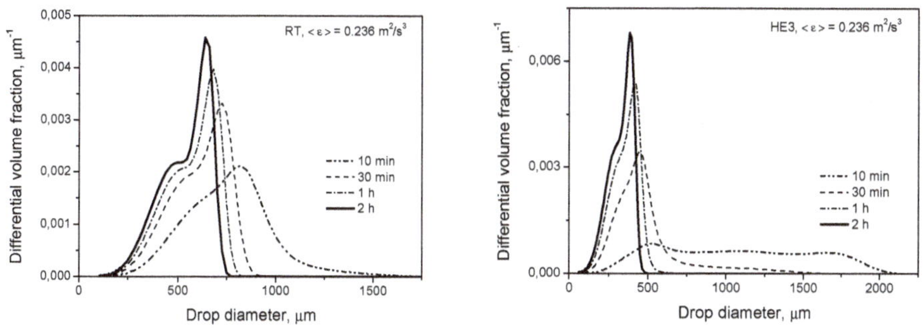


(a) (b)

**Figure 6.** Influence of the impeller type on transient drop size distributions for a dilute system ($\varphi$ = 0.001) with a dispersed phase of high viscosity ($\mu_D$ = 0.5 Pa·s): (**a**) Rushton turbine; (**b**) high efficiency impeller.

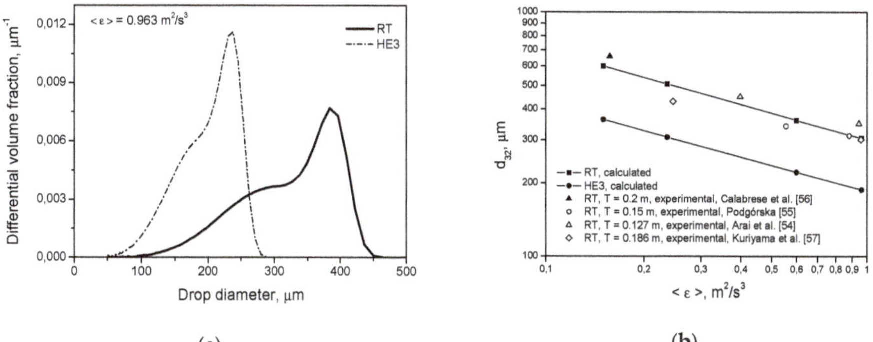


(a) (b)

**Figure 7.** Influence of the impeller type on: (**a**) drop size distribution at $t$ = 2 h (pure breakage, $\mu_D$ = 0.5 Pa·s, $\varphi$ = 0.001); (**b**) Sauter diameter.

Figure 8 shows transient drop size distributions predicted for droplets stabilized by surfactant when an additional disruptive stress due to the interfacial tension difference is generated (multifractal exponent, $\alpha_x$, is calculated from Equation (13)). Again, smaller droplets are produced in the tank equipped with a high efficiency impeller.

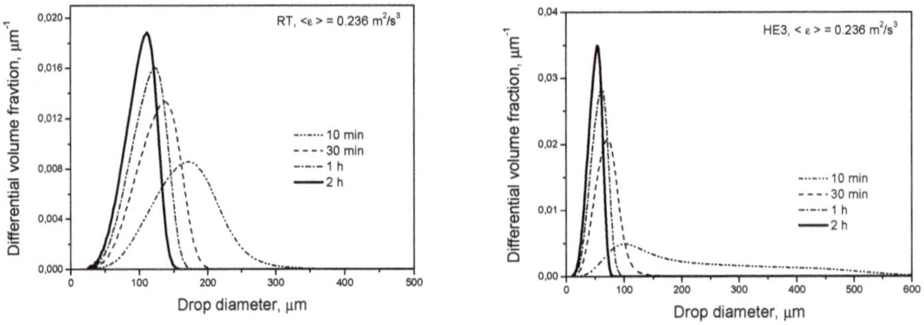


(**a**) (**b**)

**Figure 8.** Influence of the impeller type on transient drop size distributions for a dilute system ($\varphi = 0.001$) with droplets covered by surfactant: (**a**) Rushton turbine; (**b**) high efficiency impeller.

Figure 9a presents the final drop size distributions (being the result of dynamic equilibrium between breakage and coalescence) produced by different impellers at a higher dispersed phase volume fraction ($\varphi = 0.05$) for a pure liquid-liquid system (no surfactant) with a dispersed phase of low viscosity. Under such conditions, droplets have partially mobile interfaces and gentle collisions are favored (Equations (17) and (19)). Therefore, fast coalescence takes place in the bulk. The model predicts that at the same power input per unit mass, droplets produced in both systems are similar. In the case of a high dispersed phase viscosity, coalescence is highly hindered by immobilization of the drop interfaces. However, very wide drop size distributions are usually observed for such dispersions and as was shown earlier, droplets differing in size or in deformation in the contact area behave in a completely different way [32]. Coalescence of small rigid drops or small slightly deformed drops may be even faster in the zone of the high energy dissipation rate than in the bulk. However, the largest volume fraction of the population (being the result of dynamic equilibrium between breakage and coalescence) is occupied by large deformed droplets. Their behavior can be predicted using Equations (18) and (20). The predicted drop size distributions are shown in Figure 9b. Slightly smaller droplets are produced by the HE3.

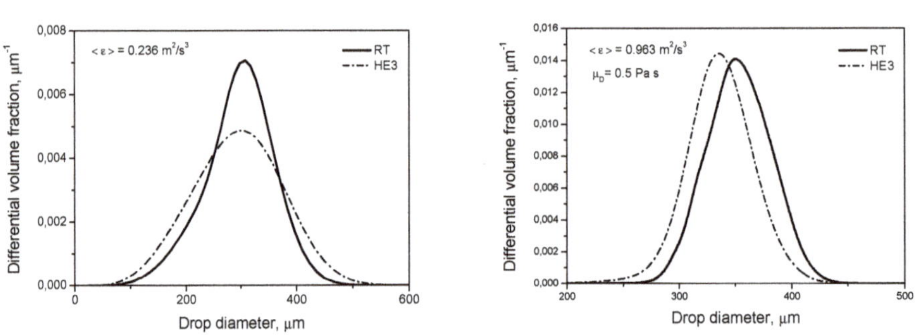


(**a**) (**b**)

**Figure 9.** Influence of the impeller type on the drop size distribution in: (**a**) fast coalescing dispersion ($\mu_D = 0.001$ Pa·s, $\varphi = 0.05$); (**b**) slowly coalescing dispersion ($\mu_D = 0.5$ Pa·s, $\varphi = 0.05$).

From the presented results, it follows that smaller droplets are produced at the same power input per unit mass by a low power number, high efficiency impeller (HE3) when drop breakage prevails.

**Funding:** This research was partly funded by Polish MNSW, grant number N N208 019534.

**Conflicts of Interest:** The author declares no conflict of interest.

## References


1. Bałdyga, J.; Bourne, J.R. Interpretation of turbulent mixing using fractals and multifractals. *Chem. Eng. Sci.* **1995**, *50*, 381–400. [CrossRef]
2. Bałdyga, J.; Podgórska, W. Drop break-up in intermittent turbulence: Maximum stable and transient sizes of drops. *Can. J. Chem. Eng.* **1998**, *76*, 456–470. [CrossRef]
3. Kraichnan, R.H. Turbulent cascade and intermittency growth. *Proc. R. Soc. Lond. Ser. A* **1991**, *434*, 65–78. [CrossRef]
4. Meneveau, C.; Sreenivasan, K.R. The multifractal nature of turbulent energy dissipation. *J. Fluid Mech.* **1991**, *224*, 429–484. [CrossRef]
5. Batchelor, G.K. *The Theory of Homogeneous Turbulence*; Cambridge University Press: Cambridge UK, 1953; pp. 183–187.
6. McComb, W.D. *The Physics of Fluid Turbulence*; Oxford University Press: Cambridge UK, 1994; pp. 100–112.
7. Benzi, R.; Biferale, L.; Paladin, G.; Vulpiani, A.; Vergassola, M. Multifractality in the statistics of the velocity gradients in turbulence. *Phys. Rev. Lett.* **1991**, *67*, 2299–2302. [CrossRef] [PubMed]
8. Frisch, U. *Turbulence: The Legacy of A.N. Kolmogorov*; Cambridge University Press: Cambridge, UK, 1995.
9. Coulaloglou, C.A.; Tavlarides, L.L. Description of interaction processes in agitated liquid-liquid dispersions. *Chem. Eng. Sci.* **1977**, *32*, 1289–1297. [CrossRef]
10. Chatzi, E.G.; Lee, J.M. Analysis of interactions for liquid-liquid dispersions in agitated vessels. *Ind. Eng. Chem. Res.* **1987**, *26*, 2263–2267. [CrossRef]
11. Chatzi, E.G.; Gavrielides, A.D.; Kiparissides, C. Generalized model for prediction of the steady-state drop size distributions in batch stirred vessels. *Ind. Eng. Chem. Res.* **1989**, *28*, 1704–1711. [CrossRef]
12. Narsimhan, G.; Gupta, J.P.; Ramkrishna, D. A model for transitional breakage probability of droplets in agitated lean liquid-liquid dispersions. *Chem. Eng. Sci.* **1979**, *34*, 257–265. [CrossRef]
13. Konno, M.; Aoki, M.; Saito, S. Scale effect on breakup process in liquid-liquid agitated tanks. *J. Chem. Eng. Jpn.* **1983**, *16*, 312–319. [CrossRef]
14. Martinez-Bazan, C.; Mintañes, J.L.; Lasheras, J.C. On the breakup of an air bubble injected into a fully developed turbulent flow. Part 1. Breakup frequency. *J. Fluid Mech.* **1999**, *401*, 157–182. [CrossRef]
15. Tsouris, C.; Tavlarides, L. Breakage and coalescence models for drops in turbulent dispersions. *AIChE J.* **1994**, *40*, 395–406. [CrossRef]
16. Luo, H.; Svendsen, H.F. Theoretical model for drop and bubble breakup in turbulent dispersion. *AIChE J.* **1996**, *42*, 1225–1233. [CrossRef]
17. Wang, T.; Wang, J.; Jin, Y. A novel theoretical breakup kernel function for bubbles/droplets in a turbulent flow. *Chem. Eng. Sci.* **2003**, *58*, 4629–4637. [CrossRef]
18. Han, L.; Gong, S.; Li, Y.; Gao, N.; Fu, J.; Luo, H.; Liu, Z. Influence of energy spectrum distribution on drop breakage in turbulent flows. *Chem. Eng. Sci.* **2014**, *117*, 55–70. [CrossRef]
19. Han, L.; Gong, S.; Ding, Y.; Fu, J.; Gao, N.; Luo, H. Consideration of low viscous droplet breakage in the framework of the wide energy spectrum and the multiple fragments. *AIChE J.* **2015**, *61*, 2147–2168. [CrossRef]
20. Solsvik, J.; Jacobsen, H.A. Development of fluid particle breakup and coalescence closure models for the complete energy spectrum of isotropic turbulence. *Ind. Eng. Chem. Res.* **2016**, *55*, 1449–1460. [CrossRef]
21. Solsvik, J.; Jacobsen, H.A. A review of statistical turbulence theory required extending the population balance closure models to the entire spectrum of turbulence. *AIChE J.* **2016**, *62*, 1795–1820. [CrossRef]
22. Hinze, J.O. Fundamentals of the hydrodynamic mechanism of splitting in dispersion process. *AIChE J.* **1955**, *1*, 289–295. [CrossRef]
23. Andersson, R.; Andersson, B. On the breakup of fluid particles in turbulent flows. *AIChE J.* **2006**, *52*, 2020–2030. [CrossRef]
24. Andersson, R.; Andersson, B. Modeling the breakup of fluid particles in turbulent flows. *AIChE J.* **2006**, *52*, 2031–2038. [CrossRef]
25. Chen, Z.; Prüss, J.; Warnecke, H.J. A population balance model for disperse systems: Drop size distribution in emulsion. *Chem. Eng. Sci.* **1998**, *53*, 1059–1066. [CrossRef]

26. Alopaeus, V.; Koskinen, J.; Keskinen, K.L.; Majander, J. Simulation of the population balances for liquid-liquid systems in a nonideal stirred tank. Part 2—parameter fitting and the use of the multiblock model for dense dispersions. *Chem. Eng. Sci.* **2002**, *57*, 1815–1825. [CrossRef]
27. Maaβ, S.; Kraume, M. Determination of breakage rates using single drop experiments. *Chem. Eng. Sci.* **2012**, *70*, 146–164.
28. Bąk, A.; Podgórska, W. Investigation of drop breakage and coalescence in the liquid-liquid system with nonionic surfactants Tween 20 and Tween 80. *Chem. Eng. Sci.* **2012**, *74*, 181–191. [CrossRef]
29. Podgórska, W.; Bałdyga, J. Drop break-up and coalescence in intermittent turbulent flow. In Proceedings of the 10th European Conference on Mixing, Delft, The Netherlands, 2–5 July 2000; pp. 141–148.
30. Podgórska, W.; Bałdyga, J. Scale-up effects on the drop size distribution of liquid-liquid dispersions in agitated vessels. *Chem. Eng. Sci.* **2001**, *56*, 741–746. [CrossRef]
31. Podgórska, W. Scale-up effects in coalescing dispersions – comparison of liquid-liquid systems differing in interface mobility. *Chem. Eng. Sci.* **2005**, *60*, 2115–2125. [CrossRef]
32. Podgórska, W. Influence of dispersed phase viscosity on drop coalescence in turbulent flow. *Chem. Eng. Res. Des.* **2007**, *85*, 721–729. [CrossRef]
33. Podgórska, W.; Marchisio, D.L. Modeling of turbulent drop coalescence in the presence of electrostatic forces. *Chem. Eng. Res. Des.* **2016**, *108*, 30–41. [CrossRef]
34. Chesters, A.K. The modeling of coalescence processes in fluid-liquid dispersions: A review of current understanding. *Trans. IChemE.* **1991**, *69*, 259–270.
35. Kuboi, R.; Komasawa, I.; Otake, T. Collision and coalescence of dispersed drops in turbulent liquid flow. *J. Chem. Eng. Jpn.* **1972**, *5*, 423–424. [CrossRef]
36. Tobin, T.; Ramkrishna, D. Modeling the effect of drop charge on coalescence in turbulent liquid-liquid dispersions. *Can. J. Chem. Eng.* **1999**, *77*, 1090–1104. [CrossRef]
37. Maindarkar, S.N.; Bongers, P.; Henson, M.A. Predicting the effects of surfactant coverage on drop size distributions of homogenized emulsions. *Chem. Eng. Sci.* **2013**, *89*, 102–114. [CrossRef]
38. Davis, R.H.; Schonberg, J.A.; Rallison, J. The lubrication force between two viscous drops. *Phys. Fluids* **1989**, *1*, 77–81. [CrossRef]
39. Atiemo-Obeng, V.A.; Roy Penny, W.; Armenante, P. Solid-liquid mixing. In *Handbook of Industrial Mixing*; Paul, E.L., Atiemo-Obeng, V.A., Kresta, S.M., Eds.; Wiley-Interscience: Hoboken, NJ, USA, 2004; pp. 543–584.
40. Ibrahim, S.; Nienow, A.W. Power curves and flow patterns for a range of impellers in Newtonian fluid-40 less-than Re-less than 5X10(5). *Chem. Eng. Res. Des.* **1995**, *73*, 485–491.
41. Pacek, A.W.; Chamsart, S.; Nienow, A.W.; Bakker, A. The influence of impeller type on mean drop size and drop size distribution in an agitated vessel. *Chem. Eng. Sci.* **1999**, *54*, 4211–4222. [CrossRef]
42. Podgórska, W. Influence of the impeller type on drop size in liquid-liquid dispersions. In Proceedings of the 13 European Conference on Mixing, London, UK, 14–17 April 2009; pp. 14–17.
43. Bąk, A.; Podgórska, W. Influence of poly(vinyl alcohol) molecular weight on drop coalescence and breakage rate. *Chem. Eng. Res. Des.* **2016**, *108*, 88–100. [CrossRef]
44. Liao, Y.; Lucas, D. A literature review of theoretical models for drop and bubble breakup in turbulent dispersions. *Chem. Eng. Sci.* **2009**, *64*, 3389–3406. [CrossRef]
45. Chemineer. Available online: https://www.chemineer.com/products/chemineer (accessed on 18 March 2019).
46. Jaworski, Z.; Nienow, A.W.; Dyster, K.N. An LDA study of the turbulent flow field in a baffled vessel agitated by an axial, down-pumping hydrofoil impeller. *Can. J. Chem. Eng.* **1996**, *74*, 3–15. [CrossRef]
47. Bujalski, W.; Nienow, A.W.; Chatwin, S.; Cooke, M. The dependency on scale of power numbers of Rushton disc turbines. *Chem. Eng. Sci.* **1987**, *42*, 317–326. [CrossRef]
48. Podgórska, W.; Bałdyga, J. Drop break-up and coalescence in a stirred tank. *Task Quarterly* **2003**, *7*, 409–424.
49. Podgórska, W.; Bałdyga, J. Effect of large- and small-scale inhomogeneity of turbulence in an agitated vessel on drop size distribution. In Proceedings of the 17th International Symposium on Chemical Reaction Engineering ISCRE, Hong Kong, China, 25–28 August 2002; pp. 1–21.
50. Ng, K.; Yianneskis, M. Observations of the distribution of energy dissipation in stirred vessels. *Chem. Eng. Res. Des.* **2000**, *78*, 334–341. [CrossRef]
51. Okamoto, Y.; Nishikawa, M.; Hashimoto, K. Energy dissipation rate distribution in mixing vessels and its effect on liquid-liquid dispersion and solid-liquid mass transfer. *Int. Chem. Eng.* **1981**, *21*, 88–94.

52. Zhou, G.; Kresta, S.M. Distribution of energy between convective and turbulent flow for three frequently used impellers. *Chem. Eng. Res. Des.* **1996**, *74*, 379–389.
53. Baldi, S.; Yianneskis, M. On the quantification of energy dissipation in the impeller stream of a stirred vessel from fluctuating velocity gradient measurements. *Chem. Eng. Sci.* **2004**, *59*, 2659–2671. [CrossRef]
54. Podgórska, W. Modeling of high viscosity oil drop breakage process in intermittent turbulence. *Chem. Eng. Sci.* **2006**, *61*, 2986–2993. [CrossRef]
55. Arai, K.; Konno, M.; Matunaga, Y.; Saito, S. Effect of dispersed-phase viscosity on the maximum stable drop size for breakup in turbulent flow. *J. Chem. Eng. Jpn.* **1977**, *10*, 325–330. [CrossRef]
56. Calabrese, R.V.; Chang, T.P.K.; Dang, P.T. Drop break-up in turbulent stirred-tank contactors. Part I: Effect of dispersed phase viscosity. *AIChE J* **1986**, *32*, 657–666. [CrossRef]
57. Kuriyama, M.; Ono, M.; Tokanai, H.; Konno, H. Correlation of transient sizes of highly viscous drops in dispersion process in liquid-liquid agitation. *Chem. Eng. Res. Des.* **1996**, *74*, 431–437.
58. Konno, M.; Saito, S. Correlation of drop sizes in liquid-liquid agitation at low dispersed phase volume fraction. *J. Chem. Eng. Jpn.* **1987**, *20*, 533–535. [CrossRef]

*entropy*

MDPI

*Article*

# Statistical Lyapunov Theory Based on Bifurcation Analysis of Energy Cascade in Isotropic Homogeneous Turbulence: A Physical–Mathematical Review

**Nicola de Divitiis**

Department of Mechanical and Aerospace Engineering, "La Sapienza" University, Via Eudossiana, 18, 00184 Rome, Italy; n.dedivitiis@gmail.com or nicola.dedivitiis@uniroma1.it; Tel.: +39-0644585268; Fax: +39-0644585750

Received: 26 April 2019; Accepted: 20 May 2019; Published: 23 May 2019

**Abstract:** This work presents a review of previous articles dealing with an original turbulence theory proposed by the author and provides new theoretical insights into some related issues. The new theoretical procedures and methodological approaches confirm and corroborate the previous results. These articles study the regime of homogeneous isotropic turbulence for incompressible fluids and propose theoretical approaches based on a specific Lyapunov theory for determining the closures of the von Kármán–Howarth and Corrsin equations and the statistics of velocity and temperature difference. While numerous works are present in the literature which concern the closures of the autocorrelation equations in the Fourier domain (i.e., Lin equation closure), few articles deal with the closures of the autocorrelation equations in the physical space. These latter, being based on the eddy–viscosity concept, describe diffusive closure models. On the other hand, the proposed Lyapunov theory leads to nondiffusive closures based on the property that, in turbulence, contiguous fluid particles trajectories continuously diverge. Therefore, the main motivation of this review is to present a theoretical formulation which does not adopt the eddy–viscosity paradigm and summarizes the results of the previous works. Next, this analysis assumes that the current fluid placements, together with velocity and temperature fields, are fluid state variables. This leads to the closures of the autocorrelation equations and helps to interpret the mechanism of energy cascade as due to the continuous divergence of the contiguous trajectories. Furthermore, novel theoretical issues are here presented among which we can mention the following ones. The bifurcation rate of the velocity gradient, calculated along fluid particles trajectories, is shown to be much larger than the corresponding maximal Lyapunov exponent. On that basis, an interpretation of the energy cascade phenomenon is given and the statistics of finite time Lyapunov exponent of the velocity gradient is shown to be represented by normal distribution functions. Next, the self–similarity produced by the proposed closures is analyzed and a proper bifurcation analysis of the closed von Kármán–Howarth equation is performed. This latter investigates the route from developed turbulence toward the non–chaotic regimes, leading to an estimate of the critical Taylor scale Reynolds number. A proper statistical decomposition based on extended distribution functions and on the Navier–Stokes equations is presented, which leads to the statistics of velocity and temperature difference.

**Keywords:** energy cascade; bifurcations; Lyapunov theory

## 1. Introduction

This article presents a review of previous works of the author regarding an original Lyapunov analysis of the developed turbulence which leads to the closures of the von Kármán–Howarth and Corrsin equations and to the statistics of both velocity and temperature difference [1–7]. This theory studies the fully developed homogeneous isotropic turbulence through the bifurcations of the incompressible Navier–Stokes equations using a specific statistical Lyapunov analysis of the fluid kinematic field. In addition, now it is introduced the energy cascade interpretation and explained some of the mathematical properties of the proposed closures. This work is organized into two parts. One is the reasoned review of previous results but with new demonstrations and theoretical procedures. The other one, presented in sections marked with asterisk symbol "*", concerns new theoretical issues of the proposed turbulence theory.

Although numerous articles were written which concern the closures of the Lin equation in the Fourier domain [8–16], few works address the closures of the autocorrelation equations in the physical space. These last ones, being based on the eddy–viscosity concept, describe diffusive closure models. Unlike the latter, the proposed Lyapunov theory provides nondiffusive closures in the physical space based on the property that, in developed turbulence, contiguous fluid particles trajectories continuously diverge. Thus, the main purpose of this review is to summarize the results of the previous works based on a theory which does not use the eddy–viscosity paradigm and to give new theoretical insights into some related issues.

The homogeneous isotropic turbulence is an ideal flow regime characterized by the energy cascade phenomenon where the diverse parts of fluid exhibit the same statistics and isotropy. On the other hand, the turbulent flows occurring in nature and in the various fields of engineering are generally much more complex than homogeneous isotropic turbulence. In such flows, spatial variations of average velocity and of other statistical flow properties can happen causing very complex simultaneous effects that add to the turbulent energy cascade and interact with the latter in a nontrivial fashion. Hence, the study of the energy cascade separately from the other phenomena requires the analysis of isotropic homogenous turbulence.

The von Kármán–Howarth and Corrsin equations are the evolution equations of longitudinal velocity and temperature correlations in homogeneous isotropic turbulence, respectively. Both the equations, being unclosed, need the adoption of proper closures [17–20]. In detail, the von Kármán–Howarth equation includes $K$, the term due to the inertia forces and directly related to the longitudinal triple velocity correlation $k$, which has to be properly modelled. The modeling of such term must take into account that, due to the inertia forces, $K$ does not modify the kinetic energy and satisfies the detailed conservation of energy [18]. This latter states that the exchange of energy between wave–numbers is only linked to the amplitudes of such wave–numbers and of their difference [21]. Different works propose for the von Kármán–Howarth equation the diffusion approximation [22–24]

$$k = 2\frac{D}{u}\frac{\partial f}{\partial r} \tag{1}$$

where $r$ and $D = D(r)$ are separation distance and turbulent diffusion parameter, respectively and $u^2 = \langle u_i u_i \rangle /3$ corresponds to the longitudinal velocity standard deviation. Following Equation (1), the turbulence can be viewed as a diffusivity phenomenon depending upon $r$, where $K$ will include a term proportional to $\partial^2 f/\partial r^2$. In the framework of Equation (1), Hasselmann [22] proposed, in 1958, a closure suggesting a link between $k$ and $f$ which expresses $k$ in function of the momentum convected through a spherical surface. His model, which incorporates a free parameter, expresses $D(r)$ by means of a complex expression. Thereafter, Millionshtchikov developed a closure of the form $D(r) = k_1 ur$, where $k_1$ represents an empirical constant [23]. Although both the models describe two possible mechanisms of energy cascade, in general, do not satisfy some physical conditions. For instance, the Hasselmann model does not verify the continuity equation for all the initial conditions, whereas the Millionshtchikov equation gives, following Equation (1), values of velocity difference skewness in contrast with experiments and energy cascade [18]. More recently, Oberlack and Peters [24] suggested a closure where $D(r) = k_2 r u \sqrt{1-f}$, being $k_2$ a constant parameter. The authors show that such closure reproduces the energy cascade and, for a proper choice of $k_2$, provides results in agreement with the experiments [24].

For what concerns the Corrsin equation, this exhibits $G$, the term responsible for the thermal energy cascade. This quantity, directly related to the triple velocity–temperature correlation $m^*$, also needs adequate modellation. As $G$ depends also on the velocity correlation, the Corrsin equation requires the knowledge of $f$, thus it must be solved together to the von Kármán–Howarth equation. Different works can be found in the literature which deal with the closure of Corrsin equation. Some of them study the self-similarity of the temperature correlation in order to analyze properties and possible expressions for $G$. Such studies are supported by the idea that the simultaneous effect of energy cascade, conductivity and viscosity, makes the temperature correlation similar in the time. This question was theoretically addressed by George (see [25,26] and references therein) which showed that the decaying isotropic turbulence reaches the self–similarity, while the temperature correlation is scaled by the Taylor microscale whose current value depends on the initial condition. More recently, Antonia et al. [27] studied the temperature structure functions in decaying homogeneous isotropic turbulence and found that the standard deviation of the temperature, as well as the turbulent kinetic energy, follows approximately the similarity over a wide range of length scales. There, the authors used this approximate similarity to estimate the third–order correlations and found satisfactory agreement between measured and calculated functions. On the other hand, the temperature correlation can be obtained using proper closures of von Kármán–Howarth and Corrsin equations suitable for the energy cascade phenomenon. On this argument, several articles has been written. For instance, Baev and Chernykh [28] (and references therein) analyzed velocity and temperature correlations by means of a closure model based on the gradient hypothesis which relates pair longitudinal second and third order correlations, by means of empirical coefficients.

Although other works regarding the von Kármán–Howarth equation were written [29–33], to the author's knowledge a physical–mathematical analysis based on basic principles which provides analytical closures of von Kármán–Howarth and Corrsin equations has not received due attention. Therefore, the aim of the this work is to present a review of the Lyapunov analysis presented in [1–7] and new theoretical insights into some related issues.

In the present formulation, based on the Navier–Stokes bifurcations, the current fluid placements, together with velocity and temperature fields, are considered to be fluid state variables. This leads to the closures of the autocorrelation equations and helps to interpret the mechanism of energy cascade as due to the continuous divergence of the contiguous trajectories.

In line with Ref. [3], the present work first addresses the problem for defining the bifurcations for incompressible Navier–Stokes equations, considering that these latter can be reduced to an opportune symbolic form of operators for which the classical bifurcation theory of differential equations can be applied [34]. In such framework, this analysis remarks that a single Navier–Stokes bifurcation will generate a doubling of the velocity field and of all its several properties, with particular reference to the characteristic length scales. If on one side the lengths are doubled due to bifurcations, on the other hand the characteristic scale for homogeneous flows in infinite domains is not defined. Hence, the problem to define the characteristic length—and therefore the flow Reynolds number—in such situation is also discussed. Such characteristic scale is here defined in terms of spatial variations of initial or current velocity field in such a way that, in fully developed homogeneous isotropic turbulence, this length coincides with the Taylor microscale. As far as the characteristic velocity is concerned, this is also defined in terms of velocity field so that, in developed turbulence, identifies the velocity standard deviation.

The trajectories bifurcations in the phase space of the velocity field are here formally dealt with using a proper Volterra integral formulation of the Navier–Stokes equations, whereas the turbulence transition is qualitatively analyzed through general properties of the bifurcations and of the route toward the fully developed chaos. This background, regarding the general bifurcations properties and the route toward the chaos, will be useful for this analysis.

The adopted statistical Lyapunov theory shows how the fluid relative kinematics can be much more rapid than velocity and temperature fields in developed turbulence, so that fluid strain and velocity fields are statistically independent with each other. Moreover, in addition to References [1–7], this analysis introduces the bifurcation rate of the velocity gradient, a quantity providing the frequency at which the velocity gradient determinant vanishes along fluid particles trajectories. The bifurcation rate, in fully developed turbulence, is shown to be much greater than the maximal Lyapunov exponent of the velocity gradient. This explains the energy cascade through the relation between material vorticity, Lyapunov vectors and bifurcation rate using the Lyapunov theory. In detail, the energy cascade can be viewed as a continuous and intensive stretching and folding process of fluid particles which involves smaller and smaller length scales during the fluid motion, where the folding frequency equals the bifurcation rate.

Next, the statistics of the Lyapunov exponents is reviewed. In agreement with Reference [6], we show that the local Lyapunov exponents are uniformely unsymmetrically distributed in their interval of variation. Unlike Reference [6] which uses the criterion of maximum entropy associated with the fluid particles placements, the isotropy and homogeneity hypotheses are here adopted. A further result with respect to the previous issues pertains the finite time Lyapunov exponents statistics: through the bifurcation analysis and the central limit theorem, we show that the finite time Lyapunov exponent tends to a fluctuating variable distributed following a normal distribution function.

Thereafter, the closure formulas of von Kármán–Howarth and Corrsin equations are derived through the Liouville equation and finite scale Lyapunov exponent statistics. These closures do not correspond to a diffusive model, being the result of the trajectories' divergence in the continuum fluid. Such formulas coincide with those just obtained in References [1,4,5] where it is shown that such closures adequately describe the energy cascade phenomenon, reproducing, negative skewness of velocity difference, the Kolmogorov law and temperature spectra in line with the theoretical argumentation of Kolmogorov, Obukhov–Corrsin and Batchelor [35–37], with experimental results [38,39] and with numerical data [40,41]. These closures are here achieved by using different mathematical procedures with respect to the other articles [1,4,5]. While the previous works derive such closures studying the local fluid act of motion in the finite scale Lyapunov basis [1,4] and adopting maximum and average finite scale Lyapunov exponents [5], here these closures are obtained by means of the local finite scale Lyapunov exponents PDF, showing that the assumptions of References [1,4,5] agree with this analysis, corroborating the previous results. Some of the properties of the proposed closures are then studied, with particular reference to the evolution times of the developed correlations and their self–similarity. In detail, as new result with respect the previous articles, this analysis shows that the proposed closures generate correlations self–similarity in proper ranges of separation distance, which is directly linked to the particles trajectories divergence.

Furthermore, a novel bifurcation analysis of the closed von Kármán–Howarth equation is proposed, which considers the route starting from the fully developed turbulence toward the non–chaotic regimes. This extends the discussion of the previous works and represents an alternative point of view for studying the turbulent transition. According to this analysis, the closed von Kármán–Howarth equation is decomposed in several ordinary differential equations through the Taylor series expansion of the longitudinal velocity correlation. This procedure, which also accounts for the aforementioned self–similarity, leads to estimating the Taylor scale Reynolds number at the transition. This latter is found to be 10, a value in good agreement with several experiments which give values around 10, and in particular with the bifurcations analysis of the energy cascade of Reference [3], which provides a critical Reynolds number of 10.13 if the route toward the turbulence follows the Feigenbaum scenario [42,43].

Finally, the statistics of velocity and temperature difference, of paramount importance for estimating the energy cascade, is reviewed. While References [1,2,4,7] determine such statistics through a concise heuristic method, this analysis uses a specific statistical decomposition of velocity and temperature which adopts appropriate stochastic variables related to the Navier–Stokes bifurcations. The novelty of the present approach with respect to the previous articles is that the random variables of such decomposition are opportunely chosen to reproduce the Navier–Stokes bifurcation effects and the isotropy: these are highly nonsymmetrically distributed stochastic variables following opportune extended distribution functions which can assume negative values. Such decomposition, able to reproduce negative skewness of longitudinal velocity difference, provides a statistics of both velocity and temperature difference in agreement with theoretical and experimental data known from the literature [44–48]. Here, in addition to References [1,2,4,7], a detailed mathematical analysis is presented which concerns the statistical properties of the aforementioned extended distribution functions in relation to the Navier–Stokes bifurcations.

In brief, the original contributions of the present work can be summarized as:

(i) The bifurcation rate associated with the velocity gradient is shown to be much larger than the maximal Lyapunov exponent of the velocity gradient.
(ii) As the consequence of (i), the energy cascade can be viewed as a succession of stretching and folding of fluid particles which involves smaller and smaller length scales, where the particle folding happens at the frequency of the bifurcation rate.
(iii) As the consequence of (i), the central limit theorem provides reasonable argumentation that the finite time Lyapunov exponent is distributed following a gaussian distribution function.
(iv) The proposed closures generate correlations self–similarity in proper ranges of variation of the separation distance which is directly caused by the continuous fluid particles trajectories divergence.
(v) A specific bifurcation analysis of the closed von Kármán–Howarth equation is proposed which allows to estimate the critical Taylor scale Reynolds number in isotropic turbulence.
(vi) A statistical decomposition of velocity and temperature is presented which is based on stochastic variables distributed following extended distribution functions. Such decomposition leads to the statistics of velocity and temperature difference, where the intermittency of these latter increases as Reynolds number and Péclet number rise.

## 2. Background

In the framework of the link between bifurcations and turbulence, this section deals with some of the fundamental elements of the Navier–Stokes equations and heat equation, useful for the present analysis. In particular, we will address the problem of defining an adequate bifurcation analysis for the Navier–Stokes equations and will analyze the meaning of the characteristic length scales when a homogeneous flow is in an infinite domain. All the considerations regarding the fluid temperature can also be applied to any passive scalar that exhibits diffusivity. A statistically homogeneous and isotropic flow with null average velocity is considered.

In order to formulate the bifurcation analysis, we start from the Navier–Stokes equations and the temperature equation

$$\begin{aligned} &\nabla_{\mathbf{x}} \cdot \mathbf{u} = 0, \\ &\frac{\partial \mathbf{u}}{\partial t} = -\nabla_{\mathbf{x}} \mathbf{u}\, \mathbf{u} - \frac{\nabla_{\mathbf{x}} p}{\rho} + \nu \nabla_{\mathbf{x}}^2 \mathbf{u} \end{aligned} \tag{2}$$

$$\frac{\partial \vartheta}{\partial t} = -\mathbf{u} \cdot \nabla_{\mathbf{x}} \vartheta + \chi \nabla_{\mathbf{x}}^2 \vartheta \tag{3}$$

where $\mathbf{u} = \mathbf{u}(t, \mathbf{x})$, $p = p(t, \mathbf{x})$ and $\vartheta = \vartheta(t, \mathbf{x})$ are velocity, pressure and temperature fields, $\nu$ and $\chi = k\rho/C_p$ are fluid kinematic viscosity and thermal diffusivity, being $\rho$ = const, $k$ and $C_p$ density, fluid thermal conductivity and specific heat at constant pressure, respectively. In this study $\nu$ and $\chi$

are supposed to be independent from the temperature, thus Equation (2) is autonomous with respect to Equation (3), whereas Equation (3) will depend on Equation (2).

To define the bifurcations of Equations (2) and (3), such equations are first expressed in the symbolic form of operators. To this end, in the momentum Navier–Stokes equations, the pressure field is eliminated by means of the continuity equation, thus Equations (2) and (3) are formally written as

$$\dot{\mathbf{u}} = \mathbf{N}(\mathbf{u}; \nu), \tag{4}$$

$$\dot{\vartheta} = \mathbf{M}(\mathbf{u}, \vartheta; \chi) \tag{5}$$

in which $\mathbf{N}$ is a nonlinear quadratic operator incorporating $-\mathbf{u} \cdot \nabla_{\mathbf{x}} \mathbf{u}$, $-\nu \nabla_{\mathbf{x}}^2 \mathbf{u}$ and the integral nonlinear operator which expresses the pressure gradient as a functional of the velocity field, being

$$p(t, \mathbf{x}) = \frac{\rho}{4\pi} \int \frac{\partial^2 u_i' u_j'}{\partial x_i' \partial x_j'} \frac{dV(\mathbf{x}')}{|\mathbf{x}' - \mathbf{x}|} \tag{6}$$

Therefore, $p$ provides nonlocal effects of the velocity field [49] and the Navier–Stokes equations are reduced to be an integro–differential equation formally expressed by Equation (4). For what concerns Equation (5), it is the evolution equation of $\vartheta$, where $\mathbf{M}$ is a linear operator of $\vartheta$. Accordingly, transition and turbulence are caused by the bifurcations of Equation (4), where $\nu^{-1}$ plays the role of the control parameter. At this stage of the analysis, it is worth to remark the following two items: (a) there is no explicit methods of bifurcation analysis for integro–differential equations such as Equation (4). (b) since the flow is statistically homogeneous in an infinite domain, characteristic scales of the problem are not defined.

The item (a) can be solved according to the analysis method proposed by Ruelle and Takens in Reference [34]: it is supposed that the infinite dimensional space of velocity field $\{\mathbf{u}\}$ can be replaced by a finite–dimensional manifold, then Equation (4) can be reduced to be the equation of the kind studied by Ruelle and Takens in Reference [34]. Therefore, the classical bifurcation theory of ordinary differential equations [34,43,50] can be formally applied to Equation (4) and the present analysis can be considered valid within the limits of the formulation proposed in Reference [34].

For what concerns the characteristic length, a homogeneous flow in infinite domain is free from boundary conditions, thus the characteristic scale, being not defined, is here chosen in function of the spatial variations of the current velocity field. Thus, for all flow regimes in infinite regions, (i.e., non–chaotic, turbulent and transition flows), characteristic length and velocity, $L$ and $U$ respectively, are here chosen in terms of volume integrals of $\mathbf{u}$ in the following manner

$$\begin{aligned} U^2 &= \lim_{\mathcal{V} \to \infty} \frac{1}{\mathcal{V}} \int_{\mathcal{V}} \mathbf{u}(t, \mathbf{x}) \cdot \mathbf{u}(t, \mathbf{x}) d\mathcal{V}(\mathbf{x}), \\ G^2 &= \lim_{\mathcal{V} \to \infty} \frac{1}{\mathcal{V}} \int_{\mathcal{V}} \nabla_{\mathbf{x}} \mathbf{u} : \nabla_{\mathbf{x}} \mathbf{u} \, d\mathcal{V}(\mathbf{x}), \\ L^2 &= c \, \frac{U^2}{G^2} \end{aligned} \tag{7}$$

where $\mathcal{V}$ is the fluid domain volume, ":" denotes the Frobenius inner product and $c = O(1)$ is a dimensionless constant which will be properly chosen. The flow Reynolds number is then defined in terms of $U$ and $L$ as

$$Re = \frac{UL}{\nu}. \tag{8}$$

Equation (8) provides an extension of the Taylor scale Reynolds number which applies for every flow regime. In particular, such definition holds also for non turbulent flows, where $U$ and $L$, although not velocity standard deviation and statistical correlation scale, provide a generalization of the latter. In fully developed homogeneous turbulence, the volume integrals appearing in Equation (7) equal statistical averages calculated over the velocity field ensemble, such as velocity standard deviation and dissipation rate. Accordingly, in isotropic homogeneous turbulence, $L$ and $U$ identify, respectively, the Taylor scale $\lambda_T$ and standard deviation $u$ of one of the velocity components and $Re = U\,L/\nu$ coincides with the Taylor scale Reynolds number $R_T$. Such definitions (7) extend the concept of velocity variance and statistical correlation scale and will be used for the bifurcation analysis proposed in this work.

### 3. Navier–Stokes Bifurcations

Before introducing the bifurcations analysis of the Navier–Stokes equations in the operatorial form (4), it is worth remarking that a given point in the space of velocity fields set $\bar{\mathbf{u}} \in \{\mathbf{u}\}$—or temperature field $\bar{\vartheta} \in \{\vartheta\}$—corresponds to a spatial distribution including all its characteristics, in particular the length scales associated with $\bar{\mathbf{u}}$.

The bifurcations of Equation (4) happen when the Jacobian $\nabla_{\mathbf{u}}\mathbf{N}$ exhibits at least an eigenvalue with zero real part (NS–bifurcations), and this occurs when

$$\det(\nabla_{\mathbf{u}}\mathbf{N}) = 0. \tag{9}$$

Such bifurcations are responsible for multiple velocity fields $\hat{\mathbf{u}}$ which provides the same field $\dot{\mathbf{u}}$. In fact, during the fluid motion, multiple solutions $\hat{\mathbf{u}}$ and $\hat{\vartheta}$ can be determined, at each instant, through inversion of Equation (4)

$$\begin{aligned} \dot{\mathbf{u}} &= \mathbf{N}(\mathbf{u}; \nu) \\ \hat{\mathbf{u}} &= \mathbf{N}^{-1}(\dot{\mathbf{u}}; \nu), \\ \hat{\vartheta} &= \mathbf{M}^{-1}(\dot{\vartheta}, \hat{\mathbf{u}}; \chi) \end{aligned} \tag{10}$$

In the framework of the trajectories bifurcations in the phase space, the fluid motion can be expressed by means of Equation (4) and initial conditions $\mathbf{u}(0)$ and $\vartheta(0)$, using the following Volterra integral formulation

$$\begin{aligned} &\mathbf{u}(t) - \mathbf{u}(0) - \int_0^t \mathbf{N}(\mathbf{u}(\tau); \nu)\, d\tau \equiv \mathcal{N}(\mathbf{u}; \nu) = 0, \\ &\vartheta(t) - \vartheta(0) - \int_0^t \mathbf{M}(\mathbf{u}(\tau), \vartheta(\tau); \chi)\, d\tau \equiv \mathcal{M}(\mathbf{u}, \vartheta; \chi) = 0 \end{aligned} \tag{11}$$

where $\mathcal{N}$ and $\mathcal{M}$ are proper operators such that

$$\begin{aligned} &\mathcal{N} : \{\mathbf{u}\} \to \mathcal{N}(\{\mathbf{u}\}), \\ &\mathcal{M} : \{\mathbf{u}\} \times \{\vartheta\} \to \mathcal{M}(\{\mathbf{u}\} \times \{\vartheta\}) \end{aligned} \tag{12}$$

Specifically, $\mathcal{N}$ is a nonlinear operator of $\mathbf{u}$, where the image $\mathcal{N}(\{\mathbf{u}\})$ has the same structure of $\{\mathbf{u}\}$, whereas $\mathcal{M}$ is linear with respect to $\vartheta$ and the image $\mathcal{M}(\{\mathbf{u}\} \times \{\vartheta\})$ is isomorphic with $\{\vartheta\}$. Thus, $\mathcal{N}$ and $\mathcal{M}$ admit in general the following jacobians

$$\begin{aligned} &\nabla_{\mathbf{u}}\mathcal{N}, \\ &\nabla_{\vartheta}\mathcal{M} \end{aligned} \tag{13}$$

According to Equation (11), a trajectory bifurcation happens when $\nabla_{\mathbf{u}}\mathcal{N}$ is singular, that is when

$$\det\left(\nabla_{\mathbf{u}}\mathcal{N}\right) = 0 \tag{14}$$

and the multiple solutions of Equation (11), say $\hat{\mathbf{u}}$ and $\hat{\vartheta}$, are given in terms of $\mathbf{u}$ and $\vartheta$ through

$$\begin{aligned} &\mathcal{N}\left(\hat{\mathbf{u}};\nu\right) = \mathcal{N}\left(\mathbf{u};\nu\right) = 0, \\ &\mathcal{M}(\hat{\mathbf{u}},\hat{\vartheta};\chi) = \mathcal{M}(\mathbf{u},\vartheta;\chi) = 0 \end{aligned} \tag{15}$$

using the implicit functions theorem. Therefore, if velocity and temperature fields are supposed to be known for $\nu$=$\nu_0$, the fields calculated for $\nu \neq \nu_0$ are formally expressed as

$$\begin{aligned} &\hat{\mathbf{u}}(\nu) = \mathbf{u}(\nu_0) - \int_{\nu_0}^{\nu} \left(\nabla_{\mathbf{u}}\mathcal{N}\right)^{-1} \frac{\partial \mathcal{N}}{\partial \nu}\, d\nu, \\ &\hat{\vartheta}(\nu) = \vartheta(\nu_0) + \int_{\nu_0}^{\nu} \left(\nabla_{\vartheta}\mathcal{M}\right)^{-1} \left(\nabla_{\mathbf{u}}\mathcal{M}\right) \left(\nabla_{\mathbf{u}}\mathcal{N}\right)^{-1} \frac{\partial \mathcal{N}}{\partial \nu}\, d\nu, \end{aligned} \tag{16}$$

## 4. Qualitative Analysis of the Route Toward the Chaos

With reference to Equation (10) or (16), when $\nu^{-1}$ is relatively small, $\mathbf{N}$ and $\mathcal{N}$ behave like linear operators and Equation (15) returns $\hat{\mathbf{u}} \equiv \mathbf{u}(t,\mathbf{x})$ as unique solution. Increasing $\nu^{-1}$, the Navier–Stokes equations encounter the first bifurcation at $\nu = \nu_1$, the jacobian $\nabla_{\mathbf{u}}\mathcal{N}$ is singular there, and thereafter Equation (15) determines different velocity fields $\hat{\mathbf{u}}$ with the corresponding length scales. A single bifurcation causes a doubling of $\mathbf{u}$, that is, a doubling of the velocity values and of the length scales. Although the route toward the chaos can be of different kinds [34,42,43,51], one common element of these latter is that the number of encountered bifurcations at the onset of the chaotic regimes is about greater than three. Hence, if $\nu^{-1}$ is quite small, the velocity field can be represented by its Fourier series of a given basic scale. The first bifurcation introduces new solutions $\hat{\mathbf{u}}$ whose Fourier characteristic lengths are independent from the previous one. Thereafter, each bifurcation adds new independent scales, and, after the third bifurcation ($\nu^{-1} = \nu_*^{-1}$), the transition occurs, the several characteristic lengths and the velocity values appear to be continuously distributed and thus the velocity field is represented by the Fourier transform there. In such situations, a huge number of such solutions are unstable; $\mathbf{u}(t,\mathbf{x})$ tends to sweep the entire velocity field set and the motion is expected to be chaotic with a high level of mixing. As for $\hat{\vartheta}$, $\mathbf{M}$ and $\mathcal{M}$ are both linear operators of $\vartheta$, thus $\hat{\vartheta}$ follows the variations of $\hat{\mathbf{u}}$.

If $\nu^{-1}$ does not exceed its critical value, say $\nu_*^{-1}$, the velocity fields satisfying Equation (15) are limited in number and this corresponds to the intermediate stages of the route toward the chaos. On the contrary, when $\nu^{-1} > \nu_*^{-1}$, the region of developed turbulence where $\lambda_{NS} > 0$ is observed, being $\lambda_{NS}$ the average maximal Lyapunov exponent of the Navier–Stokes equations, is formally calculated as

$$\lambda_{NS} = \lim_{T \to \infty} \frac{1}{T} \int_0^T \frac{y \cdot \nabla_{\mathbf{u}} \mathbf{N} y}{y \cdot y} \, dt,$$
$$\dot{y} = \nabla_{\mathbf{u}} \mathbf{N}(\mathbf{u}; \nu) y, \tag{17}$$
$$\dot{\mathbf{u}} = \mathbf{N}(\mathbf{u}; \nu),$$

and $y$ is the Lyapunov vector associated with the Navier–Stokes equations. Then, $\nu_*^{-1}$ depends on $\mathbf{u}$, and $Re_*$, calculated with Equation (7), can be roughly estimated as the minimum value of $Re$ for which $\lambda_{NS} \geq 0$.

Figure 1 qualitatively shows the route from non–chaotic regimes toward the developed turbulence. Specifically, Figure 1a,b report two bifurcation maps at a given instant, providing the velocity component $u_1$ in a point of the space and one characteristic scale $\ell$ of the velocity field in function of $\nu^{-1}$. Figure 1c–e symbolically represent, for assigned values of $\nu$, the velocity field set (points inside the dashed circle), three different solutions of the Navier–Stokes equations—say P, Q and R—and the several subsets $\sigma_1$, $\sigma_2$,... which correspond to islands that are not swept during the fluid motion. The figure also depicts $L$=$L(\nu^{-1})$ and $U$=$U(\nu^{-1})$ (Figure 1f,g), formally calculated with Equation (7). Following Equation (16), these maps are not universal, as $u_1 = u_1(\nu^{-1})$, $\ell = \ell(\nu^{-1})$, $L = L(\nu^{-1})$ and $U = U(\nu^{-1})$ do not represent universal laws and their order of magnitude will depend on velocity field at $\nu_0^{-1}$. When $\nu^{-1} > \nu_3^{-1}$, the number of solutions diverges and the bifurcation tree of $u_1$ and $\ell$ drastically changes its structure showing tongue geometries that develop from the different bifurcations. As long as $\nu^{-1}$ does not exceed much $\nu_3^{-1}$, the extension of such tongues is relatively bounded, whereas the measure of the islands $\sigma_k$ is quite large. This means that, although $u_1$ and $\ell$ exhibit chaotic behavior there, these do not sweep completely their variation interval, thus Equation (4) do not behave like an ergodic dynamic system there. This corresponds to Figure 1c, where the velocity fields P, Q and R, being differently placed with respect to $\sigma_k$, $k = 1, 2, ..$ will exhibit different values of average kinetic energy and dissipation rate in $\mathcal{V}$. As $\nu^{-1}$ rises, these tongues gradually increase their extension whereas the measures of $\sigma_k$ diminish (see Figure 1d) until reaching a situation in which the bifurcation tongues overlap with each other and the islands $\sigma_k$ vanish (Figure 1e). Such developed overlapping corresponds to the chaotic behavior of $u_1$ and $\ell$, where these latter almost entirely describe their variation interval: Equation (4) behave like an ergodic dynamic system there, whereas all the velocity fields, in particular P, Q, and R, although different to each other, give the same values of average kinetic energy and dissipation rate in $\mathcal{V}$. This is the onset of the fully developed turbulence.

As far as $L$ and $U$ are concerned, these are both functionals of $\mathbf{u}$ following Equation (7), accordingly their variations in terms of $\nu^{-1}$ are peculiar, with quite different results with respect to $u_1$ and $\ell$. In particular, the structure of the first three bifurcations do not show important differences with respect to $u_1$ and $\ell$, whereas, after the third bifurcation ($\nu^{-1} > \nu_3^{-1}$), the chaotic regime begins and the bifurcation tree of $U$ and $L$ exhibits a completely different shape to the corresponding zone of $u_1$ and $\ell$. In detail, the chaotic region extension of $U$ and $L$ appears to be more limited than that of $u_1$ and $\ell$ until to collaps in the lines A–B when $\nu^{-1} > \nu_A^{-1}$. This is because the several bifurcations in $\nu_3^{-1} < \nu^{-1} < \nu_A^{-1}$ correspond to a large number of solutions that show different levels of average kinetic energy and dissipations rate in $\mathcal{V}$ which are in some way comparable to each other, respectively. Hence, although the chaotic regime is characterized by myriad of values of $u_1$ and $\ell$ which widely sweep the corresponding ranges, $L$ and $U$, being related to average kinetic energy and dissipation rate, will exhibit smaller variations. For relatively high values of $\nu^{-1}$, when the velocity fluctuations behavior is ergodic, the averages calculated on phase trajectory tends to the spatial averages. The region of chaotic regime collaps into the line A–B there. Along such lines, for assigned $\nu$, all the solutions—in particular P, Q and R—will exhibit the same level of kinetic energy and dissipation and this represents the regime of fully developed turbulence.

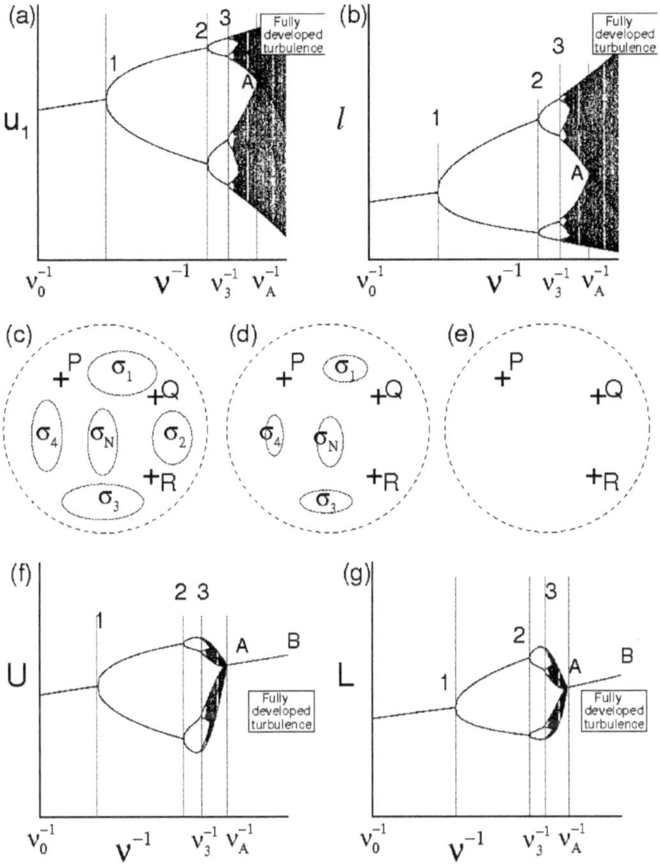


**Figure 1.** Qualitative scheme of the route toward the turbulence. (**a**,**b**): velocity and length scale in terms of kinematic viscosity. (**c**–**e**): symbolic representation of solutions in the velocity fields set. (**f**,**g**): U and L in terms of kinematic viscosity.

The Reynolds number $Re = \nu^{-1} U\, L$ is shown in terms of $\nu^{-1}$ in Figure 2. Also this map is non universal as it depends on $\nu_0^{-1}$. Nevertheless, such representation allows to identify the critical Reynolds number $Re_* = R_{T*} = \nu_*^{-1} U_*\, L_*$, the minimum value of $R_T$ for which the flow maintains statistically homogeneous and isotropic compatible with $\lambda_{NS} \geq 0$. Hence, a critical Reynolds number $Re_* = R_{T*}$ will assume a unique value, represented by the point A of Figures 1 and 2, which plays the role of an universal limit in homogeneous isotropic turbulence. Then, $\nu_* \equiv \nu_A$, $L_* \equiv L_A$, $U_* = U_A$ and the lines A–B represent regimes of fully developed homogeneous isotropic turbulence where

$$\left. \begin{array}{l} L \rightarrow \lambda_T \\ \\ U \rightarrow u \\ \\ Re \rightarrow R_T \end{array} \right\} \quad \text{along A–B} \tag{18}$$

We conclude this section by remarking that the characteristic length of the problem is an undefined quantity in infinite domain. Therefore, the length scales of **u** are used for determining the flow Reynolds number the critical value of which, $Re_* = \nu_*^{-1} U_* L_*$ has to be properly estimated. Accordingly, $L_* \equiv \lambda_{T*}$ and $U_* \equiv u_*$, linked with each other, will depend on $R_{T*}$ and $\nu$.

Such qualitative analysis is here used as background to formulate a specific bifurcation analysis of the velocity correlation equation and to determine an estimate of the critical Reynolds number $R_{T*}$.

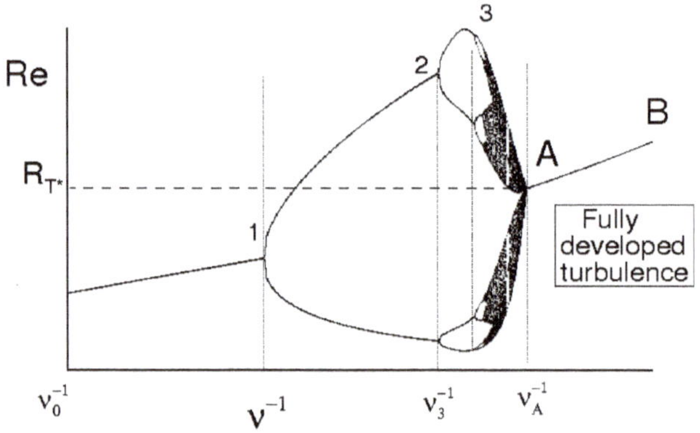


**Figure 2.** Qualitative scheme of the route toward the turbulence: Reynolds number in terms of kinematic viscosity.

## 5. Kinematic Bifurcations. Bifurcation Rate

The Navier–Stokes bifurcations have significant implications for what concerns the relative kinematics of velocity field. This kinematics is described by the separation vector $\xi$ (finite scale Lyapunov vector), which satisfies the following equations

$$\dot{x} = \mathbf{u}(t, x),$$
$$\dot{\xi} = \mathbf{u}(t, x + \xi) - \mathbf{u}(t, x), \tag{19}$$

being $x(t)$ and $\mathbf{y}(t) = x(t) + \xi(t)$ two fluid particles trajectories. In the case of contiguous trajectories, $|\xi| \to 0$, and Equation (19) read as

$$\dot{x} = \mathbf{u}(t, x),$$
$$d\dot{x} = \nabla_{\mathbf{x}} \mathbf{u}(t, x) dx, \tag{20}$$

where $dx$ and $\nabla_{\mathbf{x}} \mathbf{u}(t, \mathbf{x})$ are, respectively, elemental separation vector and velocity gradient. One point of the physical space is of bifurcation for the velocity field (kinematic bifurcation) if $\nabla_{\mathbf{x}} \mathbf{u}(t, \mathbf{x})$ has at least an eigenvalue with zero real part and this happens when its determinant vanishes, that is,

$$\det\left(\nabla_{\mathbf{x}} \mathbf{u}(t, \mathbf{x})\right) = 0. \tag{21}$$

As seen, when $R_T > R_{T*}$, due to Navier–Stokes bifurcations, the velocity field evolution will be characterized by continuous distributions of length scales and velocity values. Therefore, for $t > 0$, the velocity gradient field will exhibit nonsmooth spatial variations where $\langle \nabla_{\mathbf{x}} \mathbf{u}(t, \mathbf{x}) \rangle = 0$, and its determinant, $\det\left(\nabla_{\mathbf{x}} \mathbf{u}(t, \mathbf{x})\right)$, is expected to frequently vanish along fluid particles trajectories. To justify this, one could search a link between such property and the statistics of the eigenvalues of $\nabla_{\mathbf{x}} \mathbf{u}$ which directly arises from the fluid incompressibility [52]. In this regard, observe that an arbitrary particle trajectory $l_t : x(t)$ belongs to the surface $\Sigma_1$

$$\Sigma_1 : \Psi_1(t; x, y, z) \equiv \nabla_{\mathbf{x}} \cdot \mathbf{u}(t, \mathbf{x}) = 0 \tag{22}$$

and identically satisfies the equation

$$l_t \in \Sigma_1 : \frac{\partial \Psi_1}{\partial t} + \nabla_{\mathbf{x}} \Psi_1 \cdot \dot{x} = 0 \tag{23}$$

Thanks to Navier–Stokes bifurcations and fully developed turbulence hypothesis, for $t > 0$, $\Sigma_1$ and $l_t$ will show abrupt variations in their local placement, orientation and curvatures and will tend to sweep the entire physical space. On the other hand, the vanishing condition of velocity gradient determinant

$$\Sigma_2 : \mathcal{D}(t; x, y, z) \equiv \det\left(\nabla_{\mathbf{x}} \mathbf{u}(t, \mathbf{x})\right) = 0. \tag{24}$$

defines the surface $\Sigma_2 \neq \Sigma_1$. Thus, the points which satisfy both the conditions (22) and (24) belong to the line $l_b = \Sigma_1 \cap \Sigma_2$, and represent all the possible kinematic bifurcations which could happen along $l_t$. Because of fully developed turbulence, $l_b$ will also show nonsmooth spatial variations and will tend to describe the entire physical space. Therefore, the kinematic bifurcations that occur along $l_t$ are obtained as $l_t \cap l_b$, being $l_t, l_b \in \Sigma_1$. As $l_t$ and $l_b$ are two different curves of the same surface $\Sigma_1$ that exhibit chaotic behaviors, their intersections are expected to be very frequent, forming a highly numerous set of points on $\Sigma_1$ according to the qualitative scheme of Figure 3 wherein $l_t$ and $l_b$ are represented by solid and dashed lines. Specifically, for $R_T > R_{T*}$, $t > 0$, the Navier–Stokes bifurcations produce the regime of fully developed turbulence, where length scales and velocity values are continuously doubled and this causes situations where the number of the intersections between $l_t$ and $l_b$ (kinematic bifurcations) diverges. To show this, the kinematic bifurcation rate is now introduced. This quantity, calculated along a fluid particle trajectory, is defined as follows:

$$\begin{aligned} S_b &= \lim_{T \to \infty} \frac{1}{T} \int_0^T \delta(\mathcal{D}) \, |\frac{D\mathcal{D}}{Dt}| \, dt, \\ \frac{D\mathcal{D}}{Dt} &= \frac{\partial \mathcal{D}}{\partial t} + \nabla_{\mathbf{x}} \mathcal{D} \cdot \mathbf{u} \end{aligned} \tag{25}$$

The rate $S_b$ can be much greater than the eigenvalues modulus of $\nabla_{\mathbf{x}} \mathbf{u}$ and than its maximal Lyapunov exponent. In fact, due to the Navier–Stokes bifurcations and to the hypothesis of fully developed chaos, the characteristic scales of $\mathbf{u}$ are continuously doubled, thus $\mathcal{D} \equiv \det(\nabla_{\mathbf{x}} \mathbf{u})$ is expected to be a function of the kind

$$\begin{aligned} &\det\left(\nabla_{\mathbf{x}} \mathbf{u}\right) = \mathcal{D}\left(\mathbf{y}_1, \mathbf{y}_2, ..., \mathbf{y}_n\right), \\ &\mathbf{y}_k = \frac{\mathbf{x}}{\ell_k}, \; k = 1, 2, ..., n \\ &\ell_1 > \ell_2 > ... > \ell_n, \\ &O|\frac{\partial \mathcal{D}}{\partial \mathbf{y}_1}| \approx O|\frac{\partial \mathcal{D}}{\partial \mathbf{y}_2}| ... \approx O|\frac{\partial \mathcal{D}}{\partial \mathbf{y}_n}|, \end{aligned} \tag{26}$$

where, due to bifurcations, $n$ tends to diverge and

$$\begin{aligned} &\nabla_{\mathbf{x}} \mathcal{D} = \sum_{k=1}^{n} \frac{\partial \mathcal{D}}{\partial \mathbf{y}_k} \frac{1}{\ell_k}, \\ &O\left(\frac{1}{\ell_n} |\frac{\partial \mathcal{D}}{\partial \mathbf{y}_n} \cdot \mathbf{u}|\right) >>> O\left(|\frac{\partial \mathcal{D}}{\partial t}|\right) \end{aligned} \tag{27}$$

For one assigned velocity field, from Equations (25) and (26), the simultaneous values of $\mathbf{u}$ and $\nabla_{\mathbf{x}}(\det(\nabla_{\mathbf{x}}\mathbf{u}))$ can cause very frequent kinematic bifurcations whose rate can be significantly greater than the maximal Lyapunov exponent of Equation (19) or (20). In fact, following Equations (25) and (26), the order of magnitude of $S_b$ identifies the ratio (large scale velocity)–(small scale length)

$$S_b \approx \frac{u}{\ell_n} \tag{28}$$

where the small scale $\ell_n$ represents the minimum distance between to successive kinematic bifurcations encountered along fluid particle trajectory. This means that the changing rate of $\nabla_{\mathbf{x}}\mathbf{u}$ along $l_t$ can be much more rapid than the rate of divergence of two contiguous trajectories.

At this stage of the present study, $S_b$ is assumed to be much greater than the maximal Lyapunov exponent of Equation (19) and its estimation will be performed in the following as soon as $\ell_n$ is identified by means of this analysis.

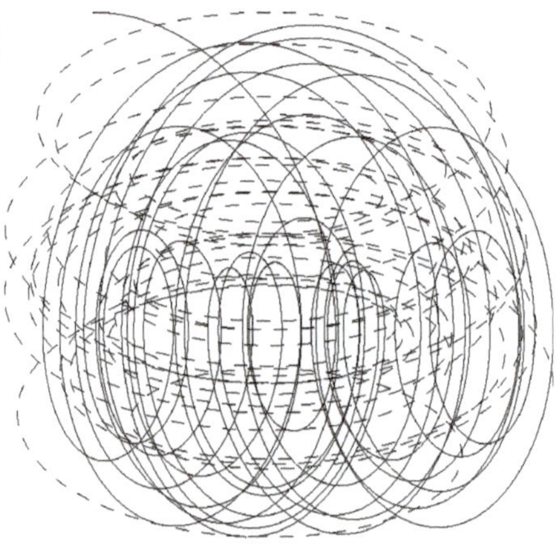


**Figure 3.** Qualitative scheme of fluid particle trajectory $l_t$, bifurcation line $l_b$ and their intersections over $\Sigma_1$.

## 6. Lyapunov Kinematic Analysis

The aim of this section is to discuss how, in fully developed turbulence, the fluctuations of fluid particles displacements and local strain can be much more rapid and statistically independent with respect to the time variations of velocity field. To analyze this, consider that, in fully developed turbulence, the Navier–Stokes bifurcations cause non smooth spatial variations of $\mathbf{u}(t, \mathbf{x})$ which in turn deternine very frequent kinematic bifurcations. Due to the fluid incompressibility, two fluid particles will describe chaotic trajectories, $\boldsymbol{x}(t)$ and $\mathbf{y}(t) = \boldsymbol{x}(t) + \boldsymbol{\xi}(t)$, which diverge with each other with a local rate of divergence quantified by the local Lyapunov exponent of finite scale $\xi$

$$\tilde{\lambda} = \frac{\dot{\boldsymbol{\xi}} \cdot \boldsymbol{\xi}}{\boldsymbol{\xi} \cdot \boldsymbol{\xi}} \tag{29}$$

According to such definition of $\tilde{\lambda}$, around to a given instant, $t_0$, $\boldsymbol{\xi}$ and $\dot{\boldsymbol{\xi}}$ can be expressed as

$$\begin{aligned} \boldsymbol{\xi} &= \mathbf{Q}(t)\boldsymbol{\xi}(t_0) \exp\left(\tilde{\lambda}\,(t - t_0)\right), \\ \dot{\boldsymbol{\xi}} &= \tilde{\lambda}\boldsymbol{\xi} + \boldsymbol{\omega}_E \times \boldsymbol{\xi} \end{aligned} \tag{30}$$

as long as $|\xi| \approx |\xi(t_0)| = r$, where $\mathbf{Q}$ is an orthogonal matrix giving the orientation of $\xi$ with respect to the inertial frame $\mathcal{R}$ and $\omega_E$ is the angular velocity of $\xi$ with respect to $\mathcal{R}$ whose determination is carried out by means of a proper orthogonalization procedure of the Lyapunov vectors described in Reference [6]. The classical local Lyapunov exponent is obtained for $|\xi| \to 0$, $\tilde{\lambda} \to \Lambda$, that is

$$\tilde{\Lambda} = \frac{dx \cdot \nabla_{\mathbf{x}} \mathbf{u} dx}{dx \cdot dx} \tag{31}$$

On the other hand, $dx$ can be expressed through Equation (20) as follows

$$dx = \exp\left(\int_0^t \nabla_{\mathbf{x}} \mathbf{u}(t', x(t'))dt'\right) dx_0 \tag{32}$$

where the exponential denotes the series expansion of operators

$$\exp\left(\int_0^t \nabla_{\mathbf{x}} \mathbf{u}(t', x(t'))dt'\right) = \mathbf{I} + \int_0^t \nabla_{\mathbf{x}} \mathbf{u}(t', x(t'))dt' + ... \tag{33}$$

Although in developed turbulence the Navier–Stokes bifurcations cause abrupt spatial variations of velocity and temperature, with $\lambda_{NS} > 0$, due to fluid dissipation, $\mathbf{u}$ and $\vartheta$ are in any case functions of slow growth of $t \in (0, \infty)$, whereas $\xi$ and $dx$, being not bounded by the dissipation effects, are functions of exponential growth of $t$. Therefore, in line with the analysis of Reference [3], and taking into account that $S_b >> \sup\{\tilde{\lambda}\}$, that $\xi$ and $dx$ are much more rapid than $\mathbf{u}(t, \mathbf{x})$ being $\sup\{\tilde{\lambda}\} >> \lambda_{NS}$, it follows that $\xi$ and $dx$ will exhibit power spectra in frequency intervals which are completely separated with respect to those of the power spectum of $\mathbf{u}$. To study this, consider now the Taylor series expansion of $\mathbf{u}$ with respect to $t$ of the trajectories equations, that is,

$$\begin{aligned}
&\dot{x} = \mathbf{u}(0, x(t)) + ..., \\
&\dot{\xi} = \mathbf{u}(0, x(t) + \xi(t)) - \mathbf{u}(0, x(t)) + ..., \quad \text{for finite scale } |\xi|, \\
&d\dot{x} = \nabla_{\mathbf{x}} \mathbf{u}(0, x(t))dx + ..., \quad \text{for contiguous trajectories}
\end{aligned} \tag{34}$$

The first terms (terms of 0 order) of such Taylor series do not correspond to time variations in velocity field, thus these do not modify the fluid kinetic energy. Furthermore, as $\sup\{\tilde{\lambda}\} >> \lambda_{NS}$ (fully developed turbulence), such terms reproduce the particles trajectories as long as $0 < t < O(1/\lambda_{NS})$, that is

$$\begin{aligned}
&\dot{x} \simeq \mathbf{u}(0, x(t)), \\
&\dot{\xi} \simeq \mathbf{u}(0, x(t) + \xi(t)) - \mathbf{u}(0, x(t)), \quad \text{for finite scale } |\xi|, \quad \forall t \in (0, a), \quad a = \mathcal{O}\left(\frac{1}{\lambda_{NS}}\right) \\
&d\dot{x} \simeq \nabla_{\mathbf{x}} \mathbf{u}(0, x(t))dx, \quad \text{for contiguous trajectories}
\end{aligned} \tag{35}$$

Following Equation (35), the fluctuations of $\xi$ and $dx$ are statistically independent with respect to the time variations of the velocity field. Next, $\sup\{\tilde{\lambda}\} >> \lambda_{NS}$, thus the number of kinematic bifurcation, which happen for $0 < t < O(1/\lambda_{NS})$, is expected to be quite high and can be considered to be significative from the statistical point of view.

Now, according to the mathematical analysis of the continuum media [53], the following map is considered

$$\chi(., t) : \mathbf{x}_0 \to x(t) \tag{36}$$

which expresses the placement of material elements at the current time $t$ in function of their referential position, say $\mathbf{x}_0 = x(0)$ [53]. From Equation (32), the local fluid strain $\partial x(t)/\partial x_0$ is then an exponential growth function of $t$ which, thanks to the above mentioned property of independence of $dx$ from $\mathbf{u}(t, \mathbf{x})$, results to be independent and much faster with respect to the time variations of the velocity field. In fact, from the Lyapunov theory of kinematic field, such strain reads as

$$\frac{\partial x}{\partial x_0} \equiv \exp\left(\int_0^t \nabla_{\mathbf{x}}\mathbf{u}(t', x(t'))dt'\right) \equiv \exp\left(\int_0^t \nabla_{\mathbf{x}}\mathbf{u}(0, x(t'))dt'\right) + ... = \mathbf{G}\exp\left(\tilde{\Lambda}\, t\right), \tag{37}$$

where $\mathbf{G}$ is a proper fluctuating matrix whose elements $G_{ij} = O(1)$ are functions of of slow growth of $t$. As long as $t \in (0, a)$ we have

$$\frac{\partial x}{\partial x_0} \simeq \exp\left(\int_0^t \nabla_{\mathbf{x}}\mathbf{u}(0, x(t'))dt'\right) = \mathbf{G}\exp\left(\tilde{\Lambda}\, t\right), \quad \forall t \in (0, a) \tag{38}$$

that is $\partial x(t)/\partial x_0$ is independent of the time variations of the velocity field.

In brief, as $\sup\{\tilde{\lambda}\} >> \lambda_{NS}$, two time scales are here considered: one associated with the velocity field and the other one related to the relative fluid kinematics. Thus, $\boldsymbol{\xi}$, $\partial x(t)/\partial x_0$ and $\tilde{\lambda}$ are statistically independent of $\mathbf{u}$. Furthermore, due to very frequent kinematic bifurcations in $(t, t + 1/\lambda_{NS})$, $\boldsymbol{\xi}$, local strain and $\tilde{\lambda}$ are expected to be continuously distributed in their variation ranges. This conclusion is supported by the arguments in References [54,55] (and references therein), where the author remarks among other things that the fields $\mathbf{u}(t, \mathbf{x})$, (and therefore also $\mathbf{u}(t, \mathbf{x} + \boldsymbol{\xi}) - \mathbf{u}(t, \mathbf{x})$) produce chaotic trajectories also for relatively simple mathematical structure of $\mathbf{u}(t, \mathbf{x})$ (also for steady fields!).

## 7. *Turbulent Energy Cascade, Material Vorticity and Link with Classical Kinematic Lyapunov Analysis

By means of theoretical considerations based on the classical Lyapunov theory and on the property that the kinematic bifurcation rate is much larger than the maximal Lyapunov exponent of the velocity gradient, an interpretation of the kinetic energy cascade phenomenon is given which shows that $\boldsymbol{\eta} \equiv dx$ is much more rapid and statistically independent with respect to $\mathbf{u}$. Following such considerations, the vorticity equation of a material element (material vorticity)—directly obtained making the curl of the incompressible Navier–Stokes equations—is compared with the evolution equation of $\boldsymbol{\eta}$ which follows the classical Lyapunov theory. These equations read as

$$\begin{aligned}
&\frac{D\omega}{Dt} \equiv \frac{\partial \omega}{\partial t} + \nabla_{\mathbf{x}}\omega\, \mathbf{u} = \nabla_{\mathbf{x}}\mathbf{u}\, \omega + \nu\nabla_{\mathbf{x}}^2\omega, \\
&\text{being } \omega = \nabla_{\mathbf{x}} \times \mathbf{u}, \\
&\frac{D\boldsymbol{\eta}}{Dt} \equiv \dot{\boldsymbol{\eta}} = \nabla_{\mathbf{x}}\mathbf{u}\, \boldsymbol{\eta}, \\
&\frac{Dx}{Dt} \equiv \dot{x} = \mathbf{u}(t, x), \\
&t \in (t_0, t_0 + a)
\end{aligned} \tag{39}$$

From such relations, it is apparent that, for inviscid fluids ($\nu = 0$), the time variations of $\boldsymbol{\eta}$ and of $\omega$ along a fluid particle trajectory $x$=$x(t)$ follow the same equation, thus $\omega$ identifies those particular Lyapunov vectors such that $\boldsymbol{\eta} \propto \nabla_{\mathbf{x}} \times \mathbf{u}$ at the initial time $t_0$. On the other hand, regardless of the initial condition $\boldsymbol{\eta}(0)$, $\boldsymbol{\eta}(t)$ tends to align with the direction of the maximum rising rate of the trajectories distance [56]. If $\omega(t_0) = k\, \boldsymbol{\eta}(t_0)$, then $\omega(t) = k\, \boldsymbol{\eta}(t)$, $\forall t > t_0$ (von Helmholtz), where $k$ does not depend on $t$, while $\boldsymbol{\eta}$ is a fast growth function of $t$. Thus, following the Lyapunov theory, for inviscid fluids, $|\omega|$, calculated along $x$=$x(t)$, tends to exponentially rise with $t$. More in general, for inviscid fluids, $\omega$ and $\eta$ are both fast growth (exponential) functions of the time, where $\omega$ tends to align to the direction of maximum growth rate of $|\eta|$ [56].

A nonzero viscosity influences the time variations of the material vorticity making this latter a slow growth function of $t \in (t_0, \infty)$, whereas $\eta$ and $\xi$ remain in any case exponential growth functions of $t$. This implies that, for $\nu \neq 0$, the characteristic time scales of $\mathbf{u}$ (and $\vartheta$) and $\eta$ are different and that after the time $t_0 + a$, the fluctuations of $\xi$ result in being statistically independent from $\mathbf{u}$. This holds also when $\nu \to 0$ for properly small length scales, except for $\nu = 0$.

Based on the previous observations, the combined effect of very frequent bifurcations and stretching term $\nabla_{\mathbf{x}}\mathbf{u}\ \omega$ produces the kinetic energy cascade. This phenomenon regards each fluid particle, where $\nabla_{\mathbf{x}}\mathbf{u}\ \omega$ acts on the material vorticity in the same way in which $\nabla_{\mathbf{x}}\mathbf{u}\ \eta$ influences $\eta$. In fact, according to Equation (39), as long as $|\nabla_{\mathbf{x}}\mathbf{u}\ \omega| >> \nu|\nabla^2_{\mathbf{x}}\omega|$, arbitrary material lines $\eta$—thus arbitrary material volumes built on different Lyapunov vectors $\eta$, that is, $\eta_1 \times \eta_2 \cdot \eta_3$—moving along $x(t)$, experience the material vorticity growth and deform according to the Lyapunov theory. According to the analysis of the previous section, such growth phenomenon, due to $\nabla_{\mathbf{x}}\mathbf{u}\ \omega$, preserves the average kinetic energy and corresponds to the continuous kinetic energy transfer from large to small scales that is, the kinetic energy cascade phenomenon. Due to the arbitrary choice of $x(t)$, this pertains to all the fluid particles. For what concerns the thermal energy cascade, $\vartheta$ is a passive scalar, the temperature follows the velocity fluctuations according to Equation (15), thus the cascade of thermal energy is direct consequence of the mechanism of kinetic energy cascade.

In brief, the energy cascade can be linked to the material vorticity tendency to be proportional to the classical Lyapunov vectors whose modulus changes according to the Lyapunov theory. Specifically, according to Equation (39) and taking into account that $S_b >> \sup\{\tilde{\lambda}\}$, $\eta$ is much faster and statistically independent with respect to the velocity field, while the energy cascade can be viewed as a continuous and intensive stretching and folding process of fluid particles which involves smaller and smaller length scales during their motion and where the particle folding process happens with a frequency given by the bifurcation rate.

## 8. Distribution Functions of u, $\vartheta$, $x$, $\xi$ and $\tilde{\lambda}$

Following the present formulation, $\mathbf{u}$, $\vartheta$, $x$ and $\xi$ are the fluid state variables. Therefore, the distribution function of $\mathbf{u}$, $\vartheta$, $x$ and $\xi$, say $P$, varies according to the Liouville theorem associated with (4), (5) and (19) [57]

$$\frac{\partial P}{\partial t} + \frac{\delta}{\delta \mathbf{u}} \cdot (P\dot{\mathbf{u}}) + \frac{\delta}{\delta \vartheta} \cdot (P\dot{\vartheta}) + \frac{\partial}{\partial x} \cdot (P\dot{x}) + \frac{\partial}{\partial \xi} \cdot (P\dot{\xi}) = 0 \tag{40}$$

where, according to the notation of Equations (4) and (5), $\delta/\delta\mathbf{u}$ and $\delta/\delta\vartheta$ are functional partial derivatives with respect to $\mathbf{u}$ and $\vartheta$, respectively and $(\partial/\partial\circ)\cdot$ stands for the divergence with respect to $\circ$. In line with the previous analysis and with References [5,6], $P$ can be factorized as follows

$$P(t, \mathbf{u}, \vartheta, x, \xi) = F(t, \mathbf{u}, \vartheta) P_\xi(t, x, \xi) \tag{41}$$

being $F$ and $P_\xi$ the distribution functions of ($\mathbf{u}$, $\vartheta$) and of ($x$, $\xi$), respectively. It is worth remarking that Equation (41) represents the crucial point of this analysis, being the hypothesis of fully developed turbulence following the present formulation. The evolution equations of $F$ and $P_\xi$ are formally obtained from Equation (40) and taking into account the aforementioned statistical independence (41). This allows to split the Liouville Equation (40) in the two following equations

$$\begin{aligned} &\frac{\partial F}{\partial t} + \frac{\delta}{\delta \mathbf{u}} \cdot (F\dot{\mathbf{u}}) + \frac{\delta}{\delta \vartheta} \cdot (F\dot{\vartheta}) = 0, \\ &\frac{\partial P_{\xi}}{\partial t} + \frac{\partial}{\partial x} \cdot (P_{\xi}\dot{x}) + \frac{\partial}{\partial \xi} \cdot (P_{\xi}\dot{\xi}) = 0 \end{aligned} \tag{42}$$

where the boundary conditions of $P_{\xi}$ read as

$$P_{\xi} = 0, \ \forall (x, \xi) \in \partial \{\{x\} \times \{\xi\}\} \tag{43}$$

In case of homogeneous and isotropic turbulence, $P_{\xi}$ does not depend on $x$ and can be expressed in function of the finite scale $r$ as follows

$$\begin{aligned} &P_{\xi} \approx \sum_k \delta(\xi - \mathbf{r}_k), \\ &|\mathbf{r}_k| = r, \forall k \end{aligned} \tag{44}$$

where $\delta$ denotes the Dirac's delta and $\mathbf{r}_k$ are uniformly distributed points on a sphere $\mathcal{S}$ of radius $r$ due to isotropy hypothesis, being $k$ a generic index indicating the several points on $\mathcal{S}$. This leads to

$$P_{\xi} = \frac{1}{4\pi r^2}\,\delta(|\xi| - r) = \begin{cases} C \to \infty \ \text{if} \ |\xi| = r \\ 0 \ \text{elsewhere} \end{cases} \tag{45}$$

Also $\tilde{\lambda}$ and $\omega_E$ are statistically independent of the velocity field and are continuously distributed in their ranges of variation. In particular, the PDF of $\tilde{\lambda}$, say $P_{\lambda}$, can be calculated by means of $P_{\xi}$ with the Frobenius–Perron equation

$$P_{\lambda}\left(\tilde{\lambda}\right) = \int_x \int_{\xi} P_{\xi}\, \delta \left(\tilde{\lambda} - \frac{\dot{\xi} \cdot \xi}{\xi \cdot \xi}\right) dx d\xi \tag{46}$$

Now, in isotropic turbulence, the longitudinal component of the velocity difference $\dot{\xi} \cdot \xi / r$ is uniformely distributed in its variation range as $\xi$ sweeps $\mathcal{S}$, while, due to the fluid incompressibility, $\tilde{\lambda}$ is expected to vary in the interval $(-\lambda_S/2, \lambda_S)$, where $\lambda_S = \sup\{\tilde{\lambda}\}$. Therefore, substituting Equation (44) in Equation (46), we found that $\tilde{\lambda}$ uniformely sweeps $(-\lambda_S/2, \lambda_S)$, according to

$$P_{\lambda} = \begin{cases} \dfrac{2}{3}\dfrac{1}{\lambda_S}, \ \text{if} \ \tilde{\lambda} \in \left(-\dfrac{\lambda_S}{2}, \lambda_S\right) \\ 0 \ \text{elsewhere} \end{cases} \tag{47}$$

Observe that Equations (45) and (47) agree with the results of Reference [6], where the author shows that $\xi$ and $\tilde{\lambda}$ are both uniformely distributed in their ranges by means of the condition $\mathcal{H}$ = max compatible with certain constraints, being $\mathcal{H}$ the entropy associated with the kinematic state $(x, \xi)$. This is because the isotropic homogeneous turbulence hypotheses, here expressed through Equations (44) and (45), correspond to the maximum of $\mathcal{H}$. The causes of the nonsymmetric distribution of $\tilde{\lambda}$ with respect to the origin, also analyzed in Reference [6], are fluid incompressibility and alignment property of $\xi$ with respect to the maximum rising rate direction. Following such property, regardless of the initial condition $\xi(0)$, $\xi(t)$ tends to align with the direction of the maximum rising rate of the trajectories distance [56]. Therefore, such a distribution function provides positive average Lyapunov exponents and gives the link between average and square mean values of the finite scale Lyapunov exponent according to

$$\langle \tilde{\lambda} \rangle_{\xi} = \frac{1}{2}\sqrt{\langle \tilde{\lambda}^2 \rangle_{\xi}} > 0. \tag{48}$$

where $\langle \circ \rangle_{\xi}$ indicates the average of $\circ$ calculated, through $P_{\xi}$ or $P_{\lambda}$.

## 9. *Finite Time Lyapunov Exponents and Their Distribution in Fully Developed Turbulence

Altough the local Lyapunov exponent $\tilde{\lambda}$ quantifies the local trajectories divergence in a point of space, in practice, the trajectory stability is evaluated by observing the particle motion in a finite time interval, say $(t_0, t_0 + \tau)$. For this reason, it is useful to define the finite time Lyapunov exponent as the average of $\tilde{\lambda}$ in such time interval, that is

$$\begin{aligned} &\tilde{\lambda}_{\tau} = \frac{1}{\tau}\int_{t_0}^{t_0+\tau} \tilde{\lambda}\, dt = \frac{1}{\tau}\int_{t_0}^{t_0+\tau} \frac{d}{dt}\ln \varrho \, dt = \frac{1}{\tau}\ln\left(\frac{\varrho(t_0+\tau)}{\varrho(t_0)}\right). \\ &\varrho = |\boldsymbol{\xi}|. \end{aligned} \tag{49}$$

This exponent trivially satisfies

$$\lim_{\tau \to 0} \tilde{\lambda}_{\tau} = \tilde{\lambda}. \tag{50}$$

If $\tau$ is properly high, a statistically significant number of kinematic bifurcations $n$ can occur for $t \in (t_0, t_0 + \tau)$, thus $\tilde{\lambda}_{\tau}$ is in general a fluctuating variable which exhibits variations whose amplitude diminishes as $\tau$ increases. Accordingly, $\tilde{\lambda}_{\tau}$ will be distributed following a Gaussian PDF in fully developed turbulence. In fact, due to the bifurcations encountered in $(t_0, t_0 + \tau)$, $\tilde{\lambda}_{\tau}$ can be written as sum of several terms, each of them related to the effects of a single bifurcation, that is,

$$\tilde{\lambda}_{\tau} = \frac{1}{\tau}\ln\left(\frac{\varrho(t_0+\tau)}{\varrho(t_0)}\right) = \frac{1}{\tau}\ln\left(\frac{\varrho(t_0+\tau)}{\varrho_{n-1}}\frac{\varrho_{n-1}}{\varrho_{n-2}}\ldots\frac{\varrho_1}{\varrho(t_0)}\right) = \frac{1}{\tau}\sum_{k=1}^{n}\ln\left(\frac{\varrho_k}{\varrho_{k-1}}\right) \tag{51}$$

where $\ln(\varrho_k/\varrho_{k-1})$ gives the contribution of the $k$th bifurcation starting from $t_0$, being $\varrho_{k-1}$ and $\varrho_k$ the Lyapunov vectors moduli calculated immediately before and after the $k$th bifurcation. On the other hand, due to fully developed chaos, each of such terms is expected to be statistically independent of all other ones and if $\tau \to \infty$, the number of encountered bifurcations $n$ diverges. Hence, a proper variant of the central limit theorem can be applied and this would guarantee that $\tilde{\lambda}_{\tau}$ tends to a Gaussian stochastic variable [58]. The novelty of the present section consists in the implication that the property $S_b >> \lambda_{\tau}$ has on Equation (51). Such property should ensure that $\lambda_{\tau}$ can be approximated to a gaussian stochastic variable also for certain finite values of $\tau$. In fact, if $\tau \approx 1/\lambda_{\tau}$ or $\tau \gtrsim 1/\lambda_{\tau}$, the time interval $(t_0, t_0 + \tau)$ should include a statistically significant number of kinematic bifurcations, thus the distribution function of $\lambda_{\tau}$ is expected to be a Gaussian PDF, expecially for relatively high values of the Taylor scale Reynolds number.

## 10. Closure of von Kármán–Howarth and Corrsin Equations

Starting from the property of statistical independence (41) and adopting the Liouville theorem, the closure formulas of von Kármán-Howarth and Corrsin equations are here determined and the effects of the chaotic trajectories divergence on these closures are discussed.

In fully developed isotropic homogeneous turbulence, the pair correlation functions of longitudinal velocity components and of temperature, defined as

$$
\begin{aligned}
f(r) &= \frac{\langle u_r(t,\mathbf{x}) u_r(t,\mathbf{x}+\mathbf{r})\rangle}{u^2} \equiv \frac{\langle u_r u_r' \rangle}{u^2}, \\
f_\theta(r) &= \frac{\langle \vartheta(t,\mathbf{x}) \vartheta(t,\mathbf{x}+\mathbf{r})\rangle}{\theta^2} \equiv \frac{\langle \vartheta \vartheta' \rangle}{\theta^2}.
\end{aligned} \tag{52}
$$

satisfy the von Kármán–Howarth equation [17] and Corrsin equation [19,20], respectively, where

$$
u_r = \mathbf{u}(t,\mathbf{x}) \cdot \frac{\mathbf{r}}{r}, \; u_r' = \mathbf{u}(t,\mathbf{x}+\mathbf{r}) \cdot \frac{\mathbf{r}}{r} \tag{53}
$$

von Kármán–Howarth and Corrsin equations are properly obtained from the Navier–Stokes and heat equations written in two points of space, say $\mathbf{x}$ and $\mathbf{x}+\mathbf{r}$. These correlation equations read as follows

$$
\begin{aligned}
\frac{\partial f}{\partial t} &= \frac{K}{u^2} + 2\nu \left( \frac{\partial^2 f}{\partial r^2} + \frac{4}{r}\frac{\partial f}{\partial r} \right) + \frac{10\nu}{\lambda_T^2} f, \\
\frac{\partial f_\theta}{\partial t} &= \frac{G}{\theta^2} + 2\chi \left( \frac{\partial^2 f_\theta}{\partial r^2} + \frac{2}{r}\frac{\partial f_\theta}{\partial r} \right) + \frac{12\chi}{\lambda_\theta^2} f_\theta,
\end{aligned} \tag{54}
$$

The boundary conditions associated with such equations are

$$
\begin{aligned}
& f(0) = 1, \quad \lim_{r\to\infty} f(r) = 0, \\
& f_\theta(0) = 1, \quad \lim_{r\to\infty} f_\theta(r) = 0,
\end{aligned} \tag{55}
$$

being $u \equiv \sqrt{\langle u_r^2 \rangle}$, $\theta \equiv \sqrt{\langle \vartheta^2 \rangle}$, where $\lambda_T \equiv \sqrt{-1/f''(0)}$ and $\lambda_\theta \equiv \sqrt{-2/f_\theta''(0)}$ are Taylor and Corrsin microscales, respectively. The quantities $K$ and $G$, arising from inertia forces and convective terms, give the energy cascade and are expressed as [17,19,20]

$$
\begin{aligned}
\left(3 + r\frac{\partial}{\partial r}\right) K &= \frac{\partial}{\partial r_k} \langle u_i u_i' \left(u_k - u_k'\right)\rangle, \\
G &= \frac{\partial}{\partial r_k} \langle \vartheta \vartheta' \left(u_k - u_k'\right)\rangle,
\end{aligned} \tag{56}
$$

where the repeated index denotes the summation convention. Following the theory [17,19,20], $K$ and $G$ are linked to the longitudinal triple velocity correlation function $k$ and to the triple correlation between $u_r$ and $\vartheta$, according to

$$
\begin{aligned}
K(r) &= u^3 \left( \frac{\partial}{\partial r} + \frac{4}{r} \right) k(r), \;\; \text{where} \;\; k(r) = \frac{\langle u_r^2 u_r' \rangle}{u^3}, \\
G(r) &= 2u\theta^2 \left( \frac{\partial}{\partial r} + \frac{2}{r} \right) m^*(r), \;\; \text{where} \;\; m^*(r) = \frac{\langle u_r \vartheta \vartheta' \rangle}{\theta^2 u},
\end{aligned} \tag{57}
$$

As well known from the literature [17,19,20], without particular hypotheses about the statistics of $\mathbf{u}$ and $\vartheta$, $K$ and $G$ are unknown quantities which can not be expressed in terms of $f$ and $f_\theta$, thus at this stage of this analysis, both the correlations Equation (54) are not closed.

In order to obtain analytical forms of $K$ and $G$, observe that these latter, representing the energy flow between length scales in the fluid, do not modify the total amount of kinetic and thermal energies [18,19]. Indeed, convective term, inertia and pressure forces determine interactions between Fourier components of velocity and temperature fields providing the transfer of kinetic and thermal energy between volume elements in the wavenumber space, whereas the global effect of such these interactions leaves $u^2$ and $\theta^2$ unaltered [18,19]. On the other hand, the proposed statistical independence property (41) allows to write the time derivative of $P$ as sum of two terms

$$\frac{\partial P}{\partial t} = P_\xi \frac{\partial F}{\partial t} + F \frac{\partial P_\xi}{\partial t} \tag{58}$$

the first one of which, being related to $\partial F / \partial t$, provides the time variations of velocity and temperature fields. The second one, linked to $\partial P_\xi / \partial t$, not producing a change of $u^2$ and $\theta^2$, identifies the energy cascade effect. Therefore, $K$ and $G$ arise from the second term of (58) and can be expressed, by means of the Liouville theorem (40) and Equation (42), in terms of material displacements $\boldsymbol{\xi}$, taking into account flow homogeneity and fluid incompressibility. Specifically, from Equations (40)–(42), $K$ and $G$, directly arising from $-F\partial(P_\xi \dot{\boldsymbol{\xi}})/\partial \boldsymbol{\xi}$, are calculated as follows

$$\begin{aligned} K &= -\int_{\mathcal{U}} \int_{\Xi} F \frac{\partial}{\partial \boldsymbol{\xi}} \cdot \left(P_\xi \dot{\boldsymbol{\xi}}\right) u_\xi u_\xi^* \, d\mathcal{U} d\Xi, \\ G &= -\int_{\mathcal{U}} \int_{\Xi} F \frac{\partial}{\partial \boldsymbol{\xi}} \cdot \left(P_\xi \dot{\boldsymbol{\xi}}\right) \vartheta \vartheta^* d\mathcal{U} d\Xi, \end{aligned} \tag{59}$$

where $\mathcal{U} = \{\mathbf{u}\} \times \{\vartheta\}$, $\Xi = \{\boldsymbol{\xi}\}$ and $d\mathcal{U}$ and $d\Xi$ are the corresponding elemental volumes, and

$$\begin{aligned} & u_\xi = \mathbf{u}(t, \boldsymbol{x}) \cdot \frac{\boldsymbol{\xi}}{\xi}, \quad u_\xi^* = \mathbf{u}(t, \boldsymbol{x} + \boldsymbol{\xi}) \cdot \frac{\boldsymbol{\xi}}{\xi}, \\ & \vartheta = \vartheta(t, \boldsymbol{x}), \quad \vartheta^* = \vartheta(t, \boldsymbol{x} + \boldsymbol{\xi}), \end{aligned} \tag{60}$$

Integrating Equation (59) with respect to $\mathcal{U}$, we obtain

$$\begin{aligned} K &= -u^2 \int_{\Xi} \frac{\partial}{\partial \boldsymbol{\xi}} \cdot \left(P_\xi \dot{\boldsymbol{\xi}}\right) f(\xi) \, d\Xi, \\ G &= -\theta^2 \int_{\Xi} \frac{\partial}{\partial \boldsymbol{\xi}} \cdot \left(P_\xi \dot{\boldsymbol{\xi}}\right) f_\theta(\xi) \, d\Xi, \end{aligned} \tag{61}$$

Again, integrating by parts Equation (61) with respect to $\Xi$, taking into account the boundary conditions (43) ($P_\xi \equiv 0, \forall \boldsymbol{\xi} \in \partial \Xi$) and the isotropy hypothesis, $K$ and $G$ are written as

$$\begin{aligned} K &= u^2 \int_{\Xi} P_\xi \frac{\partial f}{\partial \boldsymbol{\xi}} \cdot \dot{\boldsymbol{\xi}} \, d\Xi = u^2 \int_{\Xi} P_\xi \frac{\partial f}{\partial \xi} \frac{\boldsymbol{\xi}}{\xi} \cdot \dot{\boldsymbol{\xi}} \, d\Xi, \\ G &= \theta^2 \int_{\Xi} P_\xi \frac{\partial f_\theta}{\partial \boldsymbol{\xi}} \cdot \dot{\boldsymbol{\xi}} \, d\Xi = \theta^2 \int_{\Xi} P_\xi \frac{\partial f_\theta}{\partial \xi} \frac{\boldsymbol{\xi}}{\xi} \cdot \dot{\boldsymbol{\xi}} \, d\Xi, \end{aligned} \tag{62}$$

Now, the Lyapunov theory provides $\dot{\boldsymbol{\xi}} = \tilde{\lambda} \boldsymbol{\xi} + \boldsymbol{\omega}_E \times \boldsymbol{\xi}$, and in isotropic homogeneous turbulence $P_\xi = \delta(|\boldsymbol{\xi}| - r)/4\pi r^2$, thus $K$ and $G$ are

$$\begin{aligned} K &= u^2 \int_{\Xi} P_\xi \frac{\partial f}{\partial \xi} \xi \tilde{\lambda} \, d\Xi = u^2 \frac{\partial f}{\partial r} r \langle \tilde{\lambda} \rangle_\xi, \\ G &= \theta^2 \int_{\Xi} P_\xi \frac{\partial f_\theta}{\partial \xi} \xi \tilde{\lambda} \, d\Xi = \theta^2 \frac{\partial f_\theta}{\partial r} r \langle \tilde{\lambda} \rangle_\xi, \end{aligned} \tag{63}$$

Furthermore, the finite scale Lyapunov theory also gives the relationship between velocity correlation and Lyapunov exponents according to

$$\left\langle (u_{\xi}^{*} - u_{\xi})^2 \right\rangle_{\xi} = 2u^2 \left(1 - f(r)\right) = \left\langle \tilde{\lambda}^2 \right\rangle_{\xi} r^2, \tag{64}$$

where $\langle \tilde{\lambda} \rangle_{\xi}$ and $\langle \tilde{\lambda}^2 \rangle_{\xi}$ are linked with each other through Equation (48), therefore the closure formulas of $K$ and $G$ are in terms of autocorrelations and of their gradients

$$\begin{aligned} K(r) &= u^3 \sqrt{\frac{1-f}{2}} \frac{\partial f}{\partial r}, \\ G(r) &= u\theta^2 \sqrt{\frac{1-f}{2}} \frac{\partial f_\theta}{\partial r}, \end{aligned} \tag{65}$$

These closure formulas do not include second order derivatives of autocorrelations, thus Equation (65) do not correspond to a diffusive model. The energy cascade expressed by Equation (65) is not based on the eddy viscosity concept, being the result of the trajectories divergence in the continuum fluid. This cascade phenomenon and Equation (65) are here interpreted as follows:

(1) In fully developed chaos, the Navier–Stokes bifurcations determine a continuous distribution of velocity, temperature and of length scales, where one single bifurcation causes doubling of velocity, temperature, length scale and of all the properties associated with the velocity and temperature fields according to Equations (15) and (16). This leads to nonsmooth spatial variations of velocity field and very frequent kinematic bifurcations.

(2) The huge kinematic bifurcations rate generates in turn continuous distributions of $\tilde{\lambda}$ and $\boldsymbol{\xi}$, while fluid incompressibility and the mentioned alignment property of $\boldsymbol{\xi}$ make $\tilde{\lambda}$ unsymmetrically distributed with $\bar{\lambda}(r) \equiv \left\langle \tilde{\lambda} \right\rangle_{\xi} > 0$ and the relative particles trajectories to be chaotic.

(3) The tendency of the material vorticity to follow direction and variations of the Lyapunov vectors gives the phenomenon of the kinetic energy cascade.

The main asset of Equation (65) with respect to the other models is that Equation (65) are not based on phenomenological assumptions, such as, for instance, the eddy viscosity paradigm [22–24,28,29,33] but are obtained through theoretical considerations concerning the statistical independence of $\boldsymbol{\xi}$ from $\mathbf{u}$ and the Liouville theorem.

**Remark 1.** *At this stage of the present analysis, it is worth remarking on the importance of the hypothesis of the statistical independence of* $\mathbf{u}$ *and* $\boldsymbol{\xi}$ *expressed by Equation (41). This latter, expressing the hypothesis of fully developed turbulence following this study, leads to the analytical expressions of K and G separating the effects of the trajectories divergence in the physical space from those of the velocity field fluctuations in the Navier–Stokes phase space. Without such hypothesis, the energy cascade effect can not be expressed through the term* $-F\partial(P_{\xi}\dot{\boldsymbol{\xi}})/\partial\boldsymbol{\xi}$ *and using Equation (59), thus the proposed closures (65) cannot be determined.*

Thanks to their theoretical foundation, Equation (65) do not exhibit free model parameters or empirical constants which have to be identified. These closure formulas coincide with those just obtained by the author in the previous works [1,4,5]. While References [1,4] derive such closures expressing the local fluid act of motion in the finite scale Lyapunov basis and using the frame invariance property of $K$ and $G$, Reference [5] achieves the same formulas adopting maximum and average finite scale Lyapunov exponents, properly defined and the statistical independence of $\boldsymbol{\xi}$ and $\mathbf{u}$. Here, unlike References [1,4,5], Equation (65) are determined exploiting the unsymmetric distribution function of $\tilde{\lambda}$ just studied in Reference [6], showing that the assumptions of References [1,4,5] are congruent with the present analysis, corroborating the results of the previous work.

References [1,4] show that these closures adequately describe the energy cascade phenomenon and the energy spectra. In detail, $K$ reproduces the kinetic energy cascade mechanism following the Kolmogorov law and $G$ gives the thermal energy cascade in line with the theoretical argumentation of Kolmogorov, Obukhov–Corrsin and Batchelor [35–37], with experimental results [38,39] and with numerical data [40,41]. Moreover, Equation (65) allows the calculation of the skewness of $\Delta u_r$ and $\partial u_r/\partial r$ which is directly linked to the energy cascade intensity. This is [18]

$$H_3(r) \equiv \frac{\langle(\Delta u_r)^3\rangle}{\langle(\Delta u_r)^2\rangle^{3/2}} = \frac{6k(r)}{(2(1-f(r)))^{3/2}} \tag{66}$$

Then, substituting Equation (65) in Equation (66), the skewness of $\partial u_r/\partial r$ is

$$H_3(0) = -\frac{3}{7} \tag{67}$$

This constant quantifies the effect of chaotic relative trajectories on the energy cascade in isotropic turbulence and agrees with the several results obtained through direct numerical simulation of the Navier–Stokes equations (DNS) [59–61] ($-0.47 \div -0.40$) and by means of Large–eddy simulations (LES) [62–64] ($-0.42 \div -0.40$). For the sake of reader convenience, Table 1 recalls the comparison, presented in Reference [5,6], between the value of the skewness $H_3(0)$ of this analysis and those achieved by the aforementioned works. The results were that the maximum absolute difference between the proposed value and the other results were less than 10%. Therefore, the proposed hypotheses, leading to the distribution function (47) and to the closures (65), seem to be adequate assumptions for estimating turbulent energy cascade and spectra.

**Table 1.** Comparison of the results: Skewness of $\partial u_r/\partial r$ at diverse Taylor–scale Reynolds number $R_T \equiv u\lambda_T/\nu$ following different authors.

| Reference | Simulation | $R_T$ | $H_3(0)$ |
|---|---|---|---|
| Present analysis | - | - | $-3/7 = -0.428...$ |
| [59] | DNS | 202 | − 0.44 |
| [60] | DNS | 45 | − 0.47 |
| [61] | DNS | 64 | − 0.40 |
| [62] | LES | <71 | − 0.40 |
| [63] | LES | $\infty$ | − 0.40 |
| [64] | LES | 720 | − 0.42 |

We conclude this section by observing the limits of the proposed closures (65). These limits directly derive from the hypotheses under which Equation (65) are obtained: Equation (65) are valid only in a regime of fully developed chaos where the turbulence exhibit homogeneity and isotropy. Otherwise, during the transition through intermediate stages of turbulence or in more complex situations with particular boundary conditions, for instance in the presence of wall, Equation (65) cannot be applied.

## 11. Properties of the Proposed Closures

Here, some of the properties of the proposed closures (65) are renewed, with particular reference to the evolution times of the developed velocity and temperature autocorrelations. In detail, we will show that these correlations reach their developed shape in finite times which depend on the initial condition and that, after this period, the hypothesis of statistical independence could be not more verified. This result is given in Reference [5], where the author adopts a specific Lyapunov analysis using two exponents properly defined. Unlike in Reference [5], such a result is here achieved through the previously obtained local finite scale Lyapunov exponent distribution (47). To analyze this, the evolution equations of $u$, $\theta$, $\lambda_T$ and $\lambda_\theta$ are first obtained taking the coefficients of order $r^0$ and $r^2$ of Equation (54) arising from the Taylor series expansion of even powers of $f$ and $f_\theta$ [17,19,20]

$$f = 1 - \frac{1}{2}\left(\frac{r}{\lambda_T}\right)^2 + ...,$$
$$f_\theta = 1 - \left(\frac{r}{\lambda_\theta}\right)^2 + ..., \tag{68}$$

This leads to the following equations

$$\frac{du^2}{dt} = -\frac{10\nu}{\lambda_T^2}u^2,$$
$$\frac{d\theta^2}{dt} = -\frac{12\chi}{\lambda_\theta^2}\theta^2, \tag{69}$$

$$\frac{d\lambda_T}{dt} = -\frac{u}{2} + \frac{\nu}{\lambda_T}\left(\frac{7}{3}f^{IV}(0)\lambda_T^4 - 5\right),$$
$$\frac{d\lambda_\theta}{dt} = -\frac{u}{2}\frac{\lambda_\theta}{\lambda_T} + \frac{\chi}{\lambda_\theta}\left(\frac{5}{6}f_\theta^{IV}(0)\lambda_\theta^4 - 6\right) \tag{70}$$

While Equation (69) do not depend on the particular adopted closures [17,19,20], Equation (70) are obtained using the proposed closures (65). On the other hand, it is useful to consider the fluctuations of the classical Lyapunov exponent, defined as

$$\tilde{\Lambda} = \lim_{r \to 0} \tilde{\lambda} = \lim_{r \to 0} \frac{d}{dt} \ln \varrho,$$
$$\varrho = |\xi| \tag{71}$$

which are related to $f$ through Equations (64) and (68) in such a way that

$$\Lambda = \sqrt{\langle \tilde{\Lambda}^2 \rangle} = \frac{u}{\lambda_T} \propto \lim_{r \to 0} \frac{d}{dt} \langle \ln \varrho \rangle_\xi \approx |\frac{d \ln \lambda_T}{dt}|. \tag{72}$$

being $\Lambda$ the root mean square of $\tilde{\Lambda}$.

Following Equation (70), the time variations of $\lambda_T$, $\lambda_\theta$ and $\Lambda$ are now discussed. The first terms at the R.H.S. of Equation (70) provide the turbulent energy cascade, whereas the other ones arise from the fluid diffusivities. While these latter contribute to increasing both the correlation lengths, the energy cascade mechanism tends to reduce these scales and if such a mechanism is sufficiently stronger than diffusivities, then $d\lambda_T/dt < 0$ and $d\lambda_\theta/dt < 0$.

For sake of our convenience, the condition $\nu = 0$, $\chi = 0$ is first studied. In this case, $u$ and $\theta$ are both constants, whereas $\lambda_T$, $\lambda_\theta$ and $\Lambda$ vary with t. In detail, $\lambda_T$ and $\lambda_\theta$ are proportional to each other and vary linearly with time according to

$$\frac{\lambda_T(t)}{\lambda_T(0)} \equiv \frac{\lambda_\theta(t)}{\lambda_\theta(0)} = 1 - \frac{\tau}{2},$$
$$\frac{\Lambda(t)}{\Lambda(0)} = \frac{1}{1 - \tau/2}, \tag{73}$$
$$\tau = t\,\Lambda(0),$$

while $\Lambda$ monotonically rises and goes to infinity in a finite time, being $\tau$ the dimensionless time. When $\nu = \chi = 0$, the energy cascade provides that both the microscales decrease until to $\tau \to 2$, where both the correlations are considered to be fully developed, $\lambda_T \to 0$, $\lambda_\theta \to 0$ and $\Lambda \to \infty$ (see solid lines of Figure 4).

Thus, the two correlations will exhibit developed shapes in finite times whose values depend on the initial condition $\Lambda(0)$. The meaning that both the microscales are decreasing functions of $\tau$ is that kinetic and thermal energies are continuously transferred from large to small scales following the previous scheme. Next, as $\tau \to 2$, $\Lambda \to +\infty$ and this means that the velocity gradient diverges in a finite time depending on $\Lambda(0)$ and that contiguous particles trajectories diverge with a growth rate infinitely faster than velocity and temperature fields.

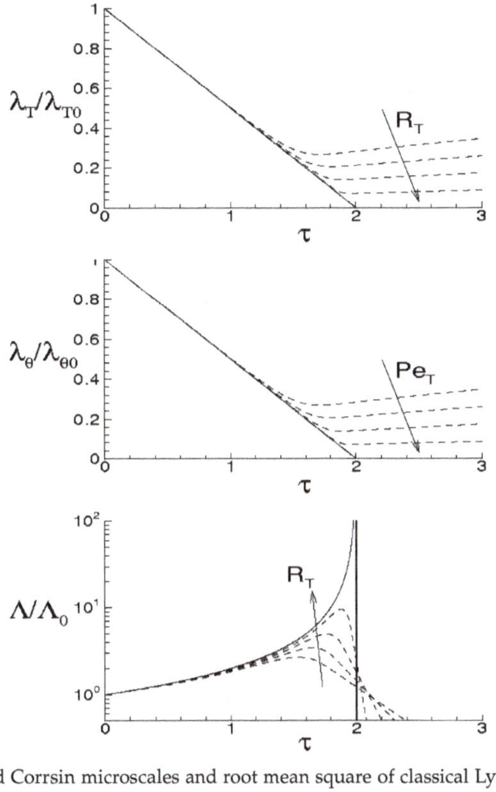


**Figure 4.** Taylor and Corrsin microscales and root mean square of classical Lyapunov exponent in function of the dimensionless time.

For $\nu > 0$, $\chi > 0$, then $du/dt < 0$ and $d\theta/dt < 0$ in any case and $f$ and $f_\theta$ are here supposed to be fully developed as soon as $d\lambda_T/dt = 0$ and $d\lambda_\theta/dt = 0$, respectively. These situations are qualitatively shown in the figure by the dashed lines for different values of $R_T$ and $Pe$, where $R_T = \lambda_T u/\nu$ and $Pe = Pr\ R_T$ are, respectively, Reynolds number and Péclet number, both referred to the Taylor microscale, being $Pr = \nu/\chi$ the Prandtl number. When the initial microscales are relatively large, the diffusivities effects are quite smaller than the convective terms, the energy cascade is initially stronger than the diffusivities effects and both the microscales exhibit about the same trend just discussed for $\nu = \chi = 0$. According to Equations (69) and (70), the interval where $\tau$ ranges can be splitted in two subregions for both $f$ and $f_\theta$. The first ones correspond to values of $\tau \in (0,2)$ such that $d\lambda_T/dt < 0$ and $d\lambda_\theta/dt < 0$, which are upper bounded by the endpoints $\tau_1 < 2$, $\tau_2 < 2$ where $d\lambda_T/dt(\tau_1) = 0$ and $d\lambda_\theta/dt(\tau_2) = 0$ (dashed lines), respectively, being in general $\tau_1 \neq \tau_2$. There, the kinetic and thermal energy cascades are momentarily balanced by viscosity and thermal diffusivity, respectively and both the autocorrelations can be considered fully developed. For both the correlations, such momentary balance happens in finite times $\tau < 2$ which depend on the initial condition. As far as $\Lambda$ is concerned, this initially coincides about with that obtained for $\nu = 0$, then reaches its maximum for $\tau \lesssim 2$ and thereafter diminishes due to viscosity. When $\Lambda$ achieves its maximum, $d\Lambda/dt = 0$, chaos and mixing reach their maximum levels, the correlations are about fully developed, thus relative kinematics and fluid strain change much more rapidly than velocity field. Thereafter, we observe regions where $d\Lambda/dt < 0$. There, due to the relatively smaller values of the microscales, the dissipation is stronger than the energy cascade and both the correlation lengths tend to rise according to Equation (70). Such a region, which occurs immediately after the condition $d\Lambda/dt = 0$, corresponds to the regime of decaying turbulence.

Observe that the proposed closures (65) are expected to be verified where $d\Lambda/dt > 0$, in which the Navier–Stokes bifurcations generate the regime of fully developed turbulence. On the contrary, in regime of decaying turbulence –$d\Lambda/dt < 0$–, after a certain time, say $\tau^+ > \tau_1 \approx 2$, it results $\Lambda/\Lambda(0) < 1$. In such situations, the relative kinematic and fluid strain could be not faster than velocity field, thus the statistical independence hypothesis (41) could be not satisfied and Equation (65) will be not defined. Therefore, the condition $\tau \approx 2$ or $\Lambda/\Lambda(0) < 1$ provides a further limit of validity for the proposed closure formulas.

## 12. *Self–Similarity and Developed Correlations of the Proposed Closures

This section analyzes self–similarity and developed shape of $f$ and $f_\theta$ produced by the proposed closures. The new result with respect to the previous works consists in to remark that the proposed closures generate correlations self–similarity in proper ranges of $r$, which is directly related to the fluid trajectories divergence. To study this question, observe that a given function of $t$ and $r$, say $\psi = \psi(t, r)$, which completely exhibits self–similarity with respect to $r$ as $t$ changes, is a function of the kind

$$\psi(t,r) = \psi\left(\frac{r}{\hat{L}(t)}\right) \tag{74}$$

and exactly satisfies the equation

$$\begin{aligned} \frac{\partial \psi}{\partial t} &= -\frac{\partial \psi}{\partial r}\frac{r}{\hat{L}}\frac{d\hat{L}}{dt} \equiv C(t) r \frac{\partial \psi}{\partial r}, \\ C(t) &= \frac{d \ln \hat{L}}{dt} \end{aligned} \tag{75}$$

wherein $\hat{L}$ is the characteristic length associated with the specific problem. From such equation, the self–similarity of $\psi$ is linked to the variation rate $d \ln \hat{L}(t)/dt$. Now, thanks to the mathematical structures of the proposed closures (65), and taking into account that $f$ and $f_\theta$ are both even functions of $r$ which near the origin behave like Equation (68), $K$ and $G$ can be expressed through even power series of $f$ as follows

$$\begin{aligned} K &= u^3\sqrt{\frac{1-f}{2}}\frac{\partial f}{\partial r} = \frac{u^3}{2}\frac{r}{\lambda_T}\frac{\partial f}{\partial r} + ... = \frac{u^2}{2}\Lambda r\frac{\partial f}{\partial r} + ... \\ G &= \theta^2 u\sqrt{\frac{1-f}{2}}\frac{\partial f_\theta}{\partial r} = \frac{\theta^2 u}{2}\frac{r}{\lambda_T}\frac{\partial f_\theta}{\partial r} + ... = \frac{\theta^2}{2}\Lambda r\frac{\partial f_\theta}{\partial r} + ..., \\ \Lambda &\propto \frac{d}{dt}\left\langle \ln \varrho \right\rangle_\xi \approx \left|\frac{d \ln \lambda_T}{dt}\right| \end{aligned} \tag{76}$$

thus, the evolution equations of both the autocorrelations can be written in the following way

$$\frac{\partial f}{\partial t} = u\sqrt{\frac{1-f}{2}}\frac{\partial f}{\partial r} + ... = \frac{u}{2\lambda_T} r \frac{\partial f}{\partial r} + ... = \frac{\Lambda}{2} r \frac{\partial f}{\partial r} + ...$$

$$\frac{\partial f_\theta}{\partial t} = u\sqrt{\frac{1-f}{2}}\frac{\partial f_\theta}{\partial r} + ... = \frac{u}{2\lambda_T} r \frac{\partial f_\theta}{\partial r} + ... = \frac{\Lambda}{2} r \frac{\partial f_\theta}{\partial r} + ..., \tag{77}$$

$$\Lambda \propto \frac{d}{dt} \langle \ln \varrho \rangle_\xi \approx |\frac{d \ln \lambda_T}{dt}|$$

Comparing Equations (75) and (77), it follows that the proposed closures (65) generate self–similarity in a range of variation of $r$ where $\Lambda/2r\partial f/\partial r$ and $\Lambda/2r\partial f_\theta/\partial r$ are dominant with respect to the other terms. As the result, such self–similarity is directly caused by the continuous fluid trajectory divergence—quantified by $\Lambda$—which happens thank to very frequent kinematic bifurcations. In such these intervals, the correlations will exhibit self–similarity during their time evolution, thus $f$ and $f_\theta$ can be expressed there as follows

$$f(t,r) \simeq f\left(\frac{r}{\lambda_T(t)}\right),$$
$$f_\theta(t,r) \simeq f_\theta\left(\frac{r}{\lambda_T(t)}\right), \tag{78}$$

In such regions, the energy cascade is intensive and much stronger than the diffusivities effects, thus following Equation (70), $\lambda_\theta(t)$ is expected to be proportional to $\lambda_T(t)$

$$\frac{\lambda_\theta(t)}{\lambda_\theta(0)} \simeq \frac{\lambda_T(t)}{\lambda_T(0)}, \tag{79}$$

Next, as $\vartheta$ is a passive scalar, energy cascade and fluid diffusivities act on $u$ and $\theta$ in such a way that their increments are proportional with each other. Therefore, far from the initial condition, we expect that

$$\frac{\theta(t)}{\theta(0)} \simeq \frac{u(t)}{u(0)}, \tag{80}$$

Now, Equation (79) provides a link between the correlation scales and $Pr$. In fact, substituting Equation (79) in Equation (69), we obtain

$$\frac{\lambda_\theta}{\lambda_T} = \sqrt{\frac{6}{5}\frac{1}{Pr}} \tag{81}$$

Furthermore, from Equation (70), also $f^{IV}(0)$ and $f_\theta^{IV}(0)$ are related to the Prandtl number

$$\frac{f_\theta^{IV}(0)}{f^{IV}(0)} = \frac{7}{3} Pr^2 \tag{82}$$

Hence, the developed autocorrelations can be estimated searching for the solutions of the closed von Kármán–Howarth and Corrsin equations in the self–similar form (78) when $d\lambda_T/dt = d\lambda_\theta/dt = 0$. This leads to the following ordinary differential equations system

$$\sqrt{\frac{1-f}{2}}\frac{df}{d\hat{r}} + \frac{2}{R_T}\left(\frac{d^2 f}{d\hat{r}^2} + \frac{4}{\hat{r}}\frac{df}{d\hat{r}}\right) + \frac{10}{R_T} f = 0,$$

$$\sqrt{\frac{1-f}{2}}\frac{df_\theta}{d\hat{r}} + \frac{2}{R_T Pr}\left(\frac{d^2 f_\theta}{d\hat{r}^2} + \frac{2}{\hat{r}}\frac{df_\theta}{d\hat{r}}\right) + \frac{12}{R_T Pr}\left(\frac{\lambda_T}{\lambda_\theta}\right)^2 f_\theta = 0, \qquad (83)$$

$$\hat{r} = \frac{r}{\lambda_T}.$$

Several solutions of these equations were numerically obtained in [2,4], where the author shows that velocity and temperature correlations agree with the Kolmogorov law, with the theoretical arguments of Obukhov–Corrsin and Batchelor and with the numerical simulations and experiments known from the literature [19,35–41].

For sake of reader convenience, Figures 5 and 6 report the velocity correlations and the corresponding spectra $E(\kappa)$, $T(\kappa)$ numerically calculated with the first equation of Equation (83) for $R_T$ = 100, 200, 300, 400, 500, 600, being

$$\begin{bmatrix} E(\kappa) \\ \\ T(\kappa) \end{bmatrix} = \frac{1}{\pi}\int_0^\infty \begin{bmatrix} u^2 f(r) \\ \\ K(r) \end{bmatrix} \kappa^2 r^2 \left(\frac{\sin \kappa r}{\kappa r} - \cos \kappa r\right) dr \qquad (84)$$

where all these cases correspond to the same level of average kinetic energy. The integral correlation scale of $f$ results to be a rising function of $R_T$, while the triple longitudinal velocity correlation $k$ maintains negative with a minimum of about $-0.04$ whose value is achieved for values of $r/\lambda_T$ which rise with the Reynolds number. For what concerns the spectra, observe that increasing $\kappa$, the kinetic energy spectra behave like $E(\kappa) \approx \kappa^4$ near the origin, then exhibit a maximum and thereafter are about parallel to the dashed line $\kappa^{-5/3}$ in a given interval of the wave–numbers. The size of this latter, which defines the inertial range of Kolmogorov, rises as $R_T$ increases. For higher values of $\kappa$, which correspond to scales less than the Kolmogorov length, $E(\kappa)$ decreases more rapidly than in the inertial range. As $K$ does not modify the kinetic energy, the proposed closure gives $\int_0^\infty T(\kappa) d\kappa \equiv 0$.

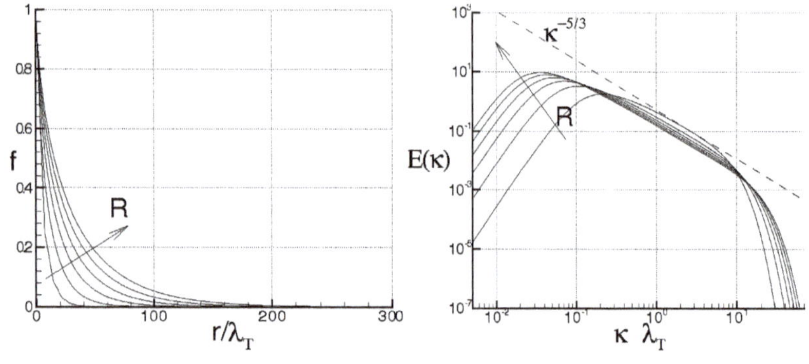


**Figure 5.** Longitudinal velocity correlations (**left**) and energy spectra (**right**) at different Taylor scale Reynolds numbers $R_T$ = 100, 200, 300, 400, 500, 600.

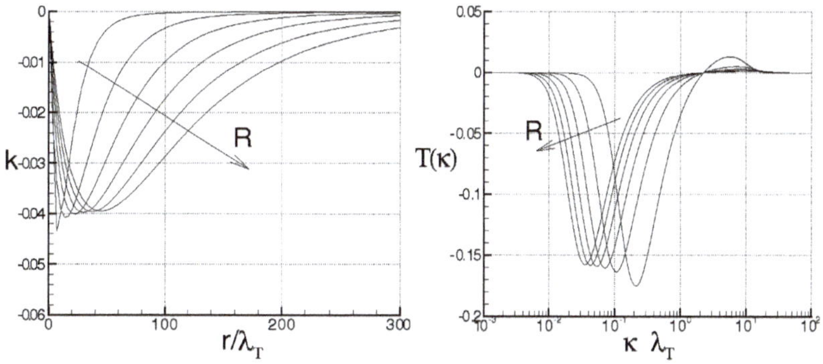


**Figure 6.** Triple longitudinal velocity correlations (**left**) and the corresponding spectra (**right**) at different Taylor scale Reynolds numbers $R_T$ = 100, 200, 300, 400, 500, 600.

From these solutions, the Kolmogorov constant $C$, here calculated as

$$C = \max_{\kappa \in (0,\infty)} \frac{E(\kappa)\kappa^{5/3}}{\varepsilon^{2/3}} \tag{85}$$

is shown in Table 2 in function of the Reynolds number, where $\varepsilon = -3/2\, du^2/dt$. The obtained values of $C \approx 2$ are in good agreement with the corresponding values known from the literature.

**Table 2.** Kolmogorov constant for different Taylor-Scale Reynolds number.

| $R_T$ | C |
|---|---|
| 100 | 1.8860 |
| 200 | 1.9451 |
| 300 | 1.9704 |
| 400 | 1.9847 |
| 500 | 1.9940 |
| 600 | 2.0005 |

Next, Figure 7 shows the temperature spectra $\Theta(\kappa)$ and the temperature transfer function $\Gamma(\kappa)$ calculated as follows [65]

$$\left[ \begin{array}{c} \Theta(\kappa) \\ \\ \Gamma(\kappa) \end{array} \right] = \frac{2}{\pi} \int_0^\infty \left[ \begin{array}{c} \theta^2 f_\theta(r) \\ \\ G(r) \end{array} \right] \kappa r \sin \kappa r \, dr \tag{86}$$

in such a way that

$$\int_0^\infty \Theta(\kappa)\, d\kappa = \theta^2, \quad \int_0^\infty \Gamma(\kappa)\, d\kappa = 0 \tag{87}$$

The variations of $\Theta(\kappa)$ with $R_T$ and $Pr$ are quite peculiar and consistent with previous studies according to which there are regions where $\Theta(\kappa)$ exhibits different scaling laws $\Theta(\kappa) \approx \kappa^n$.

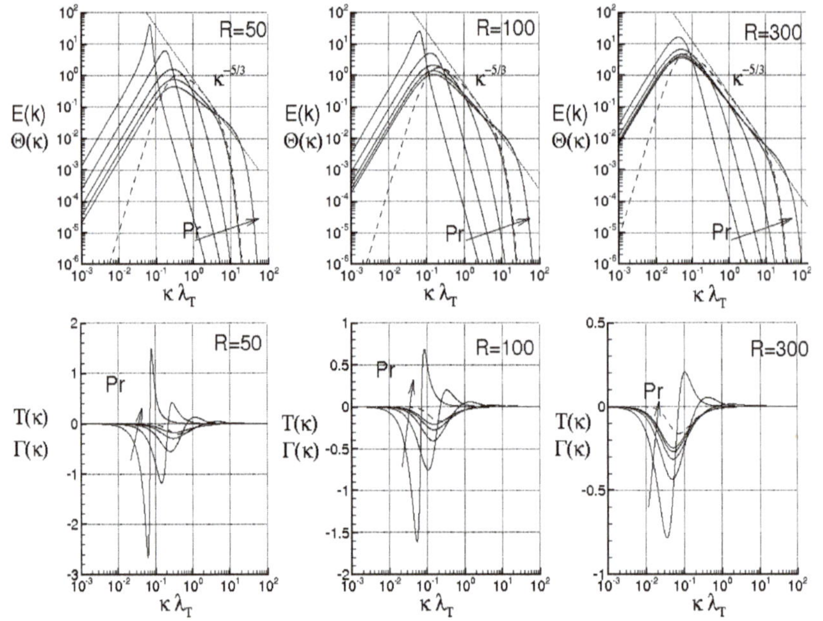


**Figure 7.** Spectra for Pr = $10^{-3}$, $10^{-2}$, 0.1, 1.0 and 10, at different Reynolds numbers. **Top**: kinetic energy spectrum $E(\kappa)$ (dashed line) and temperature spectra $\Theta(\kappa)$ (solid lines). **Bottom**: velocity transfer function $T(\kappa)$ (dashed line) and temperature transfer function $\Gamma(\kappa)$ (solid line).

Following the proposed closures, $n \simeq 2$ when $\kappa \to 0$ in any case. For $Pr = 0.001$, when $R_T$ ranges from 50 to 300, the temperature spectrum essentially exhibits two regions: one in proximity of the origin where $n \simeq 2$ and the other one, at higher values of $\kappa$, where $-17/3 < n < -11/3$, (value very close to $-13/3$). The value of $n \approx -13/3$, here obtained in an interval around to $\hat{r} \approx 1$, is in between the exponent proposed by [36] ($-17/3$) and the value determined by [40] ($-11/3$) by means of numerical simulations. Increasing $\kappa$, $n$ significantly diminishes and $\Theta(\kappa)$ does not show scaling law. When $Pr = 0.01$, an interval near $\hat{r} \approx 1$ where $-17/3 < n < -13/3$ appears and this is in agreement with [36]. Next, for $Pr = 0.1$, the previous scaling law vanishes, whereas for $R_T = 50$ and 100, $n$ changes with $\kappa$ and $\Theta(\kappa)$ does not show clear scaling laws. When $R = 300$, the birth of a small region is observed, where $n \approx -5/3$ has an inflection point. For $Pr = 0.7$ and 1, with $R_T = 300$, the width of this region is increased, whereas at $Pr = 10$ and $R = 300$, we observe two regions: one interval where $n$ has a local minimum with $n \simeq -5/3$ and the other one where $n$ exhibits a relative maximum, with $n \simeq -1$. For larger $\kappa$, $n$ diminishes and the scaling laws disappear. The presence of the scaling law $n \simeq -5/3$ agrees with the theoretical arguments of [20,37] (see also [39,41] and references therein). Figure 7 also reports (on the bottom) the spectra $\Gamma(\kappa)$ (solid lines) and $T(\kappa)$ (dashed lines) which describe the energy cascade mechanism.

## 13. *Bifurcation Analysis of Closed von Kármán–Howarth Equation: From Fully Developed Turbulence Toward Non–Chaotic Regimes

Starting from non–chaotic regimes, the transition towards the fully developed turbulence happens through intermediate stages [34,42,43,51] which correspond to bifurcations where the relative Reynolds numbers show the same order of magnitude. This section presents a specific bifurcation analysis, which, unlike the classical route toward the chaos [34,42,43,51], analyzes the inverse route: the starting condition is represented by the fully developed homogeneous isotropic turbulence and the route followed is that towards the non–chaotic regime. Such route corresponds to the path $B \to A$ of

Figures 1f,g and 2. Along the line $B \to A$, $R_T$ gradually diminishes and the bifurcations of the closed von Kármán–Howarth equation, properly defined, will be here studied. This analysis estimates $R_T^*$ through the closures (65) and their previously seen properties, where $R_T^*$ defines the minimum value of $R_T$ for which the turbulence maintains fully developed, homogeneous and isotropic. This provides the order of maginitude of $Re$ at the transition, indicating a further limit of the proposed closures.

In order to formulate a bifurcation analysis for the velocity correlation equation, consider now the various coefficients of the closed von Kármán–Howarth equation which arise from the even Taylor series expansion of $f(t,r) = \sum_k f_0^{(k)} r^k/k!$. Each of such these coefficients corresponds to one of the following equations

$$\begin{cases} \dfrac{du}{dt} = -5\nu \dfrac{u}{\lambda_T^2}, \\ \dfrac{d\lambda_T}{dt} = -\dfrac{u}{2} + \dfrac{\nu}{\lambda_T}\left(\dfrac{7}{3} f_0^{IV} \lambda_T^4 - 5\right), \\ \dfrac{d f_0^{IV}}{dt} = ..., \\ ... \\ \dfrac{d f_0^{(n)}}{dt} = ..., \\ ... \end{cases} \tag{88}$$

Such equations can be written by introducing the infinite dimensional state vector

$$\mathbf{Y} \equiv \left(u, \lambda_T, f_0^{IV}, ....f_0^{(n)}, ...\right). \tag{89}$$

which represents the state of the longitudinal velocity correlation. Therefore, Equation (88), formally written as

$$\dot{\mathbf{Y}} = \mathbf{F}(\mathbf{Y}, \nu) \tag{90}$$

are equivalent to the closed von Kármán–Howarth equation. Equation (90) defines a bifurcation problem where $\nu$ plays the role of control parameter. Thus, this bifurcation analysis studies the variations of $\mathbf{Y}$ caused by $\nu$ according to

$$\mathbf{F}(\mathbf{Y}, \nu) = \mathbf{F}(\mathbf{Y}_0, \nu_0) \tag{91}$$

For $\nu > \nu_0$, $\mathbf{Y}$ is formally calculated through the implicit functions inversion theorem

$$\mathbf{Y} = \mathbf{G}(\mathbf{Y}_0, \nu_0, \nu) \equiv \mathbf{Y}_0 - \int_{\nu_0}^{\nu} \left(\nabla_{\mathbf{Y}} \mathbf{F}\right)^{-1} \frac{\partial \mathbf{F}}{\partial \nu}\, d\nu \tag{92}$$

where $\nabla_{\mathbf{Y}} \mathbf{F}$ is the jacobian $\partial \mathbf{F}/\partial \mathbf{y}$. A bifurcation of Equation (90) happens when this jacobian is singular, that is,

$$\det\left(\nabla_{\mathbf{Y}} \mathbf{F}\right) = 0 \tag{93}$$

If $\nu_0$ is quite small ($R_T$ properly large), the energy cascade is dominant with respect to the viscosity effects and $\nabla_{\mathbf{Y}} \mathbf{F}$ is expected to be nonsingular. Increasing $\nu$, $\mathbf{Y}$ smoothly varies according to Equation (92) and thereafter the dissipation gradually becomes stronger than the energy cascade until reaching the first bifurcation where condition (93) occurs. With reference to Figure 2, this corresponds to the path $B \to A$ until to reach $A$. There, a hard loss of stability is expected for the fully developed turbulence toward non–chaotic regimes [66]. Therefore, $R_T^*$ is calculated as that value of $R_T$ at bifurcation which gives the maximum of the largest real part of the eigenvalues of $\nabla_{\mathbf{Y}}\mathbf{F}$ [66,67] compatible with the current value of the average kinetic energy $u^2$, that is,

$$\begin{aligned} & R_T^* \mid \sup_k \{\Re(l_k)\} = \max, \\ & \det(\nabla_{\mathbf{Y}}\mathbf{F}) = 0, \\ & u^2 = \text{given} \end{aligned} \tag{94}$$

where $l_k$, $k$=1, 2,... are the eigenvalues of $\nabla_{\mathbf{Y}}\mathbf{F}$.

On the other hand, as previously seen, far from the initial condition, the energy cascade acts keeping $f$ similar in the time in a given interval of variation of $r$. There, the evolution of $f$ is expected to be described—at least in first approximation—by Equation (78) and this suggests that—under such approximation—the knowledge of $u$ and $\lambda_T$ can be considered to be sufficient to describe the evolution of $f$. Hence, only the first two components of the state vector $\mathbf{Y}$ are taken which correspond to the coefficients of the order of $r^0$ and $r^2$ of Equation (88). Thus, thanks to the self–similarity, the infinite dimensional space where $\mathbf{Y}$ lies is replaced by a finite dimensional manifold and the state vector is reduced to

$$\mathbf{Y} \equiv (u, \lambda_T), \tag{95}$$

$f_0^{IV}$ plays the role of a parameter which characterizes the velocity correlation and the jacobian $\nabla_{\mathbf{Y}}\mathbf{F}$ reads as

$$\nabla_{\mathbf{Y}}\mathbf{F} = \begin{pmatrix} \dfrac{\partial \dot{u}}{\partial u} & \dfrac{\partial \dot{u}}{\partial \lambda_T} \\[2ex] \dfrac{\partial \dot{\lambda_T}}{\partial u} & \dfrac{\partial \dot{\lambda_T}}{\partial \lambda_T} \end{pmatrix} \tag{96}$$

whose determinant is

$$\det(\nabla_{\mathbf{Y}}\mathbf{F}) = -\frac{5\nu^2}{\lambda_T^2}\left(7 f_0^{IV}\lambda_T^2 + \frac{10}{\lambda_T^2}\right) + 5\nu\frac{u}{\lambda_T^3} \tag{97}$$

From Equation (97), as long as $\nu > 0$ is properly small, $\det(\nabla_{\mathbf{Y}}\mathbf{F}) > 0$. In order that a bifurcation happen, $\det(\nabla_{\mathbf{Y}}\mathbf{F})$ must vanish for a certain value of $\nu$ and this implies that $f_0^{IV}\lambda_T^4 > -10/7$. Thus, increasing $\nu$, $\det(\nabla_{\mathbf{Y}}\mathbf{F})/\nu$ diminishes and there exists a value of $\nu$ where this jacobian determinant vanishes. To determine $R_T^*$, $f_0^{IV}$ is eliminated through the bifurcation condition ($\det(\nabla_{\mathbf{Y}}\mathbf{F}) = 0$) and Equation (97), that is,

$$f_0^{IV} = \frac{1}{7\lambda_T^2\nu}\left(\frac{u}{\lambda_T} - 10\frac{\nu}{\lambda_T^2}\right) \tag{98}$$

Therefore, the singular jacobian is

$$\nabla_{\mathbf{Y}}\mathbf{F} = \begin{pmatrix} -5\nu/\lambda_T^2 & 10\nu u/\lambda_T^3 \\[2ex] -1/2 & u/\lambda_T \end{pmatrix} \tag{99}$$

and admits the following eigenvalues and eigenvectors $l_1$, $l_2$, $\mathbf{y}_1$ and $\mathbf{y}_2$, respectively

$$l_1 = 0, \quad \mathbf{y}_1 = \left(u, \frac{\lambda_T}{2}\right)$$
$$l_2 = \frac{u^2}{\nu}\left(\frac{1}{R_T} - \frac{5}{R_T^2}\right), \quad \mathbf{y}_2 = \left(u, R_T \frac{\lambda_T}{10}\right) \tag{100}$$

The eigenvalue $l_2 \in \mathbb{R}$ maintains positive for $R_T > 5$ and reaches its maximum $l_{2max} = 5\nu/\lambda_T^2$ for $R_T$ = 10. Accordingly, $R_T^*$ is estimated as

$$R_T^* = 10 \tag{101}$$

which corresponds to $f_0^{IV}$ = 0.

Another characteristic value of $R_T$ is obtained in the case where both the eigenvalues vanish. This is $R_T$ = 5 and is expected to represent the onset of the decaying turbulence regime. In fact, in such situation, it is reasonable that $f$ and $\lambda_T$ are

$$\frac{d\lambda_T}{dt} \simeq 0,$$
$$f \simeq \exp\left(-\frac{1}{2}\left(\frac{r}{\lambda_T}\right)^2\right) \tag{102}$$

Hence, $f_0^{IV}\lambda_T^4 \simeq 3$ and $R_T \simeq 4$, in agreement with the previous estimation.

**Remark 2.** *It is worth remarking that $R_T^*$ provides the minimum of $R_T$ in fully developed isotropic homogeneous turbulence, thus this gives the order of magnitude of $R_T$ at the transition. Of course, the transition toward the chaos consists in intermediate stages (bifurcations of Navier–Stokes equations) where the turbulence is not developed and the velocity statistics does not exhibit, in general, isotropy and homogeneity. Hence, the obtained results provide the order of magnitude of $R_T$ at the transition. On the basis of this analysis, during the transition, $R_T$ ranges as*

$$4 \lesssim R_T \lesssim 10 \tag{103}$$

The obtained value of $R_T^*$ = 10 is in very good agreement with the bifurcations analysis of the turbulent energy cascade [3], where the author shows that, in the transition toward the developed turbulence, if the bifurcations cascade follows the Feigenbaum scenario [42,43], the critical Taylor scale Reynolds number is about 10.13 and occurs after three bifurcations.

We conclude this section by remarking the limits under which $R_T^*$ is estimated. Such limits derive from the local self–similarity produced by the closures (65) which allow to consider only the first two equations of (88).

## 14. Velocity and Temperature Fluctuations

The purpose of this section is to obtain, by means of the previous Lyapunov analysis, formal expressions of velocity and temperature fluctuations which will be useful for estimating the statistics of these latter. For sake of our convenience, Navier–Stokes and thermal energy equations are now written in the following dimensionless divergence form

$$\frac{\partial \mathbf{u}}{\partial t} = \text{div}\, \hat{\mathbf{T}}, \qquad \text{in which} \qquad \hat{\mathbf{T}} = \mathbf{T} - \mathbf{u} \otimes \mathbf{u}, \tag{104}$$
$$\frac{\partial \vartheta}{\partial t} = -\text{div}\, \hat{\mathbf{q}} \qquad\qquad \hat{\mathbf{q}} = \mathbf{q} + \mathbf{u}\vartheta$$

where $\mathbf{T}$ and $\mathbf{q}$ denote, respectively, dimensionless stress tensor and heat flux, according to the Navier-Fourier laws

$$\mathbf{T} = -\mathbf{I}p + \mathbf{T}_v,$$
$$\mathbf{T}_v = \frac{1}{Re}\left(\nabla_{\mathbf{x}}\mathbf{u} + \nabla_{\mathbf{x}}\mathbf{u}^T\right), \tag{105}$$
$$\mathbf{q} = -\frac{1}{Pe}\nabla_{\mathbf{x}}\vartheta$$

being $\mathbf{I}$ the identity tensor, $\mathbf{T}_v$ the viscous stress tensor and the pressure $p$ is given according to Equation (6).

In order to obtain the analytical forms of velocity and temperature fluctuations, Equation (104) are first expressed in terms of referential coordinate $\mathbf{x}_0$

$$\frac{\partial u_i}{\partial t} = \left(\frac{\partial \hat{T}_{ij}}{\partial x_{0k}}\right)\left(\frac{\partial x_{0k}}{\partial x_j}\right) \equiv \left(\frac{\partial \hat{T}_{ij}}{\partial x_{0k}}\right) G^{-1}_{jk} \exp\left(-\tilde{\Lambda} t\right), \; i = 1,2,3 \tag{106}$$
$$\frac{\partial \vartheta}{\partial t} = -\left(\frac{\partial \hat{q}_j}{\partial x_{0k}}\right)\left(\frac{\partial x_{0k}}{\partial x_j}\right) \equiv -\left(\frac{\partial \hat{q}_j}{\partial x_{0k}}\right) G^{-1}_{jk} \exp\left(-\tilde{\Lambda} t\right)$$

where the repeated index denotes the summation convention. The adoption of the referential coordinates allows to factorize of $\partial \mathbf{u}/\partial t$ and $\partial \vartheta/\partial t$ as a product of two statistically uncorrelated matrices: one depending on velocity and temperature fields and the other representing the local fluid deformation. Velocity and temperature fluctuations are here obtained integrating Equation (106) in the set $(t, a)$. Due to the alignment property of the Lyapunov vectors [56], $\exp(-\tilde{\Lambda} t)$ rapidly goes to zero as $t \to \infty$ in any case, whereas $\partial \hat{T}_{ij}/\partial x_{0k}$ and $\partial \hat{q}_j/\partial x_{0k}$ are functions of slow growth of $t$. Hence, velocity and temperature fluctuations are formally calculated integrating Equation (106) in the set $(t, \infty)$ where $\partial \hat{T}_{ij}/\partial x_{0k}$ and $\partial \hat{q}_j/\partial x_{0k}$ are considered to be constant and equal to the corresponding values at the current time. Such fluctuations are then expressed in function of current velocity and temperature fields according to

$$u_i = \left(\frac{\partial \hat{T}_{ij}}{\partial x_{0k}}\right) W_{jk}, \; i = 1,2,3 \tag{107}$$
$$\vartheta = -\left(\frac{\partial \hat{q}_j}{\partial x_{0k}}\right) W_{jk}$$

being

$$W_{jk} = \int_0^\infty G^{-1}_{jk} \exp\left(-\tilde{\Lambda} t\right)\, dt \tag{108}$$

where $|W_{jk}| < \infty$ as $G^{-1}_{jk}$ is represented by slow growth functions of $t$.

It is worth to remark that Equation (107) are, in general, rough approximations of velocity and temperature fluctuations. Nevertheless, in fully developed turbulence, $dx(t)$ is considered to be much more rapid than $\mathbf{u}(t,\mathbf{x})$, thus Equation (107) provide one accurate way to express velocity and temperature in terms of referential coordinates by means of the Lyapunov theory.

## 15. *Statistics of Velocity and Temperature Difference

In developed turbulence, longitudinal velocity and temperature difference, $\Delta u_r = (\mathbf{u}(t,\mathbf{x}') - \mathbf{u}(t,\mathbf{x})) \cdot \mathbf{r}/r$ and $\Delta\vartheta = \vartheta(t,\mathbf{x}') - \vartheta(t,\mathbf{x})$, $\mathbf{r} = \mathbf{x}' - \mathbf{x}$, play a role of paramount importance as these quantities describe energy cascade, intermittency and are linked to dissipation. This section analyzes the statistics of such quantities in fully developed homogeneous isotropic turbulence through the previously seen kinematic Lyapunov analysis and using a proper statistical decomposition of velocity and temperature. In order to determine this statistics, the Navier–Stokes bifurcations effect on $\Delta u_r$ and $\Delta\vartheta$ is first analyzed. To this purpose, $\Delta u_r$ and $\Delta\vartheta$ are expressed in function of current velocity and temperature through Equation (107)

$$
\begin{aligned}
\Delta u_r &= \left(\frac{\partial \hat{T}_{ij}}{\partial x_{0k}}\right)' W'_{jk} - \left(\frac{\partial \hat{T}_{ij}}{\partial x_{0k}}\right) W_{jk} \\
\Delta\vartheta &= -\left(\frac{\partial \hat{q}_j}{\partial x_{0k}}\right)' W'_{jk} + \left(\frac{\partial \hat{q}_j}{\partial x_{0k}}\right) W_{jk}
\end{aligned}
\tag{109}
$$

The several bifurcations happening during the fluid motion determine a continuous doubling of $\mathbf{u}$ in several functions, say $\hat{\mathbf{v}}_k$, $k$ = 1, 2, ..., in the sense that each encountered bifurcation introduces new functions $\hat{\mathbf{v}}_k$ whose characteristics are independent of the velocity field at previous time. Then, due to bifurcations, $\mathbf{u}$ is of the form

$$
\mathbf{u}(t,\mathbf{x}) \approx \sum_k \hat{\mathbf{v}}_k(t,\mathbf{x}),
\tag{110}
$$

It is worth remarking that, while $\mathbf{u}(t,\mathbf{x})$ is solution of the Navier–Stokes equations, the functions $\hat{\mathbf{v}}_k$ are not. Therefore, the functions $\hat{\mathbf{v}}_k$ are the result of the mathematical segregation due to bifurcations of a fluid state variable which physically only exist in combination, thus each of them is not directly observable. This implies that $\mathbf{u}$ will be distributed, in line with the Liouville theorem, according to a classical definite positive distribution function. On the contrary, each single function $\hat{\mathbf{v}}_k$, representing mathematical segregation of the fluid state, will be distributed following extended distribution functions which can exhibit negative values [68–70] compatible with conditions linked to the specific problem. These conditions mainly arise from (a) the Navier–Stokes equations and from (b) the isotropic hypothesis. For what concerns (a), in order that pressure and inertia forces can cause sizable variations of velocity autocorrelation, each term $\hat{\mathbf{v}}_k \equiv (\hat{v}_1, \hat{v}_2, \hat{v}_3)$ will be distributed following highly nonsymmetric extended distribution function, for which

$$
\frac{|\langle \hat{v}_{ki}^3 \rangle|}{\langle \hat{v}_{ki}^2 \rangle^{3/2}} >>> 1, \quad i = 1,2,3
\tag{111}
$$

As for (b), due to isotropic hypothesis, $\mathbf{u}$ would be distributed following a gaussian PDF [18], thus, according to the Navier–Stokes equations, pressure and inertia forces will not give contribution to the time derivative of the third statistical moment of $\mathbf{u}$. Accordingly, the absolute value of odd statistical moments of order n of $\hat{\mathbf{v}}_k$ is expected to be very high in comparison with the even statistical moments of order n + 1, that is,

$$
\frac{|\langle \hat{v}_{ki}^n \rangle|}{\langle \hat{v}_{ki}^2 \rangle^{n/2}} >>> \frac{|\left\langle \hat{v}_{ki}^{n+1} \right\rangle|}{\langle \hat{v}_{ki}^2 \rangle^{(n+1)/2}}, \quad n = 3,5,7,..., \quad i = 1,2,3.
\tag{112}
$$

This suggests that $\Delta\mathbf{u}$ and $\mathbf{u}$ can be expressed through a specific statistical decomposition [71], as a linear combination of opportune stochastic variables $\xi_k$ that reproduce the doubling bifurcations effect and whose extended distribution functions satisfy Equations (111) and (112). Furthermore, as $\vartheta$ is a passive scalar, its fluctuations are the result of $\mathbf{u}$ and of thermal diffusivity, thus also $\vartheta$ is written by means of the same decomposition

$$\mathbf{u} = \sum_k \mathbf{U}_k \xi_k, \qquad \vartheta = \sum_k \Theta_k \xi_k \tag{113}$$

where $\mathbf{U}_k$ and $\Theta_k (k = 1, 2, ...)$ are coordinate functions of $t$ and $\mathbf{x}$, being $\nabla_{\mathbf{x}} \cdot \mathbf{U}_k = 0$, $\forall k$ and $\xi_k$ ($k$ = 1, 2, ...) are dimensionless independent centered stochastic variables such that

$$\langle \xi_k \rangle = 0, \quad \langle \xi_i \xi_j \rangle = \delta_{ij}, \quad \langle \xi_i \xi_j \xi_k \rangle = \begin{cases} q \neq 0, \ \forall\, i = j = k \\ 0 \ \text{else} \end{cases} \tag{114}$$

where $q$, providing the skewness of $\xi_k$ k = 1, 2..., satisfies to

$$|q| >>> 1, \ \langle \xi_i^2 \rangle, \ \langle \xi_i^4 \rangle, \ i = 1, 2, ..., \tag{115}$$

Therefore, the distribution functions of $\xi_k$ can assume negative values compatible with Equations (114) and (115).

Through the decomposition (113), we will show that the negative value of $H_u^{(3)}(r)$ has very important implications for what concerns the statistics of $\Delta u_r$ and $\Delta\vartheta$, with particular reference to the intermittency of these latter which rises as Reynolds number and Péclet number increase. To study this question, consider first the analytical forms of the fluctuations of $u_i$ and $\vartheta$ in terms of $\xi_k$ obtained by substituting Equation (113) into Equation (107)

$$u_i = \sum_j \sum_k A_{jk}^{(i)} \xi_j \xi_k + \frac{1}{R_T} \sum_k a_k^{(i)} \xi_k, \ i = 1, 2, 3 \qquad \vartheta = \sum_j \sum_k B_{jk} \xi_j \xi_k + \frac{1}{Pe} \sum_k b_k \xi_k, \tag{116}$$

where $\sum_j \sum_k A_{jk}^{(i)} \xi_j \xi_k$ and $1/R_T \sum_k a_k^{(i)} \xi_k$ are the contributions of inertia and pressure forces and of the fluid viscosity, respectively, whereas $\sum_j \sum_k B_{jk} \xi_j \xi_k$ and $1/Pe \sum_k b_k \xi_k$ arise from the convective term and fluid conduction. Because of turbulent isotropy, it is reasonable that $u_i$ and $\vartheta$ are both Gaussian stochastic variables [18,71,72], thus the various terms of Equation (116) satisfy the Lindeberg condition, a very general, necessary and sufficient condition for satisfying the central limit theorem [71,72]. Such theorem does not apply to $\Delta u_i$ and $\Delta\vartheta$ as these latter are the difference between two correlated Gaussian variables, thus their PDF are expected to be very different with respect to Gaussian distributions. To study the statistics of $\Delta u_r$ and $\Delta\vartheta$, the fluctuations of these latter are first expressed in terms of $\xi_k$

$$\Delta u_r(\mathbf{r}) = \sum_j \sum_k \Delta A_{jk} \xi_j \xi_k + \frac{1}{R_T} \sum_k \Delta a_k \xi_k, \qquad \Delta\vartheta(\mathbf{r}) = \sum_j \sum_k \Delta B_{jk} \xi_j \xi_k + \frac{1}{Pe} \sum_k \Delta b_k \xi_k, \tag{117}$$

being

$$\Delta A_{jk} = \sum_{i=1}^{3} \left( A_{jk}^{(i)}(\mathbf{x}+\mathbf{r}) - A_{jk}^{(i)}(\mathbf{x}) \right) \frac{r_i}{r} \equiv S_{ujk} + \Omega_{ujk},$$

$$\Delta a_k = \sum_{i=1}^{3} \left( a_k^{(i)}(\mathbf{x}+\mathbf{r}) - a_k^{(i)}(\mathbf{x}) \right) \frac{r_i}{r}, \tag{118}$$

$$\Delta B_{jk} = B_{jk}(\mathbf{x}+\mathbf{r}) - B_{jk}(\mathbf{x}) \equiv S_{\theta jk} + \Omega_{\theta jk},$$

$$\Delta b_k = b_k(\mathbf{x}+\mathbf{r}) - b_k(\mathbf{x}),$$

In Equation (118), the matrices $\Delta A_{jk}$ and $\Delta B_{jk}$ are decomposed following their symmetric and antisymmetric parts, respectively $S_{ujk}$, $S_{\theta jk}$ and $\Omega_{ujk}$, $\Omega_{\theta jk}$. These last ones give null contribution in Equation (117), whereas the terms arising from $S_{ujk}$ and $S_{\theta jk}$ are expressed as

$$\sum_j \sum_k S_{Xjk} \xi_j \xi_k = \sum_i S_{Xii} \xi_i^2 + \sum_{j \neq k} S_{Xjk} \xi_j \xi_k, \tag{119}$$

$$X = u, \theta$$

in which the first term of Equation (119) is decomposed in the following manner

$$\sum_i S_{Xii} \xi_i^2 = S_X^+ \left( \eta_X^2 - \sum_{j \neq k}^{+} \xi_j \xi_k \right) + \sum_i^{+} \left( S_{Xii} - S_X^+ \right) \xi_i^2 + S_X^- \left( \zeta_X^2 - \sum_{j \neq k}^{-} \xi_j \xi_k \right) + \sum_i^{-} \left( S_{Xii} - S_X^- \right) \xi_i^2, \tag{120}$$

$$X = u, \theta$$

being

$$\eta_X = \sum_i^{+} \xi_i,$$

$$\zeta_X = \sum_j^{-} \xi_j,$$

$$S_X^+ = \frac{1}{n_X^+} \sum_i^{+} S_{ii} > 0, \tag{121}$$

$$S_X^- = \frac{1}{n_X^-} \sum_i^{-} S_{ii} < 0,$$

$$X = u, \theta,$$

and

$$\begin{aligned}
\xi_X &= -S_X^+ \sum_{j\neq k}^{+} \xi_j \xi_k + \sum_i^{+} \left(S_{Xii} - S_X^+\right) \xi_i^2 - S_X^- \sum_{j\neq k}^{-} \xi_j \xi_k + \sum_i^{-} \left(S_{Xii} - S_X^-\right) \xi_i^2 + \sum_{j\neq k} S_{Xjk} \xi_j \xi_k + \sum_k \Delta a_{Xk} \xi_k \\
&\equiv \sum_{ij} M_{Xij} \xi_i \xi_j + \sum_k g_{Xk} \xi_k, \\
g_{uk} &= \frac{\Delta a_k}{R_T}, \quad g_{\theta k} = \frac{\Delta b_k}{Pe}, \quad k = 1, 2, ... \\
X &= u, \theta,
\end{aligned} \tag{122}$$

where $\sum^+$ and $\sum^-$ denote summations for $(S_{Xjj} > 0, S_{Xkk} > 0)$ and $(S_{Xjj} \le 0, S_{Xkk} \le 0)$ and $n_X^+$ and $n_X^-$ are the corresponding numbers of terms of such summations, whereas $\sum_{j\neq k}^+$ and $\sum_{j\neq k}^-$ indicate the sums of addends calculated for $j \neq k$ corresponding to $S_{Xjj} > 0$, $S_{Xkk} > 0$ and $S_{Xjj} < 0$, $S_{Xkk} < 0$, respectively. The decomposition (119) and (120) and the definitions (121) lead to the following expression of velocity and temperature difference fluctuations

$$\begin{aligned}
\Delta u_r &= \xi_u + S_u^+ \eta_u^2 + S_u^- \zeta_u^2, \\
\Delta \vartheta &= \xi_\theta + S_\theta^+ \eta_\theta^2 + S_\theta^- \zeta_\theta^2,
\end{aligned} \tag{123}$$

Now, we show that $\xi_X$, $\eta_X$ and $\zeta_X$, $X = u, \theta$ tend to uncorrelated gaussian variables. In fact, from Equation (121), $\eta_X$ and $\zeta_X$, $X = u, \theta$ are sums of random terms belonging to two different sets of uncorrelated stochastic variables (i.e., the sets for which $S_{Xii} < 0$ and $S_{Xii} > 0$), therefore $\eta_X$ and $\zeta_X$, are two uncorrelated stochastic variables such that $\langle \eta_X \rangle = \langle \zeta_X \rangle = 0$, $X = u, \theta$. Furthermore, as $\xi_k$ are statistically independent with each other, the central limit theorem applied to Equation (121) guarantees that both $\eta_X$ and $\zeta_X$ tend to two uncorrelated centered gaussian random variables. As for $\xi_X$, $X = u, \theta$, the following should be considered: due to the analytical structure of Equation (122), each term of $\xi_X$ is a centered variable, thus $\langle \xi_X \rangle = 0$. Next, in Equation (122), the following terms $-S^+ \sum_{j\neq k}^+ \xi_j \xi_k + \sum_i^+ \left(S_{Xii} - S_X^+\right) \xi_i^2$ and $-S_X^- \sum_{j\neq k}^- \xi_j \xi_k + \sum_i^- \left(S_{Xii} - S_X^-\right) \xi_i^2$ are mutually uncorrelated, as each of these is sum of random variables belonging to two different uncorrelated sets. Moreover, $\sum_{i\neq j} \xi_i \xi_j$ includes several weakly correlated terms, whereas $\sum_k g_{Xk} \xi_k$ is the sum of independent variables. On the other hand, due to hypothesis of fully developed chaos, the energy cascade, here represented by Equations (114), (115) and (117), will generate a strong mixing on the several terms of Equation (117), thus a proper variant of the central limit theorem can be applied to $\xi_X$ whose several terms are weakly dependent with each other [72]. As the result, $\xi_X$, $X = u, \theta$ will tend to centered gaussian variables statistically independent of $\eta_X$ and $\zeta_X$.

Hence, the statistics of $\Delta u_r$ and $\Delta \vartheta$ is represented by the following structure functions of the independent centered gaussian stochastic variables $\xi_X$, $\eta_X$ and $\zeta_X$ for which $\langle \xi_X^2 \rangle = \langle \eta_X^2 \rangle = \langle \zeta_X^2 \rangle = 1$.

$$\begin{aligned}
\Delta u_r &= L_u \xi_u + S_u^+ (\eta_u^2 - 1) - S_u^- (\zeta_u^2 - 1), \\
\Delta \vartheta &= L_\theta \xi_\theta + S_\theta^+ (\eta_\theta^2 - 1) - S_\theta^- (\zeta_\theta^2 - 1),
\end{aligned} \tag{124}$$

where $L_u$ and $L_\theta$ are now introduced to take into account that $\xi_X$, $\eta_X$ and $\zeta_X$ have standard deviation equal to unity. Thus

$$
\begin{aligned}
L_u\xi_u &= \sum_{ij} M_{uij}\xi_i\xi_j + \frac{1}{R_T}\sum_k \Delta a_{uk}\xi_k, \\
L_\theta\xi_\theta &= \sum_{ij} M_{\theta ij}\xi_i\xi_j + \frac{1}{Pe}\sum_k \Delta a_{\theta k}\xi_k,
\end{aligned}
\tag{125}
$$

and $L_X$, $S_X^-$ and $S_X^+$ are parameters depending upon $r$ which have to be determined. To this regard, it worth remarking that, in regime of fully developed isotropic turbulence in infinite domain, the numbers of parameters necessary to describe the statistics of $\Delta u_r$ and $\Delta\vartheta$ should be minimum compatible with assigned quantities which define the current state of fluid motion, such as average kinetic energy, temperature standard deviation and correlation functions. On the other hand, the evolution equation of $f$ [17] requires the knowledge of the correlations of the third order $k$ to be solved. Therefore, in fully developed homogeneous isotropic turbulence, the sole knowledge of $f$ and $k$ is here considered to be the necessary and sufficient information for determining the statistics of $\Delta u_r$. This implies that $S_u^+$ is proportional to $S_u^-$ through a proper quantity which does not depend on $r$, that is,

$$
S_u^+(r) = \chi S_u^-(r) \equiv \chi S_u(r)
\tag{126}
$$

where $\chi < 1$ is a function of $R_T$ giving the skewness of $\Delta u_r$, which has to be identified. Accordingly, $S_u$ and $L_u$ will be determined in function of $f$ and $k$ as soon as $\chi = \chi(Re)$ is known. For what concerns the temperature difference, observe that, due to turbulence isotropy, the skewness of $\Delta\vartheta$ should be equal to zero and this gives

$$
S_\theta^+(r) = S_\theta^-(r) \equiv S_\theta(r)
\tag{127}
$$

Therefore, the structure functions of $\Delta u_r$ and $\Delta\vartheta$ read as

$$
\begin{aligned}
\Delta u_r &= L_u\xi_u + S_u\left(\chi\left(\eta_u^2 - 1\right) - \left(\zeta_u^2 - 1\right)\right), \\
\Delta\vartheta &= L_\theta\xi_\theta + S_\theta\left(\eta_\theta^2 - \zeta_\theta^2\right),
\end{aligned}
\tag{128}
$$

Furthermore, again following the parameters minimum number, the ratio $\Psi_\theta(r) \equiv S_\theta/L_\theta$ would be proportional to $\Psi_u(r) \equiv S_u/L_u$ through a proper coefficient depending upon the Prandtl number alone, that is

$$
\Psi_\theta(r) = \sigma(Pr)\Psi_u(r)
\tag{129}
$$

where $\sigma$ is a function of the Prandtl number which has to be determined.

At this stage of the present analysis, we show that, in fully developed turbulence, $L_u$ and $L_\theta$ are, respectively, functions of $R_T$ and $Pe$, resulting in $L_u \propto R_T^{-1/2}$ and $L_\theta \propto Pe^{-1/2}$. In fact, from Equation (125) we obtain

$$
\begin{aligned}
L_u^2 &= \sum_{ijkl} M_{uij}M_{ukl}\left\langle\xi_i\xi_j\xi_k\xi_l\right\rangle + \frac{2}{R_T}\sum_k M_{ukk}\Delta a_{uk}\left\langle\xi_k^3\right\rangle + \frac{1}{R_T^2}\sum_k \Delta a_{uk}^2, \\
L_\theta^2 &= \sum_{ijkl} M_{\theta ij}M_{\theta kl}\left\langle\xi_i\xi_j\xi_k\xi_l\right\rangle + \frac{2}{Pe}\sum_k M_{\theta kk}\Delta a_{\theta k}\left\langle\xi_k^3\right\rangle + \frac{1}{Pe^2}\sum_k \Delta a_{\theta k}^2,
\end{aligned}
\tag{130}
$$

As $| \langle \xi_k^3 \rangle | >>> 1, \langle \xi_i \xi_j \xi_k \xi_l \rangle$, first and third addend of Equation (130) are negligible with respect to second one, thus $L_u$ and $L_\theta$ tend to functions of the kind

$$L_u = \frac{F_u(r)}{\sqrt{R_T}}, \qquad L_\theta = \frac{F_\theta(r)}{\sqrt{Pe}}. \tag{131}$$

where $F_u(r)$ and $F_\theta(r)$ are functions of $r$ which do not directly depend on $R_T$ and $Pe$. Hence, the dimensionless $\Delta u_r$ and $\Delta\vartheta$, normalized with respect to the corresponding standard deviations, are expressed in function of $R_T$ and $Pe$

$$\frac{\Delta u_r}{\sqrt{\langle(\Delta u_r)^2\rangle}} = \frac{\xi_u + \Psi_u(\chi(\eta_u^2 - 1) - (\zeta_u^2 - 1))}{\sqrt{1 + 2\Psi_u^2(1+\chi^2)}}, \quad \Psi_u(r) = \frac{S_u(r)}{L_u(r)} = \Phi(r)\sqrt{R_T},$$
$$\frac{\Delta \vartheta}{\sqrt{\langle(\Delta \vartheta)^2\rangle}} = \frac{\xi_\theta + \Psi_\theta(\eta_\theta^2 - \zeta_\theta^2)}{\sqrt{1 + 4\Psi_\theta^2}}, \quad \Psi_\theta(r) = \frac{S_\theta(r)}{L_\theta(r)} = \Phi(r)\sqrt{Pe} \tag{132}$$

and this identifies $\sigma = \sqrt{Pr}$. Equation (132) provide peculiar structure functions giving the statistics of $\Delta u_r$ and $\Delta\vartheta$.

Now, if $\chi = \chi(R_T)$ is considered to be known, $L_u$ and $S_u$ can be expressed in function of $\langle \Delta u_r^2 \rangle$ and $\langle \Delta u_r^3 \rangle$, where this latter is calculated adopting the proposed closure (65). In fact, $L_u$ and $S_u$ are related to $\langle \Delta u_r^2 \rangle$ and $\langle \Delta u_r^3 \rangle$ through Equation (128)

$$\left\langle (\Delta u_r)^3 \right\rangle = 6u^3 k = 8S_u^3(\chi^3 - 1),$$
$$\left\langle (\Delta u_r)^2 \right\rangle = 2u^2(1-f) = L_u^2 + 2S_u^2(\chi^2+1), \tag{133}$$

thus, $L_u$, $S_u$ and $\Phi$ are expressed in function of $f(r)$ and $k(r)$ as

$$S_u(r) = \left(\frac{3/4}{\chi^3 - 1}\right)^{1/3} u\, k(r)^{1/3},$$
$$L_u(r) = \sqrt{2}\, u \sqrt{1 - f(r) - (1+\chi^2)\left(\frac{3/4}{\chi^3-1}\right)^{2/3} k(r)^{2/3}}, \tag{134}$$
$$\Phi = \frac{S_u}{L_u}\frac{1}{\sqrt{R_T}}$$

In the expression of $L_u(r)$ of Equations (134), the argument of the square root must be greater than zero and this leads to the following implicit condition for $\chi$

$$\frac{1+\chi^2}{(\chi^3-1)^{2/3}} \le \frac{1}{2}\left(\frac{56}{3}\right)^{2/3} \tag{135}$$

where the proposed closure (65) is taken into account. Inequality (135), solved with respect to $\chi$, gives the upper limit for $\chi$

$$\chi \le \chi_\infty = 0.8659... \tag{136}$$

As far as the temperature difference is concerned, we have

$$\frac{\left\langle (\Delta u_r)^2 \right\rangle}{\left\langle (\Delta \vartheta)^2 \right\rangle} \equiv \frac{u^2}{\theta^2} \frac{1-f}{1-f_\theta} = \frac{L_u^2}{L_\theta^2} \frac{1+2\Psi_u^2(1+\chi^2)}{1+4\Psi_\theta^2} \tag{137}$$

thus Equation (137) allows to calculate $L_\theta$ in terms of the other quantities

$$L_\theta = L_u \frac{\theta}{u} \sqrt{\frac{1-f_\theta}{1-f}} \sqrt{\frac{1+2\Phi^2 R_T(1+\chi^2)}{1+4\Phi^2 Pe}} \tag{138}$$

In Equations (134) and (138), the function $\chi = \chi(R_T)$ has to be identified, and $\Phi(r)$ depends on the specific shape of $f(r)$, where, due to the constancy of $H_u^{(3)}(0)$, $\Phi(0)$ is assumed to be constant, independent of $R_T$.

The distribution functions of $\Delta u_r$ and $\Delta\vartheta$ are formally calculated through the Frobenius–Perron equation [57], taking into account that $\xi_X$, $\eta_X$ and $\zeta_X$ are independent identically distributed centered gaussian variables such that $\langle \xi_X^2 \rangle = \langle \eta_X^2 \rangle = \langle \zeta_X^2 \rangle = 1$, $X = u, \theta$

$$\begin{aligned} F_u(\Delta u_r') &= \int_\xi \int_\eta \int_\zeta P(\xi,\eta,\zeta)\, \delta(\Delta u_r' - \Delta u_r(\xi,\eta,\zeta))\, d\xi\, d\eta\, d\zeta, \\ F_\theta(\Delta\vartheta') &= \int_\xi \int_\eta \int_\zeta P(\xi,\eta,\zeta)\, \delta(\Delta\vartheta' - \Delta\vartheta(\xi,\eta,\zeta))\, d\xi\, d\eta\, d\zeta, \end{aligned} \tag{139}$$

where $\delta$ is the Dirac delta, $P(\xi,\eta,\zeta)$ is the 3D gaussian PDF

$$P(\xi,\eta,\zeta) = \frac{1}{\sqrt{(2\pi)^3}} \exp\left(-\frac{\xi^2+\eta^2+\zeta^2}{2}\right), \tag{140}$$

and $\Delta u_r(\xi,\eta,\zeta))$ and $\vartheta(\xi,\eta,\zeta))$ are determined by Equation (132).

In other words, the statistics of $\Delta u_r$ and $\Delta\vartheta$ can be inferred looking at the proposed statistical decomposition (113) which includes the bifurcations effects in isotropic turbulence. This is a non–Gaussian statistics, where the absolute value of the dimensionless statistical moments increases with $R_T$ and $Pe$. In detail, the dimensionless statistical moments of $\Delta u_r$ and $\Delta\vartheta$ are easily calculated in function of $\chi$, $\Psi_u$ and $\Psi_\theta$

$$\begin{aligned} H_u^{(n)} &\equiv \frac{\langle (\Delta u_r)^n \rangle}{\langle (\Delta u_r)^2 \rangle^{n/2}} = \frac{1}{(1+2(1+\chi^2)\Psi_u^2)^{n/2}} \sum_{k=0}^{n} \binom{n}{k} \Psi_u^k \langle \xi_u^{n-k} \rangle \langle (\chi(\eta_u^2-1) - (\zeta_u^2-1))^k \rangle, \\ H_\theta^{(n)} &\equiv \frac{\langle (\Delta\vartheta)^n \rangle}{\langle (\Delta\vartheta)^2 \rangle^{n/2}} = \frac{1}{(1+4\Psi_\theta^2)^{n/2}} \sum_{k=0}^{n} \binom{n}{k} \Psi_\theta^k \langle \xi_\theta^{n-k} \rangle \langle (\eta_\theta^2 - \zeta_\theta^2)^k \rangle, \end{aligned} \tag{141}$$

where $\Phi(0)$ and $\chi = \chi(R_T)$ have to be identified. To this end, we first analyze the statistics of $\partial u_r/\partial r$ which, following the proposed Lyapunov analysis, exhibits a constant skewness $H_u^{(3)}(0) = -3/7$. Then, $H_u^{(3)}(r)$ is first obtained from Equation (141)

$$H_u^{(3)}(r) = \frac{8\Psi_u^3(\chi^3-1)}{(1+2\Psi_u^2(1+\chi^2))^{3/2}} \tag{142}$$

and $H_u^{(3)}(0)$ is calculated for $r \to 0$

$$H_u^{(3)}(0) = \frac{8\Psi_u^3(0)(\chi^3-1)}{(1+2\Psi_u^2(0)(1+\chi^2))^{3/2}} \tag{143}$$

Accordingly, $\chi = \chi(R_T)$ is implicitly expressed in function of $\Phi(0)\sqrt{R_T}$. From Equation (143), $\chi = \chi(R_T)$ is a monotonic rising function of $R_T$ which, for $H_u^{(3)}(0) = -3/7$, admits limit

$$\chi_\infty = \lim_{R_T \to \infty} \chi(R_T) = 0.8659... \tag{144}$$

resulting in $\chi(R_T) < 0$ for properly small values of $R_T$. On the other hand, in fully developed turbulence, the PDF of $\partial u_r / \partial r$ exhibits non gaussian behavior (i.e., non gaussian tails) for $\partial u_r / \partial r \to \pm\infty$, accordingly $\chi$ must be positive. Hence, the limit condition $\chi = 0$ is supposed to be achieved for $R_T = R_T^* = 10$ which represents the minimum value of $R_T$ for which the turbulence is homogeneous isotropic. This allows to identify $\Phi(0)$ by means of Equation (143)

$$\Phi(0) = \frac{1}{\sqrt{R_T^*}} \sqrt{\frac{{H_{u0}^{(3)}}^{2/3}}{4 - 2{H_{u0}^{(3)}}^{2/3}}} = 0.1409... \tag{145}$$

Thus, Equation (143) gives, in the implicit form, the variation law $\chi = \chi(R_T)$ which is depicted in Figure 8.

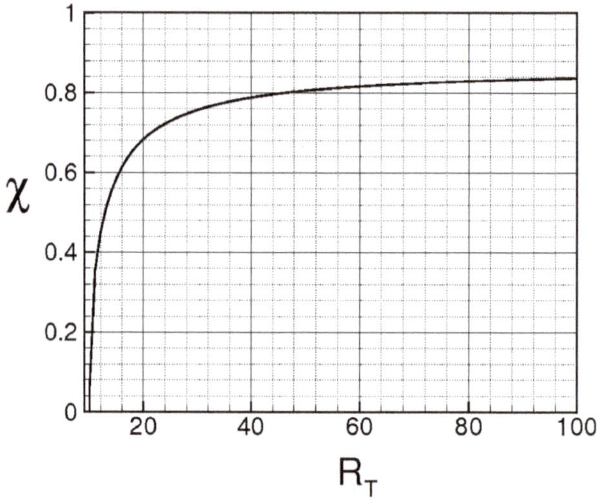


**Figure 8.** Characteristic Function $\chi = \chi(R_T)$.

We conclude this section with the following considerations regarding the proposed analysis, and summarizing some of the results just obtained in the previous works.

For non–isotropic turbulence or in more complex situations with boundary conditions or walls, the velocity will be not distributed following a normal PDF, thus Equation (112) will be not verified, and Equation (132) will change its analytical structure incorporating stronger intermittent terms [72] giving the deviation with respect to the isotropic turbulence. Hence, the absolute statistical moments of $\Delta u_r$ will be greater than those calculated through Equation (141), indicating that, in more complex cases than the isotropic turbulence, the intermittency of $\Delta u_r$ can be significantly stronger.

Next, $\Psi_u$ and $\Psi_\theta$ represent the ratios (large scale velocity)–(small scale velocity) and (large scale temperature)–(small scale temperature), respectively. In particular, $\Psi_u \propto u/u_s \approx (u^2/\lambda_T)/(u_s^2/l_s)$ being $l_s$ and $u_s$ the characteristic small scale and the corresponding velocity. This means that $u/u_s \approx \lambda_T/l_s \approx \sqrt{R_T}$ and that the Reynolds number relative to $u_s$ and $l_s$ is $u_s l_s/\nu \approx 1$, that is $l_s$ and $u_s$ identify the Kolmogorov scale and the corresponding velocity. For what concerns $\Psi_\theta$, $\vartheta$ is a passive scalar, thus $\Psi_\theta$ reads as $\Psi_\theta \propto \theta/\theta_s \approx \theta/\theta_s(u/\lambda_T)/(u_s/l_s)$and this leads to $u_s l_s/\nu \approx 1$.

At this stage of the present analysis, we can show that the kinematic bifurcation rate $S_b$, defined by Equation (25), is much larger than the kinematic Lyapunov exponents. In fact, $S_b$ can be also estimated as the ratio (large scale velocity)–(small scale length), where large scale velocity and small scale length are given by $u$ and by the Kolmogorov scale, respectively. Taking into account the Kolmogorov scale definition and Equation (69), we obtain

$$S_b \approx \frac{u}{l_s} = 15^{1/4} R_T^{1/2} \Lambda \tag{146}$$

confirming the assumption made in the relative section. In fully developed turbulence, $S_b >> \Lambda$ and is a rising function of $R_T$.

As shown in Reference [1], the statistics given by Equations (139) and (141) agree with the experimental data presented in References [47,48]. There, in experiments using low temperature helium gas between two counter–rotating cylinders (closed cell), the PDF of $\partial u_r / \partial r$ and its statistical moments are measured. Although the experiments regard wall–bounded flows, the measured PDF of velocity difference are comparable with the present results ( Equations (139) and (141)). Apart from a lightly non–monotonic evolution of $H_u^{(4)}(0)$ and $H_u^{(6)}(0)$ in [47,48], the dimensionless statistical moments of $\partial u_r / \partial r$ exhibit same trend and same order of magnitude of the corresponding quantities calculated with Equation (141). In particular, the PDFs of $\partial u_r / \partial r$ obtained with the present analysis show non gaussian tails which coincide with those measured in [47,48].

In Figure 9, the normalized PDFs of $\partial u_r / \partial r$, calculated with Equations (139) and (141), are shown in terms of $s$

$$s = \frac{\partial u_r / \partial r}{\sqrt{\left\langle (\partial u_r / \partial r)^2 \right\rangle}} \tag{147}$$

in such a way that their standard deviations are equal to the unity. The results of Figure 9a are performed for $R_T$ = 15, 30 and 60, whereas Figure 9b,c report the PDF for $R_T$ = 255, 416, 514, 1035 and 1553, where Figure 9c represents the enlarged region of Figure 9b, showing the tails of PDF for $5 < s < 8$. According to Equations (139) and (141), the tails of the PDF rise with the Reynolds number in the interval $10 < R_T < 700$, whereas for $R_T > 700$, smaller variations are observed. On the right–bottom, the results of [47] for $R_T$ = 255, 416, 514, 1035 and 1553 are shown. Despite the aforementioned non–monotonic trend (see Figure 9 (Right–bottom)), Figure 9c gives values of the PDFs and of the corresponding average slopes which agree with those obtained in [47], expecially for $5 < s < 8$. To this regards, it is worth to remark that, in certain conditions, the flow obtained in the experiments of [47] could be quite far from the isotropy hypothesis, as such experiments pertain wall–bounded flows, where the walls could significantly influence the fluid velocity in proximity of the probe.

In References [1,2,4] the scaling exponents $\zeta_V(n)$ associated with the several moments of $\Delta u_r$

$$\left\langle (\Delta u_r)^n \right\rangle \approx A_n r^{\zeta_V(n)}, \tag{148}$$

are calculated with Equation (132) through the following best fitting procedure. The statistical moments of $\Delta u_r$ are first calculated in function of $r$ using Equation (141) (see Figure 10 (Left)). Then, the scaling exponents $\zeta_V(n)$ are identified through a minimum square method which, for each statistical moment, is applied to the following optimization problem

$$J_n(\zeta_V(n), A_n) \equiv \int_{\hat{r}_1}^{\hat{r}_2} (\langle (\Delta u_r)^n \rangle - A_n r^{\zeta_V(n)})^2 dr = \min, \; n = 1, 2, ... \tag{149}$$

where $\langle\langle (\Delta u_r)^n \rangle\rangle$ are calculated with Equation (141), $\hat{r}_1$ is assumed to be equal to 0.1, whereas $\hat{r}_2$ is taken in such a way that $\zeta_V(3)$ = 1. The so obtained scaling exponents are shown in Figure 10

(Right–side) (solid symbols) where these are compared with those given by the Kolmogorov theories K41 [44] (dashed line) and K62 [45] (dotted line) and with the exponents calculated by She–Leveque [46] (continuous curve). For $n < 4$, $\zeta_V(n) \approx n/3$ and for higher values of $n$, due to the nonlinear terms of Equation (132), $\zeta_V(n)$ shows multiscaling behavior. The values of $\zeta_V(n)$ here calculated are in good agreement with the She–Leveque data, and result to be lightly greater than those obtained in [46] for $n > 8$.

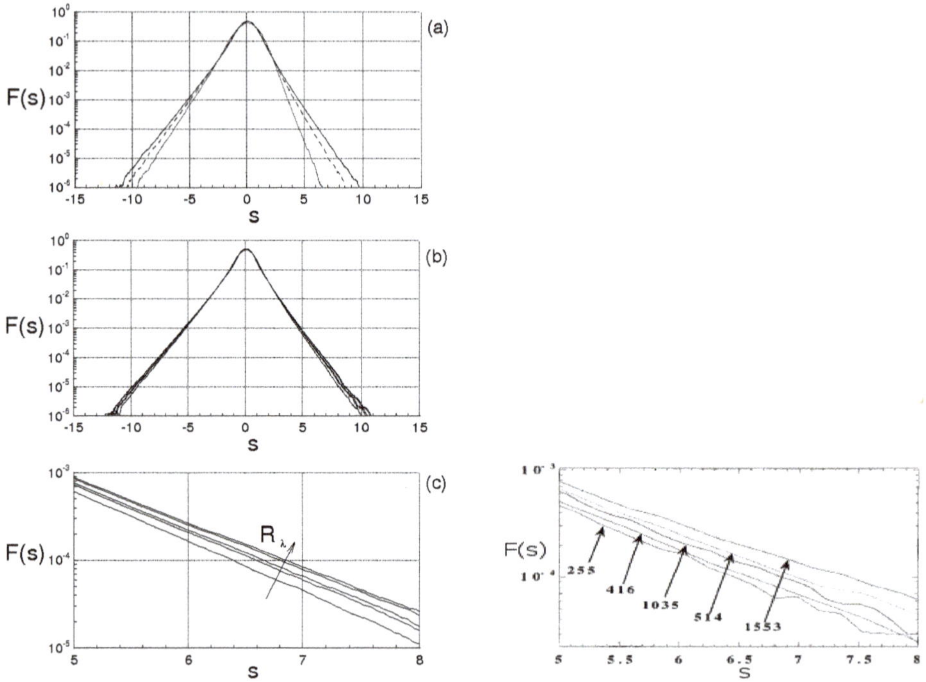


**Figure 9.** Left: PDF of $\partial u_r/\partial r$ for different values of $R_T$. (**a**) Dotted, dash–dotted and continuous lines are for $R_T$ = 15, 30 and 60, respectively. (**b**,**c**) PDFs for $R_T$ = 255, 416, 514, 1035 and 1553. (**c**) represents an enlarged part of the diagram (**b**). Right–bottom: Data from Reference [47].

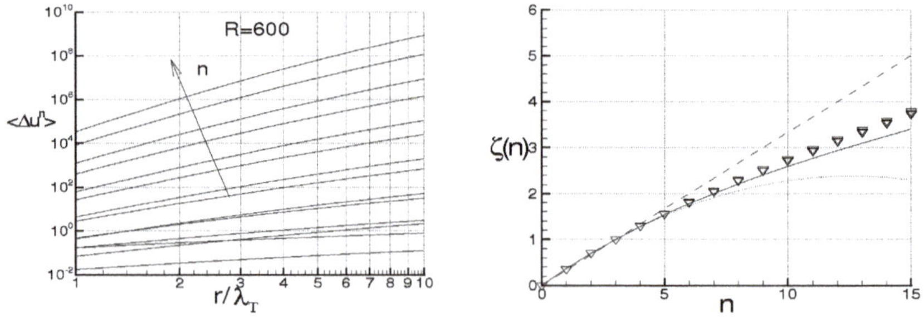


**Figure 10.** Left: Statistical moments of $u_r$ in terms of separation distance, for $R_T$ = 600. Right: Scaling exponents of $\partial u_r/\partial r$ at different $R_T$. Solid symbols are for the data calculated with the present analysis. Dashed line is for Kolmogorov K41 data [44]. Dotted line is for Kolmogorov K62 data [45]. Continuous line is for She–Leveque data [46].

As far as the temperature difference statistics is concerned, Figure 11 (Left) shows the distribution function of $\partial\vartheta/\partial r$ in terms of dimensionless abscissa

$$s = \frac{\partial\vartheta/\partial r}{\sqrt{\left\langle (\partial\vartheta/\partial r)^2 \right\rangle}} \tag{150}$$

calculated with Equations (132) and (139), for different values of $\Psi_\theta$. To show the intermittency of such PDF, the flatness $H_\theta^{(4)}$ and the hyperflatness $H_\theta^{(6)}$, defined as

$$H_\theta^{(4)} = \frac{\langle s^4 \rangle}{\langle s^2 \rangle^2}, \quad H_\theta^{(6)} = \frac{\langle s^6 \rangle}{\langle s^2 \rangle^3} \tag{151}$$

are plotted in Figure 11 (Right) in terms of $\Psi_\theta$. When $\Psi_\theta = 0$, the PDF is gaussian, thus $H_\theta^{(4)} = 3$ and $H_\theta^{(6)} = 15$. Increasing $\Psi_\theta$, the non–linear terms $\eta_\theta$ and $\zeta_\theta$ cause an increment of $H_\theta^{(4)}$ and $H_\theta^{(6)}$ and when $\Psi_\theta \to \infty$ $H_\theta^{(4)} \to 9$ and $H_\theta^{(6)} \to 225$.

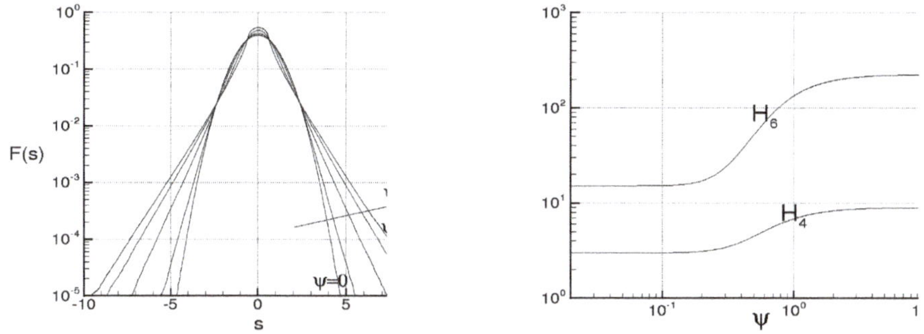


**Figure 11.** Left: Distribution function of the longitudinal temperature derivatives, at different values of $\Psi_\theta$. Right: Dimensionless statistical moments, $H_\theta^{(4)}$ and $H_\theta^{(6)}$ in function of $\Psi_\theta$.

Furthermore, the statistics of the temperature dissipation

$$\varphi = \chi \nabla\vartheta \cdot \nabla\vartheta, \tag{152}$$

is analyzed in function of $\Psi_\theta$ with particular reference to its intermittency. To this end, the Kurtosis of $\varphi$, $K_4(\varphi)$, is estimated by means of Equation (141), where, thanks to isotropy, the three components of $\nabla\vartheta \equiv (\vartheta_x, \vartheta_y, \vartheta_z)$ are identically distributed. Next, $\vartheta_x$, $\vartheta_y$ and $\vartheta_z$ are supposed to be statistically uncorrelated. This last assumption allows to estimate the Kurtosis of $\varphi$ in terms of the dimensionless statistical moments of $\partial\vartheta/\partial r$, according to

$$K_4(\varphi) = \frac{H_\theta^{(8)} - 4H_\theta^{(6)} + 6H_\theta^{(4)} - 3}{3\left(\left(H_\theta^{(4)}\right)^2 + 1 - 2H_\theta^{(4)}\right)} + 2 \tag{153}$$

where $H_\theta^{(4)}$, $H_\theta^{(6)}$ and $H_\theta^{(8)}$ are calculated using Equation (141). Figure 12 shows $K_4(\varphi)$ in function of $\Psi_\theta$, and compares the values calculated with the present theory (solid line), with those obtained by [73] through the nonlinear large–eddy simulations (symbols). The comparison shows that the data are in qualitatively good agreement. In more detail, for $\Psi_\theta \to \infty$, $K_4 \to 55$, whereas the results of Reference [73] give a value of around 60. This difference could be due to the fact that the present analysis only considers the isotropic turbulence which tends to bound the values of the dimensionless statistical moments of $\partial\vartheta/\partial r$ and of $\varphi$ and to the approximation of assuming the components of $\nabla\vartheta$ to be statistically uncorrelated.

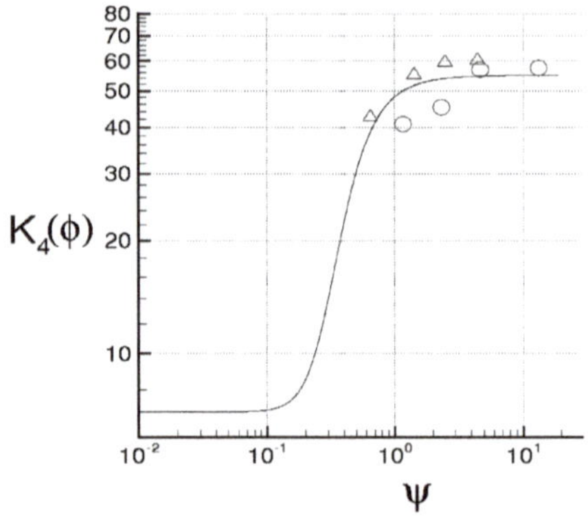


**Figure 12.** Comparison of the results: Kurtosis of temperature dissipation in function of $\Psi_\theta$. The symbols represent the results by [73].

Finally, observe that the experimental data of [47,74] allow to identify $\Phi(0)$. Table 3 reports a comparison between the value of $\Phi(0)$ calculated with the present theory and those obtained through elaboration of the experimental data of [47,74]. Form this comparison, the value of $\Phi(0)$ calculated with Equation (145) is in very good agreement with those obtained through the elaboration the data of [47,74].

**Table 3.** Identification of $\Phi(0)$ through elaboration of experimental data of [47,74] and comparison with the present analysis.

| Reference | $\Phi(0)$ |
|---|---|
| Present Analysis | 0.1409... |
| [47] | $\simeq$0.148 |
| [74] | $\simeq$0.135 |

## 16. Conclusions

A review of previous theoretical results concerning an original turbulence theory is presented. The theoretical approaches here adopted, different with respect to the other articles, confirm and corroborate the results of the previous works.

In separate sections, novel issues regarding the proposed turbulence theory are presented, and are here summarized.

- The bifurcation rate of velocity gradient, calculated along fluid particles trajectories is shown to be much larger than the maximal Lyapunov exponent of the kinematic field.
- On the basis of the previous item, the energy cascade is viewed as a stretching and folding succession of fluid particles which gradually involves smaller and smaller scales.
- The central limit theorem, in the framework of the bifurcation analysis, provides reasonable argumentation that the finite time Lyapunov exponent can be approximated by a gaussian random variable if $\tau \approx 1/\Lambda$.
- The closures of von Kármán–Howarth and Corrsin equations given by this theory determine velocity and temperature correlations which exhibit local self–similarity directly linked to the continuous particles trajectories divergence.

- The proposed bifurcation analysis of the closed von Kármán–Howarth equation studies the route from developed turbulence toward non–chaotic regimes and leads to an estimation of the critical Taylor scale Reynolds number in isotropic turbulence in agreement with the various experiments.
- Finally, a specific statistical decomposition of velocity and temperature is presented. This decomposition, adopting random variables distributed following extended distribution functions, leads to the statistics of velocity and temperature difference which agrees with the data of experiments.

**Funding:** This research received no external funding

**Acknowledgments:** This work was partially supported by the Italian Ministry for the Universities and Scientific and Technological Research (MIUR).

**Conflicts of Interest:** The author declares no conflict of interest.

## References

1. De Divitiis, N. Lyapunov Analysis for Fully Developed Homogeneous Isotropic Turbulence. *Theor. Comput. Fluid Dyn.* **2011**, *25*, 421–445. [CrossRef]
2. De Divitiis, N. Self-Similarity in Fully Developed Homogeneous Isotropic Turbulence Using the Lyapunov Analysis. *Theor. Comput. Fluid Dyn.* **2012**, *26*, 81–92. [CrossRef]
3. De Divitiis, N. Bifurcations analysis of turbulent energy cascade. *Ann. Phys.* **2015**, *354*, 604–617. [CrossRef]
4. De Divitiis, N. Finite Scale Lyapunov Analysis of Temperature Fluctuations in Homogeneous Isotropic Turbulence. *Appl. Math. Model.* **2014**, *38*, 5279–5397. [CrossRef]
5. De Divitiis, N. Von Kármán–Howarth and Corrsin equations closure based on Lagrangian description of the fluid motion. *Ann. Phys.* **2016**, *368*, 296–309. [CrossRef]
6. De Divitiis, N. Statistics of finite scale local Lyapunov exponents in fully developed homogeneous isotropic turbulence. *Adv. Math. Phys.* **2018**, *2018*, 2365602. [CrossRef]
7. De Divitiis, N. Refinement of a Previous Hypothesis of the Lyapunov Analysis of Isotropic Turbulence. *J. Eng.* **2013**, *2013*, 653027.
8. Obukhov, A.M. On the distribution of energy in the spectrum of turbulent flow. *Dokl. Akad. Nauk CCCP* **1941**, *32*, 22–24.
9. Heisenberg, W. Zur statistischen Theorie der Turbulenz. *Z. Phys.* **1948**, *124*, 628–657.
10. Kovasznay, L.S. Spectrum of Locally Isotropic Turbulence. *J. Aeronaut. Sci.* **1948**, *15*, 745–753.
11. Ellison, T.H. Zur The Universal Small-Scale Spectrum of Turbulence at High Reynolds Number. In *Mecanique de la Turbulence, Colloque International, Marseilles*; CNRS: Paris, France, 1961; pp. 113–121.
12. Pao, Y.H. Structure of turbulent velocity and scalar fields at large wave numbers. *Phys. Fluids* **1965**, *8*, 1063.
13. Leith, C. Diffusion Approximation to Inertial Energy Transfer in Isotropic Turbulence. *Phys. Fluids* **1967**, *10*, 1409–1416.
14. Clark, T.T.; Zemach, C. Symmetries and the approach to statistical equilibrium in isotropic turbulence. *Phys. Fluids* **1998**, *31*, 2395–2397.
15. Connaughton, C.; Nazarenko, S. Warm cascades and anomalous scaling in a diffusion model of turbulence. *Phys. Rev. Lett.* **2004**, *92*, 4.
16. Clark, T.T.; Rubinstein, R.; Weinstock, J. Reassessment of the classical turbilence closures: The Leith diffusion model. *J. Turbul.* **2009**, *10*, 1–23. [CrossRef]
17. Von Kármán, T.; Howarth, L. On the Statistical Theory of Isotropic Turbulence. *Proc. R. Soc. A* **1938**, *164*, 192.
18. Batchelor, G.K. *The Theory of Homogeneous Turbulence*; Cambridge University Press: Cambridge, UK, 1953.
19. Corrsin, S. The Decay of Isotropic Temperature Fluctuations in an Isotropic Turbulence. *J. Aeronaut. Sci.* **1951**, *18*, 417–423. [CrossRef]
20. Corrsin, S. On the Spectrum of Isotropic Temperature Fluctuations in an Isotropic Turbulence. *J. Appl. Phys.* **1951**, *22*, 469–473.
21. Eyink, G.L.; Sreenivasan, K.R. Onsager and the theory of hydrodynamic turbulence. *Rev. Mod. Phys.* **2006**, *78*, 87–135. [CrossRef]
22. Hasselmann, K. Zur Deutung der dreifachen Geschwindigkeitskorrelationen der isotropen Turbulenz. *Dtsch. Hydrogr. Z* **1958**, *11*, 207–217. [CrossRef]
23. Millionshtchikov, M. Isotropic turbulence in the field of turbulent viscosity. *JETP Lett.* **1969**, *8*, 406–411.

24. Oberlack, M.; Peters, N. Closure of the two-point correlation equation as a basis for Reynolds stress models *Appl. Sci. Res.* **1993**, *51*, 533–539.
25. George, W.K. *A Theory for The Self-Preservation of Temperature Fluctuations in Isotropic Turbulence*; Technical Report 117; Turbulence Research Laboratory: New York, NY, USA, January 1988.
26. George, W.K. Self-preservation of temperature fluctuations in isotropic turbulence. In *Studies in Turbulence*; Springer: Berlin, Germany, 1992.
27. Antonia, R.A.; Smalley, R.J.; Zhou, T.; Anselmet, F.; Danaila, L. Similarity solution of temperature structure functions in decaying homogeneous isotropic turbulence. *Phys. Rev. E* **2004**, *69*, 016305. [CrossRef]
28. Baev, M.K.; Chernykh, G.G. On Corrsin equation closure. *J. Eng. Thermophys.* **2010**, *19*, 154–169. [CrossRef]
29. Domaradzki, J.A.; Mellor, G.L. A simple turbulence closure hypothesis for the triple-velocity correlation functions in homogeneous isotropic turbulence. *J. Fluid Mech.* **1984**, *140*, 45–61.
30. Onufriev, A. On a model equation for probability density in semi-empirical turbulence transfer theory. In *The Notes on Turbulence*; Nauka: Moscow, Russia, 1994.
31. Grebenev, V.N.; Oberlack, M. A Chorin-Type Formula for Solutions to a Closure Model for the von Kármán-Howarth Equation. *J. Nonlinear Math. Phys.* **2005**, *12*, 1–9.
32. Grebenev, V.N.; Oberlack, M. A Geometric Interpretation of the Second-Order Structure Function Arising in Turbulence. *Math. Phys. Anal. Geometry* **2009**, *12*, 1–18. [CrossRef]
33. Thiesset, F.; Antonia, R.A.; Danaila, L.; Djenidi, L. Kármán–Howarth closure equation on the basis of a universal eddy viscosity. *Phys. Rev. E* **2013**, *88*, 011003(R). [CrossRef]
34. Ruelle, D.; Takens, F. On the nature of turbulence. *Commun. Math Phys.* **1971**, *20*, 167. [CrossRef]
35. Batchelor, G.K. Small-scale variation of convected quantities like temperature in turbulent fluid. Part 1. General discussion and the case of small conductivity. *J. Fluid Mech.* **1959**, *5*, 113–133. [CrossRef]
36. Batchelor, G.K.; Howells, I.D.; Townsend, A.A. Small-scale variation of convected quantities like temperature in turbulent fluid. Part 2. The case of large conductivity. *J. Fluid Mech.* **1959**, *5*, 134–139. [CrossRef]
37. Obukhov, A.M. The structure of the temperature field in a turbulent flow. *Dokl. Akad. Nauk.* **1949**, *39*, 391.
38. Gibson, C.H.; Schwarz, W.H. The Universal Equilibrium Spectra of Turbulent Velocity and Scalar Fields. *J. Fluid Mech.* **1963**, *16*, 365–384. [CrossRef]
39. Mydlarski, L.; Warhaft, Z. Passive scalar statistics in high-Péclet-number grid turbulence. *J. Fluid Mech.* **1998**, *358*, 135–175. [CrossRef]
40. Chasnov, J.; Canuto, V.M.; Rogallo, R.S. Turbulence spectrum of strongly conductive temperature field in a rapidly stirred fluid. *Phys. Fluids A* **1989**, *1*, 1698–1700. [CrossRef]
41. Donzis, D.A.; Sreenivasan, K.R.; Yeung, P.K. The Batchelor Spectrum for Mixing of Passive Scalars in Isotropic Turbulence. *Flow Turbul. Combust.* **2010**, *85*, 549–566. [CrossRef]
42. Feigenbaum, M.J. Quantitative Universality for a Class of Non-Linear Transformations. *J. Stat. Phys.* **1978**, *19*, 25–52. [CrossRef]
43. Eckmann, J.P. Roads to turbulence in dissipative dynamical systems. *Rev. Mod. Phys.* **1981**, *53*, 643–654. [CrossRef]
44. Kolmogorov, A.N. Dissipation of energy in locally isotropic turbulence. *Dokl. Akad. Nauk* **1941**, *32*, 19–21. [CrossRef]
45. Kolmogorov, A.N. Refinement of previous hypothesis concerning the local structure of turbulence in a viscous incompressible fluid at high Reynolds number. *J. Fluid Mech.* **1962**, *12*, 82–85. [CrossRef]
46. She, Z.S.; Leveque, E. Universal scaling laws in fully developed turbulence. *Phys. Rev. Lett.* **1994**, *72*, 336. [CrossRef]
47. Tabeling, P.; Zocchi, G.; Belin, F.; Maurer, J.; Willaime, H. Probability density functions, skewness, and flatness in large Reynolds number turbulence. *Phys. Rev. E* **1996**, *53*, 1613. [CrossRef]
48. Belin, F.; Maurer, J.; Willaime, H.; Tabeling, P. Velocity Gradient Distributions in Fully Developed Turbulence: An Experimental Study. *Phys. Fluid* **1997**, *9*, 3843–3850. [CrossRef]
49. Tsinober, A. *An Informal Conceptual Introduction to Turbulence: Second Edition of an Informal Introduction to Turbulence*; Springer Science & Business Media: Berlin, Germany, 2009.
50. Guckenheimer, J.; Holmes, P. *Nonlinear Oscillations, Dynamical Systems, and Bifurcations of Vector Fields*; Springer: Berlin, Germany, 1990.
51. Pomeau, Y.; Manneville, P. Intermittent Transition to Turbulence in Dissipative Dynamical Systems. *Commun. Math. Phys.* **1980**, *74*, 189.

52. Ashurst, W.T.; Kerstein, A.R.; Kerr, R.M.; Gibson, C.H. Alignment of vorticity and scalar gradient with strain rate in simulated Navier–Stokes turbulence. *Phys. Fluids* **1987**, *30*, 2443. [CrossRef]
53. Truesdell, C. *A First Course in Rational Continuum Mechanics*; Academic: New York, NY, USA, 1977.
54. Ottino, J.M. *The Kinematics of Mixing: Stretching, Chaos, and Transport*; Cambridge Texts in Applied Mathematics: New York, NY, USA, 1989.
55. Ottino, J.M. Mixing, Chaotic Advection, and Turbulence. *Annu. Rev. Fluid Mech.* **1990**, *22*, 207–253. [CrossRef]
56. Ott, E. *Chaos in Dynamical Systems*; Cambridge University Press: Cambridge, UK, 2002.
57. Nicolis, G. *Introduction to Nonlinear Science*; Cambridge University Press: Cambridge, UK, 1995.
58. Fujisaka, H. Statistical Dynamics Generated by Fluctuations of Local Lyapunov Exponents. *Prog. Theor. Phys.* **1983**, *70*, 1264–1275. [CrossRef]
59. Chen, S.; Doolen, G.D.; Kraichnan, R.H.; She, Z.-S. On statistical correlations between velocity increments and locally averaged dissipation in homogeneous turbulence. *Phys. Fluids A* **1992**, *5*, 458–463. [CrossRef]
60. Orszag, S.A.; Patterson, G.S. Numerical simulation of three-dimensional homogeneous isotropic turbulence. *Phys. Rev. Lett.* **1972**, *28*, 76–79. [CrossRef]
61. Panda, R.; Sonnad, V.; Clementi, E.; Orszag, S.A.; Yakhot, V. Turbulence in a randomly stirred fluid. *Phys. Fluids A* **1989**, *1*, 1045–1053. [CrossRef]
62. Anderson, R.; Meneveau, C. Effects of the similarity model in finite-difference LES of isotropic turbulence using a lagrangian dynamic mixed model. *Flow Turbul. Combust.* **1999**, *62*, 201–225. [CrossRef]
63. Carati, D.; Ghosal, S.; Moin, P. On the representation of backscatter in dynamic localization models. *Phys. Fluids* **1995**, *7*, 606–616. [CrossRef]
64. Kang, H.S.; Chester, S.; Meneveau, C. Decaying turbulence in an active–gridgenerated flow and comparisons with large–eddy simulation. *J. Fluid Mech.* **2003**, *480*, 129–160.
65. Ogura, Y. Temperature Fluctuations in an Isotropic Turbulent Flow. *J. Meteorol.* **1958**, *15*, 539–546. [CrossRef]
66. Arnold, V.I. *Catastrophe Theory*, 3rd ed.; Springer: Berlin, Germany, 1992.
67. Arnold, V.I.; Afrajmovich, V.S.; Il'yashenko, Y.S.; Shil'nikov, L.P. *Dynamical Systems V: Bifurcation Theory and Catastrophe Theory*; Springer Science & Business Media: Berlin, Germany, 2013.
68. Feynman, R.P. Negative Probability. In *Quantum Implications: Essays in Honour of David Bohm*; Peat, D., Basil, J.H., Eds.; Routledge & Kegan Paul Ltd.: Abingdon-on-Thames, UK, 1987; pp. 235–248.
69. Burgin, M. Extended Probabilities: Mathematical Foundations. *arXiv* **2009**, arXiv:0912.4767.
70. Burgin, M. Interpretations of Negative Probabilities. *arXiv* **2010**, arXiv:1008.1287.
71. Ventsel, E.S. *Theorie des Probabilites*; CCCP: Moskow, Russia, 1973.
72. Lehmann, E.L. *Elements of Large—Sample Theory*; Springer: Berlin, Germany, 1999.
73. Burton, G.C. The nonlinear large-eddy simulation method applied to and passive-scalar mixing. *Phys. Fluids* **2008**, *20*, 035103.
74. Sreenivasan, K.R.; Tavoularis, S.; Henry, R.; Corrsin, S. Temperature fluctuations and scales in grid-generated turbulence. *J. Fluid Mech.* **1980**, *100*, 597–621. [CrossRef]

MDPI
St. Alban-Anlage 66
4052 Basel
Switzerland
Tel. +41 61 683 77 34
Fax +41 61 302 89 18
www.mdpi.com

*Entropy* Editorial Office
E-mail: entropy@mdpi.com
www.mdpi.com/journal/entropy

www.ingramcontent.com/pod-product-compliance
Lightning Source LLC
Chambersburg PA
CBHW051718210326
41597CB00032B/5523
* 9 7 8 3 0 3 9 2 1 1 0 8 1 *